MyWritingLab™: Now Available for Technical Communication

MyWritingLab is an online homework, tutorial, and assessment program that provides engaging experiences for today's instructors and students.

Writing Help for Varying Skill Levels

For students who enter the course at widely varying skill levels, MyWritingLab provides unique, targeted remediation through personalized and adaptive instruction. Starting with a pre-assessment known as the Path Builder, MyWritingLab diagnoses students' strengths and weaknesses on prerequisite writing skills. The results of the pre-assessment inform each student's Learning Path, a personalized pathway for students to work on requisite skills through multimodal activities. In doing so, students feel supported and ready to succeed in class.

Respond to Student Writing with Targeted Feedback and Remediation

MyWritingLab unites instructor comments and feedback with targeted remediation via rich multimedia activities, allowing students to learn from and through their own writing.

- When giving feedback on student writing, instructors can add links to activities that address issues and strategies needed for review. Instructors may also link to multimedia resources in Pearson Writer, which include curated content from Purdue OWL.
- In the Writing Assignments, students can use instructor-created peer review rubrics to evaluate and comment on other students' writing.

NEW! Learning Tools for Student Engagement

Learning in Context

In addition to distinct, pre-loaded learning paths for writing skills practice, MyWritingLab for Technical Communication includes **modules specific to Technical Communcation**. These modules incorporate multimodal instruction and numerous model documents. Each text-specific MyWritingLab course contains readings and activities from the textbook.

Classroom Engagement

Generate classroom engagement, guide lectures, and promote peer-to-peer learning with real-time analytics. MyWritingLab now provides **Learning Catalytics**—an interactive student response tool that uses students' smartphones, tablets, or laptops to engage them in more sophisticated tasks and thinking.

Multimedia Assignments

MediaShare allows students to post multimodal assignments easily—whether they are audio, video, or visual compositions—for peer review and instructor feedback. In both face-to-face and online course settings, MediaShare saves instructors valuable time and enriches the student learning experi-ence by enabling contextual feedback to be provided quickly and easily.

Direct Access to MyLab

Users can link from any Learning Management System (LMS) to Pearson's MyWritingLab. Access MyLab assignments, rosters and resources, and synchronize MyLab grades with the LMS gradebook. New direct, single sign-on provides access to all the personalized learning MyLab resources that make studying more efficient and effective.

Visit www.mywritinglab.com for more information.

TECHNICAL COMMUNICATION

PROCESS AND PRODUCT

NINTH EDITION

SHARON J. GERSON

DeVry University (Emeritus)

STEVEN M. GERSON

Johnson County Community College

New!
2016
MLA
Updates

PEARSON

Boston   Columbus   Indianapolis   New York   San Francisco
Amsterdam   Cape Town   Dubai   London   Madrid   Milan   Munich   Paris   Montreal
Toronto   Delhi   Mexico City   São Paulo   Sydney   Hong Kong   Seoul   Singapore   Taipei   Tokyo

DEDICATION

THIS NINTH EDITION IS DEDICATED TO OUR DAUGHTERS, STACY AND STEFANI, AND OUR GRANDDAUGHTER, SOPHIA.

Senior Acquisitions Editor: Brad Potthoff
Editorial Assistant: Amanda Norelli
Program Manager: Anne Shure
Development Editor: Bruce Cantley
Product Marketing Manager: Jennifer Edwards
Field Marketing Manager: Joyce Nilsen
Media Producer: Elizabeth Bravo
Content Producer: Julia Pomann
Media Editor: Kelsey Loveday

Project Manager: Rebecca Gilpin
Text Design, Project Coordination and Electronic Page Makeup: SPi Global
Cover Design Manager: Barbara Atkinson Cover
Designer: Cenveo Publishing Services Cover
Illustration: Nesbitt
Senior Manufacturing Buyer: Roy L. Pickering, Jr.
Printer/Binder: LSC Communications/Harrisonburg
Cover Printer: LSC Communications/Harrisonburg

Acknowledgments of third-party content appear on page 560 which constitute an extension of this copyright page.

Library of Congress Cataloging-in-Publication Data
Gerson, Sharon J., 1948- author.
  Technical communication : process and product / Sharon J. Gerson,
Devry University (emeritus) ; Steven M. Gerson, Johnson County
Community College.—Ninth Edition.
     pages cm
  Includes bibliographical references and index.
  ISBN 0-13-409403-4 (alk. paper)
  1. English language—Business English. 2. English
language—Technical English. 3. Business writing—Problems, exercises,
etc. 4. Technical writing—Problems, exercises, etc. 5. Business
communication—Problems, exercises, etc. 6. Commercial correspondence—Problems, exercises, etc.
I. Gerson, Steven M., 1948- author. II. Title.
  PE1479.B87S43 2017
  808.06'66—dc23
                                                    2015031461

2 17

www.pearsonhighered.com

Student Edition ISBN 10: 0-13-467886-9
Student Edition ISBN 13: 978-0-13-467886-3

A la Carte Edition ISBN 10: 0-13-470310-3
A la Carte Edition ISBN 13: 978-0-13-470310-7

PEARSON

BRIEF CONTENTS

CONTENTS

ABOUT THE AUTHORS

SHARON J. GERSON AND STEVEN M. GERSON are dedicated career professionals who have a combined total of over 80 years teaching experience at the college and university level. They have taught technical writing, business writing, professional writing, and technical communication to thousands of students, attended and presented at dozens of conferences, written numerous articles, and published several textbooks, including *Technical Writing: Process and Product* (ninth edition), *Professional Communication in the Digital Workplace*, *The Red Bridge Reader* (third edition, co-authored by Kin Norman), *Writing That Works: A Teacher's Guide to Technical Writing* (second edition), *Workplace Communication: Process* and *Product, and Workplace Writing: Planning, Packaging, and Perfecting Communication.*

In addition to their academic work, Sharon and Steve are involved in business and industry through their business, Steve Gerson Consulting. In this business, they have worked for companies such as Sprint, AlliedSignal–Honeywell, General Electric, JCPenney, Avon, the Missouri Department of Transportation, H&R Block, Mid America Regional Council, and Commerce Bank. Their work for these businesses includes writing, editing, and proofreading many different types of technical documents, such as proposals, marketing collateral, reports, and instructions.

Steve also has presented hundreds of hands-on workshops on technical/business writing, business grammar in the workplace, oral presentations in the workplace, and business etiquette. Over 10,000 business and governmental employees have benefited from these workshops. Steve has worked closely with K–12 teachers. He has presented many well-attended, interactive workshops to give teachers useful tips about technical writing in the classroom.

Both Steve and Sharon have been awarded for teaching excellence and are listed in *Who's Who Among America's Teachers*. Steve is a Society for Technical Communication Fellow. In 2003, Steve was named Kansas Professor of the Year by the Carnegie Foundation for the Advancement of Education.

Their experience and knowledge have been gathered for you in this ninth edition of *Technical Communication: Process and Product.*

PREFACE

Welcome to the ninth edition of *Technical Communication: Process and Product*. This reader-friendly textbook combines easy-to-follow instructions for producing all forms of oral and written technical communication with interesting scenarios and examples featuring real people facing communication challenges on the job.

New to the Ninth Edition

Highlights of content new to the ninth edition of our highly readable technical communication textbook are the following:

- **Chapter 2 "Digital Communication."** Chapter 2 explores topics such as an overview of digital communication in the workplace and its benefits, social media on the job, mobile apps including writing apps, communication with QR codes, and digital collaborative tools like Google Docs/Sites, wikis, Sharepoint, Dropbox, and Skype.
- **Expanded discussion of the writing process.** A new section at the beginning of Chapter 1 transitions students from essay writing to technical communication, with helpful comparative tables. Every chapter after Chapter 3, "The Communication Process," now features a section on "The Writing Process at Work," following an individual's progress from prewriting to writing to rewriting a document tied to the chapter's topic. Approximately half of these sections are new or revised for this edition.
- **Enhanced discussion of teamwork and collaboration.** Expanded discussion of collaboration in Chapter 1 includes sections on types of teams (communities of practice, functional teams, cross-functional teams), types of team processes and how they work (segmented, sequential, prioritized, and pragmatic approaches), and planning a project.
- **Enhanced coverage of ethics.** Most chapters now feature sections on ethical considerations (ethical job interview questions; appropriate use of corporate listservs, company-provided smartphones, and the Internet on the job; ethical use of social media; copyright and plagiarism issues; and more). Every chapter now also features an "Ethical Challenge" case study in the "Apply Your Knowledge" section.
- **New digital coverage.** New technological innovations have been added to many chapters. Topics include online tasks panes with pop-up reminders, smartphone notes apps, track changes, e-mail draft modes, text messaging, groupware, online surveys, Internet job searching, using micro-video, uses of Skype, QR codes, responsive or scalable Web design, designing Web sites for small screens, online reports, and more. In addition, 16 of the 20 chapters feature "Technology Tips" that are either new or have been revised with screen captures to explain how to perform various tasks using Microsoft Office 2013.
- **Expanded chapter on routine correspondence.** This chapter now includes a discussion of letter formats (formerly in Appendix C), as well as coverage of order, recommendation, inquiry, response, and thank you correspondence.

- **Expanded chapter on proposals.** Chapter 19, "Proposals and Business Plans," now includes a section on writing business plans, exploring the audience for business plans, and components of a business plan (executive summary and the description of the business including product/service, place, personnel, price, and more).

The Communication Process in Action

For over 40 years, the writing process—prewriting, writing, and rewriting—has been the standard for teaching students how to write effectively. Many of us were taught in K–12 how to write based on this process, or we were introduced to the writing process in college. In this textbook, we build upon this process with a unique approach that applies the writing process to both oral and written communication.

Chapters begin with a "Communication at Work" scenario highlighting people communicating on the job. These employees are at businesses as diverse as the following: architectural/engineering, consulting, computer hardware, mortgage and banking, construction, biotechnology, financial planning, social media, and informatics software. Each chapter concludes with a unit entitled "The Writing Process at Work" with a businessperson using the writing process to create effective technical communication. We bring the writing process to life by using businesspeople facing real communication challenges at work. An important addition to the communication process is using technology to prewrite and write.

- In the prewriting stage, we show how the writer determines goals, audience, and communication channel. Then we show a planning technique enhanced through the use of technology tools that the writer uses to gather data for the correspondence.
- In the writing stage, we show how the writer, often using technology, organizes a rough draft and formats the text for ease of access. We also show how the writer receives suggestions for revision from peer evaluators. This illustrates the importance of teamwork in the communication process.
- In the rewriting stage, we show how the rough draft has been revised. This stage illustrates the finished product: a written document that communicates successfully with its intended audience.

The writing process allows us to accomplish key goals, including the following:

- Provide criteria for different communication channels
- Give examples of different types of communication
- Teach an effective procedure to help students succeed in the classroom and in their careers
- Use technology to prewrite and write the document
- Show the importance of revision and how it can be achieved often through the assistance of collaboration
- Reveal the thought processes and the approach to writing and revising followed by businesspeople
- Bring to life the principles of technical communication

With our approach to technical communication, this textbook provides the teacher and student with the following exciting tools:

- Real people, who are easy to relate to, encountering on-the-job communication challenges
- The actual prewriting, writing, and rewriting of documents written and created by these businesspeople

Addressing the Needs of the Evolving Workplace

Today's workplace is constantly evolving. This text addresses the important topical issues encountered in technical communication today including the following:

- Multiculturalism and the global economy
- Technological applications to prewriting and writing
- Social media, such as Twitter, Facebook, YouTube, and blogging, and its impact on technical communication
- Use of mobile communication devices like smartphones, tablets, and apps
- Criteria for effective workplace use of instant messages and text messages
- Communication in dispersed team settings
- Types of and uses of micro-videos
- Strategies for ethical technical communication
- Collaboration and teamwork especially with Wikis, Google Docs, Google Sites, and other technological tools
- Dozens of types of electronic, hard-copy, and oral communication channels
- Persuasive communication techniques for meeting the needs of the audience
- Up-to-date job search information
- Oral communication and the workplace
- Updated documentation information in Appendix B including Council of Science Editors (CSE) style format, American Psychological Association (APA), and Modern Language Association (MLA)

The Technical Communication Learning System

Technical Communication: Process and Product has been designed to provide a wealth of in-text, pedagogical features and end-of-chapter applications, as well as online resources to enhance both the learning and teaching processes.

Pedagogical Features

Aside from the "Communication at Work" and "The Writing Process at Work" sections described above, each chapter in the text contains a variety of helpful pedagogical features that enhance the narrative including the following:

LEARNING OBJECTIVES boxes provide numbered lists of key topics students will learn as they work through each chapter. The Learning Objectives also appear alongside the major headings within every chapter to reinforce learning goals.

After completing this chapter, you will be able to

- Explain how technical communication differs from other forms of communication
- Identify the various purposes of technical communication
- Adapt to written and oral communication channels
- Explain why being a successful technical communicator is important on the job
- Describe how teamwork enhances communication in the workplace
- Identify the types of workplace teams and their benefits
- Explain how the various types of team processes operate
- Collaborate effectively on diverse and dispersed teams
- Identify the challenges to effective teamwork and how to overcome them
- Consider 10 ways to resolve team conflicts

FREQUENTLY ASKED QUESTION (FAQs) boxes provide answers to questions people ask about topical issues.

FAQs: LinkedIn's Features

Q: Besides helping me with my job search, what other benefits does LinkedIn offer?

A: LinkedIn isn't just a social networking site that allows you to post a profile. Look at its other interesting features:

- LinkedIn Groups—allows users to form like-minded groups of peers within an organization or industry.
- LinkedIn DirectAds—lets you connect with a large audience by geography, job function, age, gender, industry, and company size.

TECHNOLOGY TIPS boxes show students how to use Microsoft 2013. Below is a list of the tips by chapter.

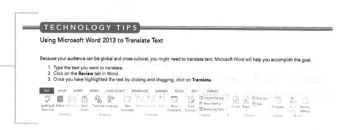

CHECKLISTS guide students through the revision stage of their writing.

CHECKLIST FOR DIGITAL COMMUNICATION

_____ 1. To enhance a job search, have you used LinkedIn?

_____ 2. Have you used a tweet to communicate with a colleague about a work-related issue?

_____ 3. Have you used social media such as Facebook or YouTube to connect with other business professionals in your field?

_____ 4. Have writing apps enhanced your communication skills?

_____ 5. Did you use QR codes to connect end users to products, services, a survey, your portfolio, or social media?

_____ 6. Have you used track changes in a team project to comment on a document and track the changes suggested?

_____ 7. Did your team use a wiki to edit text in a team project?

_____ 8. Did your team use Google Docs to upload and save files from remote work locales?

_____ 9. Did you use Google Sites with your team to organize and plan meetings and activities?

_____10. Did you use Skype to enhance communication among your colleagues?

EXAMPLE boxes provide easy-to-understand examples.

◀ EXAMPLE

GLOSSARY

BDC	Bottom dead center
CCW	Counterclockwise
Danger	This hazard alert designates the possibility of death. Be extremely careful when performing an operation.
RMS	Root mean square

BEFORE AND AFTER EXAMPLES demonstrate to students the importance of revision.

BEFORE

Wordy Prepositional Phrases

He spoke *at a rapid* rate.

She wrote *with regard to* the meeting.

I will call *in the near future.*

On two different occasions, we met.

The manager *of personnel* was hired.

AFTER

Concise Versions

He spoke rapidly (or fast).

She wrote regarding (or about) the meeting.

I will call soon.

We met twice.

The personnel manager was hired.

SAMPLE DOCUMENTS provide excellent examples of documents written for lay audiences, low-tech audiences, and high-tech audiences.

FIGURE 5.1 E-mail Message to High-Tech Audience

This e-mail, written to high-tech readers, uses terminology without explanation or definition. Its tone is businesslike and directive.

TOPICAL CHAPTER REFERENCES provide links to related topics in other chapters.

Writing Process.
See Chapter 3, "The Communication Process," for more discussion.

End of Chapter "Apply Your Knowledge" Activities

The end-of-chapter activities provide students with opportunities to apply chapter principles. In addition, because students learn in a variety of ways, we provide activities that engage students through critical thinking, collaboration, writing, oral presentations, and research.

CASE STUDIES present real-world scenarios and on-the-job communication challenges.

CASE STUDIES

1. You manage an engineering department at Acme Aerospace. Your current department supervisor is retiring. You must recommend the promotion of a new supervisor to the company's executive officer, Kelly Adams. You know that Acme seeks to promote individuals with the following traits:

   • Familiarity with modern management techniques and concerns, such as teamwork, global economics, crisis management, and the management of hazardous materials.
   • An ability to work well with colleagues (subordinates, lateral peers, and management).

ETHICAL CHALLENGES provide students with interesting, on-the-job scenarios for considering the ethics of a situation.

ETHICAL CHALLENGE

According to ethical standards, visuals should be precise. Look at the following bar chart. It shows that during one quarter, the company lost over $50,000.

Question

Is the following visual aid ethical? Why or why not?

INDIVIDUAL AND TEAM PROJECTS provide students an opportunity to apply principles discussed in the chapter drawing from personal experience, research, or collaboration.

INDIVIDUAL AND TEAM PROJECTS

1. Create a groupware site, for example, Google Sites. Use this site to assign team responsibilities. Then, use the groupware site for the team's writing and revision of a project.
2. Using online calendar software, such as Doodle, invite team members to a meeting and have them confirm their availability.

PROBLEM-SOLVING THINK PIECES help students practice critical thinking skills.

PROBLEM-SOLVING THINK PIECES

1. Using corporate-owned equipment. Tamara Jones is a receptionist at a bank. Her job is to greet clients, direct them to the appropriate bankers, and to answer questions about bank procedures. When she needs to speak to a customer or coworker, she's always available and pleasant. However, her job has periods of downtime when she waits for the next customer to arrive. During this downtime, she occasionally plays computer games. She's gotten tired of the games that are loaded on her bank computer. To enhance her "gaming" options, Tamara has downloaded numerous computer games that require a lot of computer space and memory.

WEB WORKSHOPS allow students to have an opportunity for discovery of topics related to technical communication beyond the classroom.

WEB WORKSHOP

Many governmental organizations and technical professions ask employees to follow the standards for "plain language," which focus on clarity and conciseness. To learn more about plain language, visit http://www.plainlanguage.gov. This site provides a definition of plain language, governmental mandates, and before and after comparisons. Click on any of the links in this site and report your findings either orally or in an e-mail message.

Online Resources to Enhance Technical Communication

A wide variety of online supplements for both students and instructors accompany *Technical Communication: Process and Product*, ninth edition.

The following online supplement is designed for students:

Pearson eText gives students access to *Technical Communication: Process and Product* whenever and wherever they can access the Internet. The eText pages look exactly like the printed text and include powerful interactive and customization functions. Users can create notes, highlight text in different colors, create bookmarks, zoom, click hyperlinked words and phrases to view definitions, and view as a single page or as two pages. The eText also offer a full-text search and the ability to save and export notes.

The Pearson eText app is a great companion to Pearson's eText browser-based book reader. It allows existing subscribers who view their Pearson eText titles on a Mac or PC to access their titles in a bookshelf on the iPad or an Android tablet either online or via download.

The following online supplements are available for instructors:

Instructor's Manual.   New to the ninth edition is an expanded Instructor's Manual with helpful teaching notes for the classroom. We have also added additional sample answers to case studies and other end-of-chapter assignments.

MyTest.   Pearson MyTest is a powerful assessment generation program that helps instructors easily create and print quizzes, study guides, and exams. Questions and tests are authored online, allowing instructors ultimate flexibility and the ability to efficiently manage assessments anytime, anywhere. To access MyTest, go to www.pearsonhighered.com/mytest/, log on, and follow the instructions. You must first be registered.

PowerPoint Slides.   Fully revised to accompany the ninth edition, the PowerPoint presentations provide chapter-by-chapter slides that can be projected or printed to enhance in-class instruction or simply used for review and class planning.

Acknowledgments

We would like to thank the following reviewers for helpful comments that helped us to produce this edition of *Technical Communication: Process and Product*: Jill Channing, Mitchell Community College; Sherry Lofton, Robeson Community College; Mark Mabrito, Purdue University Calumet; Eric Sentell, Southeast Missouri State University; Lawrence Sledge, Jackson State University; Kaye Temanson, North Dakota State University; Bruce Wehler, Pennslvania College of Technology; and Wanda White, Central Piedmont Community College.

In addition, we would like to thank the reviewers who helped us to shape previous editions of this book:

Patricia Boyd, Arizona State University
Floyd Brigdon, Trinity Valley Community College
Jo Ann Buck, Guilford Technical Community College
Elizabeth Christensen, Sinclair Community College
Amy Tipton Cortner, Caldwell Community College and Technical Institute
Natalie Daley, Linn-Benton Community College
Myra G. Day, North Carolina State University

Rosemary Day, Central New Mexico Community College
Treven Edwards, SUNY Plattsburgh
Paul Fattaruso, University of Denver
Bart Ganzert, Forsyth Technical Community College
Cynthia Gillispie-Johnson, North Carolina A&T State University
Joshua Hamling, New Mexico State University and Dona Ana Community College
Joyce Harp, New Mexico State University
Arthur Khaw, Kirkwood Community College
Liz Kleinfeld, Red Rocks Community College
Angela W. Lamb, Robeson Community College
Lindsay Lewan, Arapahoe Community College
Bruce R. Magee, Louisiana Tech University
Caroline Mains, Palo Alto College
Lynne Nelson Manion, Northern Maine Community College
Steve Marsden, Stephen F. Austin State University
Josie Mills, Arapahoe Community College
Heather Milton, University of California, Davis
Sharon Mouss, Oklahoma State University, Okmulgee College
Kevin Nebergall, Kirkwood Community College
Joseph Nocera, Jefferson Community College
Michele Regenold, Nicolet Area Technical College
Nancy Roberts, Griffin Technical College
Sherry Rosenthal, Community College of Southern Nevada
Terry Smith, University of Maryland – Eastern Shore
Joyce C. Staples, Patrick Henry Community College
Brian Still, Texas Tech University
Alan Tessaro, Spartanburg Community College
Chris Thaiss, University of California, Davis
Marc Wilson, Ivy Tech Community College
Esther J. Winter, Central Community College
JB Zwilling, Allen County Community College

Finally, we would like to thank the following people who worked with the book and its supplements through the publishing process:

Senior Editor Brad Potthoff
Program Manager Anne Shure
Development Editor Bruce Cantley
Editorial Assistant Amanda Norelli
Project Managers Rebecca Gilpin and Erika Jordan
Marketing Manager Jennifer Edwards
Media Supplements Editor Laura Olsen
Project Manager for permissions Joseph Croscup

Sharon J. Gerson and Steven M. Gerson

CHAPTER ONE

An Introduction to Technical Communication

After completing this chapter, you will be able to

1. Explain how technical communication differs from other forms of communication

2. Identify the various purposes of technical communication

3. Adapt to written and oral communication channels

4. Explain why being a successful technical communicator is important on the job

5. Describe how teamwork enhances communication in the workplace

6. Identify the types of workplace teams and their benefits

7. Explain how the various types of team processes operate

8. Collaborate effectively on diverse and dispersed teams

9. Identify the challenges to effective teamwork and how to overcome them

10. Consider 10 ways to resolve team conflicts

COMMUNICATION AT WORK

In the Gulfview scenario, employees on a project team reveal the importance of technical communication.

Gulfview Architectural and Engineering Services is home-based in Gulfview, Texas, with office sites in 10 U.S. cities and 5 locations throughout the world. Gulfview hopes to build a power plant in Saudi Arabia. To accomplish this task, a team of employees is working on two continents. The project requires that all team members be involved in numerous communication challenges.

Proposal. First, one team, consisting of engineers, architects, marketing specialists, accountants, lawyers, and technical communicators, put together a proposal. In this proposal, they focused on the services they could offer, the expertise of their workforce, the price they would charge for the construction, and a timeline for their work. Despite many competitors, Gulfview won the account.

E-mail, Text Messages, and Instant Messages. The construction would take Gulfview approximately two years. During that time, Gulfview personnel had to communicate with their Saudi contractors on a daily basis. E-mail, text messages, and instant messages answered this need. The team members communicated with each other by writing approximately

50 e-mail messages a day. In these transmittals, the team members focused on construction permits, negotiated costs with vendors, changed construction plans, and asked questions and received answers. They used text messages and instant messages for quick updates and to build rapport with coworkers.

Intranet Web Site and Corporate Blog. To help all parties involved (those in Saudi Arabia as well as Gulfview employees throughout the United States), Gulfview's Information Technology Department built an intranet site and a blog geared specifically toward the power plant project. This firewall-protected site, open to Gulfview employees and external vendors associated with the project, helped all construction personnel submit online forms, get corporate updates, and access answers to frequently asked questions. Many of these FAQs were managed through online help screens with pull-down menus. The blog allowed employees to provide work journals, Web logs in which they could comment on construction challenges, and get feedback from other employees working with similar issues.

Letters. To secure and revise construction permits, Gulfview personnel had to write formal letters to government officials in Saudi Arabia. In addition, Gulfview employees had to write letters to vendors, asking for quotes.

Reports. Finally, all of the employees involved in the power plant project had to report on their activities. To encourage collaboration and improve the quality of the company's writing, management created a corporate wiki where participants could write the following:

- Progress reports providing updates on the project's status
- Incident reports when job-related accidents and injuries occurred
- Feasibility reports to recommend changes to the project's plan or scope
- Meeting minutes following the many team meetings

Like all companies engaged in job-related projects, Gulfview Architectural and Engineering Services spent much of its time communicating with a diverse and dispersed audience. The challenges they faced involved teamwork, multicultural and multilingual concerns, a vast array of communication technologies, and a variety of communication channels.

---

1. Explain how technical communication differs from other forms of communication.

What Is Technical Communication?

Technical communication is oral and written communication for and about business. Technical communication focuses on products and services—how to

- Manufacture them
- Market them
- Manage them
- Deliver them
- Use them

Technical communication differs from other forms of writing with which you might be more familiar, such as creative writing and composition/rhetoric. Although all forms of writing are characterized by elements such as purpose, audience, channel of communication, content, and design, technical communication is unique in its use in the business world exclusively. Other forms of writing are applicable in classrooms and are typical parts of a

TABLE 1.1   Communication Characteristics

Type of Writing	Purpose	Audience	Channel	Content	Design
Technical Communication	Inform, recommend, persuade	Colleagues, clients, vendors, supervisors	Memo, e-mail, letter, report, social media, user manual, proposal, and more	Information about products and services	Accessible content through bullets, headings, graphics, and more
Composition/ Rhetoric	Narrate, describe, compare/contrast, argue/persuade, and more	The college professor	Essay and research paper	Any topic chosen by the professor and writer	Paragraphs
Creative Writing	Express emotions through character development, plot, literary conventions	Individuals who choose to read the document	Short story, play, poem, novel, and more	The human condition including happiness, sadness, triumph, loss, and more	Depends on the type of creative writing (stanzas, acts, chapters)

student's educational curriculum. Only technical communication, though taught in classes, is for business writers on a day-to-day basis. See Table 1.1, which reveals the contrasts among technical communication and other types of writing.

More specifically, technical communication is not freshman composition; memos, letters, and reports are not essays. The skills you might have learned in your freshman composition class will not all be applicable in a technical communication class or in the workplace. Following is Table 1.2 further explains the similarities and differences between technical communication and essays.

TABLE 1.2   Comparison/Contrast of Technical Communication and Essays

Components	Technical Communication	Essays	Summary
Development	• Uses examples, testimony, data, research	• Uses examples, anecdotes, narration, personal experiences, testimony, data, research	Has some similarities
Grammar	• Essential for professionalism in the workplace	• Essential for success in the classroom	Same for both
Organization	• Provides an introduction, body, and conclusion • Uses subject lines versus thesis statement • Uses itemization through bullets or numbers and headings versus transitional words • Can have one sentence paragraphs	• Provides an introduction, thesis statement, body paragraphs, and transitional words • Paragraphs are developed fully	Similar in some ways but different in others
Style	• Requires short words, short sentences, and short paragraphs for conciseness • Uses denotative words for clarity	• Allows for longer words, longer sentences, and longer paragraphs • Uses connotative words to convey expressive ideas	Different
Document Design	• Uses highlighting techniques, including headings, itemization, varied font sizes/types, white space, and graphics for ease of access and emphasis	• Does not have elements of design because of reliance on paragraphs	Different
Length of Document	• A successful memo, letter, report, or e-mail can be one page in length • Successful digital communication is brief, such as tweets and instant messages, which cannot exceed 140–160 characters	• To be sufficiently developed, essays are several paragraphs long and more than one page	Different

In many ways, technical communication skills and essay conventions are different. Being aware of these differences will help you to succeed in a technical communication class and later when you write on the job.

Purposes of Technical Communication

2. Identify the various purposes of technical communication.

Digital Communication

See Chapter 2, "Digital Communication," for more information about the digital workplace.

Whether you are an employer or an employee, a customer or a vendor, you will be involved with communication in the workplace. You will write business correspondence and speak to colleagues, clients, or a salesperson. Knowing how to communicate successfully in a digital work environment will help you express your point of view and influence people.

What are the purposes of technical communication? When would you be writing or speaking on the job? In the business world, you will communicate to different audiences, for different purposes, using different channels of communication. Consider the possibilities shown in Table 1.3.

TABLE 1.3 Purposes of Technical Communication

Speaking to Customers
As a computer information systems employee, you work at a 1-800 hotline helpdesk. A call comes through from a concerned client. Your job not only is to speak politely and professionally to the customer but also to follow up with an e-mail documenting your responses.

Writing a Letter
As a customer, you have just celebrated your 6-year-old child's birthday at a local pizza parlor. Unfortunately, the pizzas for the 15 guests were cold, the service was rude, and the promised entertainment was late in arriving. You now need to write a complaint letter to the store's management, recounting your experience.

Making a PowerPoint Presentation
As a trust officer in a bank, one of your jobs is to make proposals to potential clients. In doing so, you will write a proposal about your bank's services, and you will use PowerPoint to make an oral presentation for this client.

Writing Reports
As the manager of human resources, one of your major responsibilities is to document your training staff's job accomplishments. To do so, you must write year-end progress reports for the employees, which will be used to justify their raises.

Marketing
As an entrepreneur, you want to advertise your new catering business. To do so, you plan to write brochures (to be distributed locally), create a Web site (to expand your business opportunities), use your Facebook site to connect with clients, and send tweets to tell customers when your catering trucks are in the neighborhood.

Using a Wiki
As a mid-level manager, you are in charge of a team of employees who work in different cities, have varied job titles, and work different shifts. To maintain constant communication about a proposal the team is creating for distribution to the marketing department, you use a wiki so everyone is encouraged to interact and make suggestions about the proposal's content.

Seeking Employment
As a recently graduated accounting major, it is time to get a job. You need to write an effective résumé and letter of application to show corporations what an asset you will be to their work environment. Then, you will need to interview well and write a follow-up letter.

Instructing
As a technical communication professional, you write instructions to accompany a new product. The hard copy and digital manual will help customers use the product safely and efficiently.

Blogging
Your employer wants you to create and maintain a blog to reach a new market segment for the company's products.

Texting and Instant Messaging
As a manager, you need to stay in contact with your sales representatives who frequently call on customers or visit companies in the region. You use both text messages and instant messages to communicate with these employees.

3. Adapt to written and oral communication channels.

Communication Channels

Technical communication takes many different forms. Not only will you communicate both orally and in writing, but also you will rely on various types of correspondence and digital technology, dependent upon the audience, purpose, and situation.

To communicate successfully in the workplace, you must adapt to different channels of communication. Table 1.4 gives you examples of different communication channels, both oral and written.

TABLE 1.4 Communication Channels

Written Communication Channels	Oral Communication Channels
• E-mail	• Leading meetings
• Memos	• Conducting interviews
• Letters	• Making sales calls
• Reports	• Participating in teleconferences and
• Proposals	videoconferences
• Fliers	• Facilitating training sessions
• Brochures	• Participating in collaborative team projects
• Faxes	• Providing customer service
• Web sites	• Making telephone calls
• Instant messages and text messages	• Leaving voicemail messages
• Blogs	• Making presentations at conferences or to civic
• Facebook	organizations using tools such as PowerPoint
• Twitter	• Participating in interpersonal communication
• Job information (résumés, letters of	at work
application, follow-up letters, interviews)	• Conducting performance reviews

To clarify the use of different technical communication channels, look at Figure 1.1. In a survey of approximately 120 companies employing over 8 million people, the National Commission on Writing found that employees almost always use different forms of writing, including e-mail messages, PowerPoint, memos, letters, and reports ("Writing: A Ticket to Work" 11).

FIGURE 1.1 Channels Frequently Used in Technical Communication

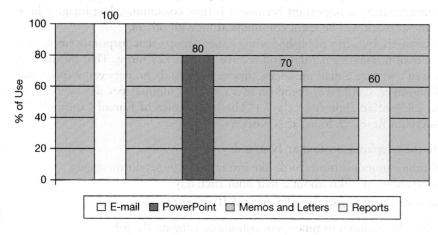

The Importance of Technical Communication

Why is communication in the workplace important? At work, your primary job is not necessarily writing—or is it? For example, you might be an accountant, computer technician, training facilitator, salesperson, buyer, supervisor, account administrator, or administrative assistant. Your employers will expect expertise from you in those areas of specialization.

> 4. Explain why being a successful technical communicator is important on the job.

As a computer technician, for instance, you have been hired due to your knowledge of hardware and software. As an accountant, you have been hired due to your knowledge of accounts payables and receivables. However, a major part of every job is an employee's ability to communicate.

To succeed on the job, you need to write and speak effectively to others—constantly. Gaston Caperton, president of the National Commission on Writing for America's Families, Schools, and Colleges, says that writing is an important skill in banking, finance, insurance, and real estate ("Writing Skills"). Technical communication is important to you for many reasons.

Operating a Business

Technical communication is not a frill or an occasional occurrence. It is a major component of your job. The National Commission on Writing for America's Families, Schools, and Colleges is a blue-ribbon group of leaders from public schools, higher education, and the business and writing communities. This commission surveyed human resource directors from 150 leading American corporations with a combined workforce of more than 10 million employees in the United States and combined annual revenues of $4 trillion ("Writing Skills"). Based on their research, the commission concluded the following:

- People need to be able to write and speak well in order to be hired. These skills are mandatory to succeed on the job. Many companies consider communication ability when promoting employees. According to one respondent to the survey, "Writing could be your ticket in . . . or it could be your ticket out" ("Writing Skills").
- Over two-thirds of employees have to write frequently on the job.
- In the fields of finance, insurance, and real estate, over 80 percent of employers consider writing skills during the interview and hiring process.

Using Time

In addition to serving valuable purposes on the job, getting a job, or meeting your needs as a customer, communication is important because it is time consuming. Just imagine how much of your time at work will be spent communicating with others.

Osterman Research, Inc., in a published white paper, reports that corporate employees send about 43 e-mail messages each day and receive around 123 more. That means that when workers aren't writing e-mail messages, they're reading them. Between writing and reading e-mail messages, employees spend on average "134 minutes per day working in email, or about 28% of an eight-hour day" ("The Importance of E-mail Continuity"). In addition, Osterman Research found that workers

- Speak on the telephone around an hour a day
- Use real-time communications tools like instant messages, text messages, videoconferences, or wikis about a half hour each day
- Communicate through social media around 10 minutes a day

See Figure 1.2 for a breakdown of time spent communicating on the job.

Though you will spend a great deal of time writing, reading, and speaking on the telephone at work, you will spend even more time communicating in other ways. Calculate the time you will spend verbally communicating in meetings, walking to and from the elevator with your coworkers, and on collaborative work teams while discussing how to complete a project. You will use oral communication skills when speaking to a customer

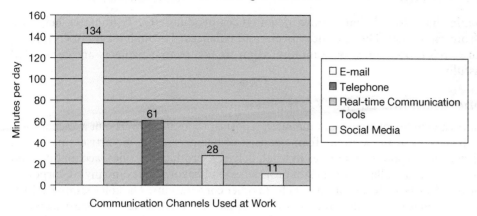

FIGURE 1.2 Time Spent Communicating on the Job

Y-axis: Minutes per day
X-axis: Communication Channels Used at Work

- E-mail: 134
- Telephone: 61
- Real-time Communication Tools: 28
- Social Media: 11

Legend:
- ☐ E-mail
- ▨ Telephone
- ☐ Real-time Communication Tools
- ☐ Social Media

in your office or in the showroom. You will need to convince your boss to let you miss two workdays while you coach your child's soccer team at an out-of-town tournament. You also will need to use effective oral communication when you represent your company at the local speaker's bureau. When you are not writing at work, you will be speaking, listening, and reading much of the time.

Costing Money

You have heard it before—time is money. Here are three simple ways of looking at the cost of your technical communication.

- **Percentage of salary.** Consider how much of your salary is being paid for your communication skills. Let's say you make $35,000 a year and spend 28 percent of your time writing e-mail (as do many employees in the workplace). Your company is paying you approximately $9,800 just to write e-mail. That does not include the additional time you spend using other written communication channels or speaking to coworkers, vendors, and clients.

    If you are not communicating effectively on the job, then you are asking your bosses to pay you a lot of money for substandard work. Your time spent communicating, both in writing and orally, is part of your salary—and part of your company's expenditures.

- **Cost of training.** Corporations spend money to improve their employees' writing skills. The National Commission on Writing for America's Families, Schools, and Colleges reported that over 40 percent of the companies responding to its survey offer training classes for employees with poor writing skills. This costs businesses approximately $3.1 billion annually.

- **Generating income.** Your communication skills do more than just cost the company money; these talents can earn money for both you and the company. A well-written sales letter, flier, brochure, proposal, or Web site can generate corporate income. Good written communication is not just part of your salary—it helps pay your wages.

5. Describe how teamwork enhances communication in the workplace.

The Importance of Teamwork in Technical Communication

Companies have found that teamwork enhances productivity. Teammates help and learn from each other. They provide checks and balances. Through teamwork, employees can develop open lines of communication to ensure that projects are completed successfully.

Collaboration in a Digital Workplace

Collaboration and Groupware

See Chapter 2, "Digital Communication," for additional information about groupware and teamwork.

In business and industry, many user manuals, reports, proposals, PowerPoint presentations, and Web sites are team written. Teams consist of engineers, graphic artists, marketing specialists, and corporate employees in legal, delivery, production, sales, accounting, and management. These collaborative team projects extend beyond the company. A corporate team also will work with subcontractors from other corporations. The collaborative efforts include communicating with companies in other cities and countries through teleconferences, faxes, and e-mail. Modern technical communication requires the participation of communities of practice: formal and informal networks of people who collaborate on projects based on common goals, interests, initiatives, and activities.

Collaboration on the job is increasing as individual work decreases because of the complexity of issues. The National Association of Colleges and Employers annually lists the skills that employers want, including oral and written communication skills and teamwork.

Collaboration ranks highly for employers because teamwork helps employees accomplish the following goals:

- Brainstorm for new ideas and consensus
- Make decisions
- Solve problems
- Determine team roles
- Assign tasks
- Complete work by team-determined deadlines

Problems with Silo Building

Working well with others requires collaboration versus silo building. The *silo* is a metaphor for departments and employees that behave as if they have no responsibilities outside their areas. They build bunkers around themselves, failing to collaborate with others. In addition, they act as if no other department's concerns or opinions are valuable.

Such stand-alone departments or people isolate themselves from the company as a whole and become inaccessible to other departments. They narrowly focus on issues, which creates problems. Poor accessibility and poor communication result in duplicate efforts and lack of cooperation. To be effective, companies need departments to maintain open lines of communication. The successful employee must be able to work collaboratively with others to share ideas. In the workplace, teamwork is essential.

Why Teamwork Is Important

Teamwork benefits employees, corporations, and consumers. By allowing all constituents a voice in project development, teamwork helps to create effective workplaces and ensures product integrity.

Diversity of Opinion. When you look at problems individually, you tend to see issues from limited perspectives—yours. In contrast, teams offer many points of view. For instance, if a team has members from accounting, public relations, customer service, engineering, and information technology, then that diverse group can offer diverse opinions. You should always look at a problem from various angles.

Checks and Balances. Diversity of opinion also provides the added benefit of checks and balances. Rarely should one individual or one department determine outcomes. When a team consists of members from different disciplines, those members can say, "Wait a minute. Your idea will negatively impact my department. We had better stop and reconsider."

Broad-Based Understanding. If decisions are made in a silo, by a small group of like-minded individuals, then these conclusions might surprise others in the company. Surprises are rarely good. You always want buy-in from the majority of your stakeholders. An excellent way to achieve this is through team projects. When multiple points of view are shared, a company benefits from broad-based knowledge.

Empowerment. Collaboration gives people from varied disciplines an opportunity to provide their input. When groups are involved in the decision-making process, they have a stake in the project. This allows for better morale and productivity.

Team Building. Everyone in a company should have the same goals—corporate success, customer satisfaction, and quality production. Team projects encourage shared visions, a better work environment, a greater sense of collegiality, and improved performance. Employees can say, "We are all in this together, working toward a common goal."

Types of Teams

Who will be on your collaborative team? This decision could come from management, who might assign team membership. Maybe the team will be formed by individuals who want to work together. Regardless of the directive, teams can take three predominant shapes: communities of practice, functional teams, or cross-functional teams. Each type of team has unique benefits and drawbacks.

> 6. Identify the types of workplace teams and their benefits.

Communities of Practice

Your team might be comprised of individuals with shared skills. These are called "Communities of Practice" or "CoPs." For example, if an entire team is made up of human resource employees, or accountants, or marketing specialists, that's a CoP. Every team member has the same knowledge base. Such groups can share best practices, a similar passion, subject matter expertise, common focus, and an innate trust in each individual's understanding of the topic.

Let's say that a company has purchased a new software program for reporting job-related travel expenses. Who would be best suited to write the instructions for employee use? If the company asks the IT staff to collaborate on this project, that team is a CoP. The team can function well because each member has the same level of knowledge about software, the company's objectives, and the audience's needs. That's the strength of a CoP. In contrast, the drawback might be the team's limited focus. Will the team have enough objectivity regarding the subject matter? A team that is too single minded could fail to recognize alternative approaches or points of view.

Functional Teams

Your team might be comprised of individuals with different but complimentary levels of expertise. This would be called a "Functional Team." For example, a company is rolling out a new product for an international market—a global positioning system watch to calculate travel distances. This watch will be accompanied by a user manual. The company's corporate communication department will develop this manual. For this project, the team will consist of writers, translators, and graphic artists. Each team member has a special talent. The writers know grammar, layout, and style; the translators know cultural habits and languages; the graphic artists put the words into images. This team can succeed because each individual has a precisely determined function. However, success depends on the team melding their skills and meeting their deadlines. A team leader might be needed to ensure that tasks are completed on time.

Cross-functional Teams

Your team might be comprised of individuals with completely different levels of expertise. This would be called a "Cross-functional Team." For example, a company wants to propose a new health and wellness program. The best approach would be to encourage input from across the entire company. The company could form a cross-functional team from multiple disciplines:

- Accounting to study the financial requirements
- Facilities to envision a location for workout classes
- Public relations to market the program
- Management to show corporate support
- Food services to focus on dietary issues
- Information technology to factor in online reporting techniques

The benefits of a cross-functional team are diversity of opinion, empowerment, buy-in, and a wide range of skills. Drawbacks could include a lack of focus, a disparity of skills, different resource levels, team members with conflicting schedules, and many more challenges. To achieve success with a cross-functional team, the majority of the team members must understand and agree to the project's objectives, timeframe, and individual responsibilities. Commitment to the team goal is mandatory.

Types of Team Processes

7. Explain how the various types of team processes operate.

Once a team is formed, how will the team proceed? What collaborative model will work best? Team processes can be as diverse as the team members or the task. No one approach is better than another, but it is good to recognize the following optional collaborative models: sequential, segmented, prioritized, or pragmatic.

Sequential Approach

Let's say you are involved in a team-written proposal. The proposal will include the following sequence of components: a title page, cover letter, table of contents, list of illustrations, abstract, introduction, body parts (including text and graphics), conclusion, and possibly a glossary and works cited page. One easy way to complete the team task is to tackle it sequentially. The team can work in concert on each part of the proposal in the order of its required parts. The team can first complete the title page, then the cover letter, then the table of contents, and so on. All members can collaborate simultaneously. This way, when one part is agreed on and completed, the team can move on to the next part.

Segmented Approach

Though the sequential approach keeps the team moving forward, it might also take more time, since everyone works together on each part. In contrast, a different approach might be to delegate responsibilities. Each team member can tackle one part of the project. Sue takes the title page, John writes the cover letter, Amir tackles the abstract, Joy-lin writes the introduction, and so on.

A segmented approach could profit from each team member's unique skills. The accountant could focus on the proposal's cost section, the graphic artist could create the tables and figures, the human resources employee could compile the staff credentials segment, and the technical writer could provide the proofreading and document design. Therefore, not only would a segmented approach save time, but also each segment would benefit from the writer's professional area of expertise.

Prioritized Approach

Sometimes, a team project contains a task that must be addressed immediately. For example, before technology requirements can be addressed, the team needs to determine budget. Before new employee development programs can be discussed, research must be completed on employee interests and needs. In situations like this, a team proposal couldn't be written sequentially or portioned out by segments. One part of the proposal would take priority. Therefore, a team's collaborative model might be directed by importance: what is most important, what is less important, and what is least important. Then, based on need, the team could proceed.

Pragmatic Approach

Finally, your team project could be organized by what you have versus what you need. If the team already has access to boilerplate information from a company's archives, the team might just start with what's practical and available. Since proposals, for example, have predesigned parts—the title page, cover letter, introduction, and so on—it might be easiest to insert what you have and then move on to more challenging parts of a project.

The approach your team takes to its task could be sequential, segmented, prioritized, or pragmatic. The collaborative model might even be a combination of these four processes. The key is to look at each approach as a tool and use whichever model or aspects of each model best works for your team and your task. Use Table 1.5 to help you organize a project.

TABLE 1.5   Project Plan

Team Member	Task	Status	Reporting Milestones	Due Date
List your team members' names.	List each team member's responsibility.	What has each team member accomplished to date? Include percentage of completion, problems encountered, and resources needed.	Determine how you want to assess team participation: • set up a wiki • establish dates for peer reviews/ editing • require a periodic e-mail progress report • call team meetings to discuss progress and problems	Determine due dates for each section of the project. Set a due date for the completed project.

Diverse Teams . . . Dispersed Teams in a Global Economy

8. Collaborate effectively on diverse and dispersed teams.

Collaborative projects will depend on diverse team members and dispersed team members.

Diverse Teams

Diversity

See Chapter 5, "Audience Recognition," for more information about diversity.

Teams will be diverse, consisting of people from different areas of expertise. Your teams will be made up of engineers, graphic artists, accountants, technical communicators, financial advisors, human resource employees, and others. In addition, the team will consist of people who are different ages, genders, cultures, and races.

Dispersed Teams

In a global economy, members of a team project might not be able to work together face to face. Team members might be located across time and space. They could work in different cities, states, time zones, countries, or shifts. For example, you might work for your company in New York, while members of your team work for the company at other sites in Chicago, Denver, and Los Angeles. This challenge to collaboration is compounded when you also must team with employees at your company's sites in India, Mexico, France, and Japan. According to a recent report, 41 percent of employees at the top international corporations live outside the borders of their company's home country (Nesbitt and Bagley-Woodward 25). As companies become more comfortable with new technologies, these companies will become mobile enterprises, allowing employees to work from diverse locations.

Choosing the Appropriate Channel for Collaboration

Technologies for collaboration are not equal. Some allow for more interactivity than others. Some are more synchronous than others. While some technologies depend solely on the written word, others allow for verbal and nonverbal communication among team members. When deciding which technology to use for your team efforts, you need to decide which collaborative tools aid frequent and effective communication. Table 1.6 provides you an overview of various technology channels for collaboration.

TABLE 1.6 Technology Channels for Collaboration

Collaboration Channel	Synchronous or Asynchronous	Written, Verbal, or Nonverbal	Interaction
Teleconference	Synchronous. Team members can have real-time, spontaneous communication.	All participants can be seen and heard. They can write, read, speak, and convey nonverbal cues (frowns, laughter, gestures, etc.).	All team members can participate equally with feedback and suggestions.
Webinar	Synchronous. Participants can engage in real-time, spontaneous communication.	All participants can talk and share written communication. Nonverbal communication is limited unless participants have Web cameras.	All team members can participate equally with feedback and suggestions.
E-mail, Text Message, Instant Message, and Twitter	Near synchronous. Participants can engage in near real-time communication if everyone is online or has his or her cell phone turned on.	All participants can share written communication. Nonverbal communication is limited to emoticons.	Participation is dependent on the team member's willingness to send and receive messages.
Blogs and Wikis	Asynchronous. Team members can access a blog or wiki at their own convenience.	All participants can share written communication. Nonverbal communication is limited to emoticons.	Participation is dependent on the team member's willingness to respond to the blog or wiki posting.

9. Identify the challenges to effective teamwork and how to overcome them.

Challenges to Effective Teamwork

Any collaborative activity is challenging to manage: Team members do not show up for class or work; one student or employee monopolizes the activity while another individual snoozes; people exert varying amounts of enthusiasm and ability; personalities clash. Some people fight over everything. Occasionally, when a boss participates on a team, employees fear speaking openly. Some team members will not stay focused on the subject. One team member will not complete an assignment. Group dynamics are difficult and can lead to performance gaps.

Human Performance Improvement

Human Performance Improvement (HPI) focuses on root cause analysis to assess and overcome the barriers inherent in teamwork. To close performance gaps, HPI analyzes the following possible causes for collaborative breakdowns:

1. **Knowledge.** Perhaps employees do not know how to perform a task. They have never acquired the knowledge or do not understand which skills are needed to complete the specific job. Varying skills of team members can impede the group's progress.

2. **Resources.** Think of these possibilities: Tools are broken or missing; the department is out of funds; you do not have enough personnel to do the job; the raw material needed for the job is below par; you ordered one piece of machinery but were shipped something different; you needed 100 items but have only 50 in stock. To complete a project, you often have to solve problems with resources.

3. **Processes.** For teams to succeed in collaborative projects, everyone must have specific responsibilities. Who reports to whom? How will these reports be handled (orally, in writing)? Who does what job? Are responsibilities shared equally? Structure, of some sort, is needed to avoid chaos, lost time, inefficiency, hurt feelings, and many other challenges to teamwork. To achieve successful collaboration, the team should set and maintain effective procedures.

4. **Information.** A team needs up-to-date and accurate information to function well. If required database information is late or incorrect, then the team will falter. If the information is too high tech for some of the team members, a lack of understanding may undermine the team effort.

5. **Support.** To succeed in any project, a team needs support. This could be financial, attitudinal, or managerial. When managers from different departments are fighting turf wars over ownership of a project, teams cannot succeed. Teams need enough money for staffing, personnel, or equipment.

6. **Wellness.** Another consideration involves the team's health and well-being. People get sick or miss work for health reasons. People have car accidents. If a teammate must miss work for a day or an extended period, this will negatively impact the team's productivity. Stress and absences can lead to arguments, missed deadlines, erratic work schedules, and poor quality.

7. **Acknowledgment.** For teams to succeed, management needs to recognize the benefits and results of collaboration. Successful workplaces reward staff with promotions, merit raises, bonuses, and awards.

8. **Trust.** A team is only as good as the trust each member has for the team effort. Team members who trust each other are more open in their brainstorming, more complimentary of each other's efforts, and work effectively toward a common goal.

9. **Leadership.** A team needs direction. Successful teams work well when their team leaders provide clear deadlines, responsibilities, task goals, and fair-minded leadership.

10. **Efficient communication.** If everyone on the team is overwhelmed with e-mail messages, text messages, tweets, or instant messages, no work will get done. Too much information is as bad as no information.

Ten HPI Intervention Techniques. After assessing root causes that challenge a team's success, HPI creates intervention options. These might include the following:

- Improved compensation packages
- Employee recognition programs
- Revised performance appraisals
- Improved employee training
- Mentoring or coaching
- Restructured work environments to enhance ergonomics

- Safety implementations
- Strategic planning changes
- Improved communication channels
- Health and wellness options—lectures, on-site fitness consultants, incentives for weight loss, and therapist and social worker interventions

People need help in order to work more effectively with each other. A progressive company recognizes these challenges and steps in to help.

Conflict Resolution in Collaborative Projects

10. Consider 10 ways to resolve team conflicts.

To ensure that team members work well together and that projects are completed successfully, consider these approaches to conflict resolution:

1. **Choose a team leader.** Sometimes, team leaders are chosen by management; sometimes, team leaders emerge from the group by consensus. However this person gains the position, the team leader becomes "point person," the individual whom all can turn to for assistance. He or she can solve problems, seek additional resources, or organize the team effort.

2. **Set guidelines.** One reason that conflicts occur is because people do not know what to expect or what is expected of them. To solve this problem, set guidelines. Hold an initial meeting (online or teleconferenced for remote, virtual teams) to define goals and establish guidelines, establish project milestones, or create schedules for synchronous dialogues. Communicate to all team members (before the meeting via e-mail or early in a project) how long the project will last. Also, clarify the team's goals, the chain of command (if one exists), and each team member's responsibilities. Use text messages, instant messages, e-mail, or Twitter messages for frequent updates.

3. **Ensure that all team members have compatible hardware and software.** This is especially important for virtual, remote teams. To communicate successfully, all team members need access to compatible e-mail platforms, computers, smartphones, and video equipment.

4. **Encourage equal discussion and involvement.** A team's success demands that everyone participate. A team leader should encourage involvement and discussion. All team members should be mutually accountable for team results, including planning, writing, editing, proofreading, and packaging the finished project. Be sure that everyone is allowed a chance to give input by participating orally, in writing, or in a wiki.

5. **Discourage taking sides.** Discussion is necessary, but conflict will arise if team members take sides. An "us against them" mentality will harm the team effort. You can avoid this pitfall by seeking consensus, tabling issues, or creating subcommittees.

6. **Seek consensus.** Not every member of the team needs to agree on a course of action. However, a team cannot go forward without majority approval.

To achieve consensus, your job as team leader is to listen to everyone's opinion, seek compromise, and value diversity. Conflict can be resolved by allowing everyone a chance to speak.

7. **Table topics when necessary.** If an issue is so controversial that it cannot be agreed upon, take a time out. Tell the team, "Let's break for a few minutes. Then we can reconvene with fresh perspectives." Maybe you need to table the topic for the next meeting. Sometimes, conflicts need a cooling-off period.

8. **Create subcommittees.** If a topic cannot be resolved, teammates are at odds, or sides are being taken, create a subcommittee to resolve the conflict. Let a smaller group tackle the issue and report back to the larger team.

9. **Find the good in the bad.** Occasionally, one team member comes to a meeting with an agenda. As team leader, seek compromise. Let the challenging team member speak. Discuss each of the points of dissension. Allow for input from the team. Some of the ideas might have more merit than you originally assumed.

10. **Deal with individuals individually.** From time to time, a team member will cause problems for the group. To handle these conflicts, avoid pointing a finger of blame at this person during the meeting. Do not react aggressively or impatiently. Doing so will lead to several problems:

   - Your reaction might call more attention to this person. Sometimes people come to meetings late or speak out in a group *just* to get attention. If you react, you might give the individual exactly what he or she wants.
   - Your reaction might embarrass this person.
   - Your reaction might make you look unprofessional.
   - Your reaction might deter others from speaking out. You want an open environment, allowing for a free exchange of ideas.

   Speak to any offending team members individually. This could be accomplished at a later date, in your office, or during a coffee break. Speaking to the person later and individually might defuse the conflict.

CHECKLIST FOR COLLABORATION

_____ 1. Have you chosen a team leader (or has a team leader been assigned)?

_____ 2. Do all participants understand the team's goal and their individual responsibilities?

_____ 3. Does the team have a schedule, complete with milestones and target due dates?

_____ 4. Does the team have compatible hardware and software for their wiki site or teleconference equipment?

_____ 5. In planning the team's project, did you seek consensus?

_____ 6. Have all participants been allowed to express themselves?

_____ 7. Have you avoided confronting people in public, choosing to meet with individuals privately to discuss concerns?

_____ 8. If conflicts occurred, did you table topics for later discussion or additional research?

_____ 9. Did you encourage diversity of opinion?

_____10. Have you remained calm in your interaction with the team?

CASE STUDY

You are the team leader of a work project at Gulfview Architectural and Engineering Services. The team has been involved in this project for a year. During the year, the team has met weekly, every Wednesday at 8:00 A.M. It is now time to assess the team's successes and areas needing improvement.

Your goal will be to recommend changes as needed before the team begins its second year on this project. You have encountered the following problems:

- One team member, Caroline Jensen, misses meetings regularly. In fact, she has missed at least one meeting a month during the past year. Occasionally, she missed two or three in a row. You have met with Caroline to discuss the problem. She says she has had child care issues that have forced her to use the company's flextime option, allowing her to come to work later than usual, at 9:00 A.M.

- Another team member, Guy Stapleton, tends to talk a lot during the meetings. He has good things to say, but he speaks his mind very loudly and interrupts others as they are speaking. He also elaborates on his points in great detail, even when the point has been made. If Guy isn't talking, he's texting.

- A third team member, Sharon Mitchell, almost never provides her input during the meetings. She will e-mail comments later or talk to people during breaks. Her comments are valid and on topic, but not everyone gets to hear what she says.

- A fourth team member, Craig Mabrito, is very impatient during the meetings. This is evident from his verbal and nonverbal communication. He grunts, slouches, drums on the table, gets up to walk around while others are speaking, and answers his cell phone.

- A fifth employee, Julie Jones, is overly aggressive. She is confrontational, both verbally and physically. Julie points her finger at people when she speaks, raises her voice to drown out others as they speak, and uses sarcasm as a weapon. Julie also crowds people, standing very close to them when speaking.

Assignment

How will you handle these challenges? Try this approach:

- Analyze the problem(s). To do so, brainstorm. What gaps might exist causing these problems?
- Invent or envision solutions. How would you solve the problems? Consider HPI issues, as discussed in this chapter.
- Plan your approach. To do so, establish verifiable measures of success (including timeframes and quantifiable actions).

Write an e-mail to your instructor sharing your findings.

ETHICAL CHALLENGE

Sophia Rose is a public relations employee for Corporate Reflector, a "makeover" company that works to improve a client's image. One of Corporate Reflector's new clients, the city of Greenfield, has asked for help. Sophia began planning by interviewing the city's residents and business owners and workers. She used a *S-W-O-T* analysis, focusing on the city's <u>S</u>trengths, <u>W</u>eaknesses, <u>O</u>pportunities for improvement, and any <u>T</u>hreats to the city's success.

Prior to the interviews, Sophia clarified her goals to the city participants. They all were made aware that their comments eventually would be used for reporting purposes. During the interviews, to ensure accuracy, Sophia either taped what the people said or took very thorough notes. Before writing her rough draft, Sophia always double-checked with her sources of information to be sure that their quotes are correct.

Sophia has contacted one individual, Burt Knoblauch, for verification. Upon hearing what Sophia plans to use as his quote, Burt first denies having said it. Sophia, however, has his comment on tape and plays it for him. Burt then tells Sophia that he does not want her to use his comments because his boss and his neighbors might be offended. Sophia reminds Burt that she had told him and all *S-W-O-T* participants that their comments would be used in a follow-up report. She even has signed release forms for Burt and others.

What should Sophia do? She knows for a fact that her quote is accurate. In addition, Burt's quote is essential to her report, since his comments highlight a key problem facing the city. Still, Burt has told her that using the quote could cause him problems at work and in his neighborhood.

Question

What is Sophia's ethical responsibility in this situation?

INDIVIDUAL AND TEAM PROJECTS

Teamwork—Business and Industry Expectations

Individually or in small groups, visit local banks, hospitals, police or fire stations, city offices, service organizations, manufacturing companies, engineering companies, or architectural firms. Once you and your teammates have visited these sites, have asked your questions (see the following assignments), and have completed your research, share your findings using one of the following methods:

- **Oral**—As a team, give a 3- to 5-minute briefing to share with your colleagues the results of your research.
- **Oral**—Invite employee representatives from other work environments to share with your class their responses to your questions.
- **Written**—Write a team e-mail message, letter, or report about your findings.

  1. Ask employees at the sites you visit if, how, and how often they are involved in team projects. In your team, assess your findings and report your discoveries.
  2. Ask employees at the sites you visit about the different communication channels they use on the job. In your team, assess your findings and report your discoveries.
  3. Research companies that rely on teamwork. Focus on which industries these companies represent and the goals of their team projects. You could also consider the challenges they encounter, their means of resolving conflicts, the number of individuals on each team, and whether the teams are diverse. Then report these findings to your professor or classmates, either orally or in writing.
  4. Visit the Society for Technical Communication (STC) Web site to learn about its membership. See which industries employ technical communicators and determine these writers' job responsibilities. Also, learn which colleges and universities have programs in technical communication and what the programs entail. What else can you learn about technical communication from the STC Web site?
  5. Research major publications of technical communication, such as *Intercom, Technical Communication,* and *The Journal of Scientific and Technical Communication.* What topics do the articles in these journals focus on?

PROBLEM-SOLVING THINK PIECES

To understand and practice conflict resolution, complete the following assignments.

1. Attend a meeting. This could be at your church, synagogue, or mosque; a city council meeting; your school, college, or university's board of trustees meeting; or a meeting at your place of employment. Was the meeting successful? Did it have room for improvement? To help answer these questions, use the following Conflict Resolution in Team Meetings Matrix. Then, report your findings to your professor or classmates, either orally or in writing. Write an e-mail message, memo, or report, for example.

CONFLICT RESOLUTION IN TEAM MEETINGS MATRIX			
Goals	Yes	No	Comments
1. Were meeting guidelines clear?			
2. Did the meeting facilitator encourage equal discussion and involvement?			
3. Were the meeting's attendees discouraged from taking sides?			
4. Did the meeting facilitator seek consensus?			
5. Did the meeting facilitator deal with individuals *individually*?			

2. Have you been involved in a team project at work or at school? Perhaps you and your classmates grouped to write a proposal, research Web sites, create a Web site, or perform mock job interviews. Maybe you were involved in a team project for another class. Did the team work well together? If so, analyze how and why the team succeeded. If the team did not function effectively, why not? Analyze the gaps between what should have been and what was. To help you with this analysis, use the following Human Performance Index Matrix. Then, report your findings to your instructor or classmates either orally or in writing. Write an e-mail message, memo, or report, for example.

HUMAN PERFORMANCE INDEX MATRIX			
Potential Gaps	Yes	No	Comments
1. Did teammates have equal and appropriate levels of knowledge to complete the task?			
2. Did teammates have equal and appropriate levels of motivation to complete the task?			
3. Did the team have sufficient resources to complete the task?			
4. Did teammates understand their roles in the process needed to complete the task?			
5. Did the team have sufficient and up-to-date information to complete the task?			
6. Did the team have sufficient support to complete the task?			
7. Did wellness issues affect the team's success?			

WEB WORKSHOP

1. How important is technical communication in the workplace? Go online to research this topic. Find five Web sites that discuss the importance of communication in the workplace, and report your discoveries to your teacher and/or class. To do so, write a brief report, memo, or e-mail message. You could also report your information orally.

2. Create a class wiki for collaborative writing. To do so, consider using one of the following sites:

   - WetPaint
   - Wikispaces
   - eGroupWare
   - Group-Office
   - TikiWiki
   - Google Docs

CHAPTER TWO

Digital Communication

After completing this chapter, you will be able to

1. Understand the effect of digital communication in today's workplace

2. Consider the impact of social media in the workplace

3. Recognize mobile apps as part of communication on the job

4. Employ Quick Response (QR) codes for ease of access in businesses

5. Use groupware to collaborate in the digital workplace

6. Create effective collaborative documents and video communication

COMMUNICATION AT WORK

In the Century Software scenario, employees in diverse locations reveal the importance of digital communication.

Shelly Smith is a product reliability team leader at **Century Software**, a company that designs, develops, sells, and maintains software systems for payroll, benefits, and insurance.

Why does Century use teams? Shelly says, "Teams are important when a company is trying to improve procedures. Change is always difficult. This is especially true when changes impact individual workers. To avoid surprises, you must have buy-in from all stakeholders. The team needs to understand the current situation, have a chance to give their input, know what's coming, and understand how any changes will affect them personally."

Shelly's goal as team leader is to ensure product reliability. Since product development depends on different departments within Century nationwide, Shelly formed a cross-functional team. Her group consists of 15 employees from software development, design, quality assurance, sales, and customer service.

Shelly spends between 15 to 20 hours a week as team leader. Her job includes acquiring and maintaining a budget, facilitating the team meetings and team tasks, and keeping the team on track. Plus, she must stay in touch with her team members. Century uses five digital communication channels to ensure a consistent, collaborative workforce: online calendaring, teleconferences, videoconferences, webinars, and wikis.

- **Online calendaring.** With her company's collaborative scheduling software, Shelly can propose meeting times for her team. Then, the team invitees can check their schedules online and agree upon times when everyone is available to participate in teleconferences or videoconferences.
- **Teleconferences.** Shelly, who works in Century's Birmingham office, has team members who work in San Francisco, Salt Lake City, Cleveland, Columbus, Rochester, and Baltimore. To communicate with her dispersed team members, she has a conference call just to touch base. Sometimes there is a formal agenda, and sometimes she asks questions to see what is going on. A casual, weekly teleconference allows her to stay up to date on issues facing each member of the team. Shelly believes that collaborating is good for problem solving.
- **Videoconferences.** You can't communicate effectively with 15 people on the telephone. While teleconferencing works well for Shelly and smaller groups, when she needs to communicate with the entire team about corporatewide issues that affect policy, budget, personnel, and strategic planning, face-to-face meetings might be the optimum solution. However, transporting numerous people to a central location is neither time-efficient nor cost-effective. A three-hour meeting might require two days of travel plus hotel, food, and air fares. To save time and money, Century uses videoconferences.
- **Webinars.** In the course of software development, training is required so team members can learn new applications. Webinars allow Shelly's team to participate in online training seminars without incurring travel expenses. Century's webinars allow Shelly's dispersed team to participate in live group discussions and training from their individual computers.
- **Wikis.** Wikis allow Shelly's team members to write collaboratively. They can add, remove, and edit content online. Plus, a wiki lets them track the history of a document as it is revised. Wikis help the dispersed team to create effective reports.

Shelly says that she spends approximately 50 percent of her work time communicating via e-mail messages, telephone calls, and teleconferences. For efficiency, cost savings, and consistent communication to a geographically dispersed workforce, Century has found that multiple, digital channels help team members achieve their communication goals.

On-the-Job Communication in a Digital Workplace

> 1. Understand the effect of digital communication in today's workplace.

Today's workplace has changed. The days of fumbling through a briefcase for loose paper and then thumbing through those papers to find relevant text are gone. Professional communication is no longer paperbound, no longer tethered to a desk or a desktop. Technical communication in a 21st-century workplace is increasingly digital, globally networked, mobile, multiplatform, and adaptable to emerging technologies.

Workplace professionals now can communicate quickly and efficiently from the palms of their hands. An employee can use his or her smartphone or tablet to

- Check an online calendar to learn which meeting is upcoming
- Scroll through e-mail messages while walking to the meeting
- Text a colleague about a change in venue for the meeting
- Download a progress report from the cloud

- Create a pie chart to include in the report
- Snap a photo of notes the meeting facilitator writes on a white board
- Leave the meeting and update the company's Facebook page while walking back to the office

Technical writers now communicate through e-mail, text messages, instant messages, Web sites, and social media. In addition, they convey content in motion. With mobile devices, workplace communication is no longer an 8-to-5 job. Employees, locally and globally, communicate to their colleagues all the time from anyplace.

The Pew Research Center states that by 2020, mobile devices will be the primary way in which people connect to the Internet (Welinske). Digital communication through Web sites, blogging, e-mail messages, text messaging, instant messaging, wikis, webinars, collaborative communication tools, and more has impacted the size, speed, and tone of business correspondence.

This trend is especially evident in the smartphone and tablet, multiuse tools for researching the Internet, taking photographs, making phone calls, and texting. Approximately 75 percent of smartphone and tablet users text consistently (Lenhart). The omnipresence of smartphones and tablets, communication channels that are both written and oral, can have a variety of values for technical communication.

Smartphone and Tablet Portability Equals Mobility

Smartphones and tablets allow people to communicate everywhere, anywhere, anytime—content in motion. The mobility of smartphones and tablets is valuable for employees involved in work-related travel or employees who telecommute. From your smartphone or tablet, you can text, tweet, and send e-mail messages to your coworkers and clients—from wherever you are to wherever they are.

Digital Communication Encourages Multitasking

Imagine this scenario: You're at your office writing a formal report. You need additional information to develop your ideas. To research, you use your smartphone to call a colleague. She verbally shares data with you over the phone, sends an e-mail message with a pie chart she just developed, suggests that you access a Web site she's just found that will provide you details for your report, and follows that up by retweeting a Twitter message she just received that talks about your topic. You can also take a photo with your smartphone and use the image in your report to document workplace issues. You're on your office computer, but simultaneously, you're using your phone to multitask.

Digital Communication Allows for Immediate Feedback

What are others thinking? With a smartphone or tablet, you can access your Twitter or Facebook accounts and find out what's what. You can read your blog or someone else's on the go. You can use your smartphone to see what your LinkedIn connections are up to (new jobs, skills, accomplishments, and more). Digital communication tools let you access a network of colleagues, clients, followers, connections, and friends. Here's another example of how your smartphone or tablet lets you get feedback. A colleague sends you a tweet and tells you to click on a hashtagged (#) word or phrase in the tweet. This allows you to access other tweets in that category or topic. The hashtag, which is community-driven metadata, provides you additional, immediate content for a report you're researching.

Apps Can Help You Find New Job Opportunities

Mobile apps provide new job opportunities. With mobile apps proliferating, employees will need to add a new skill to their skill set. In addition to hardcopy and online help, employees will need to design user help tutorials and assistance for a mobile environment that demands less screen space, a smaller work/viewing area, and fewer words. Conciseness and clarity have always been important in technical communication. With digital communication, these skills are even more essential.

Social Media

2. Consider the impact of social media in the workplace.

Perhaps the most significant technological trend in today's digital workplace is social media. Nielsen statistics show that Facebook, YouTube, and Wikipedia are the most visited social media sites. Almost 80 percent of Fortune 500 companies use blogs, Twitter, Facebook, and YouTube for communication with stakeholders (Barnes and Lescault). Through social media, businesses connect with people. Pinterest is another growing social media site, allowing people to express their creativity and create brands. LinkedIn is an invaluable social media tool in the job search, helping individuals connect with employers, network with employees, and promote one's professional profile. Twenty-two percent of all online visitors spend their time on social networks and blog sites. Note the following statistics related to social media usage.

- Facebook has over one billion users.
- LinkedIn has over 300 million users.
- Twitter has over 200 million monthly users.
- YouTube has over one billion monthly viewers.

Social Media

See Chapter 10, "The Job Search," and Chapter 12, "Social Media," for information about social media channels.

Due to the emergence of social media as a major force in technical communication, we have devoted Chapter 12 to a discussion of Facebook, Twitter, YouTube, and blogging.

Communication Advances Through Mobile Apps

3. Recognize mobile apps as part of communication on the job.

Apps, short for application software, are the perfect accompaniment for smartphone users. Workplace communicators need to be aware of and understand apps used for writing because technology alters the length and look of communication. Apps are part of this changing technology and an important part of communication today.

An Example of Apps at Work

For example, Marziah Karch is a senior instructional designer at Northwestern Evaluation Association and the author of instructional books, including *Android for Work, Droids Made Simple,* and *Android Tablets Made Simple.* The books explain how business professionals can successfully use Android and ensure workplace productivity while taking their offices on the road. Marziah uses mobile apps to

- Write and check her e-mail
- Manage her calendar
- Inform colleagues of her location (conference, airport, hotel, restaurant, customer's site, etc.)
- Create portable WiFi hotspots to use her laptop, tablet, or other electronic devices
- Access a documents dropbox
- Store and retrieve documents from the cloud
- Log her travel itinerary

Marziah tends to compose either on her laptop, tablet, or desktop PC, but when she's on the road, she uses mobile writing apps to open Word or PDF documents for editing or previewing. The key to mobile apps, for Marziah, is instantaneous access from the comfort and convenience of her smartphone (Karch).

Types of Writing Apps

Writing apps cover a wide gamut of capabilities. Some apps are limited to note taking, blogging, and simple text. Others allow for a greater scope of options, including font sizes and colors, graphics, highlighting, and organization into distinct folders. Still other apps focus on PowerPoint development. See Table 2.1 for a list of current writing apps available for technical communication and an overview of their capabilities:

TABLE 2.1 Mobile Apps

Apps Names	App Capabilities
BlogWriter	• limited to text blogging • sync, publish, edit, and delete posts • upload and post photos
Documents To Go	• view, edit, and create Microsoft Word documents • view, edit, and create PowerPoint presentations • embedded graphics and tables
Evernote	• access through Windows, Macs, and mobile devices • take notes, tag them, and organize notebooks • share content
MS OneNote	• take notes and import documents, PDFs, and images
Notesy	• create, edit, save, and synchronize notes to a Dropbox account
Pages	• create, edit, and view documents • insert tables and images • sort and organize documents into folders
Quickoffice Mobile Office Suite	• create and edit Microsoft Office files • create and edit PowerPoint slides
Simplenote	• organize and tag ideas • search notes and a revision history • share text through web-enabled devices
Workflowy	• outline with sublists presented as bullet points

> 4. Employ Quick Response (QR) codes for ease of access in businesses.

Communication Advances Through QR Codes

A Quick Response code, or QR code, is a two-dimensional image that can be scanned by a mobile device with an appropriate app. Once the end user scans the code, he or she can access various information, including social media, e-mail newsletters, event registration, prices, local deals, technical specifications, enhanced instructional content, and more.

EXAMPLE ▼

Reasons for Creating a QR Code

Here's how businesses use QR codes:

- **Help customers access your Web site.** A client can be shopping in a store, scan the QR code associated with a product, and access your Web site. This can give your customer more knowledge about the company and lead to additional sales.
- **Give participants easy access to an online survey.** Let's say you're creating research for a proposal or seeking input from end users regarding your company's products or services. Participants can scan the QR code and get access to your survey from the convenience of their mobile devices.

- **Increase e-commerce through e-mail lists.** By scanning a QR code, clients could add their e-mail addresses to your company's contact list. This would help customers stay connected to your business. You then could send e-mail updates regarding upcoming events, sales, or new product releases.
- **Link directly to your social media sites.** This way, participants can "like," "follow," or "pin" you and your products, increasing online traffic for your company.
- **Give people instant access to your corporate blog.** Through a QR code, you can grow your blog readership.
- **Lead viewers to your micro-videos.** If an audience can scan a QR code link to your micro-video sites, such as Skype's Qik, YouTube, Vine, or Instagram, you could help make a video go viral.
- **Use QR codes to enhance user manuals.** You have a new product that will be packaged in a 2-inch by 4-inch box. The size of the box limits the amount of instructional content you can include. With a QR code, the end user can access more detailed instructions online.
- **Add QR codes to product packaging.** This will let customers link to online customer service contact information (e-mail address, street address, Web sites), user manuals, and technical specifications, warranties, or product add-ons.

As an added benefit, you can access free QR code generators.

Using Groupware to Collaborate in the Digital Workplace

> 5. Use groupware to collaborate in the digital workplace.

When employees are dispersed geographically, getting all team members together would be costly in terms of time and money. Companies solve this problem by forming virtual, remote, digital teams that collaborate using electronic communication tools. Some of these tools are as simple as using Microsoft Word's track changes, which allows team members to comment on a document and to track the changes that are suggested.

TECHNOLOGY TIPS

Using Microsoft Word 2013 to Track Changes as You Edit

1. Open a document for revision.
2. Click the **Review** tab.

3. In the **Tracking** grid, click **Track Changes**.
4. Suggest revisions to the text by adding your comments.

Note: When you turn off **Track Changes**, your edits are not removed.

(Continued)

Following is an example of text that a team member has revised with suggested changes.

Date: August 13, 2014
From: Nicole Stefani
To: Greenfield City Council
Subject: ~~Proposal~~

Inserted: ¶
Nicole, improve your subject line by adding the focus of this proposal ¶

Introduction:

The City of Greenfield wants to promote the city. Greenfield city management wants to inform new homeowners and prospective businesses to move to Greenfield. ImageSkill can help Greenfield meet this problem.

~~Discussion:~~

Inserted: ¶
Could you make these headings more informative? Also, the semicolon is incorrect¶

To determine the best way to promote growth in Greenfield, ImageSkill staff followed this procedure. We met with a representative body of Greenfield citizens to ask questions. From this survey, ImageSkill determined which ~~area's~~ Greenfield citizens considered to be the city's most alluring. Our goal was to decide how best to maximize Greenfield's perceived strengths and opportunities for growth and give us insight into the best ~~area's~~ in which growth could be accomplished.

Inserted: Spelling error vs. "areas"

Inserted: Same spelling error

The citizens of Greenfield want to accomplish ~~there~~ goals without hurting the city's strengths. They don't want pollution, traffic, increased taxes, or higher housing costs. They want to maintain Greenfield's "small town ~~friendliness," and~~ the city's charm.

Inserted: Spelling error vs. "their"

Inserted: No comma needed here

ImageSkill can help Greenfield solve its problems. ~~We can create brochures for your Chamber of Commerce that will market your city's charms and intrinsic values as well as promote growth opportunities for new businesses, educational growth, entertainment, and parks/recreational prospects.~~ We can build a web site and a blog for your city, highlighting all of the values your citizens are so proud of. ~~As your city grows, citizens will resent change, which is inevitable, so packing this growth is the challenge, and ImageSkill solve this problem by providing you outstanding radio, tv, social media, and newspaper sound bites as well as periodical articles that promote the city's changes as best for the common good.~~

Inserted: 33 words—too long for conciseness

Inserted: 51 words—too long

Inserted: Consider bulleting points in this paragraph

Conclusion:

ImageSkill wants to be your full service marketing agent. We have the talent and the skill to enhance the City of Greenfield's growth opportunities.

Other collaborative tools include groupware, which consists of software and hardware that helps companies cut down on travel costs, allows for telecommuting, and facilitates communication for employees located in different cities and countries.

Groupware includes the following types of hardware and software:

- Digital conferencing tools such as webinars, listservs, chat systems, message and discussion boards, videoconferences, and teleconferences.
- Digital management tools, such as digital dashboards, and project management software that schedules, tracks, and charts the steps in a project. Other examples of electronic management tools are Microsoft Outlook's electronic calendaring or Google's Doodle. These allow you to send a meeting request to dispersed team members, check the availability of meeting attendees, reschedule meetings electronically, forward meeting requests, and cancel a meeting—without ever visiting with your team members face-to-face.
- Digital communication tools for writing and sending documents. These include tools like instant messages, e-mail, blogs, intranets and extranets, and wikis.
- Social media for communication of text and videos through Facebook, YouTube, Twitter, virtual worlds (Second Life), and more.

Routine Correspondence

See Chapter 9, "Routine Correspondence," for additional information about e-mail and instant messages.

Social Media

See Chapter 12, "Social Media," for additional information about social media channels of communication.

Collaborative Writing and Video Communication Tools

In the digital workplace, employees use collaborative writing tools to create effective documents.

6. Create effective collaborative documents and video communication.

Wikis

What's a *wiki*? A wiki is a Web site that allows writers to create and edit text collaboratively. Whenever a team member edits text in the wiki, newly created text becomes the version on the screen and previously written versions are stored.

Who's Using Wikis?

Many companies use wikis for collaborative writing projects. Yahoo uses a wiki. Eric Baldeschwieler, director of software development of Yahoo!, says Web collaboration is very important to its hundreds of employees located worldwide. Cmed runs pharmaceutical clinical trials and develops new technology. In this heavily regulated environment, wikis "improved communication and increased the quantity (and through peer review, the quality) of documentation." Cingular Wireless encourages its managers to use wikis for collaboration rather than relying on e-mail messages. Disney's engineers use wikis to post and maintain developmental specifications. Texas Instrument's India design center uses wikis to manage all project specific information, such as documenting ideas, plans, and status; sharing information with other teams across various work sites; and updating information and content to team members ("Twiki Success Stories").

How Can You Use a Wiki?

In your dispersed teams, whether virtual, remote, or mobile, you might use wikis in the following ways to create collaborative documents:

- **Create Web sites.** Wikis help team members easily add pages, insert graphics, create hyperlinks, and add simple navigation.
- **Aid project development with peer review.** A wiki makes it easy for team members to write, revise, and submit projects since all three activities can take place in the wiki.
- **Encourage group authoring.** Wikis allow group members to write and edit a document. This creates a sense of community within a group, allows group members to build on each other's work, and provides immediate, asynchronous access to all versions of a document.
- **Track group projects.** Each wiki page lets you track how group members are developing their contributions. The wiki also lets you give feedback and suggest editorial changes.

Web Sites

See Chapter 13, "Web Sites and Online Help," for additional information about Web site design.

Benefits of Using Wikis

Often in the business world, people are only familiar with wikis through Wikipedia. Wikipedia is an open-source, online encyclopedia. Because it is open-source, anyone can write and revise content. A wiki offers many benefits to collaboration in the workplace including the following:

1. **Involve all team members in the writing process.** When many individuals brainstorm ideas and participate in writing, revising, and proofreading, this can lead to an improved document.
2. **Minimize the need for face-to-face meetings.** An asynchronous wiki lets team members work on a project anytime and anywhere. Rather than having to attend a meeting, team members can receive updates to their wikis through e-mail, text messages, or instant messages. Because a wiki allows for detailed revisions and

discussions, meetings can be shorter and concentrate only on the topics that would benefit from face-to-face communication.

3. **Limit excessive and lengthy e-mail messages.** When content is available for threaded discussion within the wiki, the need for a constant flow of e-mail messages about the project diminishes.

4. **Transmit the finished product to end users.** The content in the wiki can be shared with customers or other individuals via Word documents or PDF files through cloud computing.

5. **Organize the team's work and maintain privacy.** A wiki allows you and your team to have in one location all of the material related to a project. In the wiki, you can designate which content is viewable by team members. You can also share designated information with anyone in the company.

Figure 2.1 is an illustration of a wiki.

FIGURE 2.1 Wiki Page

By clicking "Edit page," any team member can revise the text.

What the audience sees when the wiki is first opened.

Google Docs and Google Sites

Other collaborative writing tools you can use easily are Google Docs and Google Sites.

Google Docs.   Using Google Docs, you and group members can edit Word documents, RTF (rich text format), and HTML (hypertext markup language) files. Teams can be at any location on their computers and work on one document simultaneously. Changes made by one writer will be seen by all team members instantly.

Google Docs provides you these benefits:

- Upload and save files from remote work locales
- Edit and view a document
- Show changes in real time
- View a document's revision history
- Return to earlier versions
- Add new team members or delete writers

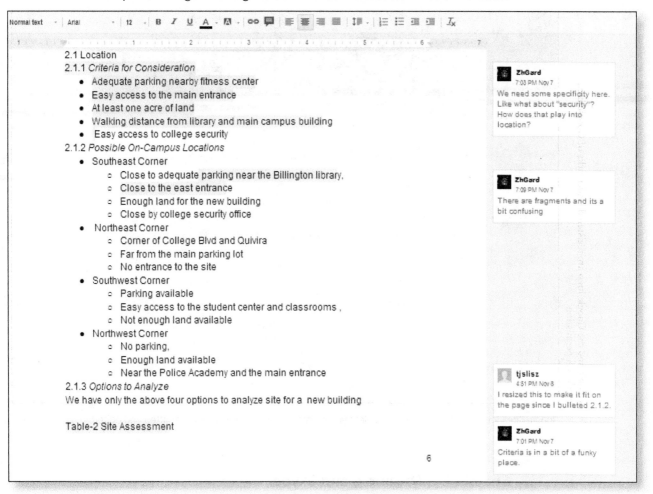

Figure 2.2 shows a page in Google Docs with comments from team members.

- Choose who can access your site
- Post documents to a blog or publish a document to a Web page

Figure 2.2 shows a page in Google Docs with comments from team members.

Google Sites. Google defines this collaborative writing tool as "a free and easy way to create and share webpages." Through Google Sites, you can

- Organize and plan meetings and activities
- Securely share information
- Collaborate in teams
- Communicate with family and friends
  (Google Sites)

See Figure 2.3 for a sample Google Sites screen capture.

FIGURE 2.3 Google Sites Task Screen

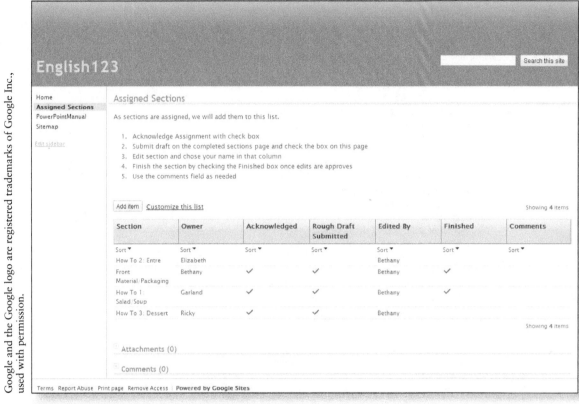

English123

Search this site

Home
Assigned Sections
PowerPointManual
Sitemap

Edit sidebar

Assigned Sections

As sections are assigned, we will add them to this list.

1. Acknowledge Assignment with check box
2. Submit draft on the completed sections page and check the box on this page
3. Edit section and chose your name in that column
4. Finish the section by checking the Finished box once edits are approves
5. Use the comments field as needed

Add item | Customize this list

Showing 4 items

Section	Owner	Acknowledged	Rough Draft Submitted	Edited By	Finished	Comments
Sort ▼	Sort ▼	Sort ▼	Sort ▼	Sort ▼	Sort ▼	Sort ▼
How To 2: Entre	Elizabeth			Bethany		
Front Material/Packaging	Bethany	✓	✓	Bethany	✓	
How To 1: Salad/Soup	Garland	✓	✓	Bethany	✓	
How To 3: Dessert	Ricky	✓	✓	Bethany		

Showing 4 items

Attachments (0)

Comments (0)

Terms Report Abuse Print page Remove Access | **Powered by Google Sites**

Google Sites lets you assign tasks. It also lets you record when work has been submitted and edited.

SharePoint, Dropbox, and Skype

Three popular ways to collaborate in today's workplace are through Microsoft SharePoint, Dropbox, and Skype.

SharePoint. Colleagues can publish writing and graphics to a SharePoint site. Then, coworkers can retrieve the text or graphics, revise the content, and review earlier versions. In addition, SharePoint is accessible on the go through smartphones and tablets. Finally, SharePoint helps teams collaborate by allowing coworkers to track meeting notes, set up calendar dates, and access a team's e-mail in one place.

Dropbox. Dropbox, as the name suggests, is an online site where employees and customers can simply drop files and photos. The information you and your team place in Dropbox can be accessed through synced computers, mobile devices, or through Dropbox's Web site. The content is secure since you determine who can access your documents.

Skype. This VOIP (voice over IP) service provides a visual, sound, and written way to collaborate. Skype allows users to make free Skype-to-Skype calls, visit with dispersed team members face-to-face, send and receive text and instant messages, and share graphics (photographs, videos, schematics, and more). Following are examples of how Skype is used for business purposes:

- One architectural engineering company uses Skype to help colleagues in Kansas City and Detroit communicate about team projects in real time. The team members not only can see and talk to each other but also can upload architectural drawings and discuss the visual's layout, design, and revisions as if the coworkers were in the same room.
- Perceptive Software is located in Kansas but is owned by Lexmark, headquartered in Lexington, Kentucky. Recently, Perceptive was asked to hire new technical writers for Lexmark's India facility. Paying candidates to fly from India to Kansas was prohibitive, so the Perceptive technical writers interviewed prospective employees in India by using Skype. This way, candidates and employers could communicate face-to-face virtually.
- A Honda dealership in Kansas no longer employs a finance manager to help complete car purchase details. Instead, the dealership subcontracts this job to a "financial producer" located in St. Louis, Missouri. The financial producer then meets virtually with car buyers via Skype. Car finance options are shown with graphics and animation on a large, flat-screen monitor; paperwork is displayed and signed on another touch-screen monitor; documents are transmitted by fax and electronically; the face-to-face interaction is completed digitally.

Other Collaborative Sites

In addition to Google Docs, Google Sites, SharePoint, and Skype, other collaborative sites include WetPaint, PBWorks, Wikispaces, Access Grid, eGroupWare, Group-Office, Mind-quarr, TikiWiki, and Trac. You also could use SkyDrive, giving you and your team access to documents, photographs, and files from a PC, Mac, phone, or tablet.

CHECKLIST FOR DIGITAL COMMUNICATION

_____ 1. To enhance a job search, have you used LinkedIn?

_____ 2. Have you used a tweet to communicate with a colleague about a work-related issue?

_____ 3. Have you used social media such as Facebook or YouTube to connect with other business professionals in your field?

_____ 4. Have writing apps enhanced your communication skills?

_____ 5. Did you use QR codes to connect end users to products, services, a survey, your portfolio, or social media?

_____ 6. Have you used track changes in a team project to comment on a document and track the changes suggested?

_____ 7. Did your team use a wiki to edit text in a team project?

_____ 8. Did your team use Google Docs to upload and save files from remote work locales?

_____ 9. Did you use Google Sites with your team to organize and plan meetings and activities?

_____10. Did you use Skype to enhance communication among your colleagues?

CASE STUDY

Shelly Smith, product reliability team leader at Century Software, manages a team of 15 employees. All of them use their smartphones for texting, calling, accessing the Internet, calendaring, and more work-related tasks. However, they also use their smartphones for personal purposes. Here's where problems occur. If a corporate smartphone is used for personal reasons, network security regarding confidential corporate data can be impaired.

Shelly needs to write a smartphone policy for her team members. It must include the following:

- Password protection
- Remote locking
- Data encryption
- Reporting policies for lost smartphone equipment
- Need to remotely wipe data
- Need to disable capabilities if an employee leaves the company
- Flexible standards for smartphone technology advances and emerging apps

Assignment

Research and write a corporate smartphone policy for Shelly Smith to emphasize the importance of digital communication in the workplace.

ETHICAL CHALLENGE

Pakash Patel is an outgoing employee at a large real estate corporation with offices in several locations. His company created a listserv for its 750 employees to ensure collegiality, to aid collaboration, and to disseminate corporate news. Pakash consistently uses the company listserv for the following reasons:

- Share jokes
- Advertise his daughter's cookie sales
- Sell his son's trash bags for soccer team travel
- Sell tickets for the religious events at his church
- Provide daily words of wisdom

Question

Is it ethical for Pakash to use his company's listserv in these ways? Why or why not?

INDIVIDUAL AND TEAM PROJECTS

1. Create a groupware site, for example, Google Sites. Use this site to assign team responsibilities. Then, use the groupware site for the team's writing and revision of a project.

2. Using online calendar software, such as Doodle, invite team members to a meeting and have them confirm their availability.

3. Use an online writing app. Then, write an e-mail message to your professor to explain the process of accessing this app, detail the app's uses, and comment on the app's pros and cons.

4. Smartphones and tablets can be used to accomplish many tasks. Make a list of ways you use your smartphone or tablet at school and personally. Then, share these uses with your classmates to learn what use might be unique and new to others.

5. What apps do you use that could have workplace value? Itemize the apps and explain your answer in either an e-mail message or memo to your professor.

6. What technologies (such as wikis, smartphones, tablets, Skype) do employees in your degree field use for collaboration? How often do employees use technology for collaborative work, what are the benefits of these technologies, and what are the problems technology creates for collaborative work?

PROBLEM-SOLVING THINK PIECE

Kate O'Brien is traveling internationally to troubleshoot a corporate problem. Her international company SportingStyle, which markets sports apparel, just purchased a smaller sporting goods company. The merger has led to challenges with labor relations, patent laws, licensing, and clashes in management styles.

During her trip, Kate needed to interview the new company's managers and workforce to determine areas for improvement. After the trip, Kate had to share her findings with her project team and report decisions to SportingStyle management.

Assignment

Review the following list of communication challenges that Kate confronted, and decide which digital communication channel she needed to use to complete her tasks. Write an e-mail message or give a short oral report to explain your decisions.

- To arrange and confirm meeting times for the various company managers
- To confer with her project team in the home office while she was traveling
- To share her company's management philosophy with the acquired company's entire staff
- To instruct the acquired company's staff on new corporate procedures
- To present the information to an international audience of dispersed stakeholders and shareholders

WEB WORKSHOP

1. Research groupware (collaborative software) products such as Google Docs, Microsoft SharePoint, IBM Lotus Notes, and Novell GroupWise. Compare and contrast the features each software product offers in terms of synchronous or asynchronous capabilities, conferencing, online proofing and revision control, project management, and so forth. Share your findings with your professor or the class in a short report.

2. Research online calendaring tools that help schedule a group meeting. Some of these products include Doodle, Microsoft Outlook, Google Calendar, Teamspace, Yahoo! Calendar, and 30 Boxes. Compare and contrast the features each software product offers in terms of access on the go from smartphones and tablets, mobile calendar synching with multiple devices, conferencing, project management, and so forth. Share your findings with your professor or the class in a short report.

CHAPTER THREE

The Communication Process

After completing this chapter, you will be able to

1. Produce successful documents using the writing process

2. Plan communication through the prewriting stage

3. Write documents by organizing and designing content

4. Rewrite documents to determine usability and correctness

COMMUNICATION AT WORK

In the following scenario, Creative International uses the writing process for effective communication with its clients.

Connie Jones (President), Mary Michelson (Project Director), and Lori Smith (Director of Sales and Marketing) have made **Creative International** a cutting edge company. Creative International works with organizations to define strategic communication goals. A key to their success is following a process "from the beginning to the end of a communication project." They prewrite, write, and rewrite.

Prewriting

- **Initial client contact.** Through telephone calls, e-mail messages, networking, their Facebook site, or a preliminary meeting, Creative gathers data to discover the client's needs. In this phase, the Creative team interviews the end users and observes them at work.

- **Clarify request meeting.** Meeting face to face with an upper-level decision maker, the Creative team collects information about the end-user's needs. Connie, Mary, and Lori don't just say, "Sure, we can do that job." Instead, they ask probing questions, such as "Why do you need that?" "Why do you want that?" "What do you want to communicate to your audience?"

- **Proposal creation.** Following the initial meeting, Creative writes a proposal, complete with schedules, project plans, the project's scope, and a description of the deliverables.

Writing

- **Drafting through collaboration.** The Creative team creates a wiki to help them write collaboratively.
- **Design, development, production, and pilot testing.** Creative designs text, graphics, audio and video training modules, and mobile apps for final beta testing.

Rewriting

- **Editing.** The Creative writing team (with input from other coworkers and the end user) revises initial drafts by making revisions through the wiki. This includes adding details, deleting unnecessary content, and correcting errors.
- **Evaluation and maintenance.** Through end user analysis, commentary on the blog, usability testing, and customer questionnaires, Creative ensures that the performance needs are met and that training materials are current and valid.

Creative International refers to its "process map" from the beginning to the end of a project. It uses process for marketing, for internal communication, and for project planning and management. The writing process that Creative follows is recursive. It includes constant sign-offs and change orders. With input from all parties, during prewriting, writing, and rewriting, Creative provides its customers "communication that provides custom solutions."

---

1. Produce successful documents using the writing process.

The Writing Process: An Overview

Technical communication is a major part of your daily work experience. It takes time to construct the correspondence, and your writing has an impact on those around you. A well-written report, e-mail message, user manual, Web site, or blog gets the job done and makes you look good. Poorly written correspondence wastes time and creates a negative image of you and your company.

However, recognizing the importance of technical communication does not ensure that your correspondence will be well written. How do you effectively write the memo, letter, blog entry, or report? How do you successfully produce the finished product? To produce successful technical communication, approach writing as a process. The process approach to writing has the following sequence:

1. **Prewrite.** Before you can write your document, you must have something to say. Prewriting allows you to spend quality time, prior to writing the correspondence, generating information, considering the needs of the audience, and choosing the communication channel.
2. **Write.** Once you have gathered your data and determined your objectives, the next step is to write a draft of your document. To do so, organize the draft, supply visual aids, and format the content so that your readers can follow easily.
3. **Rewrite.** The final step, and one that is essential to successful communication, is to rewrite your draft. Revision allows you to test for usability and to perfect your memo, Web site, letter, report, or any document so you can be a successful communicator.

The writing process is dynamic, with the three parts—prewriting, writing, and rewriting—often occurring simultaneously. You may revisit any of these parts of the process at various times as you draft your document. The writing process is illustrated in Figure 3.1.

FIGURE 3.1   The Writing Process

The Writing Process

Prewriting	Writing	Rewriting

Prewriting	Writing	Rewriting

• Determine whether your audience is internal or external.
• Write to inform, instruct, persuade, and build trust.
• Choose the correct communication channel for your audience and purpose.
• Gather your data.
	• Organize your content using modes such as problem/solution, cause/effect, comparison, argument/persuasion, analysis, chronology, etc.
	• Use figures and tables to clarify content.
	• Format the content for ease of access.
		• Test for usability.
		• Revise your draft by
		  • adding details
		  • deleting wordiness
		  • simplifying words
		  • enhancing the tone
		  • reformatting your text
		  • proofreading and correcting errors

Prewriting

Prewriting, the first stage of the process, allows you to plan your communication. If you do not know where you are going in the correspondence, you will never get there, and your audience will not get there with you. Through prewriting, you accomplish many objectives, including

> 2. Plan communication through the prewriting stage.

- Examining purposes
- Determining goals
- Considering audience
- Deciding what action you want your audience to perform
- Gathering data
- Determining the most effective communication channel

Examine Your Purposes

Before you write the document, you need to know why you are communicating. Are you planning to write because you have chosen to do so of your own accord or because you have been asked to do so by your boss? In other words, is your motivation external or internal?

External Motivation.   If someone else has requested the correspondence, then your motivation is external. Your boss, for example, expects you to write a monthly status report, a performance appraisal of your subordinate, or an e-mail report suggesting solutions to a current problem. Perhaps a vendor has requested that you write a letter documenting due dates, or a customer asks that you respond to a letter of complaint. In all of these instances, someone else has asked you to communicate.

Internal Motivation.   If you have decided to write on your own accord, then your motivation is internal. For example, you need information to perform your job more effectively, so you write a letter of inquiry. You need to meet with colleagues to plan a job, so you write an e-mail message calling a meeting and setting an agenda. Perhaps you recognize a problem in your work environment, so you create a questionnaire and transmit it via the company blog. Then, analyzing your findings, you call a meeting to report your findings. In all of these instances, you initiate the communication.

Determine Your Goals

Once you have examined why you are planning to communicate, the next step is to determine your goals in the correspondence or presentation. You might be communicating to

- Inform an audience of facts, concerns, or questions you might have
- Instruct an audience by directing actions
- Persuade an audience to accept your point of view
- Build trust and rapport by managing work relationships

These goals can overlap, of course. You might want to inform by providing an instruction. You might want to persuade by informing. You might want to build trust by persuading. Still, it is worthwhile to look at each of these goals individually to clarify their distinctions.

Communicating to Inform.   Often, you will write letters, reports, and e-mails merely to inform. In an e-mail message, for instance, you may invite your staff to an upcoming meeting. A trip report will inform your supervisor what conference presentations you attended or what your prospective client's needs are. A letter of inquiry will inform a vendor about questions you might have regarding her services. Maybe you will be asked to write a corporate blog entry or a Web site link informing your coworkers about the company's picnic, personnel birthdays, or new stock options available to employees. In these situations, your goal is to share information objectively.

Communicating to Instruct.   Instructions will play a large role in your technical communication activities. As a manager, for example, you often will need to direct action. Your job demands that you tell employees under your supervision what to do. You might need to write an e-mail providing instructions for correctly following procedures. These could include steps for filling out employee forms, researching documents in your company's intranet data bank, using new software, or writing reports according to the company's new standards.

As an employee, you also will provide instructions. Your boss might ask you to create an instructional YouTube video. As a computer information specialist, maybe you work the 1-800 hotline for customer concerns. When a customer calls about a computer's crisis, your job would be to give instructions for correcting the problem. You either will provide a written instruction in a follow-up e-mail or a verbal instruction while on the phone.

Communicating to Persuade.   If your goal in writing is to change others' opinions or a company's policies, you need to be persuasive. For example, you might want to write a proposal, a brochure, or a flier to sell a product or a service. Maybe you will write your annual progress report to justify a raise or a promotion. As a customer, you might want to write a letter of complaint or an e-mail message about poor service. Your goal in each of these cases is to persuade an audience to accept your point of view.

Communicating to Build Trust.   Building rapport (empathy, understanding, connection, and confidence) is a very important component of your communication challenge. As a manager or employee, your job is not merely to "dump data" in your written communication. You also need to realize that you are communicating with coworkers and clients, people with whom you will work every day. To maintain a successful work environment, you want to achieve the correct, positive tone in your writing. This might require nothing more than writing a tweet, saying "Thanks for the information" or "You've done a great job reporting your findings." A positive tone shows approval for work accomplished and recognition of the audience's time.

Recognizing the goals for your correspondence makes a difference. Determining your goals allows you to provide the appropriate tone and scope of detail in your communication. In contrast, failure to assess your goals can cause communication breakdowns.

Consider Your Audience

What you say and how you say it is greatly determined by your audience. Are you writing up to management, down to subordinates, or laterally to coworkers? Are you speaking to a high-tech audience (experts in your field), a low-tech audience (people with some knowledge about your field), or a lay audience (customers or people outside your work environment)? Face it—you will not write the same way to your boss as you would to your subordinates. You will not speak the same way to a customer as you would to a team member. You must provide different information to a multicultural audience than you would to individuals with the same language and cultural expectations. You must consider issues of diversity when you communicate.

Audience

See Chapter 5, "Audience Recognition," for more information about audience recognition and involvement.

Gather Your Data

Once you know why you are writing and who your audience is, the next step is deciding what to say. You have to gather data. The page or screen remains blank until you fill it with content. Your communication, therefore, will consider personnel, dates, actions required, locations, costs, methods for implementing suggestions, and so forth. As the writer, it is your obligation to flesh out the detail. After all, until you tell your readers what you want to tell them, they do not know.

Diversity

See Chapter 5, "Audience Recognition," for more information about diversity.

There are many ways to gather data. In this chapter, and throughout the textbook, we provide options for gathering information. These planning techniques include the following:

- Answering the reporter's questions
- Mind mapping
- Brainstorming or listing
- Outlining
- Storyboarding
- Creating organization charts
- Flowcharting
- Researching (online or at the library)

Documenting Research

See Chapter 16, "Research," and Appendix B, "Parenthetical Source Citations and Documentation," for more information on research techniques and documentation.

Each prewriting technique is discussed in greater detail in Table 3.1. Table 3.2 lists some good Web sites for online research.

TABLE 3.1 Prewriting Techniques

Answering the Reporter's Questions By answering *who, what, when, where, why,* and *how,* you create the content of your correspondence.	Sample Reporter's Questions	
	Who	Joe Kingsberry, Sales Rep
	What	Need to know • What our discount is if we buy in quantities • What the guarantees are • If service is provided on-site • If the installers are certified and bonded • If Acme provides 24-hour shipping
	When	Need the information by July 9 to meet our proposal deadline
	Where	Acme Radiators 11245 Armour Blvd. Oklahoma City, Oklahoma 45233 Jkings@acmerad.com
	Why	As requested by my boss, John, to help us provide more information to prospective customers
	How	Either communicate with a letter or an e-mail. I can write an e-mail inquiry to save time, but I must tell Joe to respond in a letter with his signature to verify the information he provides.

(Continued)

TABLE 3.1   Prewriting Techniques (*Continued*)

Mind Mapping
Envision a wheel. At the center is your topic. Radiating from this center, like spokes of the wheel, are different ideas about the topic. Mind mapping allows you to look at your topic from multiple perspectives and then cluster the similar ideas.

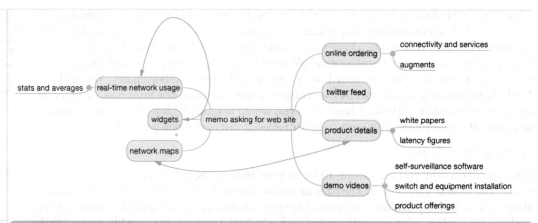

Brainstorming or Listing
Performing either individually or with a group, you can randomly suggest ideas (brainstorming) and then make a list of these suggestions. This method, which works for almost all kinds of communication, is especially valuable for team projects.

Sample Listing—Improving Employee Morale

- Before meetings, ask employees for agenda items (that way, they can feel empowered).
- Consider flextime.
- Review employee benefits packages.
- Hold yearly awards ceremony for best attendance, highest performance, most cold calls, lowest customer complaints, etc.
- Offer employee sharing for unused personal days/sick leave days.
- Roll over personal days to next calendar year.
- Include employees in decision-making process.
- Add more personal days (as a trade-off for anticipated lower employee raises).

Outlining
This traditional method of gathering and organizing information allows you to break a topic into major and minor components. This is a wonderful all-purpose planning tool.

Sample Topic Outline

1.0 The Writing Process
  1.1 Prewriting
    • Planning Techniques
  1.2 Writing
    • All-Purpose Organizational Template
    • Organizational Techniques
  1.3 Rewriting

2.0 Criteria for Effective Technical Communication
  2.1 Clarity
  2.2 Conciseness
  2.3 Document Design
  2.4 Audience Recognition
  2.5 Accuracy

Storyboarding
Storyboarding is a visual planning technique that lets you graphically sketch each page or screen of your text. This allows you to see what your document might look like.

Sample Brochure Storyboard

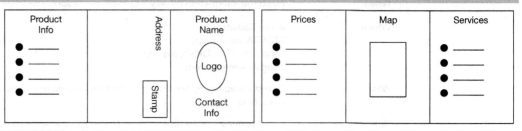

(*Continued*)

TABLE 3.1   Prewriting Techniques (*Continued*)

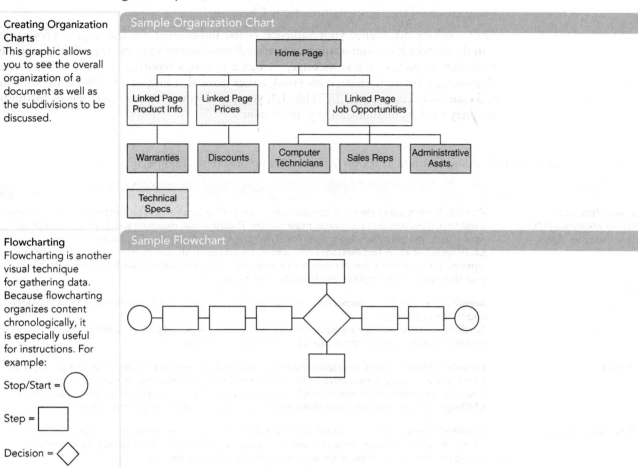

Creating Organization Charts This graphic allows you to see the overall organization of a document as well as the subdivisions to be discussed.	**Sample Organization Chart**
Flowcharting Flowcharting is another visual technique for gathering data. Because flowcharting organizes content chronologically, it is especially useful for instructions. For example: Stop/Start = ◯ Step = ▢ Decision = ◇	**Sample Flowchart**

TABLE 3.2   Internet Search Engines

Purpose	Sites
Popular Online Search Engines	Yahoo.com, Google.com, Bing.com, Ask.com
Meta-Search Engines (multithreaded engines that search several major engines at once)	MetaCrawler.com, Clusty.com, Dogpile.com
Specialty Search Engines	Findlaw.com focuses on legal resources. Webmd.com lets you access health and medical sites. MedlinePlus.gov provides consumer information. GoPubMed.com provides biomedical resources. Scirus.com provides scientific resources. ZDNet.com, EarthLink.net, and BusinessWeek.com provide resources for business.
Broad Academic Searches	Librarians' Internet Index to the Internet (http://www.ipl.org), Infomine.ucr.edu, GoogleScholar.com, FindArticle.com
Government Search Sites	USA.gov and Google's Uncle Sam
International Search Sites	Search Engine Colossus, Abyz News Links (international newspapers and magazines), Worldpress.org (international perspectives on the United States)
Directory of Search Engines	Yahoo!'s directory of search engines and Search Engine Collosus

Determine How the Content Will Be Provided—The Communication Channel

After you have determined your audience, goals, and content, the last stage in prewriting is to decide which communication channel will best convey your message. Will you write an instant message or a text message, a letter, a memo, a report, an e-mail, a Web site, a blog entry, a YouTube video, a proposal, a user manual, a flier, or a brochure, or will you make an oral presentation? In Table 3.3, you can review the many channels or methods you may use for communicating your content.

TABLE 3.3 Options for Providing Content Through Different Communication Channels

Communication Channels	Benefits and Challenges
E-mail Messages, Instant Messages (IM), Text Messages (TM)	*Benefits:* These types of electronic communication are quick and can almost be synchronous. You can have a real-time, electronic chat with one or more readers. Though e-mail messages should be short (20 or so lines of text), you can attach documents, Web links, graphics, and sound and movie files for review. *Challenges:* E-mail, instant messages (IM), and text messages (TM) tend to be less formal than other types of communication. E-mail might not be private (a company's network administrators can access your electronic communication) and have size limitations.
Letters	*Benefits:* Typed on official corporate letterhead stationery, letters are formal correspondence to readers outside your company. *Challenges:* Letters are time consuming because they must be mailed physically. Although you can enclose documents, this might demand costly or bulky envelopes.
Memos	*Benefits:* Memos—internal correspondence to one or several coworkers—allow for greater privacy than e-mail (e-mail messages are archived and can be observed by administrators within a company). Even though most memos are limited in length (one or two pages), you can attach or enclose documents. *Challenges:* Memos are both more time consuming than e-mail and less formal than letters.
Reports and Proposals	*Benefits:* Reports, internal and external, are usually very formal. They can range in length, from one page to hundreds of pages (proposals and annual corporate reports to stakeholders, for example). Because of their length, reports are appropriate for detailed information. *Challenges:* They can be time consuming to write.
Brochures and Fliers	*Benefits:* Brochures and fliers are appropriate for informational and promotional communication to large audiences. *Challenges:* Most brochures are limited to six or so panels, the equivalent of a back and front hardcopy. Similarly, a flier is usually only one page long. Thus, in-depth coverage of a topic will not occur.
Web Sites and Blogs	*Benefits:* A Web site or blog can provide informal and public communication to the entire world—anytime, anyplace. Web sites essentially have unlimited size, so you can provide lots of information, and the content can be updated instantaneously by Web designers. A Web site can include links to other sites, animation, graphics, and color. *Challenges:* Audiences need access to the Internet. Blogs could divulge sensitive corporate communication.
Oral Presentations	*Benefits:* Oral presentations can be both formal and informal and communicate directly with the audience. *Challenges:* Many people are afraid to make an oral presentation. Sometimes audience members can be poor listeners.
Social Media	*Benefits:* Social media such as Twitter, Facebook, YouTube, and LinkedIn allow you to communicate rapidly with many people. You can communicate in a friendly, informal way with your audience. *Challenges:* Social media can pose confidentiality issues and consume too much on-the-job time. Twitter allows only 140 characters of text.
Videos	*Benefits:* Instructional videos avoid language barriers and provide clear actions for people to follow. Videos can also be entertaining. *Challenges:* Creating a precise script with detailed directions for the video camera operator can be challenging, time consuming, and expensive.

Single Sourcing. Maybe you will create content that will be used in a variety of communication channels simultaneously. *Single sourcing* is the act of reusing previously designed documents for new projects and multiple deliverables. In a constantly changing marketplace, you will need to communicate your content to many different audiences using a variety of communication channels. For instance, you might need to market your product or service using the Internet, a flier, a brochure, newsletters, and a sales letter. You might have to write hard-copy user manuals and develop online help screens. To ensure that content is reusable, the best approach would be to write a "single source" of text that can be used for many documents and different media.

Writing

Writing lets you package your data. Once you have gathered your data, determined your objectives, recognized your audience, and chosen the channel of communication, the next step is writing the document. Writing the draft lets you organize your thoughts in some logical, easy-to-follow sequence. Writers usually know where they are going, but readers do not have this same insight. When readers pick up your document, they can read only one line at a time. They know what you are saying at the moment, but they don't know what your goals are. They can only hope that in your writing, you will lead them along logically and not get them lost with unnecessary data or illogical arguments.

> 3. Write documents by organizing and designing content.

Organization

To avoid confusing or misleading your audience, you need to organize your thoughts. As with prewriting, you have many organizational options. In Chapter 4, we discuss organizing according to the following traditional methods of organization:

> **Organization**
> See Chapter 4, "Objectives in Technical Communication," for more information about organization.

- Space (spatial organization)
- Chronology
- Importance
- Comparison/contrast
- Problem/solution

These organizational methods are not exclusive. Many of them can be used simultaneously within your technical communication to help your reader understand your content.

Formatting

You also must format your text to allow for ease of access. In addition to organizing your ideas, you need to consider how the text looks on the page or screen. If you give your readers a massive wall of words, they will file your document for future reading and look for the nearest exit. An unbroken page or screen of text is not reader-friendly. To invite your readers into the document, to make them want to read the technical communication, be sure to highlight key points and break up monotonous-looking text to ensure that your information is accessible.

> **Format**
> See Chapter 7, "Document Design," for information about formatting.

Rewriting

Rewriting lets you perfect your writing. After you have prewritten (to gather data, organize your thoughts, and understand your audience) and written your draft, your final step is to rewrite. Revision requires that you look over your draft to determine its usability and correctness.

> 4. Rewrite documents to determine usability and correctness.

Collaborative Evaluations

You can review your own text, but that might not be the most effective way to assess your writing. Reviewing your own text provides you a limited perspective—yours. When we read our own work, we tend to read what we think we wrote versus what we actually said. In contrast, getting help from peer reviewers could provide you more insight. You could benefit from their objectivity. They will read what's on the page instead of what you think you said. Furthermore, they might have a greater knowledge about the subject matter or grammar and syntax and be able to help you improve your content and writing. In either case, it's helpful to get feedback from other people. Throughout this book, in the "Writing Process at Work" units, we frequently show how peer evaluation assists the writer.

Usability Testing

A report, instruction, or Web site is only good if your audience can understand the content and use the information. Usability testing helps you determine the success of your draft. Through usability testing, you decide what works in the draft and what needs to be rewritten. Thus, usability focuses on the following key factors:

- **Retrievability.** Can the user find specific information quickly and easily?
- **Readability.** Can the user read and comprehend information quickly and easily?
- **Accuracy.** Is the information complete and correct?
- **User satisfaction.** Does the document present information in a way that is easy to learn and remember?

USABILITY EVALUATION CHECKLIST

Audience Recognition

_____ 1. Are technical terms defined?

_____ 2. Are examples used at the reader's level of understanding to explain difficult steps or concepts?

_____ 3. Are graphics used appropriately to help the audience understand the procedure or concept?

_____ 4. Are tone and word usage appropriate for the intended audience?

Development

_____ 5. Are steps or concepts precisely developed?

_____ 6. Is all required information provided including hazards, technical descriptions, process analyses, warranties, accessories, and required equipment or tools?

_____ 7. Is irrelevant or rarely needed information omitted?

Ease of Use

_____ 8. Can readers easily find what they want through
- A table of contents
- A navigation bar
- A glossary
- Hierarchal headings
- Headers or footers
- An index
- Links
- Frequently asked questions (FAQs)

Conciseness

_____ 9. Are words limited to 1–2 syllables when possible?

_____10. Are sentences limited to 10–15 words when possible?

_____11. Are paragraphs limited to 4–6 lines of text when possible?

Consistency

_____12. Is a consistent hierarchy of headings used?

_____13. Are graphics presented consistently (same location, same use of figure titles and numbers, similar sizes, similar style of graphics, etc.)?

_____14. Does word usage mean the same throughout the text (technical terms, hazard notations, etc.)?

_____15. Is the same style of numbering used throughout?

Document Design

_____16. Do graphics help depict how to perform steps?

_____17. Is white space used to make information accessible?

_____18. Does color emphasize hazards, key terms, or important information?

_____19. Do headings, subheadings, and talking headings aid clarity and help the readers navigate the text?

_____20. Is the text visually appealing?

Revision Techniques

After testing your document for usability, revise your text by using the following revision techniques:

- Add any missing detail for clarity.
- Delete dead words and phrases for conciseness.
- Simplify unnecessarily complex words and phrases to allow for easier understanding.
- Move around information (cut and paste) to ensure that your most important ideas are emphasized.
- Reformat (using highlighting techniques) to ensure reader-friendly ease of access.
- Enhance the tone and style of the text.
- Correct any errors to ensure accurate grammar and content.

We discuss each of these points in greater detail throughout the book.

How Important Is Proofreading?

Do employees in the workplace really care about grammar and mechanics? Is proofreading only important to teachers? Proofreading is absolutely important. Table 3.4 shows the importance of proofreading. The *National Commission on Writing* highlights what employers and employees consider to be essential skills in technical communication ("Writing: A Powerful Message from State Government" 19).

TABLE 3.4   Essential Skills in Technical Communication

Skills	Extremely Important
Accuracy	87.8%
Spelling, Grammar, Punctuation	71.4%
Clarity	69.4%
Documentation/Support	61.2%
Logic	55.1%
Conciseness	42.9%
Visual Appeal	12.2%

Revision is possibly the most important stage in the writing process. If you prewrite effectively (gathering your data, determining your objectives, and recognizing your audience) and write an effective draft, you are off to a great start. However, if you then fail to rewrite your text, you run the risk of having wasted the time you spent prewriting and writing. Rewriting is the stage in which you make sure that everything is correct. Failure to do so not only can cause confusion for your readers but also can destroy your credibility.

TECHNOLOGY TIPS

Using Microsoft Word 2013 for Rewriting

Word processing programs help you rewrite your document in many ways:

- **Spell check.** When you misspell a word, often spell check will underline the error in red (as shown in the following example with "grammer" incorrectly spelled). Spell check, unfortunately, will not catch all errors. If you use a word like to instead of too, spell check will not "no" the difference (of course, that should be "know" but spell check did not mark the error). Microsoft Word 2013's **Review** tab also provides you access to proofreading help and allows you to make comments and track changes.

(Continued)

- **Grammar check.** Word processors also can help you catch grammar errors. Grammar check underlines errors in green. When you right-click on the underlined text, the word processing package will provide an optional correction.

Grammer error look like this, for example.

- **Add/Delete.** Word processing makes adding new content and deleting unneeded text very easy. All you need to do is place your cursor where you want to add/delete. Then, to add, you type. To delete, you hit the Backspace key or the Delete key.
- **Move.** The Copy, Cut, and Paste features of word processing allow you to move text with ease.
- **Enhance/Reformat.** In addition to changing the tone of your text, you also can enhance the visual appeal of your document at a keystroke. From the **Home** tab on your toolbar, you can choose from the Word 2013 Ribbon and include bullets, italics, boldface, font changes, numbered lists, and so on.

The process approach to writing—prewriting, writing, and rewriting (including usability testing)—can help you communicate successfully in any work environment or writing situation. In fact, the greatest benefit of process is that it is generic. Process is not designed for any one profession or type of correspondence. No author of a technical communication book can anticipate exactly where you will work, what type of documents you will be required to write, or what your supervisors will expect in your writing. However, we *can* give you a methodology for tackling any communication activity. Writing as a process will help you write any kind of oral or written communication, for any boss, in any work situation.

WRITING PROCESS CHECKLIST

Prewriting

_____ 1. Have you determined the purpose for communicating?

_____ 2. Is your goal to inform, instruct, persuade, or build trust?

_____ 3. Is your audience high-tech, low-tech, or lay?

_____ 4. Have you gathered data through prewriting techniques or researching?

_____ 5. Have you chosen the correct communication channel?

Writing

_____ 6. Have you organized the information (using modes such as spatial, chronology, comparison/contrast, etc.)?

_____ 7. Is the content formatted for readability?

Rewriting

_____ 8. Have you tested for usability?

_____ 9. Have you revised by adding, deleting, simplifying, enhancing tone, and reformatting?

_____10. Have you proofread for accuracy?

The Writing Process at Work

Each company you work for over the course of your career will have its own unique approach to writing memos, letters, reports, e-mail messages, proposals, or Web sites. Your employers will want you to do it their way. Company requirements vary. Different jobs and fields of employment require different types of correspondence. However, you will succeed in tackling any writing task if you have a consistent process approach to writing.

Following is an e-mail report written using the process approach to writing. The document was produced in the workplace by Huan Zhu, an information technologist (IT) writing to his supervisor.

Prewriting

Huan needed to create a user manual and instructional video for the installation of a video card, with added functions such as video capture, MPEG-2 and MPEG-4 decoding, light pen, TV output, and the ability to connect multiple monitors. His boss asked for a follow-up status report about the creation of the manual and the instructional video. Huan's prewriting (shown in Figure 3.2) tells us who he is writing to; what types of documents

FIGURE 3.2 Using Online Tasks to Prewrite Reporter's Questions

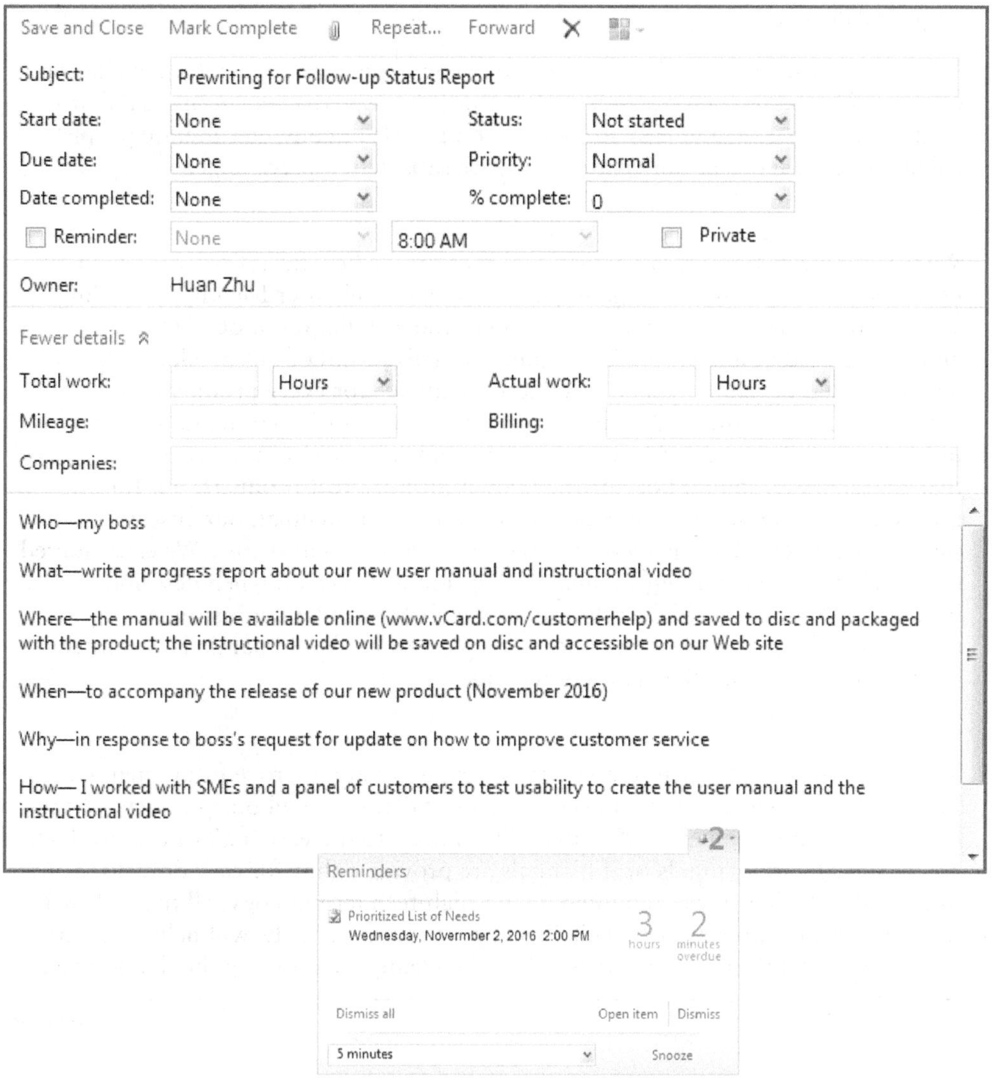

The "Reminders" pop-up helps Huan stay on track with his writing.

he worked on; where the user manual will be accessible to the customer; when the new product will be released; why the boss requested the update; how Huan gathered data for his manual and tested for usability. Huan created his prewriting in "Tasks," which would send him a reminder to complete the job.

Writing

After prewriting, Huan wrote a rough draft and took the word *rough* seriously. He did not consider document design, grammar, or sentence structure. When you draft, do not worry about errors or how the correspondence looks. It is meant to be rough, to free you from worry about making errors. You can correct errors when you revise. No writing is ever perfect. Every memo, letter, report, or e-mail message can be improved. Huan asked a colleague to review his rough draft and suggest revisions. Figure 3.3 shows the rough draft with suggestions for improvement.

FIGURE 3.3   Rough Draft with Suggested Revisions

Huan, don't put the subject line in all caps. What about the manual? You need additional information about the video.

Change the word "recently" to a specific date, clarify who you worked with on this project, and remind the reader exactly why you're writing this report.

Note that your last line in the first paragraph is a fragment (incomplete sentence). You have similar grammar errors elsewhere.

No one will read this or your last paragraph, Huan. They're both too bulky. Can you add white space, add headings, and use some bulleted lists?

Define "SMEs," Huan.

REPORT ON MANUAL

Recently, you asked us to develop an instruction manual and video for installation of our vCard 9600 pro series video card. Our focus was to create a manual that was aesthetically pleasing and easy to follow for all levels of computer literate people. All in efforts to increase our customer service satisfaction percentage.

We feel we successfully created a manual that can be beneficial to anyone who uses this product. Our installation guide is an extremely user friendly, providing an easy-to-follow, step-by-step instruction to the installation of our 9600 pro and can be used as a basis for developing manuals for all of our products. We feel we should develop detailed installation manuals for every product and service we provide, this will increase customer service satisfaction and relieve our technical support department's hold time on calls. We're sure you agree that Extreme PCs should not only be known for our great products, but for our great customer service as well. Since the release of this manual, our customer satisfaction percentage has increased greater than our competitors. We encountered two problems in writing our manual. Using computer jargon when describing installation. Our use of high-tech terminology led to reader confusion, as documented by customer complaints. Screen captures should parallel the content. We had problems with the instructional video.

The manual was a complete success. However, we could improve future manuals as follows: More instruction manuals detailing installation on all our products. We're certain that our customer satisfaction will continue to increase if all of our products are packaged with manuals or if manuals are provide online for easy download. To manage the challenge we encountered with high-tech terminology, all manuals will now include glossaries to define challenging word usage. SMEs will help with this. More time to work on the manuals. We had so many ideas during the development

(Continued)

FIGURE 3.3   Rough Draft with Suggested Revisions (*Continued*)

process, that we ran out of time to implement them all. With this first manual created, however, we can use it as a template. This will save time for future manual development. We'd also like to see an increase in staffing to take on such a massive project. Plus, we'd like to increase our tech support department to reduce hold time and ensure customer satisfaction levels.

Our boss asked for a status report about the vCard project. Your conclusion does not emphasize what was learned from the project. A status report of the vCard project will help him decide how to set up the next company project.

Your conclusion needs some "punch" for emphasis.

Rewriting

After Huan read his colleague's comments, he revised the report and sent it to his boss. Figure 3.4 shows his finished product. Once the manager received the report, he commended Huan for a job well done. The manager found the report easy to read because of the use of white space, headings, listed and bulleted information, and smaller paragraphs. The manager also noted the grammatically correct and well-constructed sentences. When you approach writing as a step-by-step process (prewriting, writing, and rewriting), your results usually are positive—and you will receive positive feedback from your supervisors.

FIGURE 3.4   Finished Report

Date: November 16, 2016
To: Sean Rogers
From: Huan Zhu
Subject: Report on Video Card Instruction Manual and Installation Video

What we hoped to achieve

On Friday, November 2, 2016, you asked our team to develop an instruction manual and video for installation of our vCard 9600 pro series video card. Our focus was to create a hard-copy manual and an instructional video that were aesthetically pleasing and easy to follow for all levels of computer literate people. We hoped to increase our customer service satisfaction percentage. We completed the user manual and video in time for the vCard 9600 rollout. This report will bring you up to date on our achievements.

What we accomplished

We successfully created a manual and a video that can be beneficial to anyone who uses our product. Our installation guide is user friendly, providing an easy-to-follow, step-by-step instruction to the installation of our 9600 pro. The manual and video are available online at www.vCard.com/customerhelp. We saved the text and accompanying video to disc and packaged them with the product.

We believe that these end-user help aids can serve as a basis for developing manuals for all our products. Our instructional video is easy to use for the consumer and can be a model for all future instructional videos. Working with our customer usability panel and SMEs (subject matter experts) proved to be an invaluable resource and a great way to achieve customer service satisfaction.

(*Continued*)

FIGURE 3.4 Finished Report (*Continued*)

Page 2
November 16, 2016
Huan Zhu

Work remaining to be done

We feel we should develop detailed installation manuals and instructional videos for every product and service we provide. This will increase customer service satisfaction and relieve our technical support department's hold time on calls. Since the release of this manual and video, our customer satisfaction percentage has increased by approximately 12 percent over our competitors, as validated by independent survey results.

Problems encountered

We encountered two problems in writing our manual:

1. We used computer jargon when describing installation. Our use of high-tech terminology led to reader confusion, as documented by customer complaints.
2. We did not include enough screen captures to help our target audience. This led to complaints about lack of understanding.

We encountered two problems when scripting our video:

1. The instructional steps were not detailed enough for the video camera operator to shoot.
2. Our "actors" had insufficient time for learning lines.

Suggestions for future projects

The manual and video were successful after we made revisions based on comments from our usability panel. However, we could improve future manuals and videos as follows:

• **More Manuals:** More instruction manuals detailing installation of our products need to be created. We're certain that our customer satisfaction will continue to increase if all products are packaged with manuals or if manuals are provided online for easy download.

• **More Instructional Videos:** Most if not all of our products could benefit from instructional videos. We could create videos in different languages to be sold in our international market. YouTube videos would be appealing to many of our customers.

• **Glossaries:** To manage the challenge we encountered with high-tech terminology, all manuals will now include glossaries to define challenging word usage. Our in-house SMEs can help with definitions of terms.

• **Time:** We need more time to work on the manuals and the videos. We had so many ideas during the development process that we ran out of time to implement them all. With this first manual and video created, however, we can use them as templates. This will save time for future manual and video development.

• **Personnel:** We'd also like to see an increase in staffing to take on such a massive project. Plus, we'd like to increase our tech support department to reduce hold time and ensure customer satisfaction levels. My team believes that hiring only two more help desk employees could successfully handle this need.

If I can answer any questions, let me know. Thank you for giving me the opportunity to work on this project.

CASE STUDIES

1. You are the co-chair of the "Mother's Weekend" at your sorority, fraternity or other school organization. Using mind mapping and listing, brainstorm the activities, menus, locations, decorations, dates, and fees for this weekend's festivities. Brainstorm the pros and cons of hosting the weekend at your sorority or fraternity house or at a hotel or restaurant.

Assignment

Write an outline showing the decisions you've made regarding the topics above. Then, write a short memo or e-mail to your organization's executive board sharing your findings.

2. You work for the Oneg, Oregon, City Planning Department. Your boss, Carol Haley, has received complaints from citizens concerned about a wastewater facility being built in their neighborhood. The homeowners are worried about odors, chemical runoff in nearby Tomahawk Creek, decreases in home values, and a generally diminished quality of life in the neighborhood. You must respond to these complaints, acting upon the citizens' issues.

   For odor abatement, the wastewater management company plans to control fumes and particulate matter. Many of the concerns regarding runoff and home values can be solved through improved land management and ecological restoration. Finally, you have learned that the wastewater company wants to be a good neighbor. To do so, it plans to become actively involved in the community by building more parks, playgrounds, hike/bike trails, and by stocking the nearby pond.

Assignment

In small teams or as individuals, decide which communication channel you will use to share your findings with the community. Will you use a Web site, a blog entry, a tweet, a letter, an e-mail, oral communication, and so forth? Brainstorm the pros and cons of each channel. Then, write an e-mail to your teacher explaining which channel you chose and why.

3. Electronic City is a retailer of DVDs, televisions, DVD players with MP3 playback/JPEG viewer, computer systems, cameras, telephones, fax machines, printers, and more. Electronic City needs to create a Web site to market its products and services. The content for this Web site should include the following: prices, store hours, warranties, service agreements, employment, technical support, and product information.

Assignment

Review the list of Web site topics for Electronic City. Using an organizational chart, decide how to group these topics. Which will be major links on the Web site's navigation bar? Which will be topics of discussion within each of the major links? Once you have organized the links, sketch the Web site by creating a storyboard.

ETHICAL CHALLENGE

Largemont Industries is rolling out a new product in 2016. It will be accompanied by a user manual. Sue Hopkins, the technical writer in charge of the project, believes that her audience of professionals will be offended by an excessive number of definitions. She also believes that using pronouns like "you" and "your" are too casual for a professional workplace. The draft manual was read by a panel of customers to test for usability. The audience was split in their reception of the manual. Many accepted the manual as written, but a percentage complained about tone and lack of clarity.

Question

What is Sue's ethical responsibility to her readers?

INDIVIDUAL AND TEAM PROJECTS

1. To practice prewriting, take one of the following topics. Then, using the suggested prewriting technique, gather data.
   a. **Reporter's questions.** To gather data for your résumé, list answers to the reporter's questions for two recent jobs you have held and for your past and present educational experiences.
   b. **Mind mapping.** Create a mind map for your options for obtaining college financial aid.
   c. **Brainstorming or listing.** List five reasons why you have selected your degree program or why you have chosen the school you are attending.
   d. **Outlining.** Outline your reasons for liking or disliking a current or previous job.
   e. **Storyboarding.** If you have a personal Web site, use storyboarding to depict graphically the various screens. If you do not have such a site, use storyboarding to depict graphically what your site's screens would include.
   f. **Creating organizational charts.** What is the hierarchy of leadership or management at your job or college organization (fraternity, sorority, club, or team)? To depict graphically who is in charge of what and who reports to whom, create an organizational chart.
   g. **Flowcharting.** Create a flowchart of the steps you followed to register for classes, buy a car, or seek employment.
   h. **Researching.** Go online or find a hard copy of the *Occupational Outlook Handbook*. Then, research a career field that interests you. Reading the *Occupational Outlook Handbook*, find out the nature of the work, working conditions, employment opportunities, educational requirements, and pay scale.

2. Using the techniques illustrated in this chapter, edit, correct, and rewrite the following flawed memo.

Date: April 3, 2016
To: William Huddleston
From: Julie Schopper
Subject: Trainning Classes

Bill, our recent training budget has increased beyond our projections. We need to solve this problem. My project team has come up with several suggestions, you need to review these and then get back to us with your input. Here is what we have come up with.

We could reduce the number of training classes, fire several trainers, but increase the number of participants allowed per class. Thus we would keep the same amount of income from participants but save a significant amount of money due to the reduction of trainer salaries and benefits. The downside might be less effective training, once the trainer to participant ratio is increased. As another option, we could outsource our training. This way we could fire all our trainers, which would mean that we would save money on benefits and salaries, as well as offer the same number of training sessions, which would keep our trainer to participant ratio low.

What do you think? We need your feedback before we can do anything so even if your busy, get on this right away. Please write me as soon as you can.

PROBLEM-SOLVING THINK PIECE

In an interview, a company Benefits Manager said that she spent over 50 percent of her workday on communication issues. These included the following:

- Consulting with staff, answering their questions about retirement, health insurance, and payroll deductions
- Meeting weekly with Human Resources (HR) colleagues
- Collaborating with project team members
- Preparing and writing quarterly reports to HR supervisors
- Teleconferencing with third-party insurance vendors regarding new services and/or costs
- E-mailing supervisors and staff, in response to questions
- Updating information about benefits on the corporate blog
- Creating a new link on the company Web site about changes to insurance and benefits
- Calling and responding to telephone calls
- Faxing information as requested
- Responding to and sending instant messages and text messages
- Writing letters to vendors and staff to document services

Though she had to use various methods of both written and oral communication, the communication channels each have benefits and drawbacks. E-mailing, for example, has pluses and minuses (convenience over depth of discussion, perhaps). Think about each of the communication options above. Using the following table, list the benefits of each particular type of communication versus the drawbacks.

Communication Channels	Benefits	Drawbacks	Possible Solutions
One-on-one discussions			
Group meetings			
Collaborative projects			
Written reports			
Teleconferences			
E-mail			
Blogging			
Web site			
Phone calls			
Faxes			
IM/TM			
Letters			

WEB WORKSHOP

1. Proofreading is a key component of successful technical communication. Access the following Web sites and read what these sites suggest as editing/proofreading hints. Compare the content to your approaches to proofreading and editing. Write an e-mail message or memo summarizing your findings.
   - St. Cloud University's Literacy Education Online
   - Purdue University's Online Writing Lab

2. TecEd provides information about Web site, laboratory, field, and remote usability testing. Research the Web site to learn more about usability. Then, report your findings to your professor in an e-mail message or memo.

CHAPTER FOUR

Objectives in Technical Communication

After completing this chapter, you will be able to

1. Achieve clarity in technical communication

2. Simplify writing through shorter words, sentences, and paragraphs

3. Create correct and accurate technical communication

4. Organize content through effective patterns

In the ImagineTec scenario, Nicole Stefani addresses the needs of her audience by writing text that is clear, concise, and accurate.

ImagineTec is a marketing firm that deals primarily with governmental and not-for-profit agencies. Its mission is to mold clients' public relations and to promote an effective overall image. The company accomplishes this through an array of services including

- Web design
- Social media development strategies
- Desktop publishing
- Editing
- Multimedia production
- Events management
- Image control
- Communications training

ImagineTec recently hired Nicole Stefani to be a member of its public relations team. Nicole graduated with honors from the University of Indiana with a degree in journalism and a specialization in public relations. After graduation, she interned in Chicago at a renowned PR company, working with such clients as Nike, Crate & Barrel, and Banana Republic. She brings to ImagineTec expertise as a communications and social media specialist.

In Chicago, she worked as an advocate for businesses and nonprofit organizations. She says, "I understand the attitudes and concerns of a community and public interest groups. I work to establish and maintain relationships with my clients, other businesses, and the media." Nicole will be called upon to draft press releases, make promotional films, plan conventions, interact with members of print and broadcast journalism, communicate via social media, and prepare annual reports and proposals.

Nicole says, "As a new-hire at ImagineTec, I need to write a proposal in response to an RFP (request for proposal) from Greenfield City Management and make a PowerPoint presentation to the city council." Nicole has spent the last three weeks gathering data for her proposal to the city council.

Nicole initially made contact with the client. She visited with them on the telephone and used e-mail correspondence to gather information that the client considered important. Nicole then reviewed the client's needs with her boss, Marc Shabbot, who suggested some additional questions she could ask in face-to-face meetings. Through two meetings with upper-level decision makers, she collected information about the end user's needs. She asked questions, such as "How can we meet your needs?" and "How will you measure success?"

After Nicole gathered data from the client, she had to write the proposal complete with schedules, project plans, the project's scope, and a description of the deliverables. This proposal also would be reviewed by Marc, so Nicole wanted to do a good job.

Nicole said, "I know the importance of this proposal for ImagineTec, the city council, and for me. This is my first big job writing a proposal for ImagineTec, and I want the proposal to be well organized, well designed, well written, clear, concise, accurate, and persuasive. I'm fortunate that Marc is willing to edit my proposal. He has worked for ImagineTec for over 10 years and knows how to write a winning proposal. He's a great team leader, and I look forward to his comments about the proposal. I had a technical writing class in college where I learned the communication process of prewriting, writing, and rewriting documents for the business world. With Marc's help and the communication process, I know that I'm going to be a successful communicator in the workplace."

Achieving Clarity in Technical Communication

> 1. Achieve clarity in technical communication.

The ultimate goal of good technical communication is clarity. If you write a letter, report, e-mail message, user manual, blog entry, or Web site that is unclear to your readers, problems can occur. Unclear technical communication can lead to missed deadlines, damaged equipment, inaccurate procedures, incorrectly filled orders, or danger to the end user. To avoid these problems and many others caused by unclear communication, write to achieve clarity.

Why is clarity important? Readers are less informed than you might assume for several reasons:

- **If there is a CC (complimentary copy).** Your primary reader might know what you are talking about. In contrast, the other four or five people who get copies of the e-mail or report could be less informed.
- **When many people will hear your oral presentation.** If you are speaking to an audience composed of many different people, they might not have the same level of knowledge about your topic. Your job is to clarify so all members of your audience basically have the same level of knowledge.

- **When time has passed.** Even if your primary reader knows the background and understands the topic, time might have passed between your initial discussion and your actual writing of the correspondence. In addition, correspondence is filed for future reference. You need to clarify for future readers who won't be familiar with the content.

Provide Specific Detail

One way to achieve clarity is by supplying specific, quantified information. If you write using vague, abstract adjectives or adverbs, such as *some* or *recently*, your readers will interpret these words in different ways. The adverb *recently* will mean 30 minutes ago to one reader, yesterday to another, and last week to a third reader. This adverb, therefore, is not clear. The same applies to an adjective like *some*. If you write, "I need some information about the budget," your readers can only guess what you mean by *some*. Do you want the desired budget increase for 2017 or the budget expenditures for 2016?

Look at the following example of vague writing caused by imprecise, unclear adjectives. (Vague words are underlined.)

BEFORE

Our latest attempt at molding preform protectors has led to some positive results. We spent several hours in Dept. 15 trying different machine settings and techniques. Several good parts were molded using two different sheet thicknesses. Here's a summary of the findings.

First, we tried the thick sheet material. At 240°F, this thickness worked well. Next, we tried the thinner sheet material. The thinner material is less forgiving, but after a few adjustments we were making good parts. Still, the thin material caused the most handling problems.

The engineer who wrote this report realized that it was unclear. To solve the problem, she rewrote the report, quantifying the vague adjectives.

AFTER

During the week of 10/4/16, we spent approximately 12 hours in Dept. 15 trying different machine settings, techniques, and thicknesses to mold preform mold protectors. Here is a report on our findings:

- 0.030" Thick Sheet At 240°F, this thickness worked well.
- 0.015" Thick Sheet This material is less forgiving, but after decreasing the heat to 200°F, we could produce good parts.

Still, material at 0.015" causes handling problems.

Your goal as a technical communicator is to express yourself clearly. To do so, state your exact meaning through specific, quantified word usage (measurements, dates, monetary amounts, and so forth).

Answer the Reporter's Questions

A second way to write clearly is to answer the reporter's questions—who, what, when, where, why, and how. The best way we can emphasize the importance of answering these reporter's questions is by sharing with you the following e-mail message, written by a highly placed executive, to a newly hired employee.

E-mail

See Chapter 9, "Routine Correspondence," for information about e-mail.

BEFORE

Date: November 16, 2016
To: Staff
From: Earl Eddings, Manager
Subject: Research

Please be prepared to plan a presentation on research. Make sure the information is very detailed. Thanks.

That's the entire e-mail. The questions are, "What doesn't the newly hired employee know?" "What additional information would that employee need to do the job?" "What needs clarifying?"

To achieve successful communication, the writer needs to answer the reporter's questions. *What* is the subject of the presentation and the research? *What* exactly is the reader supposed to do? Will the reader of this e-mail make a presentation, plan a presentation, or prepare to plan a presentation? *Who* is the audience? The word "staff" is too encompassing. Will all of the staff be involved in this project? *Why* is this presentation being made? That is, what is the rationale or motivation for this presentation? *When* will the presentation be made? *How* much detail is "very detailed"? *Where* will the presentation take place?

In contrast, the "After" e-mail message below achieves clarity by answering reporter's questions.

AFTER

Date: November 16, 2016   What

To: Melissa Hider

From: Earl Eddings   Who

Subject: Research for Homeland Security Presentation

Please make a presentation on homeland security for the Weston City Council. This meeting is planned for November 20, 2016, in Conference Room C, from 8:00 a.m. to 5:00 p.m.

When   Where

We have a budget of $6,000,000. Thus, to use these funds effectively, our city must be up to date on the following concerns:

- Bomb-detection options
- Citizen preparedness   Why
- Defense personnel training

How • Crisis management

Use PowerPoint software to make your presentation. With your help, I know Weston and the KC metro area will benefit.

Use Easily Understandable Words

Another key to clarity is using words that your readers can understand easily. Avoid obscure words and be careful when you use acronyms, abbreviations, and jargon.

Avoid Obscure Words.   Write to express, not to impress; write to communicate, not to confuse. If your reader must use a dictionary, you are not writing clearly.

Read the following unclear example.

EXAMPLE ▶

Words like "nonduplicatable" and phrases like "execute without abnormal termination" are hard to understand. For clarity, write "do not duplicate secure messages" and "JCL system testing will ensure that applications continue to work."

The following rules are to be used when determining whether or not to duplicate messages:

- Do not duplicate nonduplicatable messages.
- A message is considered nonduplicatable if it has already been duplicated.

Your job duties will be to ensure that distributed application modifications will execute without abnormal termination through the creation of production JCL system testing.

Following is a list of difficult, out-of-date terms and the modern alternatives:

BEFORE	AFTER
aforementioned	discussed above
initial	first
as per your request	as you requested
subsequent	later
cognizant	know
endeavor	try
remittance	pay
attached herewith	attached
pertain to	about
obtain	get

Simplifying Words, Sentences, and Paragraphs for Conciseness

2. Simplify writing through shorter words, sentences, and paragraphs.

A second major goal in technical communication is conciseness, providing detail in fewer words. Conciseness is important for at least three reasons.

Conciseness Saves Time

Other people's time is valuable. Your audience cannot nor should they spend too much time reading your e-mail or listening to you speak. They have their own jobs to perform. Long documents and lengthy oral presentations often waste people's time. Keep it short—for the sake of your audience as well as for your sake as the communicator.

Technology Demands Conciseness

Conciseness, however, is even more important in a digital workplace. Technology is impacting the size of your technical communication. The size of the smartphone and tablet screens and the character limitations placed on the length of the communication by Twitter and text messages, for example, make the difference. A tweet (the name for messages sent via Twitter) is limited to 140 characters. Text messages are limited to 160 characters. If a text message exceeds 160 characters, the message must be separated into a separate post. Thus, when you write these types of digital communication, you need to consider the way in which technology limits your space. Today, more and more, effective technical communication must be concise enough to *fit in a box*. Notice how the size of the "box" containing the following communication affects the way you package your content (see Table 4.1 and Figure 4.1).

TABLE 4.1 Approximate Size Characteristics of Communication Channels

Communication Channel	Overall Size	Lines per Page or Screen	Characters per Line
Hard-copy text	8½ × 11 inches	55	70–80
PC or laptop e-mail screen	About 6 × 4 inches	20–22	60–70
Smartphone screen (depending on horizontal versus vertical orientation)	About 2 × 4 inches	20 (vertical) 10 (horizontal)	45 (vertical) 80 (horizontal)

FIGURE 4.1   The Shrinking Size of Technical Communication

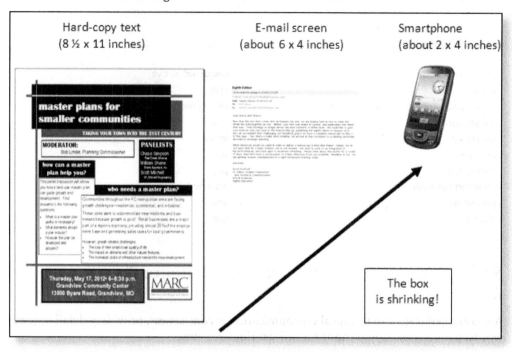

Hard-copy text
(8 ½ x 11 inches)

E-mail screen
(about 6 x 4 inches)

Smartphone
(about 2 x 4 inches)

The box
is shrinking!

Résumés.   A one-page résumé is standard. One hard-copy page measures 8½ × 11 inches. That is a box. In fact, this box (one page of text) allows for only about 55 lines of text, and each line of text allows for only about 70–80 characters (a "character" is every letter, punctuation mark, or space).

E-mail Messages.   In contrast, much of today's written communication in the workplace is accomplished through e-mail messages. Because e-mail screens displayed on computer monitors tend to be smaller (around only 4 × 6 inches) than a hard-copy page, e-mails should be concise. Yes, you can scroll an e-mail message endlessly, but no one wants to do that. In fact, the reason that readers like a one-page résumé is that one page allows for what is known as the "W-Y-S-I-W-Y-G" factor ("What You See Is What You Get"). Readers like to see what they will be getting in the correspondence. In contrast, if you make your readers scroll endlessly in e-mail, they do not see what they get. This causes problems. Therefore, a good e-mail message should fit in the box, letting the reader see the entire content at one glance. It should be limited to about 20 lines of text.

Mobile Messaging.   The technological impact is even more dramatic when you consider the screen size for handheld, mobile messaging equipment such as smartphones and tablets. Today's e-readers are accessing documents using smartphones and tablets. Technical communicators need to make content concise to fit on smartphone screens.

Online Help

See Chapter 13, "Web Sites and Online Help," for more information.

Online Help Screens.   Another example of technical communication that must fit in a box is online help screens. If you work in information technology or computer sciences, you might need to create or access online help screens. If you do so, then your text will be limited by the size of these screens.

Microsoft PowerPoint. You also must fit your technical communication within a box (or boxes) when you use Microsoft PowerPoint software.

PowerPoint

See Chapter 20, "Oral Presentations," for information about PowerPoint.

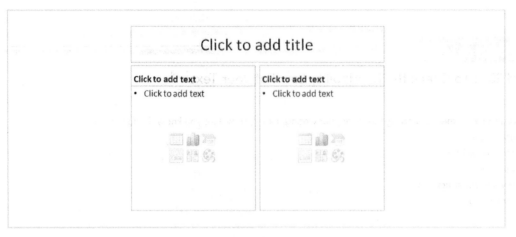

◀ **EXAMPLE**

Examples of text boxes provided in PowerPoint

Twitter and Instant Messages. A further size limitation created by technology is evident in tweets and instant messages. A tweet (the name for messages sent via Twitter) is limited to 140 characters. Instant messages are limited to 160 characters.

Conciseness Improves Readability

"Readability" is the level for reading your document. It defines whether you are writing at a 5th grade, 9th grade, or 12th grade level, for example. You can gauge your readability level by using many readability formulas, including the Linsear Write Index, the Lazy Word Index, the Flesch Reading Ease Score, and the Flesch-Kincaid Grade Level Score. A popular tool for determining readability is Robert Gunning's Fog Index. Gunning bases his readability findings on the length of your sentence and the length of your word usage.

Several online sites will help you calculate a document's readability. For example, in *Online-Utility*, just type or download text into the site and receive instant feedback on the "number of characters, words, sentences, and average number of characters per word, syllables per word, and words per sentence." Another Web site, *Readability Formulas*, provides "Free readability tools to check for Reading Levels, Reading Assessment, and Reading Grade Levels."

Readability is important for the following reasons. According to the National Center for Education Statistics, close to 22 percent of the adult population "reads at the lowest literacy level" (Labbe B9). That equals about 70 million people in the United States. You might think that this fact won't apply to you because you'll always be writing to college graduates. That's a mistake. Only about 30 percent of Americans graduate from college. Therefore, if you are writing at a college level, you might create readability challenges for approximately 70 percent of your audience (U.S. Census Bureau). Regardless of your work environment, you will need to communicate to the general public.

Reading habits are changing. Many of today's readers browse—they skim rather than read text in depth. Much of this, according to studies, is due to the emergence of social media. Look at the following statistics that highlight the change in reading habits of average college students:

- Under 10 books read in a year
- Over 2,000 Web pages browsed in a year
- Over 1,500 Facebook profiles viewed annually

- Approximately 40 pages of college assignments written annually
- Over 500 e-mail message written in a year
- Thousands of text messages sent annually

TECHNOLOGY TIPS

Using Microsoft Word 2013 to Check the Readability Level of Your Text

Microsoft Word 2013 allows you to check the "readability" level of your writing. Doing so will let you know the following:
- How many words you've used
- How many sentences you've written
- How many words per sentence
- If you've used passive voice constructions
- The grade level of your writing

1. Click on **File**, and then click **Options**.

2. Click on **Proofing**, and then select **Check grammar with spelling** and **Show readability statistics**. Then click **OK**.

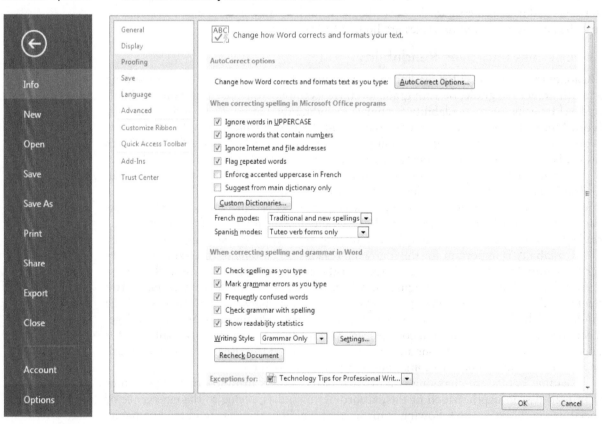

(Continued)

Once you have enabled this readability feature, open a file and check the spelling. When Word has finished checking the spelling and grammar, you will see a display similar to the one below.

Counts
Words	126
Characters	650
Paragraphs	4
Sentences	11

Averages
Sentences per Paragraph	2.7
Words per Sentence	11.4
Characters per Word	4.9

Readability
Passive Sentences	9%
Flesch Reading Ease	52.8
Flesch-Kincaid Grade Level	8.7

OK

In this example, the text consisted of 126 words, 4 paragraphs, and 11 sentences. This averaged 11.4 words per sentence, equaling about 8.7 grade level writing on the Flesch-Kincaid scale. Nine percent of the sentences were written in passive voice.

Limit Word Length for Conciseness

To achieve clarity *and* conciseness, limit your word usage length. Try to use one and two syllable words versus words with three or more syllables. You cannot nor should you avoid words like "accountant," "engineer," "telecommunications," "computer," or "nuclear." Though these words have more than two syllables, they are not difficult to understand. When you need longer words, use them. Also, try to avoid old-fashioned, legalistic words, like "pursuant," "accordance," and "aforementioned." Too often, writers and speakers use these words to *impress* their audience. In contrast, communicators should *express* their content clearly and simply. Look at the following, lengthy "Before" sentence:

BEFORE

I would like you to take into consideration the following points, which I know will assist you in better applying new HIPAA rules and regulations currently burdened by the need to execute all data manually and on paper rather than through standardized, electronic transmissions.

The above sentence is 44-words long and contains 10 words with more than three syllables. None of these words is challenging individually. Still, the mass of syllables makes the sentence hard to understand. To solve the challenges presented by the length of the sentence and the length of the words, simplify as in the "After" example.

AFTER

Please consider the following points. This will help you apply new HIPAA rules by submitting data online instead of having to type text on separate forms.

In this revision, the 44 words have been reduced to 26 words. Also, the original long sentence has been cut into two smaller sentences. The remaining two sentences contain only four multisyllabic words. The conciseness saves you and your reader time and makes the information easier to understand.

Following is a list of long words that can be simplified for conciseness and easier understanding:

BEFORE	AFTER
Long Words	**Concise Version**
utilize	use
anticipate	await (or expect)
cooperate	help
indicate	show
initially	first (or 1.)
presently	now
prohibit	stop
inconvenience	problem
pursuant	before
endeavor	try
sufficient	enough
subsequent	next

Limit Sentence Length for Conciseness

For conciseness and improved readability, limit the length of your sentences. The GNOME Documentation Style Guide (a Unix and Linux desktop suite and development platform) provides the following information about readable sentence length:

Readability	Length of Sentence
Very easy to read	Average sentence length is 12 words or less.
Plain English	Average sentence length is 15 to 20 words.
Extremely difficult to read	Average sentence length exceeds 20 words.

To limit sentence length, use the following techniques for deleting dead words and phrases.

Delete "Be" Verbs.   "Be" verbs include conjugations of the verb "be": *is, are, was, were, would, will, been,* and *am.* Often, these verbs create unnecessarily wordy sentences. For example, look at the "Before" and "After" examples.

BEFORE	AFTER
Wordiness Caused by "Be" Verbs	**Deleting for Conciseness**
Bill *is* of the opinion that stock prices will decrease.	Bill thinks stock prices will decrease.
I *am* in receipt of your bill for $1,000.	I received your $1,000 bill.
If I can *be* of any assistance to you, please call.	If I can help, please call.
They *are* planning to fax new invoices tomorrow.	They plan to fax new invoices tomorrow.
Barb *had* been hoping to move into her new office complex today.	Barb hoped to move into her new office complex today.

Use Active Voice Versus Passive Voice.   In active voice sentences, the subject performs an action. In passive voice sentences, the subject is acted upon. Sometimes, "Be" verbs create passive voice sentences, as in the "Before" example.

It *has been* decided that Joan Smith *will* head our Sales Department.

Passive voice causes two problems.

1. Passive constructions are often unclear. Who decided that Joan Smith will head the department? To solve this problem and to achieve clarity, replace the vague indefinite pronoun "it" with a precise noun, as in the "After" example.

2. Passive constructions are often wordy. Passive sentences require helping verbs, such as _has been_. The "After" example omits the helping verb _has been_ and the verb _will_.

Larry named Joan Smith head of the Sales Department.

Delete the Expletive Pattern. When you begin sentences with "there" or "it," you create the expletive pattern of sentence structure. Notice how the expletive pattern again uses "Be" verbs. The expletives (*there* and *it*) create wordy sentences, as shown in the "Before" examples in contrast to the "After" examples.

There are three people who will work for Acme.	Three people will work for Acme.
It has been decided that ten engineers will be hired.	Ten engineers will be hired.

Avoid *Shun* Words. Another way to write more concisely is to avoid words ending in *-tion* or *-sion*—*shun* sounds. For example, look at the "Before" and "After" examples.

Shun Words	Concise Versions
came to the conclusion	concluded (or decided, ended, stopped)
with the exception of	except for (or but)
make revisions	revise (or change)
investigation of the	investigate (or look at, review, assess)
consider implementation	implement (or use)
utilization of	use

Avoid Camouflaged Words. Camouflaged words are similar to *shun* words. In both instances, a key word is buried in the middle of surrounding words (usually helper verbs or unneeded prepositions). For example, in the phrase *make an amendment to*, the key word *amend* is camouflaged behind unnecessary words. Look at the "Before" and "After" examples.

BEFORE	AFTER
Camouflaged Words	Concise Versions
make an *adjustment* of	adjust (or revise, alter, change, edit, fix)
have a *meeting*	meet
thanking you in advance	thank you
for the purpose of *discussing*	discuss
arrive at an *agreement*	agree
at a *later* moment	later

Limit Prepositional Phrases.   Prepositions can be important words in your communication. They help you convey information about time and place. Occasionally, however, prepositional phrases create wordy sentences. A prepositional phrase includes a *preposition* and a *noun* or *pronoun* that serves as the object of the preposition. For example, "at a later moment" is a prepositional phrase. It includes the preposition *at* and the noun *moment.* This prepositional phrase is wordy and can be revised to read "later."

BEFORE	AFTER
Wordy Prepositional Phrases	Concise Versions
He spoke *at a rapid* rate.	He spoke rapidly (or fast).
She wrote *with regard to* the meeting.	She wrote regarding (or about) the meeting.
I will call *in the near future.*	I will call soon.
On two different occasions, we met.	We met twice.
The manager *of personnel* was hired.	The personnel manager was hired.

Use the Meat Cleaver Method of Revision.   One way to limit the number of words per sentence is to cut the sentence in half or thirds. The following sentence, which contains 43 words, is too long.

BEFORE

To maintain proper stock balances of respirators and canister elements and to ensure the identification of physical limitations that may negate an individual's previous fit-test, a GBC-16 Respirator Request and Issue Record will need to be submitted for each respirator requested for use.

AFTER

Using the meat cleaver approach makes this sentence more concise. The "Before" sentence, now rewritten as three sentences, is easier to read.

Please submit a GBC-16 Respirator Request and Issue Record for each requested respirator. We then can maintain proper respirator and canister element stock balances. We also can identify physical limitations that may negate an individual's previous fit-test.

Limit Paragraph Length

The number of lines you write in a paragraph is arbitrary. Some paragraphs, due to the complexity of the subject matter, might require development. Other paragraphs requiring less development can be shorter. Nonetheless, an excessively long paragraph is ineffective. In a long paragraph, you force your reader to wade through many words and digest large amounts of information. This hinders comprehension. In contrast, shorter paragraphs invite reading and help your readers understand your content. A paragraph in effective technical communication should consist of no more than four to six typed lines.

BEFORE

Our project management approach will provide your city clear deliverables and meet your RFP criteria. Orlin & Sons proposes the following sequence. We will assess the adequacy of your current facilities from a technology perspective, starting on January 13, to be completed by February 1. Then, beginning on February 5, O&S will meet with residents' focus groups to identify community needs, including health, culture, history, and quality-of-life issues. This will allow us to identify necessary improvements to meet your current and ongoing requirements. We will complete this project milestone on February 25. The final step of the process involves setting team goals needed to work with city, county, and state regulatory agencies. We will begin this step on March 8 and conclude by March 15.

The "Before" paragraph is neither visually attractive nor easy to understand. To invite your readers into the document and help them grasp the details, improve the document's design. Make your text open and inviting by using formatting techniques, as shown in the "After" example. Not only is the "After" example easy to access but also it is more concise than the "Before" example. The "Before" example consists of 130 words. The "After" example consists of only 72 words.

AFTER

Our project management approach will provide your city clear deliverables and meet your RFP criteria. Orlin & Sons proposes the following sequence.

- Assess the adequacy of your current facilities from a technology perspective.

  Begin/End Dates: 1/13–2/1
- Meet with residents' focus groups to identify community needs: health, culture, history, and quality-of-life issues.

  Begin/End Dates: 2/5–2/25
- Set team goals needed to work with city, county, and state regulatory agencies.

  Begin/End Dates: 3/8–3/15

Break up wall-to-wall words, margin-to-margin text with smaller paragraphs. Use bullets to make each paragraph stand out more effectively. The boldfaced, indented dates emphasize key milestones.

Achieving Accuracy in Technical Communication

3. Create correct and accurate technical communication.

Clarity and conciseness are primary objectives of effective technical communication. However, if your writing is clear and concise but incorrect—grammatically or contextually—then you have misled your audience and destroyed your credibility. To be effective, your technical communication must be accurate. Accuracy in technical communication requires that you proofread your text. The examples of inaccurate technical communication below are caused by poor proofreading (we have underlined the errors to highlight them).

First City Federal Savings and Loan
1223 Main
Oak Park, Montana

October 12, 2016
Mr. and Mrs. David Harper
2447 N. Purdom
Oak Park, Montana

Dear Mr. and Mrs. Purdom:

Note that the savings and loan incorrectly typed the customer's street rather than the last name.

National Bank
1800 Commerce Street
Houston, TX

September 9, 2016

Adler's Dog and <u>Oat</u> Shop
8893 Southside
Bellaire, TX

Dear <u>Sr:</u>

In response to your request, your account with <u>use</u> has been <u>close</u> out. We <u>are submitted</u> a check in the amount of $468.72 (your existing balance). If you have any questions, please <u>fill</u> free to contact us.

In addition to all the other errors, it should be "Dog and <u>C</u>at Shop." The errors make the writer look incompetent.

Proofreading Tips

1. **Let someone else read it.** We miss errors in our own writing for two reasons. First, we make the error because we don't know any better. Second, we read what we think we wrote, not what we actually wrote. Another reader might help you catch errors.
2. **Let it sit.** Write your correspondence and then read it later. When you read it later in the day, you'll be more objective.
3. **Print it out.** You can proofread more effectively by printing the document and then reading it for errors, line by line.
4. **Use technology.** Computer spell checks are useful for catching most errors. They might miss proper names, homonyms (*their*, *they're*, or *there*) or incorrectly used words, such as *device* to mean *devise*.
5. **Read it out loud.** Sometimes we can hear errors that we cannot see. For example, we know that *a outline* is incorrect. It just sounds wrong. *An outline* sounds better and is correct.

FAQs The Importance of Proofreading

Q: Is anyone in the workplace concerned about proofreading? Isn't proofreading just something that only English professors care about?

A: Proofreading is essential in technical writing. Here are two stories to prove our point.

- "A misplaced decimal point resulted in one company paying . . . $120,000 in taxes on a piece of industrial equipment, instead of the $1,200 the firm rightfully owed."
- A Chicago-based company purchased an industrial sander for $54,589.62. Unfortunately, when listing the purchase on their year-end taxes, the company reported the purchase price as $5,458,962. This misplaced decimal point and commas equaled a difference of over 5 million dollars. The issue is now in court, of course, costing even more money.

A single mark of punctuation can be important.

Source: Tony Rizzo, Kansas City Star

Organizing Technical Communication

4. Organize content through effective patterns.

If you are clear, concise, and accurate, but no one can follow your train of thought because your text rambles, you still haven't communicated effectively. Successful technical communication also must be well organized. No one method of organization always works. Following are patterns of organization that you can use to help clarify content.

Analysis. Topics can be difficult for audiences to understand. For example, what does your employee benefits package include? What does your homeowner's insurance cover? What are the benefits of your cell phone calling plan? What courses do you need to take to complete your degree program? Analyzing a topic helps you focus on the smaller pieces that make up the whole. The following example uses analysis to provide information about smartphone rates.

◀ EXAMPLE

SmartPhone Rates
Look at what AaBbTelecom offers you as part of your smartphone services:

Plan Prices	$29.99	$49.99	$59.99
Minutes	450	450	Unlimited
Nights and Weekends	5,000 minutes	Unlimited	Unlimited
Mobile to Mobile	Unlimited	Unlimited	Unlimited
Roaming and Long Distance	None	None	None
Optional Features	Text messaging	Text messaging Pre-installed games Ringtones	Text messaging Pre-installed games Ringtones Bluetooth wireless technology

Spatial Organization. To organize a topic spatially, you discuss how the topic looks if viewed from left to right, right to left, inside to outside, or bottom to top. Spatial organization is useful when providing physical descriptions of products or perhaps reporting on work-related accidents or events. For example, an accident report might factor in geographical location—north, east, south, and west. The home burglary claims report shown in Figure 4.2 uses compass directions for spatial organization.

Organization

See Chapter 14, "Technical Descriptions and Process Analyses," for discussion of spatial and chronological organization.

Chronology. You might use chronological order to organize many types of communication. If you are a recording secretary for a board, agency, sorority, fraternity, or city council, you could use chronology to report meeting minutes. This would entail noting who said what first, who responded next, and so forth. Reverse chronology is used for many résumés. In a reverse chronological résumé, you discuss your current job or educational status first. Then, you list your prior employment and educational accomplishments. Chronology is mandatory if you are discussing the steps in a procedure. Using a chronological order also can help your audience follow trends. The "Before" example on page 71, taken from a company's Web site advertising its clothing lines, is not organized effectively, while the "After" is reorganized chronologically.

FIGURE 4.2   Report Using Spatial Organization

Claims Report

Date: January 16, 2016
To: Larry Lerner, Regional Manager
From: Susan McGarvey
Subject: Claims Report on Burglary at 1600 Oaklawn

Introduction

Time/Date When Claim Was Filed: 8:45 p.m./January 15, 2016
Policy Number: 3209-6491
Effective Date of Insurance Coverage: May 15, 2015
Policy Holder(s): Mr. John Stamper and Mrs. Carol Stamper
Mailing Address: 1600 Oaklawn
City/State/Zip: Caligon, MS 34267
Phone: 314-555-2424

Description of the Burglary

Narrative: The residents (Mr. John Stamper and Mrs. Carol Stamper), returning from an evening out, found their house broken into, vandalism, and missing items.

Exit/Entry: Entry appears to have been made by cutting an L-shaped hole in the northwest bedroom (BR) window. The perpetrators then apparently left the BR and traveled due south to the southwest BR, where vandalism occurred. Then, the culprits walked east down the hall to the family room (FAM). When the Stampers returned home, they found their garage (GR) door open, suggesting that the perpetrators exited south from their home.

Missing Items/Estimated Costs:

- Sony 46″ high definition television ($1,400)
- $200 cash that had been laying on the family room desk
- A Nintendo Wii Console ($259)
- A Blu-ray player ($159)

Vandalism/Costs:

- The southwest BR had random spray paint on the ceiling and walls ($200—materials and labor). Please see the attached photographs.
- Window repair ($125—material and labor)

Status of the Claim

This claim has been given to claims adjuster Mary Rivera for disposition. I have informed Mary that turnaround time on claim clearance must be two weeks maximum to meet our company's new mandates for customer satisfaction.

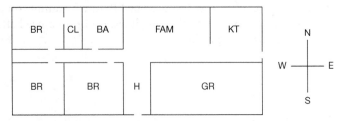

Compass directions can be used in reports to help you organize content and help the reader visualize the event.

BEFORE

We have clothing and other things from the 50s to the 90s. Our 90s' clothing includes everything from early 90s' grunge (raggedy cut-off shorts and flannel plaid shirts) to late 90s' hip-hop clothes and add-ons (oversized jewelry, sports jerseys, hooded sweatshirts, and fancy club duds). Our 50s' goodies cover the "beats" to the "blues brothers" clothing. We have black suits, black skinny ties, as well as black berets and turtlenecks. Our 70s' clothing and items include all kinds of disco clothing including leisure suits and everything from Ban-lon to skin-tight pants, shirts, white belts, and boots. We even have disco dance hall balls and posters for your own disco dance floor. Our 80s' clothing focuses on what we call "business-nerd" including pocket protectors, wing-tip shoes, horn-rimmed glasses (with the nose piece already pre-taped). We also cover "Flash Dance" clothes like leg warmers and off-the-shoulder sweatshirts. Finally, our 60s' hippie stuff is way cool, such as wide, hallucinogenic ties, Nehru jackets, short-short mini-dresses, long-long granny dresses, way-out-there boots, peace symbols, fringe belts, and vests.

The "Before" example discusses clothing options in the 90s, 50s, 70s, 80s, and 60s. Organizing the topic so randomly makes the text hard to follow.

AFTER

We have clothing and other things from the 50s to the 90s.

- **The 50s**—Our goodies cover the "beats" to the "blues brothers" clothing. We have black suits, black skinny ties, as well as black berets and turtlenecks.
- **The 60s**—Our hippie stuff is way cool. We sell wide, hallucinogenic ties, Nehru jackets, short-short mini-dresses, long-long granny dresses, way-out-there boots, peace symbols, fringe belts, and vests.
- **The 70s**—Our merchandise includes all kinds of disco clothing including leisure suits and everything from Ban-lon to skin-tight pants, shirts, white belts, and boots. We even have disco dance hall balls and posters for your own disco dance floor.
- **The 80s**—Check out our "business-nerdwear" including pocket protectors, wing-tip shoes, and horn-rimmed glasses (with the nose piece already pre-taped). We also cover "Flash Dance" clothes like leg warmers and off-the-shoulder sweatshirts.
- **The 90s**—We have clothing from early 90s' grunge (raggedy cut-off shorts and flannel plaid shirts) to late 90s' hip-hop clothes and add-ons (oversized bling bling jewelery, sports jerseys, hooded sweatshirts, and fancy club duds).

The "After" example is improved in at least three ways: The text has been organized chronologically by decade; the bullets make the text easier to access; the boldface headings highlight each era.

Importance. If you organize your text by importance, you tell readers which parts of the discussion are most important and which are less important. Figure 4.3 is a letter written by the chairperson of the board of a publicly traded utility company. In this letter to the other board members, the chair suggests ways in which the organization can meet stockholder concerns regarding changes that are taking place in the industry.

Letters

See Chapter 9, "Routine Correspondence," for information about letters.

Comparison/Contrast. One way to make decisions is by comparing and contrasting options. Which car should you buy? In which degree program should you enroll? Which job offer should you take? By comparing and contrasting topics or options, you can help your audience see the pros and cons or the choices available. The example following on page 73 compares and contrasts three social media sites to clarify service options.

Problem/Solution. Another way to organize your content is to focus on problems and solutions. For instance, if you work in customer service, you might have to respond to consumer complaints. To do so, you might focus on the problem your customer has identified and then present your company's solutions. Proposals often use a problem/solution method of organization. If you are giving an oral presentation as part of a

Executive Summary

See Chapter 18, "Long, Formal Reports," and Chapter 19, "Proposals and Business Plans," for information about executive summaries and organization.

FIGURE 4.3   Letter Organizing Content by Importance

Arrowhead Utilities
1209 Arrowhead Dr.
Like Washington, IA 39921

September 12, 2016

Ms. Stacy Helgoe
1982 Evening Star Rd.
Lawrence, KS 78721

Dear Ms. Helgoe:

We have experienced rough times lately in the utilities industry. Prices for oil, gas, water, and coal have gone up by over 50 percent, but statewide regulations have disallowed us from raising rates to meet these costs. This has led the Board to consider laying off workers, reducing our geographic area of coverage, limiting our customers' options for service, merging with a utilities competitor, and providing fewer hours of service ("enforced brownouts").

To ensure continued good relations, we should inform our stakeholders of these decisions. How should we proceed? Here are our options:

1. Implement the Board suggestions without notifying the stakeholders directly. A follow-up article in the local newspaper's business section could then report the activity.
2. Implement the Board suggestions and provide a personalized letter to each stockholder detailing the causes and our goals.
3. Present information to the stakeholders at an abbreviated annual meeting, asking for questions and answers, and then taking a vote on which options to pursue.
4. Hold four "small town" meetings prior to any vote or implementation. This will allow stakeholders ample opportunity for discussion.

My suggestion is #4, the best choice to ensure large scale stakeholder buy-in and empowerment. Any other approach, I believe, will create distrust on the part of our primary audience. I want to hear from each of you regarding your thoughts. Please call me at 914-555-7676, ext. 234, or e-mail me at cpieburn@au.org by September 21.

Sincerely,

Christy Pieburn

Christy Pieburn
Board Chair

The itemized body in this letter uses analysis to provide the reader options. In the last paragraph, the body points are itemized from *least important to most important*. Notice that point #4 is considered "best choice," thus the most important point.

Social Media Sites			
Service Options	Facebook	Friendster	Orkut
Multilingual	✓	✓	✓
Profile Editor	✓	✓	
Customize		✓	
Personalize URL		✓	
Photos	✓	✓	✓
Blog Journal		✓	
Safety Tips	✓		✓
Instant Messaging	✓		
Tags	✓	✓	
Mobile	✓		
Music	✓		

proposal, for example, you might want to mention the customer's problem and then highlight the many ways in which your company will make improvements. The following example is a proposal's executive summary, which uses problem/solution as a means of organization.

Executive Summary

Problem

Results from the employee satisfaction survey indicate that the Northwest Group needs to improve current leadership training. Our analysis reveals that managers want to hire staff from outside our Supervision Identification Program's (SIP) pool of "SIP Certified" personnel.

Managers do not currently believe that the SIP pool contains personnel with the skills needed to succeed on the job. They highlighted three areas specifically: Diversity Management, Communication Skills, and Knowledge of ISO Standardization.

Solution

We propose solving this problem as follows:

- Researching supervisory software vendors for improved online and computer-aided individual instruction
- Implementing improved ongoing and post-assessment techniques

Cause and Effect.   Another method of organizing your draft is cause and effect. In this instance, you would focus on what caused a specific situation or its results—the effects. This method of organization would be useful in writing reports. Figure 4.4 is a report organized by cause and effect.

FIGURE 4.4   Report Organized by Cause and Effect

Date: March 16, 2016
To: Edie Kreisler
From: Carlos De La Torre
Subject: National Savings and Loan Employee Interview Results

Purpose of Report

The introduction explains why this report was being written—to determine the causes of departmental problems.

In response to your request on February 23, 2016, ExecuMeasure has interviewed the National Savings and Loan's Trust Department employees. Our goal was to determine what might have been causing the department's poor morale and decreasing sales. The following is a report on our procedure and findings.

Analysis of Employee Morale Problems and Decreasing Sales

Procedures—ExecuMeasure's research consultants met with all of the Trust Department's staff and interviewed them on the basis of their job performance and interpersonal communication skills. We used a scale of 1 to 10, with 1 representing "poor" and 10 representing "outstanding."

The questions regarding job performance are "Does *name of person* a) represent the company/department in a professional manner, b) return customer calls/e-mail, c) close sales deals efficiently, d) meet his/her quotas?

For interpersonal communication, I asked the employees to assess themselves and other staff, using the same 1 to 10 point scale. The questions were "Does *name of person* a) work well with other team members, b) respond in a timely manner to team member calls/e-mail, c) conduct him/herself professionally in departmental meetings, d) act collegially (with honesty and sincerity)?

Findings—As Figure 1 below indicates, all six staff received very good ratings on job performance with scores between 7.6 and 8.7. However, two employees scored significantly lower on interpersonal communication skills than the other workers. Barb scored 4 and Dan 3.6. The norm for the other four employees was 8.3.

Figure 1:   Overall Employee Rankings

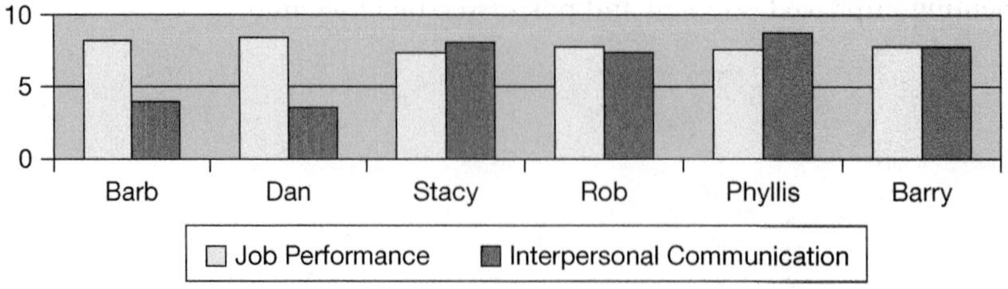

(Continued)

FIGURE 4.4   *(Continued)*

Figure 2 clarifies why Barb and Dan received their low scores, causing low employee morale.

Figure 2: Causes for Low Scores

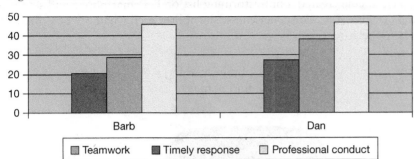

The figure specifies the causes for this office's employee problems.

When we asked the other staff members to explain their scores, they gave us the following causes: "Barb will tell me one thing but then say something else to another employee." "Dan always comes late to meetings and either acts bored or says contrary things." "Barb never returns my e-mail messages or my phone calls." "Dan will mock other coworkers." "Barb thinks that she is the only one who works here." "It's obvious that Dan has no respect for Stacy—or anyone else for that matter."

The employees' comments further clarify the causes behind the company's problems.

Summary of Findings

As is evident from our survey, Stacy, Rob, Phyllis, and Barry trust and respect each other and their manager. However, they do not have the same trust and respect for Barb and Dan, nor do they believe Barb or Dan treat them professionally. The effect of this office dynamic seems to be poor morale and office inefficiency.

The report's conclusion highlights the effect of the office's poor employee relationships.

ExecuMeasure specializes not only in determining the causes for office-environment challenges but also in providing solutions to improve office morale and efficiency. If you would like us to work with you and your colleagues on creative solutions, please contact me at 212-555-9856, ext. 234, or cdltorre@execumeasure.com.

CHECKLIST FOR CLARITY, CONCISENESS, ACCURACY, AND ORGANIZATION IN TECHNICAL COMMUNICATION

Clarity

_____ **1.** Have you answered the reporter's questions (who, what, when, where, why, and how)?

_____ **2.** Have you provided specific information, avoiding vague word usage?

Conciseness

_____ **3.** Have you limited the number of syllables per word?

_____ **4.** Have you limited the number of words per sentence?

_____ **5.** Have you limited the number of lines per paragraph?

Accuracy

_____ **6.** Have you proofread your technical communication?

Organization

_____ **7.** Have you used modes, such as analysis, spatial, importance, chronology, comparison/contrast, and problem/solution to organize your technical communication?

The Writing Process at Work

Nicole Stefani, a recently hired public relations employee at ImagineTec, relies on prewriting, writing, and rewriting to create a report for her boss, Marc.

Prewriting

To plan her project, Nicole created a brainstorming list on her smartphone and sent it to her e-mail address (Figure 4.5). For conciseness, she abbreviated Chamber of Commerce (CC) and Homes Association (HA).

FIGURE 4.5 Brainstorming List for Prewriting Sent via Smartphone

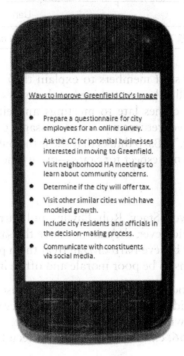

Ways to Improve Greenfield City's Image

- Prepare a questionnaire for city employees for an online survey.
- Ask the CC for potential businesses interested in moving to Greenfield.
- Visit neighborhood HA meetings to learn about community concerns.
- Determine if the city will offer tax.
- Visit other similar cities which have modeled growth.
- Include city residents and officials in the decision-making process.
- Communicate with constituents via social media.

Writing

Nicole says, "As I draft my proposal, I'm going to consider organization, writing style, clarity, layout, visual aids, and which communication channel will work best for this proposal. Should I consider using blogs, Twitter, or Facebook, for example, to engage my audience?

To draft the proposal, I prioritized the answers from the interviews and focused on which responses were most prevalent. My goal was to organize both written and oral comments according to a least important/most important presentation. That is, I offered the clients options for improving their public relations: the first options were the most cost-effective but the least promising; the latter options were more costly but more rewarding for the City." Figure 4.6 shows Nicole's rough draft for review.

Rewriting

According to Nicole, "Marc says that revising the report will ensure its readability and its persuasiveness. With Marc's help, I'm pleased with the proposal I wrote for the City of Greenfield. Following the writing process let me develop the entire report and be proud of the result." Figure 4.7 presents Nicole's revised text.

FIGURE 4.6 Rough Draft with Track Changes

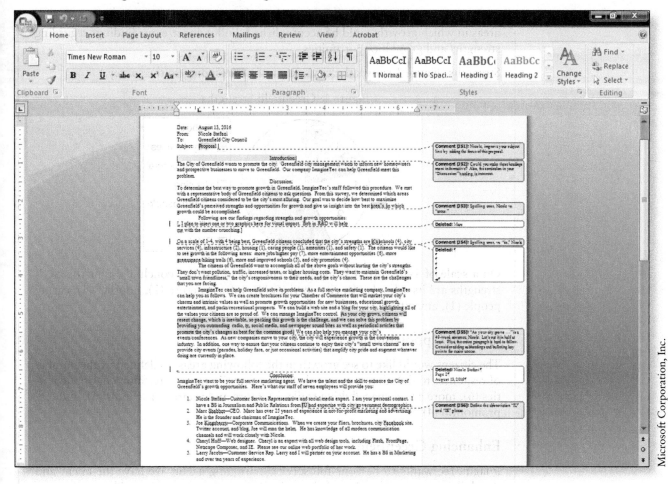

FIGURE 4.7 Revised Report Presented to the Client

Date: August 13, 2016
From: Nicole Stefani
To: Greenfield City Council
Subject: Report to Market the City of Greenfield

How We Can Meet Your Needs

The City of Greenfield wants to promote itself. Greenfield city management wants to inform new homeowners and prospective businesses to move to Greenfield. ImagineTec can help Greenfield meet this problem.

Solutions to Greenfield's Public Relations Challenges

We met with a representative body of Greenfield citizens who participated in a survey. From this survey, we determined which areas Greenfield citizens considered to be the city's most alluring. An analysis of the survey gave us insight into the best

(Continued)

FIGURE 4.7 *(Continued)*

areas in which growth could be accomplished. Following are our findings regarding strengths and growth opportunities.

Strengths (Figure 1 shows the city's strengths in response to the survey.)

Figure 1: Greenfield Strengths

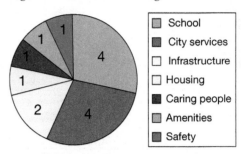

On a scale of 1–4, with 4 being best, Greenfield citizens concluded that the city's strengths are its schools (4), city services (4), infrastructure (2), housing (1), caring people (1), amenities (1), and safety (1).

Growth Opportunities

The citizens would like to see growth in the following areas, as defined by the number of responses to the survey: more jobs/higher pay (7), more entertainment opportunities (6), more green space/hiking trails (6), more and improved schools (5), and city promotion (4).

Enhancing Greenfield's Growth

ImagineTec wants to be your full service marketing agent. We have the talent and the skill to enhance the City of Greenfield's growth opportunities. Please allow me to meet with your City Council to make a PowerPoint presentation, showing you examples of ways in which our talents perfectly meet your needs.

CASE STUDY

The Blue Valley Wastewater Treatment Plant processes water running to and from Frog Creek, a water reservoir in North Upton. This water has low alkalinity (generally, <30 mg/l), low hardness (generally, <40 mg/l), and minimal water discoloration. Inorganic fertilizer nutrients (phosphorus and nitrogen) are also generally low, with limited algae growth.

Despite the normal low readings, algae-related tastes and odors occur occasionally. Although threshold odors range from 3 to 6, they have risen to 10 in summer months. Alkalinity rises to <50 mg/l, hardness to <60 mg/l, and discoloration intensifies. Taste and odor problems can be controlled by powdered activated carbon (PAC); nutrient-related algae growth can be controlled by filtrated ammonia. Both options are costly.

The odors and tastes are bothersome to outdoor enthusiasts who hike and bike the trails along Frog Creek. Residents also worry about the impact of increased alkalinity on fish and turtles. Frog Creek is treasured for its wildlife and recreational opportunities.

The Blue Valley Wastewater Treatment Plant is under no legal obligation to solve these problems. The alkalinity, hardness, color, and nutrient readings are all within regulated legal ranges. However, Blue Valley Wastewater Treatment Plant values the community residents and wants to improve the environment.

In response to this dilemma, divide into small groups to write one of the following documents. These documents could be organized by the problem/solution or comparison/contrast methods.

- You are Blue Valley Wastewater Treatment Plant's director of public relations. Write a letter to the City Commission stating your plant's point of view. (We discuss routine correspondence in Chapter 9.)
- You are North Upton community's resident representative. Write a letter to the City Commission stating your community's point of view.
- You are an employee for the Blue Valley Wastewater Treatment Plant. Write an e-mail to the plant director suggesting ways in which the plant could solve this environmental and community relations problem.

ETHICAL CHALLENGE

Kim Ngyuen is a manager at a nationwide communications company, XConnect. Because of a declining economy, company revenues have fallen 29 percent since last quarter. On Friday, Kim sent the following press release regarding the company:

"As you all know, our economy has slowed down nationwide, but prices for all services and goods are rising rapidly. Plus, the communications industry is changing. Most communications companies are streamlining their operations to meet economic challenges. Many operations are moving offshore. Investors demand that our company show profitability. Soon, with improved strategies, we will be able to increase our business. However, in the short term, to offset losses, we will cut approximately 1,000 jobs. The company will offer voluntary separation packages followed by involuntary cuts as needed."

Question

Although the information about job loss is in Kim's press release, Kim has placed this information at the end of a long paragraph. Is his placement of content ethical since this is how his employees learned the bad news? Has Kim used the best method of organization to convey his content?

INDIVIDUAL AND TEAM PROJECTS

Achieving Clarity

The following sentences are unclear. They will be interpreted differently by different readers. Revise these sentences by replacing the vague words with more specific information.

1. We need this information as soon as possible.
2. The machinery will replace a flawed piece of equipment in our department.
3. Failure to purchase this will have a negative impact.
4. Weather problems in the area resulted in damage to the computer systems.
5. The most recent occurrences were caused by insufficient personnel.

Using the Active Voice Versus the Passive Voice

Passive voice often leads to vague, wordy sentences. Revise the following sentences by writing them in the active voice.

1. Implementation of this procedure is to be carried out by the Accounting Department.
2. Benefits derived by attending the conference were twofold.
3. The information was demonstrated and explained in great detail by the training supervisor.
4. Discussions were held with representatives from Allied, who supplied analytical equipment for automatic upgrades.
5. Also attended was the symposium on polymerization.

Limiting Paragraph Length to Achieve Conciseness

You can achieve clarity and conciseness if you limit the length of your paragraphs. An excessively long paragraph (beyond six typed lines) requires too much work for your reader.

Revise the following paragraph to make it more reader friendly.

As you know, we use electronics to process freight and documentation. We are in the process of having terminals placed in the export departments of some of our major customers around the country so they may keep track of all their shipments within our system. I would like to propose a similar tracking mechanism for your company. We could handle all of your export traffic from your locations around the country and monitor these exports with a terminal located in your home office. This could have many advantages for you. You could generate an export invoice in your export department, which could be transmitted via the computer to our office. You could trace your shipments more readily. This would allow you to determine rating fees more accurately. Finally, your accounting department would benefit. All in all, your export operations would achieve greater efficiency.

Reducing Word Length to Achieve Conciseness

Multisyllabic words can create long sentences. To limit sentence length, limit word length. Find shorter words to replace the following words:

1. advise
2. anticipate
3. ascertain
4. cooperate
5. determine

6. endeavor
7. inconvenience
8. indicate
9. initially
10. presently

Reducing Sentence Length to Achieve Conciseness

Each of the following sentences is too long. Revise them using the techniques suggested in this chapter:

1. In regard to the progress reports, they should be absolutely complete by the fifteenth of each month.
2. I wonder if you would be so kind as to answer a few questions about your proposal.
3. I am in receipt of your e-mail requesting an increase in pay and am of the opinion that it is not merited at this time due to the fact that you have worked here for only one month.
4. On two different occasions, I have made an investigation of your residence, and I believe that your sump pump might result in damage to your neighbor's adjacent property. I have come to the conclusion that you must take action to rectify this potential dilemma, or your neighbor might seek to sue you in a court of law.
5. If there are any questions that you might have, please feel free to contact me by phone.

Using Organization

1. **Spatial.**

   a. Using spatial organization, write a paragraph describing your classroom, your office, your work environment, your dorm room, your apartment, or any room in your house.

   b. Using spatial organization, write an advertisement describing the interior of a car, the exterior of a mechanism or tool, a piece of clothing, or an iPod.

2. **Chronological.** Organizing your text chronologically, write a report documenting your drive to school or work, your activities accomplished in class or at work, your discoveries at a conference or on vacation, or your activities at a sporting event.

3. **Importance.** A fashion merchandising retailer asked her buyers to purchase a new line of clothing. In her e-mail, she provided them the following list to help them accomplish their task. Reorganize the list by importance, and justify your decisions.

Date: January 15, 2016
To: Buyers
From: Sharon Baker
Subject: Clothing Purchases

It is time again for our spring purchases. This year, let's consider a new line of clothing. When you go to the clothing market, focus on the following:

- Colors
- Materials
- Our customers' buying habits
- Price versus markup potential
- Quantity discounts
- Wholesaler delivery schedules

Good luck. Your purchases at the market are what make our annual sales successful.

---

4. **Comparison/contrast.** Visit two auto dealerships, two clothing stores, two restaurants, two online dating sites, two prospective employers, two colleges, or so on. Based on your discoveries, write a report using comparison/contrast to make a value judgment. Which of the two cars would you buy, which of the two restaurants would you frequent, and which of the two online dating sites would you use?

5. **Problem/solution.** You work for Acme Electronics as an electrical engineer. Dejuan Simpson, your boss, informs you that your department's electronic scales are measuring tolerances inaccurately. You are asked to study the problem and determine solutions. In your study, you find that one scale (ID #1893) is measuring within 90 percent of tolerance; another scale (ID #1887) is measuring within 75 percent of tolerance; a third scale (ID #1890) is measuring within 60 percent of tolerance; a final scale (ID #1885) is measuring within 80 percent of tolerance. Standards suggest that 80 percent is acceptable. To solve this problem, the company could purchase new scales ($2,000 per scale), reduce the vibration on the scales by mounting them to the floor ($1,500 per scale), or reduce the vibration around the scales by enclosing them in Plexiglas boxes ($1,000 per scale).

Write an e-mail to your boss detailing your findings (the problems) and suggesting the solutions.

PROBLEM-SOLVING THINK PIECE

Correcting Errors for Accuracy

The following memo contains grammatical errors. The errors include problems with spelling and punctuation, as well as agreement, capitalization, and number usage. Correct the errors to ensure professionalism.

Date: February 12, 2016
To: Martha Collins
From: Richard Davis
Subject: 2016 Digital Carriers

Attached is the supplemental 2016 Digital Carriers reports that is required to support this years growth patterns. As we have discussed in previous phone conversations the January numbers show a decrease in traffic but forecasts still suggest increased traffic therefore we are issuing plans for this contingency.

If the January forecasts prove to be accurate the carriers being placed in the network via these plans will support our future growth accept for areas where growth cannot be predicted. Some areas for example are to densely populated for forecasting. Because the company did not hire enough survey personal to do a thorough job.

Following is an update of our suggested revisions;

Digital Carriers Needed	Capitol Costs
52,304	$3,590,625

If your going to hire anyone to provide follow-up forecasts, they should have sufficient lead time. The survey teams, if you want a successful forecast, needs at least three months. 25 team members should be sufficient. If we can provide farther information please let us know.

WEB WORKSHOP

Many governmental organizations and technical professions ask employees to follow the standards for "plain language," which focus on clarity and conciseness. To learn more about plain language, visit http://www.plainlanguage.gov. This site provides a definition of plain language, governmental mandates, and before and after comparisons. Click on any of the links in this site and report your findings either orally or in an e-mail message.

CHAPTER FIVE

Audience Recognition

After completing this chapter, you will be able to

1. Recognize the audience
2. Define the terms high-tech, low-tech, lay, and multiple audiences
3. Consider an audience's personality traits
4. Understand issues of diversity
5. Recognize the importance of a multicultural audience
6. Achieve effective multicultural communication by following guidelines
7. Avoid biased language
8. Achieve audience involvement through personalized tone and positive words

---

COMMUNICATION AT WORK

In this scenario, Home and Business Mortgage works to meet the company's commitment to diversity through its technical communication.

Home and Business Mortgage (HBM), a Phoenix, Arizona, company, prides itself on being a good neighbor and an asset to the community. Its goal is to ensure that the company meets all governmental personnel regulations, including those required by the Equal Employment Opportunity Commission (EEOC), the Family Medical Leave Act (FMLA), and the Americans with Disabilities Act (ADA). HBM is committed to achieving diversity in its workplace.

In fact, the company is so focused on its appreciation of and commitment to diversity that HBM's technical communicators have created the following:

- Bilingual brochures, newsletters, and annual reports
- Brochures geared toward different stakeholders—senior citizens (age 55 and up), African-American, Hispanic, and Native-American homebuyers; women, and young singles (18–25)
- A Web site with photographs depicting Phoenix's diverse population

HBM's Web site is unique in another important way. It has sought to abide by W3C WAI mandates. The World Wide Web Consortium (W3C) hosts the Web Accessibility Initiative (WAI). This initiative states that "the power of the Web is its universality.

Access by everyone regardless of disability is an essential aspect." HBM's webmasters, in agreement with the WAI, realize that graphic-laden Web sites, as well as those with audio components, are not accessible to people with disabilities, such as hearing and visual impairments.

To meet this WAI platform, HBM's webmasters have achieved the following on the company's Web site:

- Provided links to access information in different languages
- Avoided online audio
- Limited online video and used the Alt attribute to describe each visual's function
- Increased the font of Web text to 20 points versus the standard 12 to 14 points
- Avoided all designer fonts, like cursive, which are hard to read online
- Avoided red and green for color emphasis because these colors are hard to read for audiences with visual impairments
- Avoided frames that not only fail to load on all computers but also create readability challenges online

HBM is honored to serve all of its customers, regardless of race, age, religion, or physical challenges. One way to achieve the company's commitment to diversity is through its technical communication.

When you write a memo, e-mail message, blog entry, Web site, instant message, tweet, letter, or report, someone reads it. That individual or group of readers is your audience. To compose effective technical communication, you should achieve audience recognition and audience involvement.

Audience Recognition

> 1. Recognize the audience.

When you write, give an oral presentation, convene a meeting, communicate with customers in a salesroom, or make a speech at a conference, consider the following questions:

- Who is your audience?
- What does this reader or listener know?
- What does this reader or listener not know?
- What must you write or say to ensure that your audience understands your point?
- How do you communicate to more than one person (multiple audiences)?
- What is this person's position in relation to your job title?
- What diversity issues (gender, sexual orientation, cultural, multicultural) must you consider?

If you do not know the answers to these questions, your communication might contain jargon or acronyms the reader will not understand. The tone of the e-mail may be inappropriate for management (too dictatorial) or for your subordinates (too relaxed). Your communication might not consider your audience's unique culture and language. To communicate successfully, you must recognize your audience's level of understanding. You also should factor in your audience's unique personality and traits, which could impact how successfully you communicate.

See Table 5.1 for a summary of audience variables to consider when communicating.

TABLE 5.1 Audience Variables

Knowledge of Subject Matter	Personality Traits	Issues of Diversity
• High tech • Low tech • Lay • Multiple	• Audience's perceived personality • Audience's attitude or position regarding the topic • Audience's preference regarding style • Audience's response to the topic	• Age • Gender • Race and/or religion • Sexual orientation • Language and/or culture of origin—multicultural or cross-cultural

Knowledge of Subject Matter

What does your audience know about the subject matter? Does the person work closely with you on the project? That would make the audience a high-tech peer. Does the audience have general knowledge of the subject matter but a different area of expertise? That would make the audience a low-tech peer. Is the audience totally uninvolved in the subject matter? That would make the person a lay audience. Finally, could your audience be a combination of these types? Then, you would be communicating with an audience with multiple levels of expertise.

High-Tech Audience. High-tech readers work in your field of expertise. They might work directly with you in your department, or they might work in a similar capacity for another company. Wherever they work, they are your colleagues because they share your educational background, work experience, or level of understanding.

If you are a computer programmer, for example, another computer programmer who is working on the same system is your high-tech peer. If you are an environmental engineer working with hazardous wastes, other environmental engineers focusing on the same concerns are your high-tech peers.

Once you recognize that your reader is high tech, what does this tell you? High-tech readers have the following characteristics:

- They are experts in the field you are writing about. If you write an e-mail message to a department colleague about a project you two are working on, your associate is a high-tech peer. If you write a letter to a vendor requesting specifications for a system she markets, that reader is a high-tech expert. If you write a proposal geared toward your colleagues, they are high tech.
- Because their work experience or education is comparable to yours, high-tech readers share your level of understanding. Therefore, they will understand high-tech jargon, acronyms, and abbreviations. You do not have to explain to an electronic technician, for example, what MHz means. Defining megahertz for this high-tech reader would be unnecessary.
- High-tech readers require minimal detail regarding standard procedures or scientific, mathematical, or technical theories. Two biomedical technologists, when discussing a metabolic condition in which the body cannot buffer changes in pH, do not need to define pH or acidosis (the terms being discussed).
- High-tech readers need little background information regarding a project's history or objectives unless the specific subject matter of the correspondence is new to them. If, for example, you are writing a status report to your supervisor, who has been involved in a project since its inception, then you will not need to flesh out the history of the project. On the other hand, if a new supervisor is hired or if you are updating a colleague new to your department, even though these readers are high tech, you will need to provide background data.

Figure 5.1 is an e-mail message to high-tech HVAC employees, informing them of the capabilities of a new HVAC security system. In contrast, Figure 5.2 discusses the same topic for a low-tech audience.

FIGURE 5.1 E-mail Message to High-Tech Audience

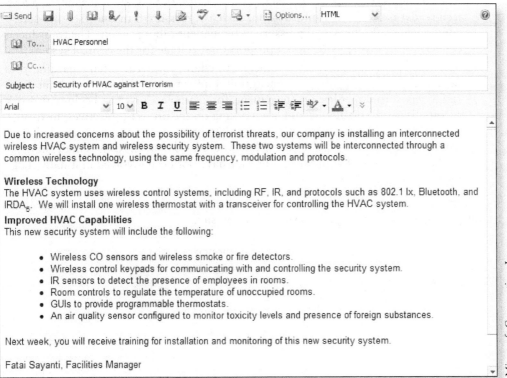

This e-mail, written to high-tech readers, uses terminology without explanation or definition. Its tone is businesslike and directive.

Microsoft Corporation, Inc.

Low-Tech Audience. Writing in the workplace, however, rarely allows you to communicate only with high-tech readers. You also will write to low-tech readers. Low-tech readers include your coworkers in other departments. Low-tech readers also might include your bosses, subordinates, or colleagues who work for other companies. For instance, if you are a biomedical equipment technician, the accountant or personnel director or graphic artists in your company are low-tech peers. These individuals have worked around your company's equipment and, therefore, are familiar with your technology. However, they do not understand the intricacies of this technology.

Although your colleagues at other companies might have your level of education or work experience, they could be low tech if they are not familiar with your company's procedures or in-house jargon, acronyms, and abbreviations. When you write e-mail, letters, and reports to low-tech readers, remember that they share the following characteristics:

- Low-tech readers are familiar with the technology you are writing about, but their job responsibilities are peripheral to the subject matter. They either work in another department, manage you, work under your supervision, or work outside your company.

- Because low-tech readers are familiar with your subject matter, they understand some abbreviations, jargon, and technical concepts. To ensure that readers understand your content, therefore, define your terms. An abbreviation like BCA can't stand alone. Define it parenthetically: BCA (Burst Cutting Area).

- Abbreviations, acronyms, and technical jargon like "Burst Cutting Area" sometimes need follow-up explanations. You should write, "BCA (Burst Cutting Area) is a circular section located near the center of a DVD disc where manufacturing information can be inscribed in a barcode."

- In addition, technical concepts must be defined for low-tech readers. For example, whereas high-tech readers understand the function of pressure transducers, a low-tech reader needs further information, such as "Pressure transducers: Solid-state components sense proximal pressure in the patient tubing circuit. The transducers convert this pressure value into a proportional voltage for the control system."
- Since the low-tech reader is not someone to whom you write often regarding your field of expertise, provide more background information. When you submit a status report to upper-level management, for example, don't just begin with work accomplished. Explain why you are working on the project (objectives, history), who is involved (other personnel), when the project began and its scheduled end date, and how you are accomplishing your goals. Low-tech readers understand the basic concepts of your work, but they have not been involved in it daily. Fill them in on past history.

Figure 5.2 is an e-mail to low-tech security personnel, informing them of procedures to follow if they discover suspicious substances on site.

FIGURE 5.2   E-mail Message to Low-Tech Audience

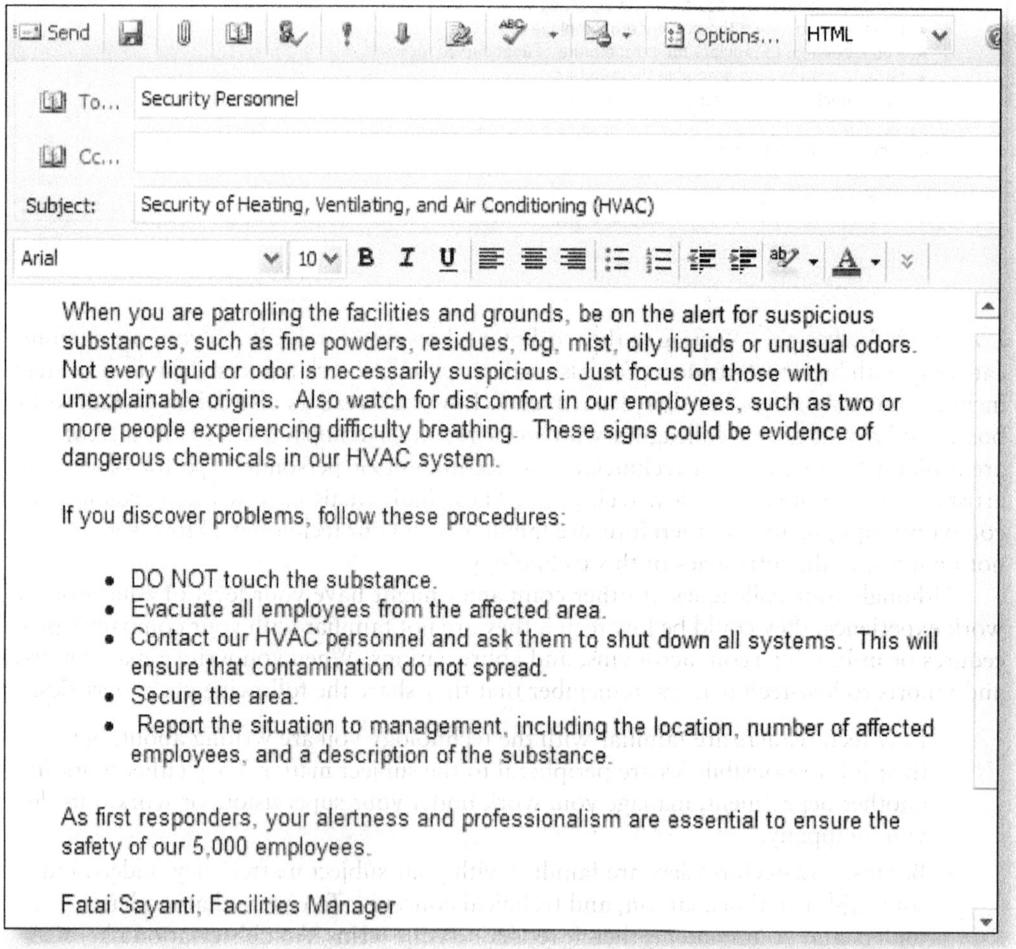

This e-mail to low-tech readers deals with a very important subject (security in the face of potential terrorism). The writer presents only what the audience needs to know in order to complete their job responsibilities. The writer conveys this content simply, using short sentences, a list, and easy-to-understand words. HVAC is defined.

Microsoft Corporation, Inc.

| Send | 🖫 🔗 📖 ⚙ ❗ ⬇ ▨ ✍ ▾ ✉ ▾ | Options... | HTML ▾ |

To...   Security Personnel

Cc...

Subject: Security of Heating, Ventilating, and Air Conditioning (HVAC)

Arial   ▾ 10 ▾ **B** *I* U ≡ ≡ ≡ ≡ ⋮≡ ⋮≡ 年 年 ᵃᵇ⁄ ▾ **A** ▾ ⌄

When you are patrolling the facilities and grounds, be on the alert for suspicious substances, such as fine powders, residues, fog, mist, oily liquids or unusual odors. Not every liquid or odor is necessarily suspicious. Just focus on those with unexplainable origins. Also watch for discomfort in our employees, such as two or more people experiencing difficulty breathing. These signs could be evidence of dangerous chemicals in our HVAC system.

If you discover problems, follow these procedures:

- DO NOT touch the substance.
- Evacuate all employees from the affected area.
- Contact our HVAC personnel and ask them to shut down all systems. This will ensure that contamination do not spread.
- Secure the area.
- Report the situation to management, including the location, number of affected employees, and a description of the substance.

As first responders, your alertness and professionalism are essential to ensure the safety of our 5,000 employees.

Fatai Sayanti, Facilities Manager

Lay Audience.   Customers and clients who neither work for your company nor have any knowledge about your field of expertise are the lay audience. If you work in network communications for a cable company, for example, and you write an e-mail to a client regarding a problem with the high-speed Internet connection, your audience is a lay reader.

If your field of expertise is biomedical equipment and you write a user manual for the patient, you are writing to a lay audience. If you are an automotive technician writing a service report for a customer, that customer is a lay reader.

Although you understand your technology, your readers who use the high-speed Internet service or the medical equipment or the car are not experts in the field. These readers are using your equipment or require your services, but the technology you are writing about is not within their daily realm of experience.

This makes writing for a lay audience difficult. To write successfully to a lay audience, remember that these readers share the following characteristics:

- Lay readers are unfamiliar with your subject matter and terminology. Therefore, you should write simply. That's not to say that you should insult your lay reader with a remedial discussion or with a patronizing tone. You must, however, explain your topic clearly through precise word usage, depth of detail, and simple graphics.

- Because your lay readers do not understand your terminology or work environment, they won't understand your in-house jargon, abbreviations, or acronyms. Avoid high-tech terms or define them thoroughly.

- Lay readers will need background information. If you leap into a discussion about a procedure without explaining to your lay readers why they should perform each step, they will not understand the causes or rationale. High-tech and possibly even low-tech readers might not need such explanations, but the lay reader needs clarity.

For example, look at the letter about green construction principles from a high-tech engineer to a lay audience, a school superintendent (Figure 5.3).

Composing effective technical communication requires that you recognize the differences among high-tech, low-tech, and lay audiences. If you incorrectly assume that all readers are experts in your field, you will create problems for yourself as well as for your readers. If you write using high-tech terms to low-tech or lay audiences, your readers will be confused and anxious. You will waste time on the phone or answering e-mail messages to clarify the points that you did not make clear in the technical document.

Multiple Audiences.   Correspondence is not always sent to just one type of audience. Your correspondence may have an audience with multiple levels of expertise. Sometimes you write to high-tech, low-tech, and lay audiences simultaneously. For example, when writing a report, most people assume that their supervisor will be the only reader. This might not be the case, however. The supervisor could send a copy of your report to the manager, who could then submit the same report to the executive officer. Similarly, your supervisor might send the report to your colleagues or to your subordinates. Your report might be sent out to other lateral departments. Figure 5.4 shows the possibilities.

Writing correspondence for multiple readers with different levels of understanding and reasons for reading creates a challenge. When you add the necessity of using a tone that will be appropriate for all of these varied readers, the writing challenge becomes even greater.

How do you meet such a challenge? The first key to success is recognizing that multiple audiences exist and that they share the following characteristics:

- Your intended audience will not necessarily be your only readers. Others might receive copies of your correspondence.

- Some of the multiple readers will be unfamiliar with the subject matter. Proper background data (objectives, overviews) will clarify the history for these readers. In a letter, report, e-mail message, Web site, or blog, this background information cannot be too elaborate. Often, all you do is provide a reference line, subject line, or link suggesting where the readers can find out more about the subject matter

FIGURE 5.3   Letter from a High-Tech Writer to a Lay Audience

This letter from a high-tech writer to a lay audience defines all terms and avoids abbreviations.

RTA Engineering
5117 East 31st Street
Kansas City, MO 64128

May 6, 2016

Dr. Jamie Wilson
Superintendent, Parkton Independent School District
205 Main
Parkton, MO 65110

Dear Dr. Wilson:

Thank you for inviting RTA Engineering and GreenLive to work with you on plans for new construction and retrofitting of Parkton Independent schools. Dale Askew, from GreenLive, and I have completed our review of the Parkton Independent School District. To ensure that your facilities meet green construction principles, we are proposing the following energy-saving building changes:

1. **Net-zero Energy.** Currently, your facilities draw energy from the city's grid. The goal of net-zero energy is to ensure that buildings generate, on their own sites, at least as much energy as they use. By installing only four wind turbines on the southwest corner of your high school (currently used for overflow parking), your entire district can produce as much power from renewable sources as from the city's grid.

2. **Retrofitting.** Our studies show that your buildings lose approximately 36 percent heat and cooling from air seepage. By retrofitting the current sites (no new construction needed), you can achieve what's known in the industry as "a tight envelope." This means that by sealing floor-to-wall joints, insulating roofs, and wrapping pipes—all easy and cost-effective fixes—Parkton can earn the coveted Platinum certification. Only businesses that are at least 80 percent energy efficient earn this award.

3. **Solar Panel Installation.** GreenLive provided two great options for solar power. Cyrstalline cell modules, common to flat panel installation, are somewhat costly but very efficient. Thin-film modules are less expensive but also less efficient. Still, they are a good option for building retrofits like you're considering. We'll be happy to provide you more details on each option.

4. **Biophilic Design.** This green technology incorporates nature in building construction. We can create vegetated roofs (partially covering your roof with grass, reeds, and bushes), rammed earth walls (building walls out of sustainable materials like earth, chalk, lime, and gravel), operable windows that open to let in light and air, skylights, and ventilation systems that let fresh air into your buildings. Not only will these techniques improve energy consumption, but also biophilic designs have proven to positively impact morale.

Once you have reviewed the attached construction and retrofit plans, Dr. Wilson, please call or e-mail me (682-0011 or mwalker@RTAEngineering.com). I look forward to answering your questions and working with you on ways to make Parkton schools green.

Sincerely,

Mike Walker

Mike Walker, CEO RTA Engineering

FIGURE 5.4 Examples of Possible Audiences

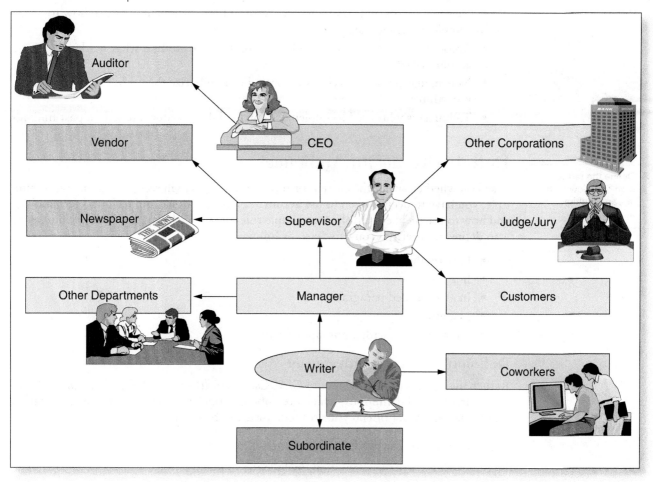

if they wish—"Reference: Operations Procedure 321 dated 9/21/16." In longer reports, background data will appear in the summary or abstract as well as in the report's introduction.

- Multiple readers have diverse understandings of your technology. This requires that you define jargon, abbreviations, and acronyms.
- You might be writing correspondence in a collaborative team. If you're the technical writer, you could work with SMEs (Subject Matter Experts). These could include software developers, customer service personnel, project managers, business analysts, and more. To communicate effectively within your team, everyone needs to understand each other's terminology.

> **Audience for Social Media, Web Sites, and Long, Formal Reports**
>
> See Chapter 12, "Social Media," Chapter 13, "Web Sites and Online Help," and Chapter 18, "Long, Formal Reports."

Writing for Future Audiences

Time creates another challenge regarding audience recognition. Technical communication usually is archived for future reference. You might write an e-mail, report, proposal, or instruction using terms or focusing on content that your primary audience understands . . . at the moment. However, months (or years) later when the correspondence is retrieved from the files, will your readers still be familiar with the topic? Will you still have the same readers? Many people, some of whom didn't even work for your company at the time of the original writing, will read your correspondence. These future readers need clarity, background information, and terms defined. Future readers could include

- Judges and juries who depend on past reports to decide cases
- New employees at a company who retrieve filed documents to learn how to complete current tasks
- Accountants and auditors who look at past records to find what work was accomplished at a prior date
- New managers or supervisors who need to familiarize themselves with workplace procedures
- Technicians who need to implement current lab procedures based on past findings

Defining Terms for Audiences

2. Define the terms high-tech, low-tech, lay, and multiple audiences.

Based on whether or not the current or future audience is high tech, low tech, lay, or multiple, you have to decide when to use acronyms, abbreviations, and technical terminology and how to use them effectively. One simple rule for low-tech, lay, and multiple audiences is to define your terms. You can do so as follows:

- In parentheses
- In a sentence
- In an extended paragraph
- In a glossary
- As online help with a pop-up definition

Defining Terms Parenthetically

Rather than just writing *CIA*, write *CIA* (*Cash in Advance*) or *Cash in Advance* (*CIA*). Such parenthetical definitions, which are only used once per correspondence, don't take a lot of time and won't offend your readers. Instead, the result will be clarity.

Defining Your Terms in a Sentence

If you provide a sentence definition, include the following:

Term + Type + Distinguishing characteristics

For example, using a sentence to define *HTTP,* you would write the following:

Term   Type

EXAMPLE ▶

HTTP (hypertext transfer protocol) is a set of rules for the secure transfer of files, including text, graphics, videos, and sound on the Internet.

Distinguishing
characteristics

Using Extended Definitions of One or More Paragraphs

When you need to provide an extended definition of a paragraph or more, in addition to providing the term, type, and distinguishing characteristics, also consider including examples, procedures, and descriptions. Look at the following paragraph definition of a video card.

EXAMPLE ▶

A video card (graphics accelerator card, display adapter, or graphics card) is a type of printed circuit board that can be inserted into an expansion slot of a computer's motherboard. The video card creates and outputs images to a display. Functions of a video card can include video capture, MPEG-2 and MPEG-4 decoding, TV output, or the ability to connect multiple monitors.

Using a Glossary

If you have not defined your terms parenthetically, in a sentence, or in a paragraph, you should use a glossary. A glossary is an alphabetized list of terms placed after your conclusion/recommendation.

◀ EXAMPLE

Glossary	
ADC	analog to digital converter
Analog	a variable signal continuous in both time and amplitude
DAC	digital to analog converter
LAN	local area network
PSTN	public switched telephone network—analog phone systems
ROI	return on investment
Voice signal	allows users to access phone options with a single voice command
VoIP	voice over Internet protocol
WAN	wide area network

Providing Pop-Ups and Links with Definitions

Use pop-up screens for online help. (See Figure 5.5.)

FIGURE 5.5 Online Help Screen with Pop-up Definition

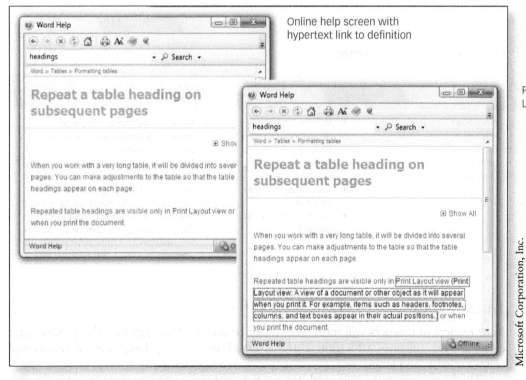

Online help screen with hypertext link to definition

Pop-up definition of "Print Layout view"

Microsoft Corporation, Inc.

Audience Personality Traits

3. Consider an audience's personality traits.

In addition to considering an audience's level of knowledge, a technical communicator should determine aspects of an audience's personality. By considering personality traits, the writer then can write appropriately using the proper tone, visual aids, and writing style.

This will help the writer meet the audience's needs. Of equal importance, by recognizing the audience's personality traits, the writer can more effectively get the desired response from the reader.

Let us start with a caveat. You obviously cannot always know your audience's personality traits. This would be a difficult goal to achieve when working with customers you've never met, a dispersed team of coworkers located in different cities, members of an audience at a large conference, or the widely diverse audience that reads your Web site or blog postings. However, many times you will write or speak to an audience that you know. This could include customers you have served in the past and vendors you often work with as well as coworkers in your office and your immediate supervisors. To communicate successfully with an audience whom you know, factor in your knowledge of their personalities, attitudes, and preferences.

For example, we work with a company that offers its employees a pay incentive. The company will pay employees to identify a problem in the workplace and then suggest solutions to this problem. The goal is to improve the company's services, efficiency, budget, and profits. To earn this monetary award, the employees must write a report analyzing the perceived problem and proposing the solutions, create PowerPoint slides regarding the topic, and make a brief oral presentation to their supervisors.

We have found from interviews that employees who successfully earn the financial incentive assess their audience's personality. An audience's personality traits and possible responses to a topic of discussion can include issues such as those shown in Table 5.2. Reviewing these considerations will help you determine the type of content you want to provide, how persuasive you need to be, and what tone and style your technical communication must achieve.

TABLE 5.2 Audience Personality Traits and Responses

Audience's Perceived Personality	Audience's Attitude or Position Regarding the Topic	Audience's Preference Regarding Style	Desired Audience Response
• Slow to act? • Eager? • Receptive? • Questioning? • Organized? • Disorganized? • Oppositional?	• Negative? • Positive? • Noncommittal? • Informed? • Uninformed?	• Will he, she, or they accept contractions? • Should you use first names, last names, or position titles? • Is short and to-the-point better? • Is long and detailed better?	• Do you want the audience to consider this idea and pass on a recommendation? • Do you want the audience to act now? • Do you want the audience to confirm what has been suggested? • Do you want the audience to reject some options but act on others? • Do you want the audience to file this information for future reference?

4. Understand issues of diversity.

Biased Language—Issues of Diversity

Your readers or listeners will more than likely be diverse. Diversity includes gender, race, ethnicity, age, religious affiliation, culture, sexual orientation, class, physical and mental characteristics, and language.

Think of it this way: You work for a city government, and your audience is the citizenry. Who comprises your city's populace? Or you work in a hospital. Who will visit your facility? Or you work in a retail establishment. Who will shop there? They are people of many different interests, levels of knowledge, and backgrounds—and they are all valuable to your business.

Why should you be concerned about a diverse audience?

1. **Diversity is protected by the law.** Prejudicial behavior and discrimination on the job will not be tolerated. Failure to comply with equal opportunity and affirmative action legislation can lead to fines and lost business.

2. **Respecting diversity is the right thing to do.** People should be treated equally, regardless of their age, gender, sexual orientation, culture, or religion.

3. **Diversity is good for business.** In a workplace where employees feel valued and respected, they feel more committed to the company. Clients and customers prefer shopping in an environment devoid of prejudice. In addition, people from diverse backgrounds spend money and buy products. Diverse hiring practices are good for business.

    General Mills is an example of a company that values diversity. According to its Web site, the company recruits employees from a variety of networks, including the following:

    - American Indian Council
    - Asian Heritage Network
    - Betty's Family (GLBT Network)
    - Black Champions Network
    - Hispanic Network
    - Middle East and North Africa Network
    - Women in Leadership Network

4. **A diverse workforce keeps companies competitive.** Talent does not come in one color, nationality, or belief system. Diversity management is such an important concern that American businesses spend millions of dollars training employees to understand multiculturalism and diversity.

The importance of diversity is verified further in a Society for Human Resource Management Survey of Diversity Initiatives, which stated that "84 percent of human resource professionals at Fortune 500 companies say their top-level executives think diversity management is important" ("What Is The 'Business Case' For Diversity?").

Multiculturalism

5. Recognize the importance of a multicultural audience.

Your company might market its products or services worldwide. International business requires multicultural communication, the sharing of written and oral information between businesspeople from different countries.

The Global Economy

According to Edward Jones, an international investment company, "S&P 500 companies generate about half of their sales outside the U.S." (Bannister). For example,

- Two-thirds of Nike's sales are outside the United States.
- Yum! Brands (Taco Bell, KFC, and Pizza Hut) sells more in China than in the United States.
- General Motors sells more cars in China than in the United States.
- Coca Cola, selling 300 different brands in 200 countries, generates 80 percent of its income outside the United States.
- Microsoft translates its operating system into more than 100 different languages.
- Black & Veatch, a global engineering company, has office locations in Australia, Canada, Chile, China, India, Indonesia, Malaysia, Nepal, Puerto Rico, Russia, Singapore, Taiwan, Thailand, Turkey, United Arab Emirates, United Kingdom, United States, and Vietnam.

Margaret Keating, vice president-operations for Hallmark Cards Inc., oversees Hallmark's North American global operations. Her workforce of 6,000 employees is geographically dispersed. Much of her time is spent writing clear and concise communication to be translated into different languages (Cardarella D19).

The Challenges of Multicultural Communication. The Internet, e-mail, Facebook, Twitter, and YouTube affect global communication and global commerce constantly. With these technologies, companies can market their products internationally and communicate with multicultural clients and coworkers at a keystroke. An international market is great for companies because a global economy increases sales opportunities. However, international commerce also creates written and oral communication challenges. Companies that work internationally must communicate with all of their employees and clients; therefore, communication must be multilingual.

For example, Medtronic, a leading medical technology company, does business in 120 countries. Many of those countries mandate that product documentation be written in the local language. To meet these countries' demands, Medtronic translates its manuals into 11 languages: French, Italian, German, Spanish, Swedish, Dutch, Danish, Greek, Portuguese, Japanese, and Chinese (Walmer 230).

Multilingual reports create unique communication challenges:

- Will each language version be identical in content and readability?
- Will the first language version suggest advantages to investors over translations?
- Are all translations carefully prepared according to tone, style, and content?
- Is each translation tested for usability and accuracy?

Multicultural Team Projects. What about international, multilingual project work teams? If, for example, your U.S. company is planning to build a power plant in China, you will work with Chinese engineers, financial planners, and regulatory officials. To do so effectively, you will need to understand that country's verbal and nonverbal communication norms. You also will have to know that country's management styles, decision-making procedures, sense of time and place, and local values, beliefs, and attitudes.

Our natural instinct is to evaluate people and situations according to our sense of values, our cultural perspectives. That is called *ethnocentrism*—a belief that one's own culture represents the norm. Such is not the case. The world's citizenry does not share the same perspectives, beliefs, values, political systems, social orders, languages, or habits. Successful technical communication takes into consideration language differences, nonverbal communication differences, and cultural differences.

Due to the multicultural makeup of your audience, you must ensure that your writing, speaking, and nonverbal communication skills accommodate language barriers and cultural customs. The classic example of one company's failure to recognize the importance of translation concerns a car that was named Nova. In English, *nova* is defined as a star that spectacularly flares up. In contrast, *no va* in Spanish is translated as "no go," a poor advertisement for an automobile.

Cross-cultural Workplace Communication

Multiculturalism will affect you not just when you communicate globally. You will be confronted with multicultural communication challenges even in your own city and state. Another term for this challenge is *cross-cultural communication,* writing and speaking between businesspeople of two or more different cultures within the same country (Nethery). Look at the following statistics regarding America's melting pot, as shown in Figures 5.6 and 5.7 ("Demographics of the United States").

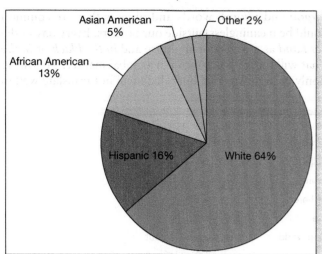

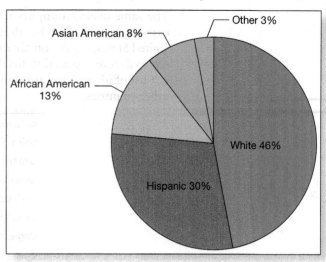

U.S. Census projects that by 2050, the White population will fall in percentage to 46.3 percent, and the Asian population will rise to around 8 percent. The Hispanic population will almost double to 30.3 percent as shown in Figure 5.7.

One of the challenges presented by the increasingly multicultural nature of our society and workplace is language. Language barriers are especially challenging for hospitals, police and fire personnel, and governmental agencies where failure to communicate effectively can have dangerous repercussions.

One hospital reported that it has staff who can translate into Spanish, Arabic, and Somali languages. However, this hospital's successful use of translators to help with doctor-patient communication is rare. Fewer than 15 percent of hospitals in the United States train translators. Most hospitals depend on the patient's relatives, an undependable source of translations. One hospital reported that an 11-year-old made over 50 mistakes (Sanchez A4). Imagine how such errors can negatively impact healthcare and medical records.

The communication challenges are not only evident for employees in healthcare and other community infrastructures. Industries as diverse as banking, hospitality (restaurants and hotels), construction, agriculture, meat production and packing, and insurance also face difficulties when communicating with clients and employees for whom English is a second language.

Guidelines for Effective Multicultural Communication

To achieve effective multicultural technical communication, follow these guidelines:

Define Acronyms and Abbreviations

Acronyms and abbreviations cause most readers a problem. Although you and your immediate colleagues might understand such high-tech usage, many readers won't. This is especially true when your audience is not native to the United States. For example, corporate employees often abbreviate the job title "system manager" as *sysmgr*. However, in German, the title "system manager" is called the *system leiter*; in French, it's *le responsable*. The abbreviation *sysmgr.* would make no sense in either of these countries (Swenson WE-193).

> 6. Achieve effective multicultural communication by following guidelines.

Avoid Jargon and Idioms

The same dilemma applies to *jargon* and *idioms*, words and phrases that are common expressions in English but that could be meaningless outside our borders. Every day in the United States, we use *on the other hand* as a transitional phrase and *in the black* or *in the red* to denote financial status. What will these idioms mean in a global market? Following is a list of idioms that are commonly used in America but which will not translate well in other countries:

ballpark figure	eye on the prize
belt-tightening	face the music
bite the bullet	guesstimate
brownie points	jump through hoops
buck stops here	pass the buck
crunch time	pull the plug
dog-eat-dog world	shape up or ship out
drum up business	through the roof

Distinguish Between Nouns and Verbs

Many words in English act as both nouns and verbs. This is especially true with computer terms, such as *file*, *scroll*, *paste*, *boot*, *code*, and *help*. If your text will be translated, make sure that your reader can tell whether you're using the word as a noun or a verb.

Watch for Cultural Biases/Expectations

Your text will include words and graphics. As a technical communicator, you need to realize that many colors and images that connote one thing in the United States will have different meanings elsewhere. For example, the idioms "in the red" and "in the black" will not necessarily communicate your intent when they are translated. Even worse, the colors black and red have different meanings in different cultures. Red in the United States connotes danger; therefore, "in the red" suggests a financial problem. In China, however, the word *red* has a positive connotation, which would skew your intended meaning. The word *black* often implies death and danger, yet "in the black" suggests financial stability. Such contradictions could confuse readers in various countries.

Animals represent another multicultural challenge. In the United States, we say you're a "turkey" if you make a mistake, but success will make you "soar like an eagle." The same meanings don't translate in other cultures. Take the friendly piggy bank, for example. It represents a perfect image for savings accounts in the United States, but pork is a negative symbol in the Mideast. If you are "cowed" by your competition in the United States, you lose. Cows, in contrast, represent a positive and sacred image in India.

Be Careful When Using Slash Marks

Does the slash mark mean "and," "or," or both "and/or"? The word *and* means "both," but the word *or* means "one or the other," not necessarily both. If your text will be translated into another language, will the translator know what you meant by using a slash mark? To avoid this problem, determine what you want to say and then say it.

Avoid Humor and Puns

Humor is not universal. In the United States, people talk about regional humor. If a joke is good in the South but not in the North, how could that same joke be effective overseas? Microsoft's software package Excel is promoted by a logo that looks like an *X* superimposed over an *L*. This visual pun works in the United States because we pronounce the letters *X* and *L* just as we would the name of the software package. If your readers are not familiar with English, however, they might miss this clever sound-alike image.

Realize That Translations May Take More or Less Space

Paper Size.   If your writing will be conveyed not on paper but on disk or on the Internet, you must consider software's line-length and screen-length restrictions. For example, a page of hard-copy text in the United States will consist of approximately 55 lines that average 80 characters per line. How could this present a problem? The standard sheet of paper in the United States measures 8½ × 11 inches. In contrast, the norm in Europe for standard-sized paper is A4—210 × 297 millimeters, or 8.27 × 11.69 inches.

Why is the size of paper important? Here is the problem: If you format your text and graphics for an 8½ × 11-inch piece of paper in Atlanta, travel to London, download the files on a computer there, and hit Print, you might find that what you get is not what you hoped for. The line breaks will not be the same. You will not be able to three-hole punch and bind the text. The margins will be off, and so will the spacing on your flowchart or table.

Web Sites.   An even greater problem occurs when you are writing for the Internet. On a Web site, you will provide a navigation bar and several frames. Why is this a problem? The word count of a document written in English will expand more than 30 percent when translated into some European languages. In Table 5.3, notice how English words become longer when translated into other languages.

TABLE 5.3   Translations Increase Word Length

English Word	Translations into Other Languages
Print	*Impression* (French)
File	*Archivo* (Spanish)
View	*Visualizzare* (Italian)
Help	*Assistance* (French)
E-mail	*Courriel* (Swiss)

The Swiss government is trying to curb what it defines as "the encroachment of English." To do so, the government's French Linguistics Service asked its citizens to avoid using the word *spam,* instead opting for *courier de masse non sollicite,* meaning "unsolicited bulk mail" ("Swiss Fight Encroachment of English" A16).

Whereas the above examples add length to a document, Chinese characters are succinct, usually about 60 percent shorter than English. You must consider length to accommodate translations (Courtis and Hassan 397).

<table>
<tr><td>

- **English.** Wash your hands before returning to work = 7 words, 10 syllables
- **Spanish.** Lávese las manos antes de volver al trabajo = 8 words, 15 syllables
- **Chinese.** 返回工作岗位前要洗手 = 10 characters

</td><td>

◄ EXAMPLE

</td></tr>
</table>

TECHNOLOGY TIPS

Using Microsoft Word 2013 to Translate Text

Because your audience can be global and cross-cultural, you might need to translate text. Microsoft Word will help you accomplish this goal.

1. Type the text you want to translate.
2. Click on the **Review** tab in Word.
3. Once you have highlighted the text by clicking and dragging, click on **Translate**.

Note: A window will pop up at the right margin. From here, you can choose to translate your text from English into at least 16 different languages, including Spanish, French, Hebrew, German, Arabic, and Chinese.

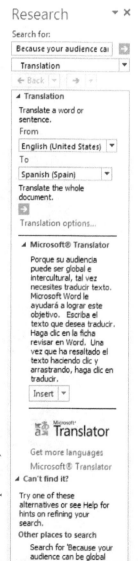

Avoid Figurative Language

Many of us use sports images to figuratively illustrate our points. We "tackle" a chore; in business, a "good defense is the best offense"; we "huddle" to make decisions; if a sale isn't made, you might have "booted" the job; if a sale is made, you "hit a home run." Each of these sports images might mean something to native speakers, but they may not communicate worldwide. Instead, say what you mean, using precise words.

Be Careful with Numbers, Measurements, Dates, and Times

Numbers and Measurements.   If your text uses measurements, you are probably using standard American inches, feet, and yards. However, most of the world measures in metrics. Thus, if you write 18 high × 20 wide × 30 deep, what are the measurements? There is a huge difference between 18 × 20 × 30 inches and 18 × 20 × 30 millimeters.

Dates.   In the United States, we tend to abbreviate dates as MM/DD/YY: 05/03/16. In the United Kingdom, however, this could be perceived as March 5, 2016, instead of May 3, 2016. See Table 5.4 for additional examples.

TABLE 5.4   Different Ways of Understanding and Writing the U.S. Date 05/03/16

Country	Date
United States	May 3, 2016
United Kingdom	March 5, 2016
France	5 mars 2016
Germany	5. Marz 2016
Sweden	16–05–03
Italy	5.3.16

Time.   Time is another challenge. Table 5.5 shows how different countries write times.

TABLE 5.5   Different Ways of Writing the U.S. Time 5:15 P.M.

Country	Time
United States	5:15 P.M.
France	17:15
Germany	17.15
Quebec, Canada	17 h 15

In addition to different ways of writing time, you must also remember that even within the United States, 1:00 P.M. does not mean the same thing to everyone. Is that central time, Pacific time, mountain time, or eastern time?

Another challenge with time occurs when we incorrectly assume that everyone everywhere abides by the same work hours. In the United States, the average workweek is 40 hours, and the typical workday is from 8:00 A.M. to 5:00 P.M. However, this is not the norm globally. French laws have reduced the workweek to 35 hours. Many Middle Eastern countries close work on Friday, the beginning of Sabbath. Offices in parts of southern Europe shut down for a traditional two-hour lunch (noon to 2 P.M.). Therefore, if you write an e-mail telling a coworker in Spain or Jordan that you will call at 1:00 P.M., Friday, his or her time, that could be an inappropriate time for your coworker.

Finally, even simple words like *today, yesterday,* or *tomorrow* can cause problems. Japan is 14 hours ahead of U.S. eastern standard time. Thus, if you need a report "tomorrow," do you mean tomorrow—the next day *your time*—or tomorrow—two days from *your reader's time?* To solve these problems, determine your audience and make changes accordingly. That might mean

- Writing out the date completely (January 12, 2016)
- Telling the reader what standard of measurement you will use ("This document provides all measurements in metrics.")
- Telling the reader what scheme of time presentation you will use ("This document relates time using a 24-hour clock rather than a 12-hour clock.")
- Using multiple formats ("Let's meet at 2:30 P.M./14:30.")
- Avoiding vague words like *today, tomorrow,* or *yesterday*
- Recognizing that people have different work schedules globally

Use Stylized Graphics to Represent People

A photograph or realistic drawing of people will probably offend someone and create a cultural conflict. You want to avoid depicting race, skin color, hairstyles, and even gender. To solve this problem, avoid shades of skin color, choosing instead pure white or black to represent generic skin. Use simple, abstract, even stick figures to represent people. Stylize hands so they are neither male nor female—and show a right hand rather than a left hand, if possible (a left hand is perceived as "unclean" in some countries) (Flint). Recognizing the importance of the global marketplace is smart business and a wise move on the part of the technical communicator.

Following is an example of poor communication in an e-mail message for a multilingual audience.

E-mail Messages

See Chapter 9, "Routine Correspondence," for more information about e-mail messages.

BEFORE

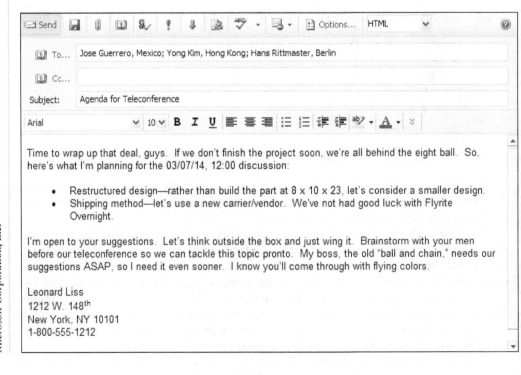

Microsoft Corporation, Inc.

In contrast, the following example corrects these communication problems:

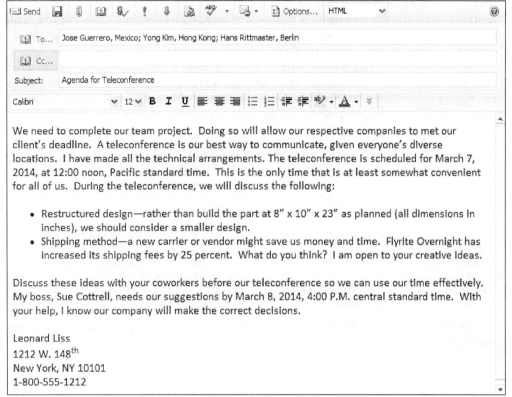

We need to complete our team project. Doing so will allow our respective companies to met our client's deadline. A teleconference is our best way to communicate, given everyone's diverse locations. I have made all the technical arrangements. The teleconference is scheduled for March 7, 2014, at 12:00 noon, Pacific standard time. This is the only time that is at least somewhat convenient for all of us. During the teleconference, we will discuss the following:

- Restructured design—rather than build the part at 8" x 10" x 23" as planned (all dimensions in inches), we should consider a smaller design.
- Shipping method—a new carrier or vendor might save us money and time. Flyrite Overnight has increased its shipping fees by 25 percent. What do you think? I am open to your creative ideas.

Discuss these ideas with your coworkers before our teleconference so we can use our time effectively. My boss, Sue Cottrell, needs our suggestions by March 8, 2014, 4:00 P.M. central standard time. With your help, I know our company will make the correct decisions.

Leonard Liss
1212 W. 148th
New York, NY 10101
1-800-555-1212

Microsoft Corporation, Inc.

Avoiding Biased Language

> 7. Avoid biased language.

In addition to recognizing your audience's level of knowledge, roles, and cultural diversity, you also must consider your audience's age, physical limitations, and gender.

Ageist Language

A word like "elderly" could imply feebleness. The words "old folks" create a negative image. To avoid these biases, write "people over seventy" or "retirees." Better yet, avoid reference to age.

BEFORE	AFTER
Professor Jones, an elderly teacher at State University, is publishing a textbook despite his age.	Professor Jones, a State University teacher, is publishing a textbook.

Biased Language About People with Disabilities

The word "handicap" creates a negative image. "Disability" is generally preferred. However, any euphemism can be offensive. You should avoid reference to a person's disability. If you need to refer to a physical problem, do so without negative characterizations.

BEFORE	AFTER
Debbie Brown, a blind market researcher, won employee of the month.	Debbie Brown, a market researcher, won employee of the month.
The AIDS victim changed insurance carriers.	The AIDS patient changed insurance carriers.
John suffers from diabetes.	John is diabetic.
Sheila is confined to a wheelchair.	Sheila uses a wheelchair.

Sexist Language

Many of your readers will be women. This does not constitute a separate audience category. Women readers will be high tech or low tech, management or subordinate. Thus, you don't need to evaluate a woman's level of understanding or position in the chain of command any differently than you do for readers in general.

When you write, you should avoid *sexist language*, which is offensive and discriminatory to all readers. Let's focus specifically on ways in which sexism is expressed and techniques for avoiding this problem. Sexism creates problems through omission, unequal treatment, and stereotyping, as well as through word choice.

Ignoring Women or Treating Them as Secondary.   When your writing ignores women or refers to them as secondary, you are expressing sexist sentiments. The following are examples of biased comments and their nonsexist alternatives:

BEFORE	AFTER
The former CEO of Yahoo is a woman named Carol A. Bartz.	The former CEO of Yahoo is Carol A. Bartz.
When setting up his experiment, the researcher must always check for errors.	When setting up experiments, the researcher must always check for errors.
As we acquired scientific knowledge, men began to examine long-held ideas more critically.	As we acquired scientific knowledge, people began to examine long-held ideas more critically.

Modifiers that describe women in physical terms not applied to men treat women unequally.

BEFORE	AFTER
The poor women could no longer go on; the exhausted men...	The exhausted men and women could no longer go on.
Mrs. Acton, a gorgeous blonde, is Joe Granger's assistant.	Jan Acton is Joe Granger's assistant.

Stereotyping.   If your writing implies that only men do one kind of job and only women do another kind of job, you are stereotyping. For example, if all management jobs are held by men and all subordinate positions are held by women, this is sexist stereotyping.

BEFORE	AFTER
Current tax regulations allow a head of household to deduct for the support of his children.	Current tax regulations allow a head of household to deduct for child support.
The manager is responsible for the productivity of his department; the foreman is responsible for the work of his linemen.	Management is responsible for departmental productivity. Supervisors are responsible for their personnel.
The administrative assistant brought her boss his coffee.	The administrative assistant brought the boss's coffee.
The teacher must be sure her lesson plans are filed.	The teacher must file all lesson plans.

Sexist language disappears when you use pronouns and nouns that treat all people equally.

Pronouns.   Pronouns such as *he, him,* or *his* are masculine. Sometimes you read disclaimers by manufacturers stating that although these masculine pronouns are used, they are not intended to be sexist. They're only used for convenience. This is an unacceptable statement. When *he, him,* and *his* are used, a masculine image is created, whether or not such companies want to admit it.

To avoid this sexist image, avoid masculine pronouns. Instead, use the plural, generic *they* or *their*. You also can use *he or she* and *his or her*. Sometimes you can solve the problem by omitting all pronouns.

BEFORE	AFTER
Sometimes the doctor calls on his patients in their homes.	Sometimes the doctor calls on patients in their homes.
The typical child does his homework after school.	Most children do their homework after school.
A good lawyer will make sure that his clients are aware of their rights.	A good lawyer will make sure that clients are aware of their rights.

Gender-Tagged Nouns.   Use nouns that are nonsexist. To achieve this, avoid nouns that exclude women and denote that only men are involved.

BEFORE	AFTER
mankind	people
manpower	workers/personnel
the common man	the average citizen
wise men	leaders
businessmen	businesspeople
policemen	police officers
fireman	firefighter
foreman	supervisor
chairman	chairperson/chair
stewardess/steward	flight attendant
waitress/waiter	server

Consider the following examples of sexist and nonsexist writing in a letter advertising landscaping services. In the "Before" example, underlined words are gender specific. In the second example, the sexist terms are replaced with underlined nonsexist words.

BEFORE

> Dear <u>Sir</u>:
>
> Interested in beautifying your company property, with no care or worries? We have been in business for over 25 years helping <u>men</u> just like you. Here's what we can offer:
>
> 1. Shrub Care—your <u>workmen</u> will no longer need to water shrubs. We'll take care of that with a sprinkler system geared toward your business's unique needs.
> 2. Seasonal System Checks—don't worry about your <u>foreman</u> having to ask <u>his</u> workers to turn on the sprinkler system in the Spring or turn it off in the Winter. Our <u>repairmen</u> take care of that for you as part of our contract.
> 3. New Annual Plantings—we plant rose bushes and daffodils that are so pretty, your secretary will want to leave <u>her</u> desk and pick a bunch for <u>her</u> office. Of course, you might want to do that for <u>her</u> yourself ☺.
>
> Call today for prices. Let our skilled <u>craftsmen</u> work for you!

Subject: Sales Information

Interested in beautifying your company property, with no care or worries? We have been in business for over 25 years helping <u>business owners</u> just like you. Here's what we can offer:

- Shrub Care—your <u>employees</u> will no longer need to water shrubs. We'll take care of that with a sprinkler system geared toward your business's unique needs.

- Seasonal System Checks—don't worry about your <u>supervisor</u> having to ask <u>his or her</u> workers to turn on the sprinkler system in the Spring or turn it off in the Winter. Our <u>staff</u> takes care of that for you as part of our contract.

- New Annual Plantings—we plant rose bushes and daffodils that are so pretty, <u>all of your employees</u> will want vase-filled flowers on <u>their</u> desks.

Call today for prices. Let our skilled <u>experts</u> work for you!

---

8. Achieve audience involvement through personalized tone and positive words.

Audience Involvement

In addition to audience recognition, effective technical communication demands audience involvement. You not only need to know whom you are writing to in your technical correspondence (audience recognition), but also you need to involve your readers—draw them into your writing and keep them interested. Achieving audience involvement requires that you strive for personalized tone and reader benefit.

Personalized Tone

Companies do not write to companies; people write to people. Remember when you write your e-mail, letter, report, or procedure, another person will read it, so achieve a personalized, collaborative tone to involve your reader. Personalized tone is even more important in social media (Facebook, YouTube, Twitter, blogs, and so on) since this communication channel seeks to build community.

Pronouns. The best way to personalize correspondence is through pronoun usage. If you omit pronouns in your technical communication, the text will read as if it has been computer generated, absent of human contact. When you use pronouns in your technical communication, you humanize the text. You reveal that the memo, letter, report, or procedure is written by people, for people.

The generally accepted hierarchy of pronoun usage is as follows:

Pronoun	Focus	Pronoun	Focus	Pronoun	Focus
You Your	The reader	We Us Our	The team	I Me My	The ego

The first group of pronouns, *you* and *your*, is preferred. When you use *you* or *your*, you are speaking directly to your reader(s) on a one-to-one basis. The readers, whoever they are, read the word *you* or *your* and see themselves in the pronoun, envisioning that they are being spoken to, focused on, and singled out. In other words, *you* or *your* appeals to the reader's sense of self-worth.

The second group of pronouns (*we, us,* and *our*) uses team words to connote group involvement. These pronouns are especially valuable when writing to multiple audiences or when writing to subordinates. *We, us,* or *our* implies to the readers that "we're all in this together." Such a team concept helps motivate by making the readers feel an integral part of the whole.

The third group (*I, me,* and *my*) denotes the writer's involvement. If overused, however, these pronouns can connote egocentricity ("All I care about is *me, me, me*"). Because of this potential danger, emphasize *you* and *your* and downplay *I, me,* and *my*. Strive for a two-to-one ratio. For every *I, me,* or *my*, double your use of *you* or *your*.

Let's look at an automobile manufacturer's user manual, which omits pronouns and, thus, reads as if it is computer generated.

BEFORE

Claims Procedure

To obtain service under the Emissions Performance Warranty, take the vehicle to the company dealer as soon as possible after it fails an I/M test along with documentation showing that the vehicle failed an EPA-approved emissions test.

> Without any pronouns, the version has no personality. It's dull and stiff, and it reads as if no human beings are involved.

Compare this flat, dry, dehumanized example with the more personalized version.

AFTER

Claims Procedure

How do you get service under the Emissions Performance Warranty? To get service under this warranty, take your car to the dealer as soon as possible after it has failed an EPA-approved test. Be sure to bring along the document that shows your car failed the test.

> When pronouns are added in this version, the writing involves the reader on a personal level.

Many writers believe that the first version is more professional. However, professionalism does not require that you write without personality. The pronoun-based writing in the second version is more friendly. Friendliness and humanism are positive attributes in technical communication. The second version also ensures reader involvement. As a professional, you want your readers (customers, clients, colleagues) to be involved, for without these readers no transaction occurs.

Names.   Another way to personalize and achieve audience involvement is through the use of names—incorporating the reader's name in your technical communication. By doing so, you create a friendly reading environment in which you speak directly to your audience.

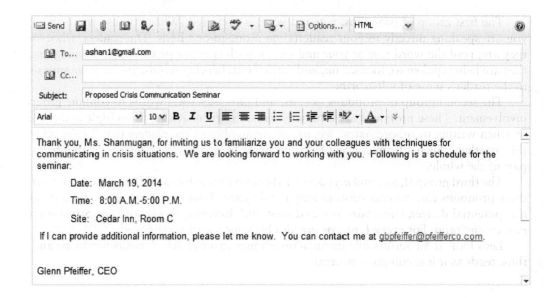

When do you use first names versus last names? If you know your reader well, have worked with him or her for a while, and know your reader will not be offended, then use the first name. However, if you do not know the reader well, have not worked with this reader before, or worry that the reader might be offended, then use the surname instead. In either instance, calling a reader by name will involve that reader and personalize your writing.

<table>
<tr><td>Audience Benefit</td></tr>
</table>

See Chapter 9, "Routine Correspondence," Chapter 15, "Instructions, User Manuals, and Standard Operating Procedures," and Chapter 18, "Long, Formal Reports," for additional discussion of audience benefit.

Reader Benefit

A final way to achieve audience involvement is to motivate your readers by giving them what they want or need. Show your audience how they will benefit from your technical communication.

Explain the Benefit.   Until you tell your readers how they'll benefit, they may not know. Therefore, in your letter, e-mail, report, or manual, state the benefit clearly. You can do this anywhere, but you're probably wise to place the statement of benefit either early (first paragraph or abstract/summary) or late (last paragraph or conclusion). Placing the benefit early in the writing will interest readers and help ensure that they continue to read. Placing the benefit at the end could provide a motivational close, leaving readers with a positive impression.

EXAMPLE ▶

This instruction conveys reader benefit.

Instructions for Poured Foundations
A poured foundation will provide a level surface for mounting both the pump and motor. Carefully aligned equipment will provide you a longer and more easily maintained operation.

E-mail messages can also develop reader benefit to involve audiences, as in the following example.

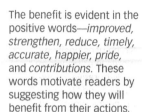

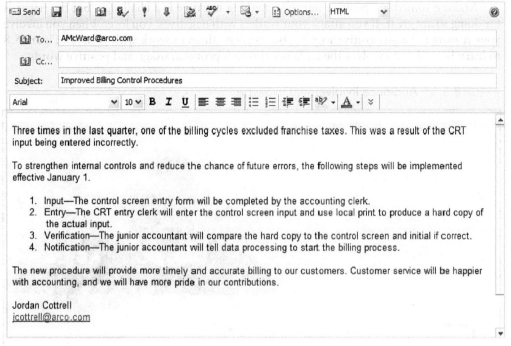

Three times in the last quarter, one of the billing cycles excluded franchise taxes. This was a result of the CRT input being entered incorrectly.

To strengthen internal controls and reduce the chance of future errors, the following steps will be implemented effective January 1.

1. Input—The control screen entry form will be completed by the accounting clerk.
2. Entry—The CRT entry clerk will enter the control screen input and use local print to produce a hard copy of the actual input.
3. Verification—The junior accountant will compare the hard copy to the control screen and initial if correct.
4. Notification—The junior accountant will tell data processing to start the billing process.

The new procedure will provide more timely and accurate billing to our customers. Customer service will be happier with accounting, and we will have more pride in our contributions.

Jordan Cottrell
jcottrell@arco.com

The benefit is evident in the positive words—*improved, strengthen, reduce, timely, accurate, happier, pride,* and *contributions*. These words motivate readers by suggesting how they will benefit from their actions.

Reader benefit is emphasized in the last paragraph. The pronouns *we* and *our* in the last sentence involve the readers in the activity. A team concept is implied.

Use Positive Words.   In the preceding examples achieving reader benefit, the motivation or value is revealed through positive words. One way to involve your audience is to use positive words throughout your correspondence.

POSITIVE WORDS				
accomplish	coordinate	favorable	plan	serve
achieve	create	guide	please	succeed
advantage	develop	happy	produce	support
appreciate	educate	help	profit	thank you
asset	effective	improve	promote	value
assist	efficient	increase	raise	
benefit	enjoy	lead	recommend	
build	ensure	maintain	reduce	
confident	establish	organize	satisfy	

To clarify how important such words are, let's look at some examples of negative writing, followed by positive revisions.

BEFORE	AFTER
• We cannot process your request. You failed to follow the printed instructions.	• So that we may process your request rapidly, please fill in line 6 on the printed form.
• The error is your fault. You keep your books incorrectly and cannot complain about our deliveries. If you would cooperate with us, we could solve your problem.	• To ensure prompt deliveries, let's get together to review our bookkeeping practices. Would next Tuesday be convenient?
• Your bill is now three weeks overdue. Failure to pay immediately will result in lower credit ratings.	• If you're as busy as we are, you've probably misplaced our recent bill (mailed three weeks ago). Please send it in soon to maintain your high credit ratings.

Making something positive out of something negative is a challenge, but the rewards for doing so are great. If you attack your readers with negatives, you lose. When you involve your readers through positive words, you motivate them to work with you. See Figure 5.8 to learn how Skype involves the audience through pronoun usage and positive words.

FIGURE 5.8   Audience Involvement in Skype

Skype uses positive words like "transform," "collaborative," "love," "trust," and "together." Pronouns "your" and "you" help to involve the audience. Finally, a smiling person makes the site visually appealing and welcoming.

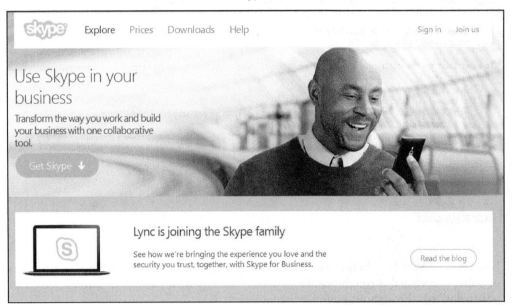

AUDIENCE CHECKLIST

_____ 1. What is my audience's level of understanding regarding the subject matter?
- High tech
- Low tech
- Lay
- Multiple

_____ 2. Given my audience's level of understanding, have I written accordingly?
- Have I defined my acronyms, abbreviations, and jargon?
- Have I supplied enough background data?
- Have I used the appropriate types of graphics?

_____ 3. Who else might read my correspondence?
- How many people?
- What are their levels of understanding?
  - High tech
  - Low tech
  - Lay
- Will my audience be multicultural?

_____ 4. What is my role in relation to my audience?
- Do I work for the reader?
- Does the reader work for me?

- Is the reader a peer?
- Is the reader a client?

_____ 5. What response do I want from my audience? Do I want my audience to act, respond, confirm, consider, decide, or file the information for future reference?

_____ 6. Will my audience act according to my wishes? What is my audience's attitude toward the subject (and me)?
- Negative
- Positive
- Noncommittal
- Uninformed

_____ 7. Is my audience in a position of authority to act according to my wishes? If not, who will make the decision?

_____ 8. What are my audience's personality traits?
- Slow to act
- Eager
- Receptive
- Questioning
- Organized
- Disorganized
- Oppositional

_____ 9. Have I motivated my audience to act?

- Have I involved my reader in the correspondence by using pronouns and his or her name?
- Have I shown my reader the benefit of the proposal by using positive words?
- Have I considered any questions or objections my audience might have?

_____ 10. Have I considered my audience's preferences regarding style?

- Is use of a first name appropriate?

_____ 11. Have I avoided sexist language?

- Have I used "their" or "his or her" to avoid sexist singular pronouns such as "his"?
- Have I used generic words such as "police officer" versus "policeman"?
- Have I avoided excluding women, writing sentences such as "All present voted to accept the proposal" versus "All the men voted to accept the proposal"?

- Have I avoided patronizing, writing sentences such as "Mr. Smith and Mrs. Brown wrote the proposal" versus "Mr. Smith and Judy wrote the proposal"?

_____ 12. Have I considered diversity?

- Have I avoided any language that could offend various age groups, people of different sexual orientations, people with disabilities, or people of different cultures and religions?
- Have I considered that people from different countries and people for whom English is a second language will be involved in the communication? This might mean that I should do the following:
  - Clarify time and measurements
  - Define abbreviations and acronyms
  - Avoid figurative language and idiomatic phrases unique to one culture
  - Avoid humor and puns
  - Consider each country's cultural norms.

The Writing Process at Work

Phil Baron, training manager at CareerBusiness Services, used the techniques of prewriting, writing, and rewriting to prepare a letter to a prospective client.

Prewriting

To prewrite his letter, Phil wrote an outline, saved in draft form. See Figure 5.9.

FIGURE 5.9   Outline for Prewriting in Draft Mode

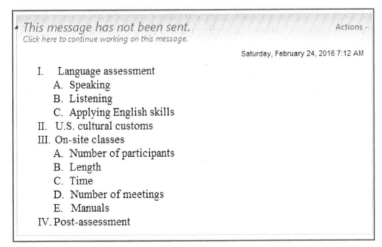

> ▸ This message has not been sent.
> Click here to continue working on this message.
>
> Actions ‑
>
> Saturday, February 24, 2016 7:12 AM
>
> I.   Language assessment
>   A. Speaking
>   B. Listening
>   C. Applying English skills
> II. U.S. cultural customs
> III. On-site classes
>   A. Number of participants
>   B. Length
>   C. Time
>   D. Number of meetings
>   E. Manuals
> IV. Post-assessment

Writing

After Phil researched his potential client's language needs, he wrote a rough draft of the cover letter. In the cover letter, Phil hoped to show his client how training sessions would be structured and what benefits the company would derive. He sent the draft of the cover letter (see Figure 5.10) to a colleague for Track Changes feedback.

February 24th, 2016

> Comment [JS1]: Remove the "th" from the date.

Joan T. Osborn
Manager, Learning and Development Americas
Arctic Cooling Technologies
7401 W. 129th Street
Overland Park, KS 66213

SUBJECT: [TRAINING PROPOSAL]

> Comment [JS2]: Expand the subject line by adding "Workplace English" for clarity.

[We are pleased to present to you a proposal to offer on-site Workplace English program to limited English speaking Arctic employees] Following a language assessment to determine the appropriate level of instruction, we will develop a comprehensive curriculum to improve each participant's English language skills and knowledge.

> Comment [JS3]: Wordy. Revise this to read "We are pleased to present a proposal for . . . ."

The program will offer an easy and quick way to learn limited amounts of everyday workplace English. Instruction will be divided into three components: Speaking in English, Listening in English and Workplace Application English. Aspects of U.S. culture will also be included in each session to discuss some of the most fundamental aspects of everyday U.S. culture that are often misunderstood.

Speaking in English teaches participants how to say practical, common phrases and questions in American English. We will focus on pronunciation, expressions, and questions for use in everyday interactions. Listening in English teaches the participants how to comprehend many basic and common expressions, phrases and questions used in everyday English. Workplace specific English teaches participants to comprehend and respond to workplace specific language including phrases and terms that are regularly used in the workplace.

Each participant will receive a manual with [CD's] that can be used to maintain the language skills they have acquired.

> Comment [JS4]: Omit the apostrophe in "CDs" for correct spelling.

We propose offering on-site classes limited to a maximum of 15 participants for 6 weeks to meet two hours, twice a week for a total of 24 hours. A once per week meeting format for 12 weeks could also be provided. All instructional material will be provided to participants. The $3,300estimated budget for this program includes curriculum development, instructional manuals, and on-site training. Following completion of the program a follow-up plan will be developed to help each participant maintain and improve his or her English language skills.

We look forward to finalizing this training program. Thank you for the opportunity to present this proposal to you for your careful consideration. We pride ourselves on providing quality training, language development, career planning, consulting, and economic development services.

Sincerely,

> Comment [JS5]: Rather than full justification, let's use ragged right margins.

Rewriting

Phil says, "My revised cover letter offers the specific ways in which we can meet the client's needs to improve multicultural communication." See Figure 5.11.

FIGURE 5.11   Rewritten Letter for Arctic Training Workshop

12345 College Boulevard
St. Louis, MO 76210
913-469-3845
Fax 913-469-4415

February 24, 2016

Joan T. Osborn
Manager, Learning and Development
Arctic Cooling Technologies
7401 W. 129th Street
Overland Park, KS 66213

Subject: Training Proposal for Arctic Workplace English

We are pleased to propose an on-site Workplace English program to Arctic employees. To determine the appropriate level of instruction, we developed a curriculum to improve each participant's English language skills and knowledge.

The program will offer an easy and quick way to learn limited amounts of everyday workplace English. Instruction will be divided into three components: Speaking in English, Listening in English and Workplace-specific English. Aspects of U.S. culture will also be included. We will discuss often misunderstood fundamental aspects of everyday U.S. culture. Following is an overview of what we can offer Arctic:

- **Speaking in English:** This training component teaches participants how to say practical, common phrases and to ask questions in American English. We will focus on pronunciation, expressions, and questions for use in everyday interactions.

- **Listening in English:** In this training unit, participants learn many common expressions, phrases and questions used in everyday English.

- **Workplace-specific English:** Here, participants learn how to comprehend and respond to common workplace phrases and terms.

Each participant will receive a manual with CDs that can be used to maintain the language skills they have acquired.

We propose offering on-site classes limited to a maximum of 15 participants for 6 weeks to meet 2 hours, twice a week for a total of 24 hours. A once-per-week meeting format for 12 weeks could also be provided. All instructional material will be provided to participants. The $3,300 estimated budget for this program includes curriculum development, instructional manuals, and on-site training. When the training program is complete, I will develop a follow-up plan to help each participant maintain and improve English language skills.

We look forward to finalizing this training program. Thank you for the opportunity to present this proposal. We pride ourselves on providing quality training and language development.

Phil Baron

Phil Baron, Program Director

Phil says, "Focusing on the needs of the audience is an important part of my job. The writing process allows me to create the best possible written document to address my audience's needs."

CASE STUDY

Home and Business Mortgage (HBM) has suffered several lawsuits recently. A former employee sued the company, contending that it practiced "ageism" by promoting a younger employee over him. In an unrelated case, another employee said she was denied a raise due to her ethnicity. To combat these concerns, HBM has instituted new human resources practices to meet Equal Employment Opportunity Commission (EEOC), the Family Medical Leave Act (FMLA), and the Americans with Disabilities Act (ADA) regulations.

HBM now needs to hire a new office manager for one of its branch operations. It has three outstanding candidates.

- **Carlos Gutierrez** Carlos is a 27-year-old recent recipient of an MBA (Masters in business administration) from an acclaimed business school where he learned many modern business applications, including human resources management, organizational behavior, diversity management, and team management strategies. He has been out of graduate school for only 2 years, but his work during that time has been outstanding. As an employee at one of HBM's branches, he has impressed his bosses by increasing the branch's market share by 28 percent through innovative marketing. In addition, his colleagues enjoy working with him and praise his team-building skills. Carlos has never managed a staff, but he has promise.

- **Cheryl Huff** Cheryl is a 37-year-old employee of a rival mortgage company, Farm-Ranch Equity. She has a BGS (Bachelor's degree in general studies) from a local university, which she acquired while working full time in the mortgage/real estate business. At Farm-Ranch, she has over 15 years' experience, working as an executive assistant for 5 years, a mortgage payoff clerk for 2 years, a loan processor for 2 years, a mortgage sales manager for 4 years, and, most recently, a residential mortgage closing coordinator. In the last two positions, she managed a staff of five employees.

  According to her references, she has been dependable on the job. Though she has accomplished much on the job, her subordinates have complained about her demanding expectations. Her references suggest that she could profit from some improved people skills.

- **Rose Massin** Rose is 47 years old and has been out of the workforce for 12 years. During that absence, she raised a family of three children. Now, Rose wants to reenter the job market. She has a Bachelor's degree in business and was the former office manager of this HBM branch.

  While she was branch manager, she did an outstanding job. This included increasing business, working well with employees and clients, and maintaining excellent relationships with lending banks and realtors. She was a highly respected employee and involved in civic activities and community volunteerism. In fact, she is immediate past-president of the local Rotary club. Though she has been out of the workforce for years, she has kept active in the city and maintained excellent business contacts. Still, she's a bit rusty on modern business practices.

Assignment

Based on the information provided, make your hiring decision. Be sure to prove your decision. Then share this finding as follows:

- **Oral Presentations**—Give a 3–5 minute presentation to share with your colleagues the results of your decision.
- **Written**—Write an e-mail, memo, or report about your findings.

ETHICAL CHALLENGE

George Penrose works for a hospital in its health information management department. He loves his job and has enjoyed the hospital's work environment for 10 years. Recently, a new supervisor was hired. The manager has treated all employees fairly, and George is impressed with the supervisor's knowledge of health information management. However, last week, the supervisor wrote a memo to all of the department's employees regarding a change in departmental policy. Starting next month, the supervisor will begin hosting nondenominational Bible study sessions. Regardless of an employee's religious beliefs or cultural background, all employees were encouraged to sign up for these Bible classes. The classes would be voluntary, and employees would be given time off their jobs to attend the sessions.

Question

Is it ethical for the company to offer religious study classes? Why or why not?

INDIVIDUAL AND TEAM PROJECTS

Sexist Language

1. Revise the following sentences to avoid sexist language.
   a. All the software development specialists and their wives attended the conference.
   b. The foremen met to discuss techniques for handling union grievances.
   c. Every technician must keep accurate records for his monthly activity reports.
   d. The president of the corporation, a woman, met with her sales staff.
   e. Throughout the history of mankind, each scientist has tried to make his mark with a discovery of significant intellectual worth.

Achieving Audience Involvement

1. Revise the following sentences to achieve audience involvement through personalization and by adding pronouns and names.
   a. The company will require further information before processing this request.
   b. It has been decided that a new procedure must be implemented to avoid further mechanical failures.
   c. The department supervisor wants to extend heartfelt thanks for the fine efforts expended.
   d. I think you have done a great job. I want you to know that you have surpassed this month's quota by 12 percent. I believe I can speak for the entire department by saying thank you.
   e. If the computer overloads, simultaneously press Reset and Control. Wait for the screen command. If it reads "Data Recovered," continue operations. If it reads "I/O Error," call the computer resource center.
2. Revise the following sentences to achieve reader involvement and benefit, using positive words and pronouns.
   a. We cannot lay your cable until you sign the attached waiver.
   b. John, don't purchase the wrong program. If we continue to keep inefficient records, our customers will continue to complain.
   c. You have not paid your bill yet. Failure to do so might result in termination of services.
   d. If you incorrectly quote and paraphrase, you will receive an *F* on the assignment.
   e. Your team has lost 6 of their last 12 games.

3. Rewrite the following flawed correspondence. Its tone is too negative and commanding, regardless of the audience. Soften the tone to achieve better audience involvement and motivation.

To: Staff
Subject: Fourth Quarter Goals

Due to a severe lack of discipline, the company failed to meet third quarter goals. To avoid repeating this disaster for the fourth quarter, this is what I think you all must do—ASAP.

1. Demand that the sales department increase cold calls by 15 percent.
2. Require weekly progress reports by all sales staff.
3. Penalize employees when reports are not provided on time.
4. Tell managers to keep on top of their staff, pushing them to meet these goals.

Remember, when one link is weak in the chain, the entire company suffers. DON'T BE THE WEAK LINK!

Defining Terms for Different Audience Levels

1. Find examples of definitions provided in computer word processing programs, e-mail packages, Internet dictionaries, and online help screens. Determine whether these examples provide the term, its type, and its distinguishing characteristics. Are the definitions effective? If so, explain why. If not, rewrite the definitions for clarity.

2. Find examples of definitions in manuals such as your car's owner's manual, a computer user manual, or manuals packaged with your MP3 player, Wii, coffeemaker, Bluetooth, iPod, or lawn mower. Note where these definitions are located. Are they placed parenthetically within the text or in a glossary? Next, determine whether the definitions provide the term, its type, and its distinguishing characteristics. Are the definitions effective? If so, explain why. If not, rewrite the definitions for clarity.

3. In small groups composed of individuals from like majors, list 10 high-tech terms (jargon, acronyms, or abbreviations) unique to your degree programs. Then, envisioning a lay audience, parenthetically define and briefly explain these terms. To test the success of your communication abilities, orally share these high-tech terms with other students who have different majors.
   - State the high-tech term to see if they understand it.
   - If they don't, provide the parenthetical definition. How much does this help? Do they understand now?
   - If they still do not understand the definition, add the third step—the brief explanation. How much information do the readers need before they understand your high-tech terms?

Recognizing Issues of Diversity

1. Rewrite the following sentences for multicultural, cross-cultural audiences.
   a. Let's meet at 8:30 P.M.
   b. The best size for this new component is 16 × 23 × 41.
   c. To keep us out of the red, we need to round up employees who can put their pedal to the metal and get us out of this hole.
   d. We need to produce fliers/brochures to increase business.
   e. The meeting is planned for 07/09/16.

2. Rewrite the following flawed correspondence. Be sure to consider your multicultural or cross-cultural audience's needs.

PROBLEM-SOLVING THINK PIECES

1. Dove Hill, GA, experienced severe thunderstorms on March 15. GAI (Goodwin & Associates Insurance Co.) insured much of the territory affected by this storm. Over 1,200 houses insured by GAI had water damage. As written, the homeowner's policy provided full-replacement coverage for water damage. In the homeowner's policies, water damage was limited to situations such as the following:

   - A hailstorm smashes a window, permitting hail and rain to enter a home.
   - A heavy rain soaks through a roof, allowing water to drip through a ceiling.
   - A broken water pipe spews water into a home.

   In contrast, homes which suffered water damage due to flooding were not covered unless the homeowners also had taken out a separate flood damage policy from the National Flood Insurance Program. For insurance purposes, "flood" is defined as "the rising of a body of water onto normally dry land." For example, flood damage can include a river overflowing its banks or a heavy rain seeping into a basement. Of the 1,200 homes suffering water damage, only 680 homes qualified for full-replacement coverage. The remaining homes suffered flood damage and were thus not covered.

Assignment

Write a letter or e-mail message from GAI to a homeowner denying coverage for a water damage claim. You are writing as a high-tech insurance professional to a lay reader. You want to maintain a good business relationship, despite the bad news. Build rapport while specifying in lay terms the denial.

2. Find examples of writing directed to high-tech, low-tech, and lay audiences. To do so, read professional journals, find instructions, look at marketing brochures, read trade magazines, or ask your colleagues and coworkers for documents.

Assignment

Once you have found these examples, bring them to class. In groups, discuss your findings to determine whether the documents are written for high-tech, low-tech, or lay readers. Use the Audience Evaluation Form that follows to share your decisions.

Audience Evaluation Form

Example Number	1	2	3	4	5
Type (Circle One)	High Low Lay	High Low Lay	High Low Lay	High Low Lay	High Low Lay
Criteria					
Language • Abbreviations • Acronyms • Jargon					
Content • General • Specific • Background					
Tone • Formal • Informal					
Format • Highlighting • Graphics					
Type of Visuals Used • Complexity • Color • Labeling					

WEB WORKSHOP

1. You are head of international relations at your corporation. Your company is preparing to go global. To ensure that your company is sensitive to multicultural concerns, research the cultural traits and business practices in five countries of your choice. To do so, access an online search engine and type in "multicultural business practices in _____" (specify the country's name). Report your findings either orally or in an e-mail.

2. Companies occasionally are sued for unfair hiring practices, wrongful termination of employees, or other discriminatory practices on the job. These lawsuits can depend on the company's perceived biases related to age, gender, physical disabilities, or ethnicity. Access an online search engine to learn about corporate challenges with lawsuits based on biases. What was the cause of the lawsuit, and how was the legal issue resolved? Report your findings either orally or in an e-mail.

3. Research the following online articles and sources for information about diversity and multiculturalism in the workplace. Then, report your findings in an e-mail message or memo to your professor.

    • Business.com. "Diversity in the Workplace" provides information about affirmative action, diversity training, minority-owned businesses, and related cultural issues.

    • Fink, Kenneth. "International Business Multiculturalism and the Internet" is an article about language and the global economy.

CHAPTER SIX

Ethical Considerations

After completing this chapter, you will be able to

1. Understand the importance of ethics in the workplace

2. Apply strategies to ensure ethical communication

3. Use language and visuals with precision

4. Prefer simple, direct expression of ideas

5. Satisfy the audience's need for information, not your own need for self-expression

6. Hold yourself responsible for how well the audience understands the message

7. Respect the audience's privacy

8. Understand the ethics of intellectual property laws

9. Recognize the importance of ethics in a digital workplace

COMMUNICATION AT WORK

In the scenario below, financial advisor Sean Colter reveals the importance of ethical communication with clients.

Sean Colter became a financial advisor for two key reasons: He was always interested in "the miracle of compounding—how money makes money," and he had a passion to "help others achieve financial independence." Sean knows that the key to success as a financial planner is to perform his job ethically. This is important for practical, legal, and ethical reasons.

Practical. The client's welfare is paramount to Sean. Failure to meet his clients' financial needs will harm his reputation, and that will have practical, negative ramifications on his business. If Sean is not a good partner with his clients, they will not continue to seek his advice. In contrast, Sean knows that doing business ethically leads to continued success for him and for his clients.

Legal. Sean must meet ethical standards set by state and federal mandates and regulated by his nationally known financial advisory company. First, Sean must receive continuing education credit from the state when tax laws change. For example, if the state passes new long-term care legislation, Sean needs to provide confirmation to his company that he has taken and passed state-required training in this area. Next, the Securities Exchange Commission (SEC) mandates that all financial advisors take and pass an annual ethics course. Finally, Sean participates in periodic online classes and lectures offered by his company.

In fact, Sean's company monitors all of its financial advisors in multiple ways. As Sean says, his company's Field Supervision and Compliance departments provide "mutually enforceable and overlaying protection every day and for almost every trade."

First, the Field Supervision and Compliance departments conduct "annually unannounced inspections." They arrive at Sean's office and "stay as long as needed to perform full introspection. They look at files, review records, and ask why certain actions have been taken with people's accounts over a three-, six-, or nine-month period." Sean, then, must justify his performance, proving that he is meeting his customer's specific financial needs.

Next, Sean must learn basic information about his client and input that data into his company's computer system before he can even open the customer's account. Then, his company's Field Supervision sends a letter to the client confirming the data. When Sean makes an investment for the client, regardless of the amount, his company can send a "pending message" to ensure that the investment is in the client's best interest. Sean must respond to this message, even if everything has been handled legally and ethically. In this way, the Field Supervision and Compliance departments check and double check to make sure that the financial advisor is legally meeting state and federal laws and the client's goals.

Ethical. Practical and legal issues are usually easy to define. Ethical issues are more challenging. To accomplish this challenging goal, Sean "lives by a core of guiding values." He puts the client's needs first, striving to always "preserve principal while growing the investment to keep pace with inflation."

<table>
<tr><td>1. Understand the importance of ethics in the workplace.</td></tr>
</table>

Ethics in the Workplace

Ethical considerations—doing the right thing—are an important part of business. You must consider the appropriateness of your behavior and words. As you communicate in the workplace, you are representing your company whether you are in the building or off-site with a vendor or a customer. You might travel to a conference in a different city to represent your company; you might have to make presentations to the city council. At work, you communicate with subordinates or other coworkers. In any situation, you should follow the standards for business decorum and comport yourself appropriately. Inappropriate behavior, words, or dress could be considered unethical and lead to workplace issues, business losses, or legal problems.

Decisions based on ethical considerations often are a "grey area" dependent on how the subsequent actions, words, or behavior are perceived by colleagues, clients, stakeholders, and legal personnel. If you find yourself in this grey area and need to determine whether something is ethical, consider the following:

1. Where the action takes place (in the workplace, online, or off-site)
2. When the action takes place (during working hours, after work, or when you are away from the office but representing your company)
3. Who sees the action or hears the words (a person in authority at the company or a subordinate, coworker, vendor, customer, or acquaintance)
4. Whether or not the action or words "subjectively" offend (jokes or other personal comments e-mailed to all employees will offend some people but not others)
5. If the action is an isolated incident
6. If the company has a policy prohibiting the action or words

7. Whether or not the person holds a position of power over other people involved
8. If the action or words are expressed when the person is a "public" face of the company
9. Whether or not the appropriate action was egregious (the degree or level of inappropriateness)

Problems Caused by Unethical Behavior

Business schools and management experts stress the importance of ethics. Many problems can occur when businesses fail to maintain ethical standards. These problems can include dissatisfied customers, large legal judgments, prison terms, antitrust litigation, loss of goodwill, lost sales, assessment of fines, and bankruptcies. In addition to these external results of poor ethical standards, poor business ethics can result in internal failures. Organizational problems include high employee turnover, poor work performance, and a stressful work environment. Employees do not want to work in corporations where bosses ask them to perform actions which are illegal, immoral, or unethical. Knowing this, however, does not make working or communicating in the workplace easy. Ethical dilemmas exist in corporations. What should you do when confronted with ethical challenges?

One way to solve this dilemma is by checking your actions against three concerns: legal, practical, and ethical. For example, if you plan to communicate sales literature for a new product, will your text be

- Legal, focusing on liability, negligence, and consumer protection laws?
- Practical, since dishonest business communication backfires and can cause the company to lose sales or to suffer legal expenses?
- Ethical, written to promote customer welfare and avoid deceiving the end user?

These are not necessarily three separate issues. Each interacts with the other. Laws are based on ethics and practical applications.

Legalities

When asked to communicate information that profits the company but deceives the customer, you might question where your loyalties lie. The boss pays the bills, but your customers also might be your neighbors. Such conflicts exist and challenge all employees. What do you do? Trust your instincts, and trust the laws.

Laws are written to protect the customer, the company, and you—the employee. Corporate, employee, and customer rights are covered by many laws and organizations. Your company must adhere legally to laws such as the Family Educational Rights and Privacy Act (FERPA), the Family and Medical Leave Act (FMLA), and the Health Insurance Portability and Accountability Act (HIPAA). Your company must meet the legal expectations of the Equal Employment Opportunity Commission (EEOC) and of the Occupational Safety and Health Administration (OSHA). If you believe you are being asked to do something illegal that will harm you or your community, seek legal counsel.

Practicalities

Though it might appear to be in the best interests of the company to hide potentially damaging information from customers, such is not the case. As a technical communicator, your goal is candor. That means that you must be truthful, stating the facts. It also means that you must not keep silent about facts that are potentially dangerous. The ultimate goal of a company is not just making a profit but making money the right way—"good ethics is good business" (Guy 9). What good is it to earn money from a customer who will never buy from you again or who will sue for reparation? That's not practical.

One simple example of the importance of practicality in technical communication is your résumé. Let's say that a job posting requires the applicant to be OSHA certified in the handling of hazardous materials. You don't have this certification, but you want the job. To meet the company's requirements, you state in your résumé that you are OSHA certified. Your actions are not only unethical because you have lied on the résumé, but also your assertion is impractical. If you commit an OSHA violation that leads to a customer illness or harm to a client, you would be hurting the company because it will be held liable for damages. You will also probably lose your job.

Ethicalities

Every industry and professional organization faces the challenges of defining and abiding by ethical standards. You can search the Internet and find ethical codes of conduct for many organizations, including the Direct Marketing Association, the Institute of Electrical and Electronics Engineers, the Association for Computing Machinery, the American Institute of Certified Public Accountants, and the International Society for Performance Improvement, to name a few. See Figure 6.1 for the National Society of Professional Engineers' information on ethics.

FIGURE 6.1   The National Society of Professional Engineers' Ethics Web Page

The National Society of Professional Engineers' Ethics Web page clarifies the importance of ethics in the workplace. It provides a code of ethics, access to a board to review cases, training, and opportunities for professionals to ask ethics questions.

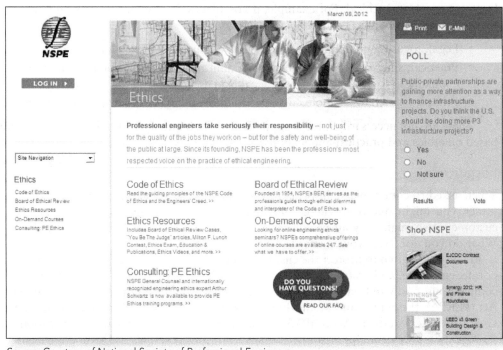

Source: Courtesy of National Society of Professional Engineers.

Ten Questions to Ask When Confronting an Ethical Dilemma

You have encountered a situation at work which does not have a clear-cut solution. You first asked yourself whether or not the situation is legal, practical, or ethical, but you still do not have an answer. Like many employees, you might feel isolated when confronted by apparently unethical behavior. What should you do next? Consider these questions:

1. Does your company have a policy regarding the situation?
2. Have you discussed the situation with your boss?
3. Have you met with the director of human resources for guidance?

4. Does your company have an ombudsman who provides support or training?
5. What do your colleagues say about the situation?
6. Do you belong to a professional organization with published policies regarding ethical behavior?
7. Do you have a moral authority with whom you can confer?
8. How has your organization dealt with similar situations in the past?
9. Does your company have a way to communicate anonymously about ethical issues, such as a drop box or secure e-mail address?
10. Has your company offered training about ethical behavior in the workplace?

How to Make Ethical Decisions

When confronted with ethical challenges, try following these writing strategies.

Define the Problem.   Is the dilemma legal, practical, ethical, or a combination of all three?

Determine Your Audience.   Who will be affected by the problem (clients, coworkers, management)? What is their involvement, what are their individual needs, and what is your responsibility—either to the company or to the community?

Maximize Values; Minimize Problems.   Ethical dilemmas always involve options. Your challenge is to select the option that promotes the greatest worth for all stakeholders involved. You won't be able to avoid all problems. The best you can hope for is to minimize those problems for both your company and your readers while you maximize the benefits for the same stakeholders.

Consider the Big Picture.   Don't just focus on short-term benefits when making your ethical decisions. Don't just consider how much money the company will make now or how easy the text will be to write now. Focus on long-term consequences as well. Will what you write please your readers so that they will be clients for years to come? Will what you write have a long-term positive impact on the economy or the environment?

A Sample Challenge to Ethical Behavior in the Workplace

You are a biomedical equipment salesperson. Your company has created a new piece of equipment to be marketed worldwide. Part of the sales literature that your boss tells you to share with potential customers contains the following sentence.

| NOTE: Our product has been tested for defects and safety by trained technicians. |

◀ EXAMPLE

When read literally, this sentence is true. The product has been tested, and the technicians are trained. However, you know that the product has been tested for only 24 hours by technicians trained on-site without knowledge of international regulations. As a loyal employee, are you required to do as your boss requests? Even though the statement is not completely true, can you legally include it in your sales literature?

The answer to both questions is no. You have an ethical responsibility to write the truth. Your customers expect it, and it is in the best interests of your company. Of equal importance, including the sentence in your sales literature is illegal. Though the sentence is essentially true, it implies something that is false. Readers will assume that the product has been thoroughly tested by technicians who have been correctly trained. Thus, the sentence deceives the readers. Such comments are actionable under law if they lead to false impressions. If you fail to properly disclose information, including dangers, warnings, cautions, or notes like the sentence in question, then your company is legally liable.

Strategies for Ethical Technical Communication

If you encounter situations like the sample challenge to ethical behavior in the workplace, the Rocky Mountain Chapter of the Society for Technical Communication (STC) (see Figure 6.2) provides a source for ethical standards.

FIGURE 6.2 Society for Technical Communication Code for Communicators

As a technical communicator, I am the bridge between those who create ideas and those who use them. Because I recognize that the quality of my services directly affects how well ideas are understood, I am committed to excellence in performance and the highest standards of ethical behavior.

I value the worth of the ideas I am transmitting and the cost of developing and communicating those ideas. I also value the time and effort spent by those who read or see or hear my communication.

I therefore recognize my responsibility to communicate technical information truthfully, clearly, and economically.

My commitment to professional excellence and ethical behavior means that I will

- Use language and visuals with precision.
- Prefer simple, direct expression of ideas.
- Satisfy the audience's need for information, not my own need for self-expression.
- Hold myself responsible for how well my audience understands my message.
- Respect the work of colleagues, knowing that a communication problem may have more than one solution.
- Strive continually to improve my professional competence.
- Promote a climate that encourages the exercise of professional judgment and that attracts talented individuals to careers in technical communication.

Source: "Code for Communicators & Ethical Principles."

Use Language and Visuals with Precision

In a survey we conducted comparing technical writers and teachers of technical communication, we discovered an amazing finding: Professional writers rate grammar and mechanics higher than teachers do (Gerson and Gerson). On a 5-point scale (5 equaling "very important"), writers rated grammar and mechanics 4.67, whereas teachers rated grammar and mechanics only 3.54. That equals a difference of 1.13, which represents a 22.6 percent divergence of opinion.

Given these numbers, would we be precise in writing "Teachers do not take grammar and mechanics seriously"? The numbers accurately depict a difference of opinion, and the 22.6 percent divergence is substantial. However, these figures do not assert that teachers ignore grammar. To say so is imprecise and would constitute an ethical failure to present data accurately. Even though writers are expected to highlight their client's values and downplay their client's shortcomings, technical communicators ethically cannot skew numbers to accomplish these goals. Information must be presented accurately, and writers must use language precisely.

Precision also is required when you use visuals to convey information. Look at the Before and After examples of visual aids showing sales growth. Both examples show that a company's sales have risen. However, the Before suggests that sales have risen consistently and dramatically. The weight of the line even exclaims success. The After is more accurate and honest. It shows sales growth, but it also depicts peaks and valleys. As with language, the writer is ethically responsible for presenting visual information precisely.

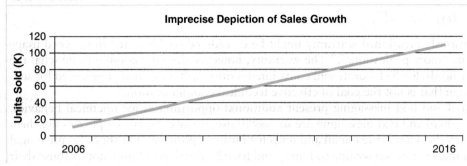

Imprecise Depiction of Sales Growth

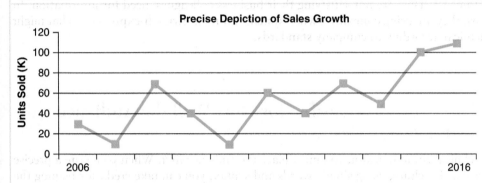

Precise Depiction of Sales Growth

Prefer Simple, Direct Expression of Ideas

While writing a proposal, you might be asked to use legal language to help your company avoid legal problems. Following is a typical warranty that legally protects your company:

> **Company Warranty Information**
>
> Acme's liability for damages from any cause whatsoever, including fundamental breach, arising out of this Statement of Limited Warranty, or for any other claim related to this product, shall be limited to the greater of $10,000 or the amount paid for this product at the time of the original purchase, and shall not apply to claims for personal injury or damages to personal property caused by Acme's negligence, and in no event shall Acme be liable for any damages caused by your failure to perform your responsibilities under this Statement of Limited Warranty, or for loss of profits, lost savings, or other consequential damages, or for any third-party claims.

4. Prefer simple, direct expression of ideas.

◀ EXAMPLE

Will your audience be able to understand this warranty's 109-word sentence and complex word usage? Readers will become frustrated and fail to recognize what is covered under the warranty. Although you are following your boss's directions, you are not communicating clearly. Writing can be legally binding and easy to understand. The first requirement does not negate the second. As a successful writer, you should help your audience understand the message by using simple words and direct expression of ideas whenever possible.

Satisfy the Audience's Need for Information, Not Your Own Need for Self-Expression

The previously mentioned warranty might be considered sophisticated in its word usage and, thus, "more professional." The warranty, however, does not communicate what the audience needs. Such elaborate and convoluted writing will only satisfy one's need for self-expression; that is not the goal of effective technical communication.

E-mail and text messaging present additional opportunities for unethical behavior. Because e-mail and text messaging are so easy to use, many company employees use them too frequently. They write e-mail and texts for business purposes, but they also use e-mail and text messages when writing to family and friends. Employees know not to abuse their company's metered mail by using corporate envelopes and stamps to send in their gas bills or write thank-you notes to Aunt Rose. These same employees, however, abuse the company's e-mail system or text messaging (on company time) by writing messages to relatives coast-to-coast. They are not satisfying their business colleagues' need for information. In contrast, they are using company-owned systems for their own self-expression. That might be unethical according to company standards.

Hold Yourself Responsible for How Well the Audience Understands the Message

As a technical communicator, you must place the audience first. When you write a precise proposal or brochure, using simple words and syntax, you can take credit for helping the reader understand your text. Conversely, you must also accept responsibility if the reader fails to understand. Ethical standards for successful communication require the writer to always remember the readers—the people who read corporate annual reports to understand how their stocks are doing, who read brochures before making travel arrangements, or who read proposals to determine whether to purchase a service. An ethical writer knows that these people will be confused by inaccessible stock reports and misled by inaccurate proposals or brochures. Take the time to check your facts. Present your information precisely, and communicate clearly so that your audience is safe and satisfied. You are responsible for your message.

Observe Liability Laws

Keeping your audience safe is a key concern for technical communicators. Liability laws cover injuries, such as accidents, and exposure to hazardous materials caused by faulty construction or dangerous products. An ethical technical communicator must warn the audience about dangers. Doing so requires the following:

- Provide clear instructions for use of the product.
- Warn users about potential risks.
- Ensure that the warning is clearly and concisely stated.
- Ensure that the warning is brought to the user's attention through headings, color, and illustrations.

Figure 6.3 is an illustration of an effective way to communicate potential product liability issues.

FIGURE 6.3 Illustration of a Warning to Prevent Liability Problems

Respect Your Audience's Privacy

7. Respect the audience's privacy.

Since so much of today's communication to coworkers, colleagues, and clients takes place online and through smartphones, you need to consider the ethical dilemmas presented by electronic communications. When writing e-mail, texting, tweeting, or distributing information through a Web site, consider confidentiality and courtesy.

- *Confidentiality.* The 1974 Privacy Act allows "individuals to control information about themselves and to prevent its use without consent" (Turner 59). The Electronic Communication Privacy Act of 1986, which applies all federal wiretap laws to electronic communication, states that e-mail messages can be disclosed only "with the consent of the senders or recipients" (Turner 60). However, both of these laws can be abused easily on the Internet. For example, neither law specifically defines consent. Data such as your credit records can be accessed without your knowledge by anyone with the right hardware and software. Your confidentiality can be breached easily.

  In addition, the 1986 Electronic Communication Privacy Act fails to define the sender. In the workplace, a company owns the e-mail system and smartphones, just as it owns other more tangible items such as desks, computers, and file cabinets. Companies have been held liable for electronic messages sent by employees. Thus, many corporations consider the contents of one's e-mail and one's e-mailbox company property, not the property of the employee. With ownership comes the right to inspect an employee's messages. Although you might write an e-mail message assuming that your thoughts are confidential, this message can be monitored without your knowledge.

  These instances might be legal, but are they ethical? Though a company can eavesdrop on your e-mail or access your life's history through databases, that does not mean they should. As an employee or corporate manager, you should respect another's right to confidentiality. Ethically, you should avoid the temptation to read someone else's e-mail or to access data about an individual without his or her consent unless you are confronted by a performance or personnel issue. Companies should inform their employees that e-mail could be monitored.

- *Courtesy.* As noted, e-mail is not as private as you might believe. Whatever you write in your e-mail message can be read by others. This is especially true if you hit the "reply to all" button, which will send an e-mail message to all readers copied in the cc: line. Therefore, be very careful about what you say in e-mail. Avoid offending coworkers by "flaming" (writing discourteous messages, usually typed in all caps). Before you assess a coworker's ideas or ability, and before you criticize your employer, remember that common courtesy, respect for others, is ethical.

Be Ethical When Using Social Media for Business

Though many companies are using social media, this communication channel has risks. What you write on a Facebook wall or depict visually in a YouTube video can have significantly negative results. For example, your initial reaction to the mention of YouTube and business might be a memory of the two employees at a company who posted a video of themselves doing unsanitary things with food that was then to be served to customers. As a result of this video, the two employees were fired. Neither of them had considered the ethical ramifications of their actions when they created and posted the video on YouTube. Similar problems can occur when you write a blog entry or tweet. Always assess the risk of online communication. Tweets, blogs, and Facebook text rarely are read by only one person; your social media messages could circulate globally.

To avoid problems with social media in the workplace, consider these ethical guidelines:

- Follow your employer's rules regarding Internet conduct.
- Never publish negative, confidential, or proprietary comments about your employer.
- Link to your sources so you can avoid plagiarizing content.
- Be accountable for your actions—to your organization, stakeholders, clients, coworkers, and yourself.
- State only what you know is true, distinguishing opinion from fact.
- Be clear about who you are, who you're representing online, and the identity of bloggers, e-mail authors, and other social media authors.
- Make sure that what you're posting is relevant to the audience, the context, and the company.

Strive Continually to Improve Professional Competence and Exercise of Professional Judgment

Professional competence, for example, includes one's ability to avoid plagiarizing. Competence also should include professional courtesy, respect for confidentiality, a sense of responsibility for one's work, and the ability to write clearly and precisely. Ultimately, your competence is dependent upon professional judgment. Do you know the difference between right and wrong? Recognizing the distinction is what ethics is all about.

As a businessperson, you always will be confronted by options, such as loyalty to your company, responsible citizenship, need for a salary, accountability to your client and your coworkers, and personal integrity. You must weigh the issues—ethically, legally, and practically—and then write and speak according to your conscience.

The Ethics of Intellectual Property Laws

> 8. Understand the ethics of intellectual property laws.

The creator of a unique product, term, image, or invention is protected by intellectual property laws. Types of intellectual property include copyrights and trademarks.

Copyright Laws

Every English teacher you have ever had has told you to avoid plagiarism (stealing another writer's words, ideas, or images). Plagiarism, unfortunately, is a problem in the workplace. If you were an unethical technical communicator, you could access text and images from many sources and easily print them as your own. An ethical communicator will not fall prey to such temptation. If you have not written it, give the other author credit. Words are like any other possession. Taking words and ideas without attributing your source through a footnote or parenthetical citation is wrong. You should respect copyright laws.

A Web page's words, graphics, and coding are protected by U.S. and international copyright laws even without a copyright application. The U.S. Copyright Office's "Circular 66, Copyright Registration for Online Works" states the following:

> Copyright protects original authorship fixed in tangible form. For works transmitted online, the copyrightable authorship may consist of text, artwork, music, audiovisual material (including any sounds), sound recordings, etc. Copyright does not protect ideas, procedures, systems, or methods of operation. Under U.S. law, copyright protection subsists from the time the work is fixed in any tangible medium of expression from which it can be perceived, reproduced, or otherwise communicated, either directly or with the aid of a machine or device.

If you and your company "borrow" from an existing Internet site, thus infringing on that site's copyright, you can be assessed actual or statutory damages. The Copyright Act allows the owner of Web material to stop infringement and obtain damages and attorney fees. In addition to financial damages, your company, if violating intellectual property laws, could lose customers, damage its reputation, and lose future capital investments.

Trademark Laws

In addition to laws protecting copyrighted words and phrases, the same holds true for a company's name; logo, symbol, design, or image that identifies the company; or combinations of these elements. In these instances, you must respect trademark laws. A company can protect its intellectual property by branding products or services with the following trademark symbols:

Visual Aids.

See Chapter 8, "Visual Aids," for more information.

1. ™ (unregistered trademark)
2. ˢᴹ (unregistered service mark)
3. ® (registered trademark)

If you copy images or text designated as trademarks by these symbols, the individual or company from whom you have taken this information can sue for trademark infringement.

Boilerplate Content

However, avoiding intellectual property theft and plagiarism in the workplace is not always clearcut. According to Jessica Reyman, noted intellectual property law author, "communicators commonly perform a variety . . . of composing activities that could be considered plagiarism in the context of the classroom" but are not necessarily plagiarism in the workplace (61). For example, technical communicators often use boilerplate information and templates—text, design, and layout that has been used before. Do you need to cite content that has already been used in your company? The answer is yes . . . and no.

If you are writing a proposal to a new client and you draw from your company's library of existing proposals for content, that is acceptable use of boilerplate content and templates. You do not need to cite the source of your information. However, if you find a proposal from some other company on the Internet and use its content or images, that is plagiarism and unethical behavior. You must cite the source of this original information or get the company's approval.

Guidelines for Protecting Intellectual Property

To protect your rights and those of others, you should

- Assume that information is covered under copyright and trademark protection laws unless proven otherwise.
- Obtain permission for use from the original creator of graphics or text.

- Cite the source of your information.
- Create your own graphics and text.
- Copyright any information you create.
- Place a copyright notice at the bottom of your Web site.

Ethical Issues in a Digital Workplace

9. Recognize the importance of ethics in a digital workplace.

Digital communication opens new ethical considerations for workplace employees, such as the following:

Privacy

When companies track your likes, tastes, interests, and more, do they invade your privacy? Do you give up your right to privacy when using a Web cam or divulging information in your social media site?

Ownership

When you alter an image through digital manipulation, is the new image yours, or does it belong to the creator of the original image? Is copying from an online source that everyone has access to considered plagiarism? Who owns what text when you share files online through groupware?

Credibility

Do social media (blogs, Facebook postings, tweets) create verifiable content? Are online comments, likes, follows, pins, and so on credible?

ETHICS CHECKLIST

_____ 1. Have you considered if your communication is legal, practical, and ethical?

_____ 2. Have you considered how your actions, words, or behavior will be perceived by colleagues, clients, stakeholders, and legal personnel?

_____ 3. Have you used visuals and words that are clear, accurate, and precisely state your meaning?

_____ 4. Have you avoided plagiarism by citing sources or creating your own text and visuals?

_____ 5. Have you observed liability laws?

_____ 6. Have you avoided trademark and copyright infringement?

_____ 7. Have you followed your company's rules for use of social media?

_____ 8. Have you respected your colleagues' rights to confidentiality and courtesy?

_____ 9. If you have encountered a situation at work which does not have a clearcut solution, have you asked yourself the 10 questions outlined in this chapter?

_____10. Have you followed the strategies in this chapter for making ethical decisions?

The Writing Process at Work

Writing Process.

See Chapter 3, "The Communication Process," for more discussion.

Effective writing follows a process of prewriting, writing, and rewriting. The writing process is dynamic with the three steps frequently overlapping. To clarify the importance of the writing process, look at how Sean Colter, the financial advisor profiled in the chapter's "Communication at Work" opening scenario, used prewriting, writing, and rewriting to prepare a presentation for clients that ethically presents information.

Prewriting

To prewrite, Sean uses the following acronym: E G A D I MADEIT! He

1. Establishes goals for his clients
2. Gathers information about their financial status and plans

3. <u>A</u>nalyzes the information so he can organize his strategy
4. <u>D</u>evelops a course of action
5. <u>I</u>mplements the best course of action for clients
6. <u>M</u>onitors the results so changes can be made where needed

This technique is recursive and flexible. When his clients' lives change (due to illness, birth of grandchildren, new expenses, new income, reduced income, etc.), the information in the prewriting changes. Following is an example of the listing that Sean took during the initial meeting with prospective clients (Figure 6.4).

FIGURE 6.4   Sean's Prewriting

George and Ida Smithson 1ˢᵗ Client Meeting (10/11/2016)

Client goals:

1. Save and build income for retirement
2. Invest for their family
3. Ensure health concerns are met

Current financial status and plans:

1. Home will be paid off in 2 years
2. Have pension plan that will pay $56,000 annually
3. Has $225,000 in state retirement plan earning 3.3%
4. Have annuity worth $248,000 at other company, earning 3.9%
5. Have current investments worth $788,000 (80% in CDs earning on average 6% but which will come due within 5 years)
6. Plan to retire in 3 years
7. George has limited home health care/Ida has none

Writing

For his meeting with the clients, Sean prepared an oral presentation with some handouts. After reviewing the presentation with his boss, Sean realized that his clients might not understand his plans for their investments and that the visual aid he had prepared was not informative (Figure 6.5). It did not clearly show his plans for their use of CDs. In addition,

FIGURE 6.5   Portfolio Distribution Line Graph George and Ida Smithson (first draft)

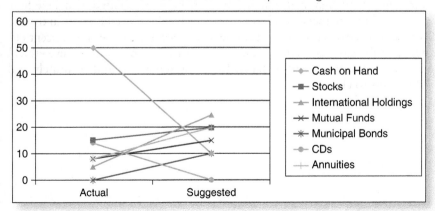

his line graph seemed to incorrectly highlight the importance of cash on hand, which he wanted to minimize. Therefore, he revised his approach by adding PowerPoint slides and met with the clients.

Rewriting

Sean used a bar chart in PowerPoint to illustrate the plan for diversified investments of the couples' potential retirement income (Figure 6.6). This revised visual aid ethically highlighted how Sean planned to distribute investments for the potential clients.

FIGURE 6.6   Sean's PowerPoint Slides Comparing Current and Planned Investments (revised version)

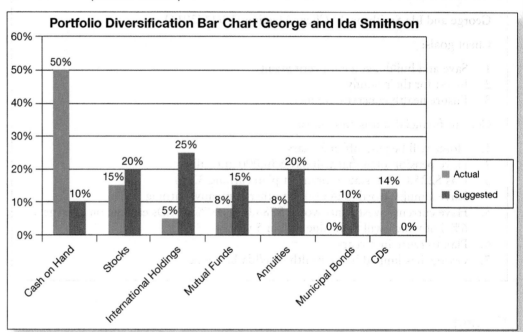

In his oral presentation, Sean explained that the Smithson's large percentage of cash on hand was not earning them money. His plan was for the clients to move some of their income to a more diversified investment where they could enhance their earnings and surpass the cost of living increases.

Sean is a very successful financial planner. In fact, he is designated as a Five Star Wealth Manager, one of the financial services professionals who scored highest in overall satisfaction. Because of his expertise and ethical behavior, Sean won over his reticent clients, who are now members of his financial family. His constant goal is to achieve ethical communication with clients. Visual aids enhance Sean's presentations and accurately and honestly reveal his plans to clients.

APPLY YOUR KNOWLEDGE

CASE STUDY

CompToday, a computer hardware company, must abide by the Sarbanes-Oxley Act, passed by Congress in response to accounting scandals. This Act specifically mandates the following related to documentation standards:

- Section 103: Auditing, Quality Control, and Independence Standards and Rules.
- Companies must "prepare, and maintain for a period of not less than 7 years, audit work papers, and other information related to any audit report, in sufficient detail to support the conclusions reached in such report."
- Section 401(a): Disclosures in Periodic Reports; Disclosures Required.
  "Each annual and quarterly financial report . . . must be presented so as not to contain an untrue statement or omit to state a material fact necessary in order to make the pro forma financial information not misleading."

Beverly Warden, documentation specialist at CompToday, is responsible for managing the Sarbanes-Oxley reports. She is being confronted by the following ethical issues:

a. To help Beverly prepare the first annual report, her Chief Financial Officer (CFO) has given her six months of audits (January through June). These prove that the company is meeting its accounting responsibilities. However, Beverly's report covers the entire year, including July through December. Section 103 states that the report must provide "sufficient detail to support the conclusions reached in [the] report."

Ethical Questions

Are the company's first six months' audits sufficient? If Beverly writes a report stating that her company is in compliance, is she abiding by her Code for Communicators & Ethical Principles, which state that a writer's work is "consistent with laws and regulations"?

What are her ethical, practical, and legal responsibilities?

- Share your findings in an oral presentation.
- Write a letter, memo, report, or e-mail to your teacher stating your opinion regarding this issue.

b. Beverly's CFO also has told her that during the year, the company fired an outside accounting firm and hired a new one to audit the company books. The first firm expressed concerns about several bookkeeping practices. The newly hired firm, providing a second opinion after reviewing the books, concluded that all bookkeeping practices were acceptable. The CFO sees no reason to mention the first firm.

Ethical Questions

Section 401(a) states that reports must contain no untrue statements or omit to state a material fact. Ethical principles also say that her writing must be truthful and accurate to the best of her ability. Beverly can report factually that the new accounting firm finds no bookkeeping errors. Should she also report the first accounting firm's assessment? Is that a material fact? If she omits any mention of the first accounting firm, as her boss suggests, is she meeting both her writer's responsibilities and the needs of Sarbanes-Oxley?

What are her ethical, practical, and legal responsibilities?

- Share your findings in an oral presentation.
- Write a letter, memo, report, or e-mail to your teacher stating your opinion regarding this issue.

ETHICAL CHALLENGE

Mary Madero has been employed at Commercial Savings and Loans for many years. She frequently communicates via e-mail with colleagues in the bank and at other savings and loans organizations. Recently, Mary was elected president of the Association for Savings and Loans Teller Supervisors (ASLTS). As ASLTS president, she is involved in fund-raising for charity, community service outreach projects, organization picnics, and the ASLTS monthly bowling league.

Sometimes she gets e-mail during the workday that relates to her new position as president. In addition, at home, she must answer ASLTS e-mail communication. In the evening and on weekends, she uses her company laptop for additional work related to her position as teller supervisor at the savings and loan.

Question

Is it ethical for Mary to use her company's e-mail account, or should she use her personal e-mail account for ASLTS business? Is it ethical for Mary to use the company laptop, or should she be using her own laptop for ASLTS business? Does her behavior promote a climate that encourages the exercise of professional judgment? Why or why not?

INDIVIDUAL AND TEAM PROJECTS

1. List five or more examples of ethical responsibilities you believe technical communicators should have when writing letters, reports, proposals, Web sites, e-mail messages, or instructions (other than those already discussed in this chapter).

2. List five or more examples of ethical responsibilities you believe employees should have when using social media in the workplace. Consider ethical lapses you've seen with the use of instant messages, text messages, blog entries, tweets, Facebook sites, or YouTube videos.

3. List five or more examples of failures to abide by ethical responsibilities you have seen either in writing or in other types of media (recordings, movies, television programs, newspapers, magazines, news reports, etc.).

4. Technical communication is factual, as are newspaper, magazine, radio, and television news reports. In contrast, recordings, movies, and television programs are art forms. Do the same ethical considerations apply for all types of media? If there are differences, explain your answer.

5. Bring to class examples of ethically flawed communication. These could include poorly written warranties (which are too difficult to understand) or dangers, warnings, and cautions (which do not clearly identify the potential for harm). You might find misleading annual reports or unethical advertisements. You might have seen videos on YouTube that do not ethically represent a company. Poor examples could even include graphics that are visually misleading. Then in small groups, rewrite these types of communication or redraw the graphics to make them ethical.

PROBLEM-SOLVING THINK PIECES

1. **Using corporate-owned equipment.** Tamara Jones is a receptionist at a bank. Her job is to greet clients, direct them to the appropriate bankers, and to answer questions about bank procedures. When she needs to speak to a customer or coworker, she's always available and pleasant. However, her job has periods of downtime when she waits for the next customer to arrive. During this downtime, she occasionally plays computer games. She's gotten tired of the games that are loaded on her bank computer. To enhance her "gaming" options, Tamara has downloaded numerous computer games that require a lot of computer space and memory.

She has been complaining about the slowness of her computer. As the manager of information of technology at the bank, you were doing an inventory of bank computers and found that Tamara's computer memory was overloaded with games.

Question

Is it ethical for Tamara to be using bank time and equipment to play computer games? Also, is it ethical for Tamara to be downloading external software? Justify your answer based on information provided in this chapter.

2. **Use of company listservs.** Elaine Williams is an employee at a large real estate corporation with offices in several locations. Her company created a listserv for its 750 employees to ensure collegiality and to disseminate corporate news. Elaine consistently uses the company listserv for the following reasons: to share jokes, to advertise her daughter's Girl Scout cookie sales, to sell her son's trash bags for soccer team travel, to sell tickets to the Easter pageant at her church, and to distribute chain letters.

Question

Is it ethical for Elaine to use the company's listserv in these ways? Why or why not?

3. **Trademark issues and name usage.** The word *xerox* is commonly used to mean "photocopy" (both as a verb and as a noun). For example, as a verb, people write, "I xeroxed the document and placed it in your inbox." As a noun, people write, "Make a xerox of the report and let me see it." Xerox Corporation tries to protect its trademark, but the company is challenged to do so. Even dictionaries use "xerox" to represent the act of making copies or copies themselves.

   The same trademark issue affects Coca Cola. Many people use the word *coke* to represent any soft drink, regardless of brand or flavor. Similarly, is Kleenex a company name or a generic tissue? For example, consider "Hand me a kleenex, please."

Question

What other examples can you think of where a company's brand name is used as a noun or verb, infringing upon the company's trademark rights?

WEB WORKSHOP

1. Go online and research codes of ethics at various corporations. Using any search engine, for example, type in phrases such as "Harley-Davidson code of ethics." You can substitute any company's name, including Ford, H&R Block, Gap, McDonalds, Google, and more. Once you have researched these sites, compare and contrast your findings.
   • What does one company focus on that others do not?
   • What do all of the companies have in common regarding their codes of ethics?
   • Do the companies' codes of ethics specifically relate to employees, social media, investors, accounting, or marketing procedures, for example, or are the codes more generic?
   • How would you revise or update any of the companies' ethical codes if asked to do so?

2. Visit Wikipedia and search for information about "creative accounting," "workplace surveillance," "whistle blowing," or "product liability." Write a short e-mail or memo reporting your findings or present a short oral presentation focusing on ethical issues related to these terms.

CHAPTER SEVEN

Document Design

After completing this chapter, you will be able to

1. Appreciate the importance of document design

2. Organize content by breaking text into smaller chunks

3. Create order through a hierarchy of headings

4. Achieve easy access of content through design elements

5. Vary design elements to create effective documents

6. Recognize how Web pages use design effectively

COMMUNICATION AT WORK

In this scenario, Roger Traver relies on meeting minutes to record information for his company.

DesignGlobal Incorporated (DGI), an engineering consulting company, holds monthly meetings at the Carriage Club, a privately owned banquet facility. To communicate to its 3,750 employees located internationally, these meetings are simulcast through video conferences. Roger Traver, CEO of DGI, invites speakers to make presentations related to news of interest to his employees. Last month, Roger invited George Smith, a university chancellor, to speak about the connection between industry and academia. George hopes to attract more professionals such as the engineers at DGI to mentor students.

Figure 7.1 shows you the meeting minutes taken at the monthly meeting where George Smith spoke. These minutes are clear, answering reporter's questions such as who, what, when, where, why, and how. The minutes also are concise in terms of word usage and sentence length. However, the minutes are unsuccessful technical communication. The wall-to-wall words not only are visually unappealing but also disallow easy access of information. An improved document design would help readers understand the meeting minutes. This flawed document design causes readability challenges.

FIGURE 7.1   Flawed Document Design

MINUTES

The meeting at the Carriage Club was attended by 30 members and guests. After the dinner, Roger Traver introduced the guest speaker, George Smith, university chancellor, and noted his accomplishments and experiences prior to education—U.S. Navy commander, Oak Ridge Laboratory researcher, and politician. Dr. Smith's talk, "Industry and Education Collaboration," was very interesting and included a history of special projects enjoyed by both academics and corporate heads. Dr. Smith suggested that we engineers could work with education to (1) provide training seminars, (2) help in urban development, and (3) provide intern opportunities. Recent industry–education collaborations include training seminars in computers, fiber optics, and human resource options. The chancellor's primary thrust was a request for $100,000 in financial aid for urban development. He said money had already been donated from three sources: a large realty firm, Capital Homes, had given $20,000; a philanthropic group, We Care, had donated a matching $20,000; Dr. Smith's university gave a matching $20,000. The remaining $40,000, Dr. Smith hoped, would come from industry donations. Finally, the chancellor noted that industry could help itself, as well as the community, by providing internships for university under-graduate majors. These internships could either be semester- or year-long arrangements, whereby students would work for minimum wage to learn more about the day-to-day aspects of their chosen fields. The chancellor said that these internships would not only increase the students' theoretical knowledge of engineering by giving them hands-on experience but also make them better future employees for the host engineering companies. Everyone would benefit. Dr. Smith noted that the students would receive a grade and credit for their work. After the speech, our VP introduced new business, calling for nomina-tions for next year's officers; gave us the agenda for our next meeting; and adjourned the meeting.

Importance of Document Design

> 1. Appreciate the importance of document design.

In technical communication, words are not your only concern. What you write is impor-tant, but how the text looks on the page is equally important. If you give your readers excessively long paragraphs or pages full of wall-to-wall words, you have made a mistake. Ugly blocks of unappealing text negatively impact readability. In contrast, effective techni-cal communication allows readers rapid access to the information, highlights important information, and graphically expresses your company's identity.

The Technical Communication Context

Why do people read correspondence? Although individuals read poetry, short stories, novels, and drama for enjoyment, few people read memos, letters, reports, or instructions for fun. They read these types of technical communication for information about a product or service. They read this correspondence while they talk on the telephone, commute to work, or walk to meetings.

Given these contexts, readers want information quickly, information they can under-stand at a glance. Reading word after word, paragraph after paragraph takes time and effort, which most readers cannot spare. Therefore, if your technical communication is visually unappealing, your audience might not even read your words. Readers will either give up before they have begun or be unable to remember what they have read. You cannot assume they will labor over your text to uncover its worth. Good technical communication allows readers rapid access to information.

Damages and Dangers

If your intended readers fail to read your text because it is visually inaccessible, imagine the possible repercussions. They could damage equipment by not recognizing important information which you have buried in dense blocks of text. The readers could give up on the text and call your company's toll-free hotline or send an e-mail message for assistance. This wastes your readers' and your coworkers' time and energy. Worse, your readers might hurt themselves and sue your company for failing to highlight potential dangers in a user manual.

Corporate Identity

Your document—whether a memo, letter, report, instruction, Web site, or brochure—is a visual representation of your company, graphically expressing the company's identity. It might be the only way you meet your clients. If your text is unappealing, that is the corporate image your company conveys to the customer. If your text is not reader-friendly, that is how your company will appear to your client. Visually unappealing and inaccessible correspondence can negatively affect your company's sales and reputation. In today's competitive workplace, any leverage you can provide your company is a plus. Document design is one way to appeal to a client.

Time and Money

Successful technical communication involves the audience. Your goal is to help the readers understand the organization of your text, recognize its order, and access information at a glance. You also want to use varied types of communication, including graphics, for readers who absorb information better visually rather than through words. What's the payoff? Studies tell us that effective document design saves money and time. One international customs department, after revising the design of its lost-baggage forms, reduced its error rate by over 50 percent. When a utilities company changed the look of its billing statements, customers asked fewer questions, saving the company approximately $250,000 per year. The U.S. Department of Commerce, Office of Consumer Affairs, reported that when several companies improved their documents' visual appeal, the companies increased business and reduced customer complaints (Schriver 250–51). Document design isn't a costly frill. Effective document design is good for your company's business.

To clarify how important document design is, look at the inaccessible meeting minutes in this chapter's opening "Communication at Work" scenario (see Figure 7.1). The meeting minutes are neither clear nor concise. You are given so much data in such an unappealing format that your first response upon seeing the correspondence probably is to say, "I'm not going to read that."

How can you make these minutes more inviting? How can you break up the wall-to-wall words and make key points more accessible? To achieve effective document design, you should provide your readers visual

- Organization
- Order
- Access
- Variety

Organization

2. Organize content by breaking text into smaller chunks.

The easiest way to organize your document's design is to break text into small chunks of information, a technique called *chunking*. When you use chunking to separate blocks of text, you help your readers understand the overall organization of your correspondence. They can see which topics go together and which are distinct.

Chunking to organize your text is accomplished by using any of the following techniques:

- **White space** (horizontal spacing between paragraphs created by double- or triple-spacing)
- **Rules** (horizontal lines typed across the page to separate units of information)
- **Section dividers and tabs** (used in longer reports to create smaller units)
- **Headings and talking headings**

Headings and Talking Headings

To improve your page layout and make content accessible, use headings and talking headings. Headings—words or phrases such as "Introduction," "Discussion," "Conclusion," "Problems with Employees," or "Background Information"—highlight the content in a particular section of a document. When you begin a new section, you should use a new heading. In addition, use subheadings if you have a long section under one heading. This will help you break up a topic into smaller, more readable units of text.

Talking headings, in contrast, are more informative than headings. A heading helps your readers navigate the text by guiding them to key parts of a document. However, headings such as "Introduction," "Discussion," and "Conclusion," do not tell the readers what content is included in the section. Talking headings, such as "Human Resources Committee Reviews 2016 Benefits Packages," informatively clarify the content that follows.

One way to create a talking heading is to use a subject (someone or something performing the action), a verb (the action), and an object (something acted upon).

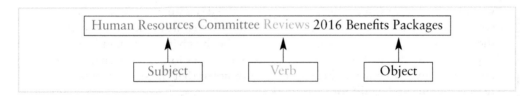

◀ **EXAMPLE**

Another way to create talking headings is to use informative phrases, such as "Problems Leading to Employee Dissatisfaction," "Uses of Company Cars for Personal Errands," and "Cost Analysis of Technology Options for the Accounting Department."

Table 7.1 provides examples of informative talking headings.

Notice how using some of these techniques improves the document design in Figure 7.2.

TABLE 7.1   Talking Headings

Sentences Used as Talking Heads	Phrases Used as Talking Heads
Rude Customer Service Leads to Sales Losses	Budget Increases to Be Frozen until 2016
Accounting Department Requests Feedback on Benefits Package	Outsourced Workers Leading to Corporate Layoffs
Corporate Profit Sharing Decreases to 27 Percent	Harlan Cisneros—New Departmental Supervisor Announced
Parking Lot Congestion Angers Employees	EEOC: Questions about Company Hiring Practices

FIGURE 7.2   Document Design to Organize the Information

First-level heading

MINUTES

The meeting at the Carriage Club was attended by 30 members and guests. After the dinner, Roger Traver introduced the guest speaker, George Smith, university chancellor, and noted his accomplishments and experiences prior to education— U.S. Navy commander, Oak Ridge Laboratory researcher, and politician. Dr. Smith's talk "Industry and Education Collaboration" was very interesting and included a history of special projects enjoyed by both academics and corporate heads. Dr. Smith suggested that we engineers could work with education to accomplish three goals.

Horizontal rule

Training Seminars

Recent industry–education collaborations include training seminars in computers, fiber optics, and human resource options.

Second-level heading

Urban Development

The chancellor's primary thrust was a request for $100,000 in financial aid for urban development. He said money had already been donated from three sources: a large realty firm, Capital Homes, had given $20,000; a philanthropic group, We Care, had donated a matching $20,000; Dr. Smith's university gave a matching $20,000. The remaining $40,000, Dr. Smith hoped, would come from industry donations.

Internships

The chancellor noted that industry could help itself, as well as the community, by providing internships for university undergraduate majors. These internships could either be semester- or year-long arrangements, whereby students would work for minimum wage to learn more about the day-to-day aspects of their chosen fields. The chancellor said that these internships would not only increase the students' theoretical knowledge of engineering by giving them hands-on experience but also make them better future employees for the host engineering companies. Everyone would benefit. Dr. Smith noted that the students would receive a grade and credit for their work.

Conclusion

After the speech, our VP introduced new business, calling for nominations for next year's officers; gave us the agenda for our next meeting; and adjourned the meeting.

Order

3. Create order through a hierarchy of headings.

Once a wall of unbroken words has been separated through chunking to help the reader understand the text's organization, the next thing a reader wants from your text is a sense of order. What's most important on the page? What's less important? What's least important? You can help your audience prioritize information by ordering—or queuing—ideas. The primary way to accomplish this goal is through a hierarchy of headings set apart from each other through various techniques.

- **Typeface.** There are many different *typefaces* (or *fonts*), including Times New Roman, `Courier`, Verdana, **Helvetica**, Arial, Bauhaus 93, Comic Sans MS, *Lucida Calligraphy,* **Cooper Black,** and **STENCIL**. Whichever typeface you choose, it will either be a *serif* or *sans serif* typeface. Serif type has "feet" or decorative strokes at the edges of each letter. This typeface is commonly used in text because it is easy to read, allowing the reader's eyes to glide across the page.

Serif ←— [decorative feet]

Sans serif is a block typeface that omits the feet or decorative lines. This typeface is best used for headings.

Sans Serif ←— [no decorative feet]

Though you have many font typefaces to choose from, all are not appropriate for every technical document. Times New Roman, Arial, and Calibri are best to use for letters, memos, e-mail, reports, resumes, and proposals because these font types are most professional looking and are easiest to read. Arial and Verdana are considered best for Web sites since these fonts are very readable online. If you want to use "designer fonts," limit them to brochures and sales letters, for example.

- **Type size.** Another way of queuing for your readers is through the size of your type. A primary, first-level heading should be larger than subsequent, less important headings: second level, third level, and so forth. For example, a first-level heading could be in 18-point type. The second-level heading would then be set in 16-point type, the third-level heading in 14-point type, and the fourth-level heading in 12-point type.

Figure 7.3 shows examples of different typefaces and type sizes.

FIGURE 7.3   Examples of Typefaces and Type Sizes

Sans Serif Typefaces	Serif Typefaces
Avant Garde 12 point	Courier 12 point
Avant Garde 14 point	Courier 14 point
Avant Garde 18 point	Courier 18 point
Futura 12 point	Bookman 12 point
Futura 14 point	Bookman 14 point
Futura 18 point	Bookman 18 point
Helvetica 12 point	Goudy 12 point
Helvetica 14 point	Goudy 14 point
Helvetica 18 point	Goudy 18 point

- **Density.** The weight of the type also prioritizes your text. Type density is created by boldfacing words.
- **Spacing.** Another queuing technique to help your readers order their thoughts is the amount of horizontal space used after each heading. You can emphasize headings with white space.
- **Position.** Your headings can be centered, aligned with the left margin, indented, or outdented (hung heads). No one approach is more valuable or more correct than another. The key is consistency. If you center your first-level heading, for example, and then place subsequent heads at the left margin, this should be your model for all chapters or sections of that report.

Figure 7.4 shows an outdented first-level heading with indented subsequent headings. Figure 7.5 shows a centered heading with subsequent headings aligned with the left margin.

Figure 7.6 reformats the meeting minutes seen in Figure 7.1 and uses queuing to order the hierarchy of ideas. The outdented first-level heading is set in a 12-point bold sans serif typeface, all caps. The second-level heading is set in a 10-point bold serif typeface and is separated from the preceding text by horizontal white space. The third-level heading is set in a 10-point bold serif typeface and is separated from the preceding text by double-spacing. It is also set on the same line as the following text. Hierarchical heading levels shown in Figure 7.6 allow the readers to visualize the order of information to see clearly how the writer has prioritized text.

FIGURE 7.4 Outdented and Indented Headings

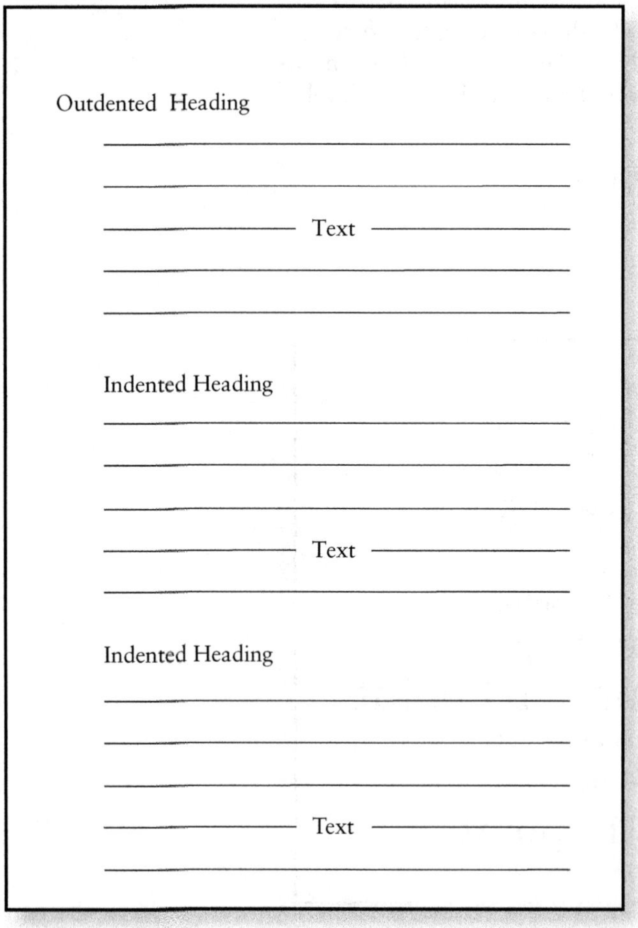

FIGURE 7.5 Centered and Left-Margin Aligned Headings

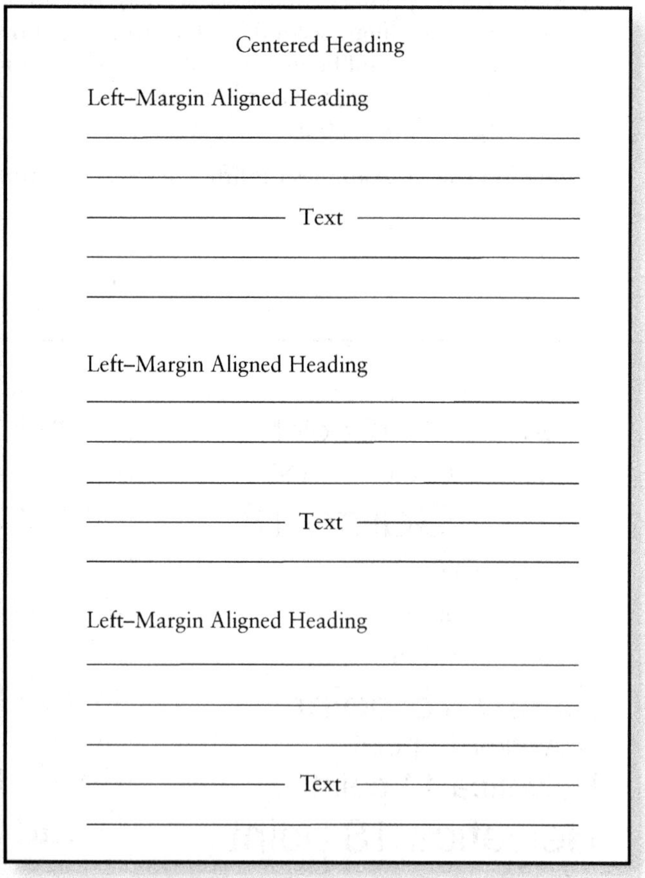

MINUTES The meeting at the Carriage Club was attended by 30 members and guests. After the dinner, Roger Traver introduced the guest speaker, George Smith, university chancellor, and noted his accomplishments and experiences prior to education—U.S. Navy commander, Oak Ridge Laboratory researcher, and politician. Dr. Smith's talk, "Industry and Education Collaboration," was very interesting and included a history of special projects enjoyed by both academics and corporate heads. Dr. Smith suggested that we engineers could work with education to accomplish three goals.

Urban Development
The chancellor's primary thrust was a request for $100,000 in financial aid for urban development. He said money had already been donated from three sources: a large realty firm, Capital Homes, had given $20,000; a philanthropic group, We Care, had donated a matching $20,000; Dr. Smith's university gave a matching $20,000. The remaining $40,000, Dr. Smith hoped, would come from industry donations.

Internships
The chancellor noted that industry could help itself, as well as the community, by providing internships for university undergraduate majors. These internships could either be semester- or year-long arrangements, whereby students would work for minimum wage to learn more about the day-to-day aspects of their chosen fields. The chancellor said that these internships would not only increase the students' theoretical knowledge of engineering by giving them hands-on experience but also make them better future employees for the host engineering companies. Everyone would benefit. Dr. Smith noted that the students would receive a grade and credit for their work.

Training Seminars. Recent industry–education collaborations include training seminars in computers, fiber optics, and human resource options.

Conclusion
After the speech, our VP introduced new business, calling for nominations for next year's officers; gave us the agenda for our next meeting; and adjourned the meeting.

This hanging head is set in all caps and a sans serif font.

Create a hierarchy of headings by changing font size and style.

Headings create accessible content.

Access

4. Achieve easy access of content through design elements.

Chunking helps the reader see which ideas go together, and a hierarchy of headings helps the reader understand the relative importance of each unit of information. Nonetheless, the document design in Figure 7.6 needs improvement. The reader still must read every word carefully to see the key points within each chunk of text. Readers are not that generous with their time. As the writer, you should make your reader's task easier.

A third way to assist your audience is by helping them *access* information rapidly—at a glance. You can use any of the following highlighting techniques to help the readers filter out extraneous or tangential information and focus on key ideas.

- **White space.** In addition to horizontal space, created by double- or triple-spacing, you also can create vertical space by indenting. This vertical white space breaks up the monotony of wall-to-wall words and gives your readers breathing room. White space invites your readers into the text and helps the audience focus on the indented points you want to emphasize.
- **Bullets.** Bullets, used to emphasize items within an indented list, are created by using asterisks (*), hyphens (-), a lowercase *o*, degree signs (°), typographic symbols (■, ❑, •, or ◆), or iconic webdings and wingdings (☞, ☺, or ✓).
- **Numbering.** Enumeration creates itemized lists that can show sequence or importance and allow for easy reference.
- **Boldface.** Boldface text emphasizes a key word or phrase.
- **All caps.** The technique of capitalizing text is an excellent way to highlight a WARNING, DANGER, CAUTION, or NOTE. However, capitalizing other types of information is not suggested because reading lowercase words is easier for your audience. All caps creates a block of letters in which individual letters aren't easily distinguished from each other. Avoid using all caps in correspondence since typing in all caps can create a negative tone. Readers might think you are yelling at them or "flaming."
- **Underlining.** Underlining should be used cautiously. If you underline too frequently, none of your information will be emphatic. One underlined word or phrase will call attention to itself and achieve reader access. Several underlined words or phrases will overwhelm your readers.
- **Italics.** Italics and underlining are used similarly as highlighting techniques.
- **Text boxes.** Place key points in a text box for emphasis.

EXAMPLE ▶

> NOTE: Only place glass containers in the *red recycle bins*.
>
> Use *green bins* for paper and plastic.

You can also italicize and use all caps within the text box, as we have in this example.

- **Fills.** You can further highlight text boxes through fills (color, gradients, and shadings).
- **Inverse type.** You also can help readers access information by using inverse type—printing white on black versus the usual black on white.
- **Color.** Another way to make key words and phrases leap off the page is to color them. *Danger* would be red, for example, *Warning* orange, and *Caution* yellow. You also can use color to help a reader access the first-level heading, a header, or a footer. (*Headers* contain information placed along the top margin of text; *footers* contain information placed along the bottom margin of text.) For instance, if your text is typed in a black font, then headings typed in a blue font would stand out more effectively. However, as with all highlighting techniques, a little bit goes a long way. Do not overuse color. Do not type several headings in different colors. Doing so could produce a very unprofessional impression.

Effective Use of Colored Headings

Committee Action
The Budget and Personnel Committee will vote to approve the audit report at the July meeting.

Recommendation
The committee will recommend that a proposal be submitted to improve roadway construction.

Staff Contacts
Mel Henderson
Sean Thomson

Ineffective Use of Colored Headings

Committee Action
The Budget and Personnel Committee will vote to approve the audit report at the July meeting.

Recommendation
The committee will recommend that a proposal be submitted to improve roadway construction.

Staff Contacts
Mel Henderson
Sean Thomson

All colors are not equal in visual value. Generally, dark-colored fonts provide the most contrast against light-colored backgrounds or vice versa. For example, a black font on a white background (or a white font on a black background) creates optimum contrast. On the other hand, a light-colored font on a light background does not improve access.

Inverse print (white on black) creates optimum contrast.

Committee Action
The Budget and Personnel Committee will vote to approve the audit report at the July meeting.

Recommendation
The committee will recommend that a proposal be submitted to improve roadway construction.

Staff Contacts
Mel Henderson
Sean Thomson

◄ EXAMPLE

Good Contrast Helps Access

Light-colored text on a light-colored background harms readability.

Committee Action
The Budget and Personnel Committee will vote to approve the audit report at the July meeting.

Recommendation
The committee will recommend that a proposal be submitted to improve roadway construction.

Staff Contacts
Mel Henderson
Sean Thomson

◄ EXAMPLE

Bad Contrast Hurts Access

FIGURE 7.7   Highlighting Techniques for Access

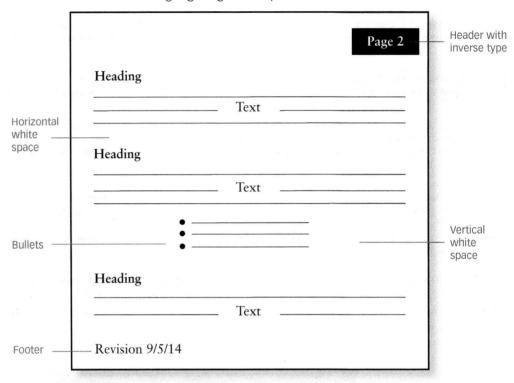

You also should use colors tastefully. Avoid garish color combinations in graphics or backgrounds (red and orange, pink and green, or purple and yellow, and so forth). Colors that clash will distract the reader more than aid access.

Here is a very important consideration: When it comes to using highlighting techniques, more is not better. A few highlighting techniques help your readers filter out background data and focus on key points. Too many highlighting techniques are distractions and clutter the document design. Be careful not to overdo a good thing. Figure 7.7 gives examples of several highlighting methods.

Variety

> 5. Vary design elements to create effective documents.

Each of the document designs in Figures 7.2 and 7.6 uses one column and is printed vertically on a traditional 8½ × 11″ page (a type of printing called *portrait;* see Figure 7.8).

This is not your only option. Your reader might profit from more variety. For example, you might want to use smaller or larger paper; vary the weight of your paper (for example 10-pound, 12-pound, or heavier card stock), or even print your text on colored paper.

More important, you can vary the document design as follows:

- **Choose a different page orientation.** Rather than use *portrait orientation* (8½ × 11″), you could choose *landscape orientation* (11 × 8½″).
- **Use more columns.** Provide your reader two to five columns of text.
- **Vary gutter width.** Separate columns of text by vertical white space called the *gutter.*
- **Use ragged-right margins.** You can use ragged-right (the right margin is not justified) to vary page layout. Some text is fully justified (both right and left

margins are aligned). This was once considered professional, giving the text a clean look. Now, however, studies confirm that right-margin-justified text is harder for the audience to read. It's too rigid. In contrast, ragged-right type is easier to read and more pleasing to the eye.

Figure 7.9 shows how you can use columns, landscape orientation, and ragged-right margins to vary your document design.

FIGURE 7.8 Portrait Orientation with One Column Fully Justified

FIGURE 7.9 Landscape Orientation with Two Columns and Ragged-Right Margins

Although you can vary your document design through page orientation and columns, the audience is still confronted by words, words, and more words. The majority of readers do not want to wade through text. Luckily, words are not your only means of communication. You can reach a larger audience with different learning styles by varying your method of communication. Graphics are an excellent alternative. Many people are more comfortable grasping information visually than verbally. Although it's a cliché, a picture is often worth a thousand words.

To clarify our point about the value of variety, see Figure 7.10, which incorporates a graphic.

> **Graphics**
> See Chapter 8, "Visual Aids," for information about graphics.

Symmetry in Design

When designing documents, achieve a balance of text, white space, and visuals. Lack of balance can confuse your audience and prevent easy access to your message. In contrast, a symmetrically balanced document is visually appealing and allows the audience to grasp your message at a glance. There's no one way to achieve symmetry. The examples in Figures 7.11 and 7.12 show how two companies' Web sites are symmetrically designed.

FIGURE 7.10   Document Design Using a Pie Chart to Add Variety

MINUTES   Thirty members and guests attended the Carriage Club meeting. After the dinner, Roger Traver introduced the guest speaker, George Smith, university chancellor, and noted his accomplishments and experiences prior to education:

- U.S. Navy commander
- Oak Ridge Laboratory researcher
- Politician

Dr. Smith's talk, "Industry and Education Collaboration," included a history of special projects enjoyed by both academics and corporate heads. Dr. Smith suggested that we engineers could work with education to accomplish three goals:

1. **Dr. Smith's Urban Development**
Primary thrust was to request $40,000 for urban development.
Although three sources have already donated money, business and industry can still help significantly, as illustrated in the following pie chart:

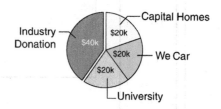

Figure 1 Donations

2. **Internships**
The chancellor noted that industry could help itself, as well as the community, by providing semester-long and year-long internships for university undergraduate majors. These internships would not only increase the students' theoretical knowledge of engineering but also make them better future employees for the host engineering companies.

3. **Training Seminars**
Recent industry–education collaborations include training seminars in computers, fiber optics, and human resources.

Conclusion
After the speech, our VP introduced new business, calling for nominations for next year's officers; gave us the agenda for our next meeting; and adjourned the meeting.

(Continued)

FIGURE 7.11 Symmetrically Balanced Web Site Divided into Three Panels

The NewType Web site has bilateral symmetry both horizontally and vertically. The site also divides into three main panels with the left and right panels balanced in size in contrast to the more prominent center panel.

FIGURE 7.12 Symmetrically Balanced Web Site

The GBA Web site has symmetry both horizontally and vertically. The site divides into two main units with the top two thirds devoted to a graphic and the bottom third divided into a panel consisting of four equally separated fields of information
Source: Courtesy of GBA.

Web Sites Showing Design Elements

Figure 7.13, a Web page from GBA, shows the importance of graphics, white space, bullets, headings, and subheadings to enhance page layout.

FIGURE 7.13   GBA Web Page

The GBA Web site uses a readable sans serif font on a white background with red headings to match the color of the company logo. Note the effective use of short paragraphs, bullets, and visual aids to enhance readability.
Source: Courtesy of GBA.

TECHNOLOGY TIPS

Document Design Using Microsoft Word

When it comes to document design, you have a world of options at your fingertips. Not only does your word processing software offer you possibilities to enhance page layout, but also the Internet provides unlimited resources.

Word Processing

In a word processing software program like Microsoft Word, you enhance your document's design in many ways.
From the **Home** tab ribbon, you can make changes to **Font**, **Paragraphs**, and **Styles**.

(Continued)

Font category

- Choose your **FONT TYPE** and **SIZE**

  Times New Roman ▾ 12 ▾

- **Bold face,** *italicize,* and <u>underline</u> the text

  B *I* <u>U</u> ▾

- ~~strikethrough~~, create ₛᵤᵦₛᵤᵣᵢₚₜₛ and ˢᵘᵖᵉʳˢᶜʳⁱᵖᵗˢ

  abc X₂ X²

- Highlight the color of text, change the font color, or change the case of selected text to uppercase or lowercase

  ab▾ A▾ A▴

- Increase or decrease the font size

  A˄ A˅

Paragraph category

- Create bulleted, numbered, or multilevel lists (as with outlines)

- Decrease or increase an indentation

- Change the margins (ragged right, centered, block right, or full block)

- Change the spacing between lines

- Color background behind selected text, create borders, alphabetize, and show or hide paragraphing

Styles category

- Format text (Quick Styles) and change colors and fonts throughout a document (Change Styles)

  Quick Styles ▾   Change Styles ▾
  Styles

From the following **Insert** tab ribbon, you can add **Tables, Illustrations, Links, Header, Footer, Text,** and include mathematical **Symbols:**

The **Illustrations** category lets you insert pictures, clip art, and charts (pie charts, bar charts, line charts, etc.). **SmartArt** provides access to more complex charts such as Venn diagrams and organizational charts. The **Shapes** pull-down menu lets you insert lines, arrows, flowchart symbols, call-outs, stars, and banners.

The **Text** category lets you insert the following:

- 36 different kinds of text boxes including sidebars

  Text boxes can be enhanced further by changing the color of the lines or fill. You also can create shadows, as shown in this example.

- **WordArt**

The Writing Process at Work

Karen Carlin, claims adjuster for Home, Health, and Hearth Insurance, is returning home from work-related travel. A client, Randall Bell, calls her to report a claim. Karen uses prewriting, writing, and rewriting to create a follow-up e-mail.

Prewriting

Because Karen is riding in a shuttle from the airport, she only has time to make a few prewriting notes on her smartphone. She knows she will revise the text and document design when she returns to her office. Figure 7.14 shows her prewriting.

Writing

The goal of Karen's prewriting was to document the facts of the claim. When she returned to her office, she drafted an e-mail to her clients and sent a copy to her supervisor. See Figure 7.15 for an e-mail draft saved in draft mode.

FIGURE 7.14 Prewriting on a Smartphone

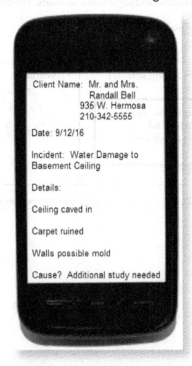

Client Name: Mr. and Mrs.
Randall Bell
935 W. Hermosa
210-342-5555

Date: 9/12/16

Incident: Water Damage to Basement Ceiling

Details:

Ceiling caved in

Carpet ruined

Walls possible mold

Cause? Additional study needed

FIGURE 7.15 E-mail in Draft Mode

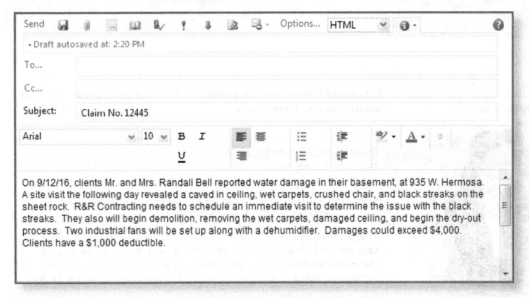

Rewriting

Editing and revising a draft e-mail will help you make it look professional. By paying attention to the e-mail's document design, Karen not only improved the look of the message but also, more importantly, she made the text more accessible for her audience. See Figure 7.16.

FIGURE 7.16 Revised E-mail with Headings and Bullets

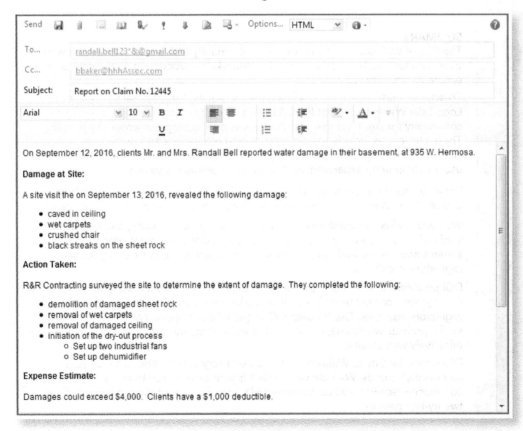

CASE STUDIES

1. In teams of three to five class members, reformat the following memo to improve its document design.

Date: November 30, 2016
To: Jan Hunt
From: Tom Langford
Subject: Cleaning Procedures for Manufacturing Walk-in Ovens #98731, #98732, and #98733

The above-mentioned ovens need extensive cleaning. To do so, vacuum and wipe all doors, walls, roofs, and floors. All vents and dampers need to be removed, and a tack cloth must be used to remove all loose dust and dirt. Also, all filters need to be replaced.

I am requesting this because loose particles of dust and dirt are blown onto wet parts when placed in the air-circulating ovens to dry. This causes extensive rework. Please perform this procedure twice per week to ensure clean production.

---

2. In teams of three to five class members, reformat the following summary to improve its document design.

SUMMARY

The City of Waluska wants to provide its community with a safe and reliable water treatment facility. The goal is to protect Waluska's environmental resources and to ensure community values.

To achieve these goals, the city has issued a request for proposal to update the Loon Lake Water Treatment Plant (LLWTP). The city recognizes that meeting its community's water treatment needs requires overcoming numerous challenges. These challenges include managing changing regulations and protection standards, developing financially responsible treatment services, planning land use for community expansion, and upholding community values.

For all of the above reasons, DesignGlobal, Inc. (DGI) Engineering is your best choice. We understand the project scope and recognize your community's needs.

We have worked successfully with your community for a decade, creating feasibility studies for Loon Lake toxic control, developing odor-abatement procedures for your streams and creeks, and assessing your water treatment plant's ability to meet regulatory standards.

DGI personnel are not just engineering experts. We are members of your community. Our dynamic project team has a close working relationship with your community's regulatory agencies. Our Partner in Charge, Julie Schopper, has experience with similar projects worldwide, demonstrated leadership, and the ability to communicate effectively with clients.

DGI offers the City of Waluska an integrated program that addresses all your community's needs. We believe that DGI is your best choice to ensure that your community receives a water treatment plant ready to meet the challenges of the twenty-first century.

ETHICAL CHALLENGE

Flow Inc., a water purification company that has been in business for over 40 years, is known throughout the Midwest for its logo and slogan, "Go With The Flow."

Nurani Rama grew up in the Midwest and is familiar with Flow Inc. She now works for a software development company named GO—Gaming Online. Nurani is in charge of marketing and public relations and was asked to create a logo and slogan for her company. She fondly remembers Flow Inc.'s slogan and logo and used it as her inspiration. At a board meeting, Nurani promoted the phrase "Flow With the GO" and showed her creation.

Question

Though Nurani's company is in a different city and industry than Flow Inc., is it ethical for her to pattern her slogan and logo after Flow Inc.'s? Why or why not?

INDIVIDUAL AND TEAM PROJECTS

1. Bring samples of technical communication to class. These could include letters, brochures, fliers, instructions, proposals, or reports. In small groups, assess the document design of each sample. Determine which samples have successful document designs and which samples have poor document designs. Base your decisions on the criteria provided in this chapter: organization, order, access, and variety. Either orally or in writing, share your findings with other teams in the class. Rewrite any samples that you can improve.

2. Read the following headings and make them more accessible by creating a hierarchy using different font sizes and font types:
   Meeting Minutes
   Agenda
   Discussion of Ongoing Projects
   Recommendations
   Pricing
   Cost of Equipment
   Cost of Facilities Update
   Cost of Insurance Benefits
   New Hiring Policies
   Job Requirements
   Employee Credentials
   Licensing

3. Revise the following headings into "talking headings" by changing them to phrases or complete sentences. To do so, add any information you choose to clarify your content.
   - Computer Problems
   - WiFi Compatibility
   - Biotechnology Advances
   - Accounting Regulations
   - Facilities Update

PROBLEM-SOLVING THINK PIECES

1. The information in the following two lists needs to be separated and highlighted more clearly. But which highlighting technique should you use for each list—bullets or numbers? Explain why you would use bullets versus numbers, or vice versa for both lists.

List 1	List 2
To access your online course, follow these steps:	To choose the right car for your family and business needs, consider these factors:
Turn on the computer.	Price
Double click the Internet icon on your desktop.	Options
Type in the following URL:	Fuel economy
http://webct.acc.edu.	Cost of repairs
Type your username and password.	Availability of dealerships
Click on the online course of your choice.	Financing
Complete assignment 1.	Capacity

2. Look at the following two lists with headings. Both use Comic Sans and colored headings, but should they? Assess the two lists and explain when and why it's OK to use designer fonts and color or when and why you shouldn't.

EXAMPLE 1   RÉSUMÉ

Objective   Use my information technology skills to improve a company's network capabilities, computer security, and troubleshooting.

Qualifications

- Four years experience in information technology
- BS in Computer Science
- Security certified

Work History

Information Technologist Pantheon Corp. St. Louis, MO 2010–present
- Networked 24 computer stations
- Created computer passwords for all employees
- Installed Spam and Virus protection on all systems

EXAMPLE 2   BROCHURE

Prices
- Buy one, get one free
- Guaranteed lowest in the market
- Last year's prices—*today!*

Options
- Sizes—4" × 2", 3" × 2", and 2" × 2"
- Colors—red, green, black, and silver
- WiFi compatible

Service
- 24/7
- On-site or online help

WEB WORKSHOP

On the Internet, access corporate Web sites. Study them and make a list of the techniques used for visual communication. Which Web sites are successful, and why? Which Web sites are unsuccessful, and why? How would you redesign the less successful Web sites to achieve better visual communication?

CHAPTER EIGHT

Visual Aids

After completing this chapter, you will be able to

1. Appreciate the benefits of visual aids

2. Understand the benefits and drawbacks of color in graphics

3. Follow criteria to create successful graphics

4. Distinguish between the two types of graphics—tables and figures

5. Design effective tables using criteria

6. Design effective figures using criteria

COMMUNICATION AT WORK

To help his clients understand complex figures, Bert Lang includes visual aids in his proposals.

Bert Lang is an investment banker at **Country Commercial Bank.** He is writing a proposal to a potential client, Sylvia Light, a retired public health nurse. Sylvia is 68 years old and worked for the Texas Public Health Department for 36 years. She has earned her State of Texas retirement and Social Security benefits. She now has $315,500 in savings allocated as follows: $78,000 in an Individual Retirement Account (IRA), $234,000 in a low-earning certificate of deposit (CD), and $3,500 in her checking account.

Sylvia contacted Bert, asking him to help her organize her portfolio for a comfortable retirement. Bert has studied Sylvia's various accounts and considered her lifestyle and expenditures. Now, he is ready to write the proposal.

In this proposal, Bert wants to show Sylvia how to reallocate her funds. She should keep some ready money available and invest a portion of capital for long-term returns. Currently, too much of her money is tied up in a CD earning 1.1 percent. Bert plans to propose that Sylvia could reallocate her funds as follows:

- $110,000 in an annuity
- $78,000 in an IRA
- $45,000 in municipal bonds
- $37,000 in a stock fund

- $42,000 in CDs
- $3,500 in a checking account

Like most people, Sylvia is unfamiliar with financial planning. Though she was an expert in her health field, tuberculosis treatment, money matters confuse her. Numbers alone will not explain Bert's vision for her money management.

To make this proposal visually appealing and more readily understandable to Sylvia, Bert will use visual aids. He will provide Sylvia a pie chart to show how he wants to invest her money. He will use a line graph to predict how much more money she can earn by reallocating her assets. Finally, he will create a table to clarify the types of investments, the amount in each investment, the interest to be earned, and the fees.

Although Sylvia has always been fiscally conservative, Bert hopes that he can explain the need for growth of capital even in retirement. The graphic aids will visually enhance his written explanation.

The Benefits of Visual Aids

1. Appreciate the benefits of visual aids.

Although your writing may have no grammatical or mechanical errors and you may present valuable information, you won't communicate effectively if your information is inaccessible. Consider the following paragraph (Figure 8.1).

FIGURE 8.1   Inaccessible Content

> According to research from electronicdot.com, two out of three Americans in 2016 used social networking sites, like Facebook, Twitter, and YouTube. Over the last four years, this percentage has grown for all age cohorts. For example, in 2013, approximately 83 percent of people aged 12 to 24 used social networking sites. This percentage grew to 86 percent in 2014, 88 percent in 2015, and 89 percent in 2016. In 2013, 71 percent of Americans between the ages of 25 to 44 visited social media sites. This percentage grew to 77 percent in 2014, 80 percent in 2015, and 81 percent in 2016. For the 45 to 64 year old cohort, social media visitation was 51 percent in 2013. This percentage grew to 55 percent in 2014. In 2015, 59 percent of people aged 45 to 64 visited social media sites. Finally, in 2016, this percentage grew to 61 percent. Overall, in 2016, 67 percent of Americans visited social networking sites.

If you read the preceding paragraph in its entirety, you are an unusually dedicated reader. Such wall-to-wall words mixed with statistics do not create readable writing.

The goal of effective technical communication is accessible information. The example paragraph fails to meet this goal. No reader can digest the data easily or see clearly the comparative changes in social media usage from year to year.

To present large blocks of data or reveal comparisons, you can supplement, if not replace, your text with graphics. In technical communication, visual aids accomplish several goals. Graphics will help you achieve conciseness, clarity, and cosmetic appeal.

Conciseness

Visual aids allow you to provide large amounts of information in a small space. Words used to convey data (such as Figure 8.1) double, triple, or even quadruple the space needed to report information. By using graphics, you can also delete many unnecessary words and phrases.

Clarity

Visual aids can clarify complex information, such as trends, comparisons, percentages, and facts and figures. See Table 8.1 for examples of how you can use graphics to clarify content.

TABLE 8.1   Use of Graphics to Clarify Content

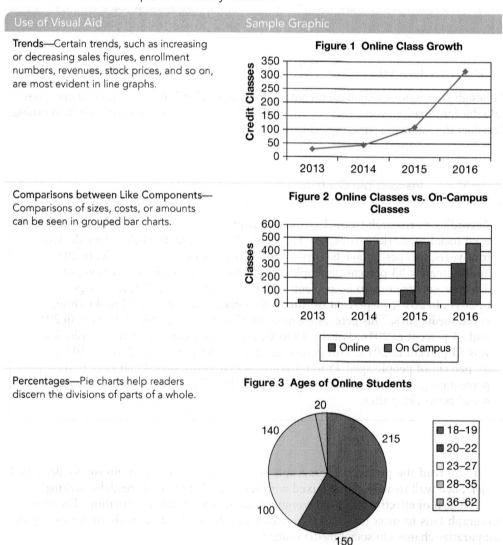

Use of Visual Aid	Sample Graphic
Trends—Certain trends, such as increasing or decreasing sales figures, enrollment numbers, revenues, stock prices, and so on, are most evident in line graphs.	**Figure 1 Online Class Growth**
Comparisons between Like Components—Comparisons of sizes, costs, or amounts can be seen in grouped bar charts.	**Figure 2 Online Classes vs. On-Campus Classes**
Percentages—Pie charts help readers discern the divisions of parts of a whole.	**Figure 3 Ages of Online Students**

(Continued)

TABLE 8.1 Use of Graphics to Clarify Content (*Continued*)

Use of Visual Aid	Sample Graphic
Facts and Figures—A table states statistics/numbers more clearly than a wordy paragraph.	(see table below)

TABLE 1 Online Student Facts

Student Age Cohort	Number of Online Students	Number of Online Credit Hours
18–19	215	1290
20–22	150	900
23–27	100	600
28–35	140	840
36–62	20	120
Total	625	3750

Cosmetic Appeal

Visual aids help you break up the monotony of wall-to-wall words. If you only give unbroken text, your reader might tire, lose interest, and overlook key concerns. Graphics help you sustain your reader's interest. Let's face it; readers like to look at pictures.

Color

All graphics look best in color, don't they? Not necessarily. Without a doubt, a graphic depicted in vivid colors will attract your reader's attention. However, the colors might not aid communication. For example, colored graphics could have these drawbacks:

1. The colors might be distracting (glaring orange, red, and yellow combinations on a bar chart would do more harm than good).
2. Colors that look good today might go out of style in time.
3. Colored graphics increase production costs.
4. Colored graphics consume more disk space and computer memory than black-and-white graphics.
5. If your reader prints the text in black and white, your original colors will have no meaning. In fact, colors, when printed in black and white, will not be distinguishable.

> 2. Understand the benefits and drawbacks of color in graphics.

FAQs: Three-Dimensional Graphics

Q: Three-dimensional graphics look more interesting than one-dimensional graphics, don't they? Shouldn't we use them to add excitement to our text?

A: Many people are attracted to three-dimensional (3-D) graphics. After all, they have obvious appeal. Three-dimensional graphics are more interesting and vivid than flat, one-dimensional graphics. However, 3-D graphics have drawbacks. A 3-D graphic is visually appealing, but it does not convey information quantifiably. The reader has difficulty distinguishing amounts because the gridlines do not "touch" the bars. A word of caution: Use 3-D graphics sparingly. Better yet, if you use the 3-D graphic to create an impression, include a table to quantify your data.

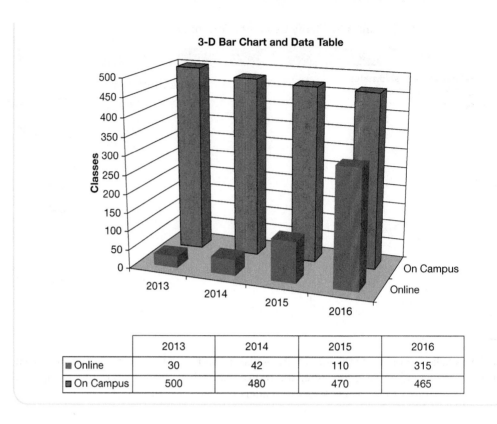

3-D Bar Chart and Data Table

	2013	2014	2015	2016
▪ Online	30	42	110	315
▪ On Campus	500	480	470	465

3. Follow criteria to create successful graphics.

Criteria for Effective Graphics

Figure 8.2 is an example of a cosmetically appealing, clear, and concise graphic. At a glance, the reader can pinpoint the comparative prices per barrel of crude oil between 2010 and 2016. Thus, the line graph is clear and concise. In addition, the writer has included an interesting artistic touch. The oil gushing out of the tower shades parts of the graph to emphasize the dollar amounts. Envision this graph without the shading. Only the line would exist. The shading provides the right touch of artistry to enhance the information communicated.

The graph shown in Figure 8.2 includes the traits common to effective visual aids. Successful tables and figures have these characteristics:

1. Integrated with the text (i.e., the graphic complements the text; the text explains the graphic.)
2. Appropriately located (preferably immediately following the text referring to the graphic, not a page or pages later)
3. Enhance the material explained in the text (without being redundant)
4. Communicate important information that could not be conveyed easily in a paragraph or longer text
5. Do not contain details that detract from rather than enhance the information
6. Sized effectively (large enough to be readable but not so large as to overwhelm the page)
7. Correctly labeled (with numbers, titles, legends, and headings)
8. Follow the style of other figures or tables in the text (same font size, font style, color, size of the graphic, and so on)

FIGURE 8.2   Line Graph with Shading

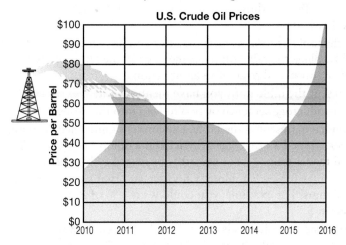

Types of Graphics

Graphics can be broken down into two basic types: tables and figures. Tables provide columns and rows of information. You should use a table to make factual information, such as numbers, percentages, and monetary amounts, easily accessible and understandable. Figures, in contrast, are varied and include bar charts, line graphs, photographs, pie charts, schematics, line drawings, and more.

4. Distinguish between the two types of graphics, tables and figures.

Tables

Let's tabulate the information about social media usage from 2013 to 2016 presented earlier in Figure 8.1. Because effective technical communication integrates text and graphics, you will want to provide an introductory sentence prefacing Table 8.2, as follows:

5. Design effective tables using criteria.

◄ EXAMPLE

Table 8.2 compares the use of social media by age cohort from 2013 to 2016.

TABLE 8.2   Percentage of Social Media Usage

Age Cohort	2013	2014	2015	2016
12–24	83%	86%	88%	89%
25–44	71%	77%	80%	81%
45–64	51%	55%	59%	61%

Table 8.2 has advantages for both the writer and the reader. First, the headings eliminate needless repetition of words, thereby making the text more readable. The example paragraph in Figure 8.1 contains 160 words; the table consists of only 31 words and numbers. Second, the audience can see easily the comparison between the use of social media by age. Thus, the table highlights the content's significant differences. Finally, if this information is included in a report, the writer will reference the table in the list of illustrations. This creates ease of access for the reader.

Criteria for Effective Tables

To construct tables correctly, do the following:

1. Number tables in order of presentation (i.e., Table 1, Table 2, Table 3).

2. Title every table. In your writing, refer to the table by its number, not its title. Simply say, "Table 1 shows . . . ," "As seen in Table 1," or "The information in Table 1 reveals . . . ."

3. Present the table as soon as possible after you have mentioned it in your text. Preferably, place the table on the same page as the appropriate text, not on a subsequent, unrelated page or in an appendix.

4. Don't present the table until you have mentioned it.

5. Use an introductory sentence or two to lead into the table.

6. After you have presented the table, explain its significance. You might write, "Thus, social media usage has grown exponentially since 2013. The peak year for social media usage by age cohort is projected to be in 2016."

7. Write headings for each column. Choose terms that summarize the information in the columns. For example, you could write "% of Error," "Length in Ft.," or "Amount in $."

8. Because the size of columns is determined by the width of the data or headings, you may want to abbreviate terms ("%" for "percent," "$" for "dollars," etc.). If you use abbreviations, however, be sure your audience understands your terminology.

9. Center tables between right and left margins when possible. Don't crowd them on the page.

10. Separate columns with ample white space, vertical lines, or dashes.

11. Show that you have omitted information by printing two or three periods or a hyphen or dash in an empty column.

12. Be consistent when using numbers. Use either decimals or numerators and denominators for fractions. You could write 3¼ and 3¾ or 3.25 and 3.75. If you use decimal points for some numbers but other numbers are whole, include zeroes. For example, write 9.00 for 9.

13. If you do not conclude a table on one page, on the second page write *Continued* in parentheses after the number of the table and the table's title.

14. Cite the source of your information if you have taken content from primary or secondary sources.

Table 8.3 is an excellent example of a correctly prepared table.

Table number and title

TABLE 8.3 Student Headcount Enrollment by Age Group and Student Status, Fall 2016

Column headings

Age Group	New Students	Continuing Students	Readmitted	Other	Total
15–17	453	33	2	2	490
18–20	1,404	1,125	132	—	2,661
21–23	339	819	269	—	1,427
24–26	263	596	213	—	1,072
27–29	250	436	134	—	820
30–39	524	1,168	372	—	2,064
40–49	271	510	186	—	967
50–59	76	121	54	—	251
60+	19	48	16	—	83
Unknown	109	92	27	2	230
Total	**3,708**	**4,948**	**1,405**	**4**	**10,065**

Indicate if no data is available by using a dash, periods, or hyphens.

Provide a final row for totals when necessary. Include a source of information when possible.

Source: "City Community College Enrollment Statistics." Office of Institutional Research. Dec. 2016.

Creating Graphics in Microsoft Word 2013 (Pie Charts, Bar Charts, Line Graphs, etc.)

You can create customized graphics in Microsoft Word as follows:

1. Click on **Insert** on the menu bar.

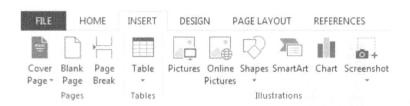

2. Click on **Chart**, and then select the type of graphic you want to insert from the graphics' template.

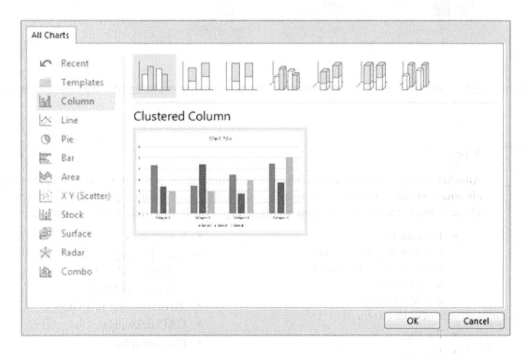

Once you choose your graphic by double-clicking on the icon ("Column" for example), Word will open a data sheet and sample graphic.

3. Customize the graphic by adding the appropriate data in the Excel spreadsheet. Changes made on the spreadsheet at the top will automatically be seen on the graphic.

	A	B	C	D	E	F	G	H	I
1		Series 1	Series 2	Series 3					
2	Category 1	4.3	2.4	2					
3	Category 2	2.5	4.4	2					
4	Category 3	3.5	1.8	3					
5	Category 4	4.5	2.8	5					
6									
7									

(Continued)

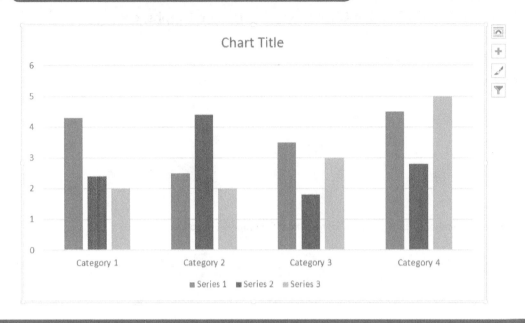

Figures

6. Design effective figures using criteria.

Another way to enhance your technical communication is to use figures. Whereas tables eliminate needless repetition of words and clarify quantifiably, figures highlight and supplement important points in your writing. Types of figures include the following:

- Bar charts
  - Grouped bar charts
  - Horizontal bar charts
  - Vertical bar charts
  - Pictographs
  - Gantt charts
- Pie charts
- Line charts
  - Broken line charts
  - Curved line charts
- Combination charts
- Flowcharts
- Organizational charts

- Schematics
- Line drawings
  - Exploded views
  - Cutaway views
- Virtual reality drawings
- CAD drawings
- Photographs
- Icons
- Internet downloadable graphics

Criteria for Effective Figures

To construct figures correctly, do the following:

1. Number figures in order of presentation (i.e., Figure 1, Figure 2, Figure 3).
2. Title each figure. When you refer to the figure, use its number rather than its title: "Figure 1 shows the relation between the average estimated price for houses and the actual sales prices."

3. Preface each figure with an introductory sentence.

4. Avoid using a figure until you have mentioned it in the text.

5. Present the figure as soon as possible after mentioning it instead of several paragraphs or pages later.

6. After you have presented the figure, explain its significance. Don't let the figure speak for itself. Remind the reader of the important facts you want to highlight.

7. Label the figure's components. For example, if you are using a bar or line chart, label the horizontal and vertical axes clearly. If you're using line drawings, pie charts, or photographs, use clear *callouts* (names or numbers that indicate particular parts) to label each component.

8. When necessary, provide a legend or key at the bottom of the figure to explain information. For example, a key in a bar or line chart will explain what each differently colored line or bar means. In line drawings and photographs, you can use numbered callouts in place of names. If you do so, you will need a legend at the bottom of the figure explaining what each number means.

9. If you abbreviate any labels, define these in a footnote. Place an asterisk (*) or a superscript number (^{1}, ^{2}, ^{3}) after the term and then at the bottom of the figure where you explain your terminology.

10. If you have drawn information from another source, note this at the bottom of the figure.

11. Frame the figure. Center it between the left and right margins or place it in a text box.

12. Size figures appropriately. Don't make them too small or too large.

Bar Charts. Bar charts show either vertical bars (as in Figure 8.3) or horizontal bars (as in Figure 8.4). These bars are scaled to reveal quantities and comparative values. You can shade, color, or crosshatch the bars to emphasize the contrasts. If you do so, include a legend explaining what each bar represents, as in Figure 8.4. Pictographs (as in Figure 8.5) use picture symbols instead of bars to show quantities. To create effective pictographs, do the following:

1. The picture should be representative of the topic discussed.

2. Each symbol equals a unit of measurement. The size of the units depends on your value selection as noted in the key or on the *x*- and *y*-axes.

3. Use more symbols of the same size to indicate a higher quantity; do not use larger symbols.

FIGURE 8.3 Vertical Grouped Bar Chart

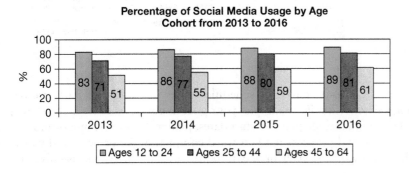

FIGURE 8.4 Horizontal Grouped Bar Chart

This bar chart, geared toward office supply sales managers, is as factual as the pictograph in Figure 8.5. However, Figure 8.4, which omits the drawings of the PCs, keeps the graphic more businesslike for the intended audience of peers who are specialists.

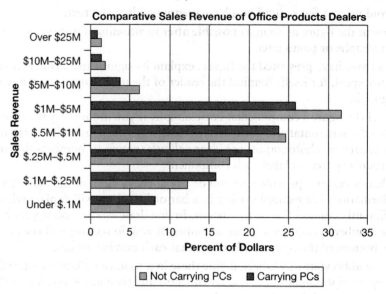

The pictograph is as factual as the bar chart in Figure 8.4. However, since the pictograph is designed for lay readers, it uses symbols of PCs to enhance the visual appeal of this topic. In addition, the PCs make the subject matter more interesting for the audience.

FIGURE 8.5 Pictograph

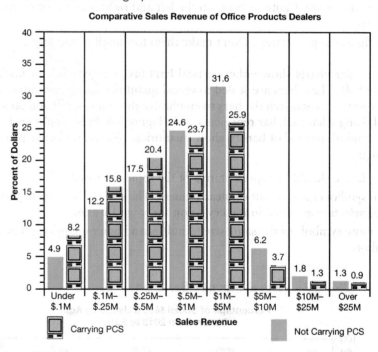

Gantt Charts. Gantt charts, or schedule charts (as in Figure 8.6), use bars to show chronological activities. For example, your goal might be to show a client the phases of a project. This could include planned start dates, planned reporting milestones, planned completion dates, actual progress made toward completing the project, and work remaining. Gantt charts provide an excellent way to represent these activities visually. They are often

FIGURE 8.6   Gantt Chart

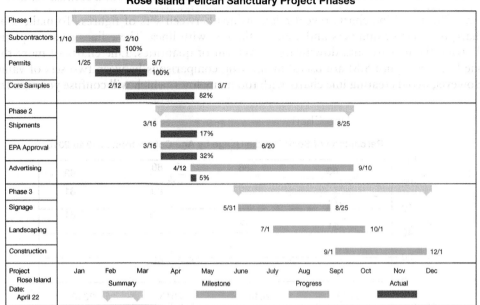

Rose Island Pelican Sanctuary Project Phases

The Gantt chart shows the schedule of activities, including start and stop dates, plus percentage of completion.

included in proposals to project schedules or in reports to show work completed. To create successful Gantt charts, do the following:

1. Label your horizontal and vertical axes. For example, the vertical axis represents the various activities scheduled, and then the horizontal axis represents time (either days, weeks, months, or years).
2. Provide gridlines (either horizontal or vertical) to help your readers pinpoint the time accurately.
3. Label your bars with exact dates for start or completion.
4. Quantify the percentages of work accomplished and work remaining.
5. Provide a legend to differentiate between planned activities and actual progress.

Pie Charts.   Use pie charts (as in Figure 8.7) to illustrate portions of a whole. The pie chart represents information as pie-shaped parts of a circle. The entire circle equals 100 percent or 360 degrees. The pie pieces (the wedges) show the various divisions of the whole.

To create effective pie charts, do the following:

1. Be sure that the complete circle equals 100 percent or 360 degrees.
2. Begin spacing wedges at the twelve o'clock position.
3. Use shading, color, or crosshatching to emphasize wedge distributions.
4. Use horizontal writing to label wedges.
5. If you don't have enough room for a label within each wedge, provide a legend defining what each shade, color, or crosshatch symbolizes.

FIGURE 8.7   Pie Chart

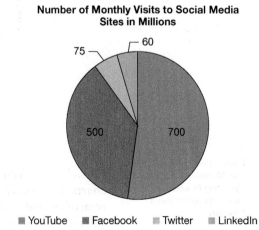

Number of Monthly Visits to Social Media Sites in Millions

■ YouTube   ■ Facebook   ■ Twitter   ■ LinkedIn

6. Provide percentages for wedges when possible.
7. Do not use too many wedges—this would crowd the chart and confuse readers.

Line Charts. Line charts reveal relationships between sets of figures. To make a line chart, plot sets of numbers and connect the sets with lines. These lines create a picture showing the upward and downward movement of quantities. Line charts of more than one line (see Figure 8.8) are useful in showing comparisons between two sets of values. However, avoid creating line charts with too many lines, which will confuse your readers.

FIGURE 8.8   Line Chart

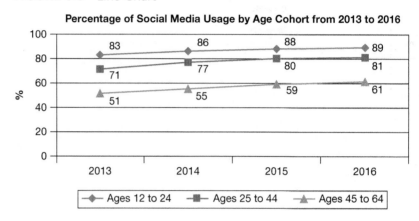

Combination Charts. A combination chart reveals relationships between two sets of figures. To do so, it uses a combination of figure styles, such as a bar chart and a line chart (as shown in Figure 8.9). The value of a combination chart is that it adds interest and distinguishes the two sets of figures by depicting them differently.

FIGURE 8.9   Combination Bar and Line Chart

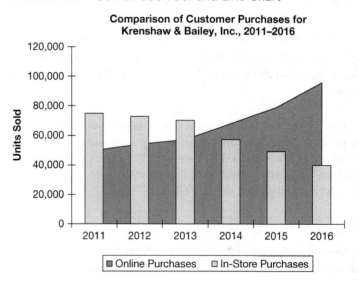

Flowcharts

See Chapter 15, "Instructions, User Manuals, and Standard Operating Procedures," for more information about flowcharts.

Flowcharts. You can show chronological sequence of activities using a flowchart. Flowcharts are especially useful for writing technical instructions. When using a flowchart, remember that ovals represent starts and stops, rectangles represent steps, and diamonds equal decisions (see Figure 8.10).

FIGURE 8.10   Flowchart

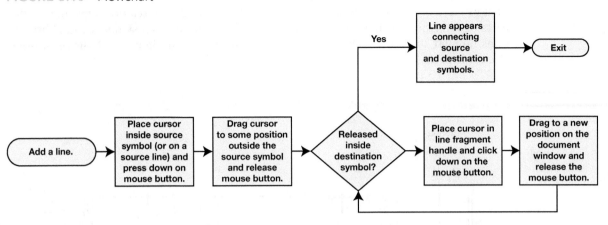

Organizational Charts.   The chart in Figure 8.11 shows the chain of command in an organization. An organizational chart helps your readers see where individuals work within a business and their relation to other workers.

FIGURE 8.11   Organizational Chart

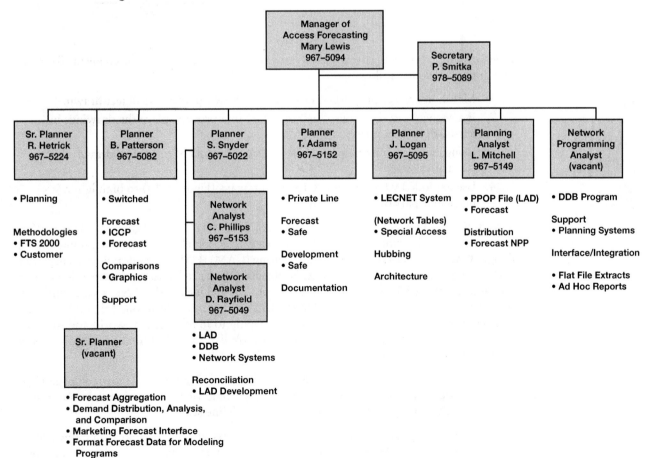

Schematics. Schematics are useful for presenting abstract information in technical fields such as electronics and engineering. A schematic diagrams the relationships among the parts of something such as an electrical circuit. The diagram uses symbols and abbreviations familiar to highly technical readers. The schematic in Figure 8.12 shows various electronic parts (resistors, diodes, condensers) in a radio.

FIGURE 8.12 Schematic of a Radio

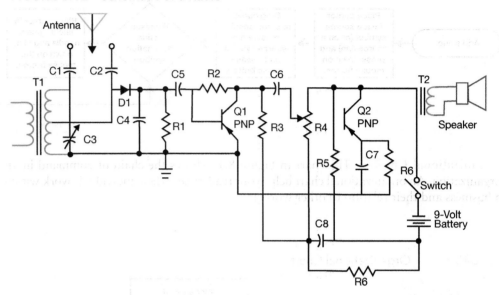

Line Drawings. Use line drawings to show the important parts of a mechanism. To create line drawings, do the following:

1. Maintain correct proportions in relation to each part of the object drawn.
2. If a sequence of drawings illustrates steps in a process, place the drawings in left-to-right or top-to-bottom order.
3. Using callouts to name parts, label the components of the object drawn (see Figure 8.13).
4. Use exploded views (Figure 8.13) or cutaways (Figure 8.14) to highlight a particular part of the drawing.

Virtual Reality Drawings, Section Elevations, and CAD/CAM Drawing. Three different types of line drawings are virtual reality views, section elevations, and computer-aided design (CAD) or computer-aided manufacturing (CAM) (CAD/CAM drawings). They offer 3-D representations of buildings, sites, or objects. Often used in the architectural/engineering industry, these 3-D drawings help clients get a visual idea of the services your company provides. Virtual reality drawings (Figure 8.15) and section elevations (Figure 8.16) add lighting, materials, and shadow reflection mapping to mimic the real world and to allow customers to see what a building or site will look like in a photorealistic setting. CAD/CAM drawings give more precise details.

CAD Drawings. CAD drawings, such as the site plan in Figure 8.17 (shown on page 175), use geometric shapes and symbols to provide a graphic view of a setting drawn to a particular scale. CAD drawings include *notations* to define scale.

Photographs. A photograph can illustrate your text effectively. Like a line drawing, a photograph can show the components of a mechanism. If you use a photo for this purpose,

FIGURE 8.13   Line Drawing of Exploded View with Callouts

Valve Parts List		
Item	Description	Product Number
1	Nut	054859
2	Cap	098997
3	O–Ring	066584
4	Diaphragm	023337
5	Body Valve	V45665
6	Connector Elbow	C33678

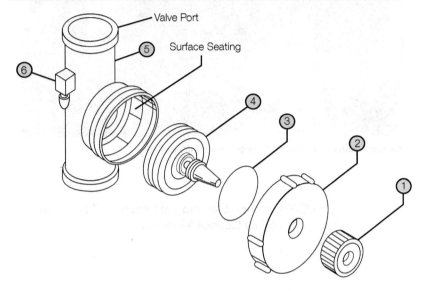

FIGURE 8.14   Line Drawing of Nut and Bolt
(Cutaway View)

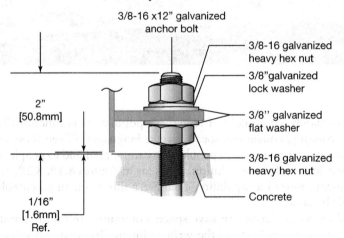

you will need to label (name), number, or letter parts and provide a key. Photographs are
excellent visual aids because they emphasize all parts equally. Their primary advantage is
that they show something as it truly is. See Figure 8.18 (shown on page 176).

FIGURE 8.15   Virtual Reality Drawing

Source: Courtesy of Johnson County Community College.

FIGURE 8.16   CAD/CAM Section Elevation of Office Facility for Lay Audience of City Council Members

The 3-D section elevation drawing provides an impressionistic view of what the site will potentially look like. It not only factors in elevation and landscaping but also allows the audience to relate to and envision the design.

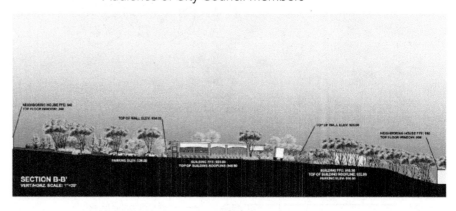

Icons.   Approximately 20 percent of America's population is functionally illiterate. In today's global economy, consumers speak diverse languages. Given these two facts, how can technical writers communicate to people who cannot read and to people who speak different languages? Icons offer one solution. Icons (as in Figures 8.19, 8.20, 8.21, and 8.22) are visual representations of a capability, a danger, a direction, an acceptable behavior, or an unacceptable behavior.

When used correctly, icons can save space, communicate rapidly, and help readers with language problems understand the writer's intent. To create effective icons, follow these suggestions:

1. **Keep it simple.** You should try to communicate a single idea. Icons are not appropriate for long discourse.
2. **Create a realistic image.** This could be accomplished by representing the idea as a photograph, drawing, caricature, outline, or silhouette.

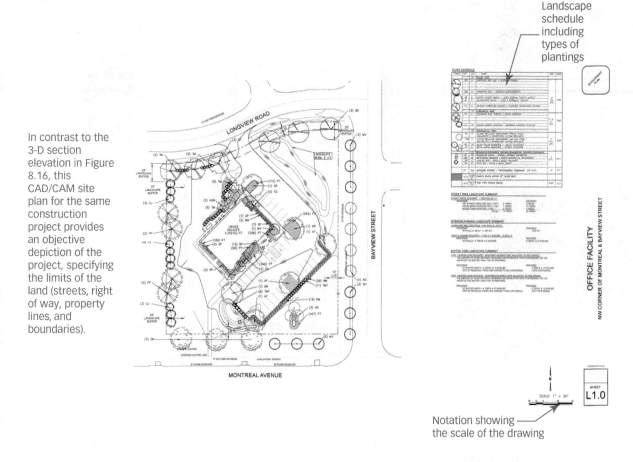

Landscape schedule including types of plantings

In contrast to the 3-D section elevation in Figure 8.16, this CAD/CAM site plan for the same construction project provides an objective depiction of the project, specifying the limits of the land (streets, right of way, property lines, and boundaries).

Notation showing the scale of the drawing

3. **Make the image recognizable.** A top view of a computer monitor or a side view of a smartphone is uninformative. Select the view of the object that best communicates your intent.

4. **Avoid cultural and gender stereotyping.** If you are drawing a hand, you should avoid showing any skin color, and you should stylize the hand so it is neither clearly male nor female.

5. **Strive for universality.** Stick figures of men and women are recognizable worldwide. In contrast, letters—such as *P* for *parking*—will mean very little in China, Africa, or the Mideast. Even colors can cause trouble. In North America, red represents danger, but red is a joyous color in China. Yellow calls for caution in North America, but this color equals happiness and prosperity in the Arab culture.

Downloading Existing Online Graphics

The Internet contains thousands of Web sites that contain graphics, including photographs, line drawings, cartoons, icons, animated images, arrows, buttons, horizontal lines, balls, letters, bullets, hazard signs, and more. You can download images from any Web site. To download graphics, place the cursor on the graphic and right-click on the mouse. A pop-up menu will appear. Either click on **Copy** or click on **Save Picture.** You can save your image in the file of your choice. The images from the Internet will already be graphics interchange format (GIF) or joint photographic experts group (JPEG) files.

> **Copyright**
> See Chapter 6, "Ethical Considerations," for information about copyright.

FIGURE 8.18 Photograph of Mechanical Piping

FIGURE 8.19 Icon of Explosives

FIGURE 8.20 Icon of Dangerous Machinery

FIGURE 8.21 Icon of Electric Shock

FIGURE 8.22 Icon of Corrosive Material

Source: Courtesy of George Butler Associates, Inc.

However, when downloading images, you must abide by copyright laws and ethical considerations. Can you ethically download any image? Can you make changes to a graphic taken from your own company's archives or from boilerplate documents? Do these graphics need a source citation? If your company owns these images, you can modify them and use the changed graphics without citing the source.

Modifying or Creating New Graphics

To modify graphics, you can download them in two ways. First, you can print the screen by pressing the Print Screen key (usually found on the upper right of your keyboard). This captures the entire screen image in a clipboard. Then you can open a graphics program and paste the captured image. Second, you can save the image in a file and then open the graphic in a graphics package.

Most graphics programs will allow you to customize a graphic. Popular programs include Paint, Paint Shop Pro, PhotoShop, Corel Draw, and Adobe Illustrator. In these graphics programs, you can manipulate the images by changing colors, adding text, reversing the images, cropping, resizing, redimensioning, rotating, retouching, deleting or erasing parts of the images, overlaying multiple images, joining multiple images, and so forth. See Figures 8.23 and 8.24 for examples of original and modified photographs taken from a company's archives.

Another option is to create your own graphic. If you are artistic, draw your graphic in a graphics program. This option might be more challenging and time consuming. However, creating your own graphic gives you more control over the finished product, provides a graphic precisely suited to your company's needs, and helps avoid infringement of copyright laws.

FIGURE 8.23 Original Photograph with Two Men and Pipe

Source: Courtesy of George Butler Associates, Inc.

FIGURE 8.24 Modified Photograph (with the two men and *PLO5* removed)

Source: Courtesy of George Butler Associates, Inc.

VISUAL AID CHECKLIST

_____ 1. Will a visual aid add to your technical communication and make the document concise and clear and add cosmetic appeal?

_____ 2. Should you use color or 3-D graphics in your visual aids?

_____ 3. Are visuals integrated with the text?

_____ 4. Do the visuals add to the text and enhance it without being redundant?

_____ 5. Do the graphics communicate information visually that could not easily be conveyed in text?

_____ 6. Are the visuals the correct size, labeled, readable, and similar in style?

_____ 7. Did you use a table when you presented factual information, such as numbers, percentages, or monetary amounts?

_____ 8. Have you used a figure to highlight and supplement important points in your writing?

_____ 9. Have you included a figure or table number and a title for your graphics and data labels?

_____10. Have you cited your source of information?

The Writing Process at Work

Now that you know the criteria for creating different types of visual aids, you next need to create the visual aid. How do you begin? As always, prewrite, write, and rewrite. Remember that the process is dynamic and the steps frequently overlap.

> **Communication Process**
>
> See Chapter 3, "The Communication Process," to learn more about prewriting, writing, and rewriting.

Prewriting

Prewrite to gather data and determine your goals and audience. Bert Lang, highlighted in this chapter's "Communication at Work" scenario, worked for the trust department of a bank and was going to make a presentation to a client, Sylvia Light, about asset allocations. Bert considered the following:

- Audience—specialists (bank employees) and lay reader (client)
- Channels—financial proposal to be reviewed by both the client and the bank
- Data—specific client financial information, including assets

Figure 8.25 shows Bert's list for planning asset allocations.

FIGURE 8.25   Planning List of Asset Allocations

Reallocate funds as follows:

- $110,000 in an annuity
- $78,000 in an Individual Retirement Account (IRA)
- $45,000 in municipal bonds
- $37,000 in a stock fund
- $42,000 in certificates of deposit (CD)
- $3,500 in a checking account

Writing

Draft the document. Bert had to decide how to present his information to Sylvia. He wrote the following rough draft of the asset allocation in paragraph form (Figure 8.26):

FIGURE 8.26   Rough Draft of Asset Allocation

After reviewing your financial situation and considering your lifestyle, you should consider reallocating your assets in the following way. Put $110,000 in an annuity, $78,000 in an Individual Retirement Account (IRA), $45,000 in municipal bonds, $37,000 in a stock fund, $42,000 in certificates of deposit (CD), and $3,500 in a checking account. This will allow you to have a higher overall rate of return, yet remain almost risk free in regards to your capital. I know that preservation of capital is of paramount importance at this time in your life.

Rewriting

Revise and rewrite the rough draft. Bert realized that a paragraph would not communicate asset allocations easily to the customer, so he revised the paragraph into a pie chart for the presentation. Figure 8.27 shows Bert's pie chart of asset allocation.

FIGURE 8.27   Revised Pie Chart of Asset Allocation

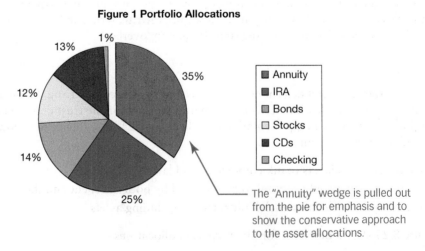

Figure 1 Portfolio Allocations

The "Annuity" wedge is pulled out from the pie for emphasis and to show the conservative approach to the asset allocations.

CASE STUDY

Bert Lang is an investment banker at Country Commercial Bank, as noted in this chapter's beginning scenario. He is writing a proposal to a potential client, Sylvia Light, a retired public health nurse. She now has $315,500 in savings allocated as follows: $78,000 in an Individual Retirement Account (IRA); $234,000 in a low-earning certificate of deposit (CD); and $3,500 in her checking account.

Sylvia contacted Bert, asking him to help her organize her portfolio for a comfortable retirement. Bert has studied Sylvia's various accounts and considered her lifestyle and expenditures. Now, he is ready to write the proposal. Bert plans to propose that Sylvia could reallocate her funds as follows:

- $110,000 in an annuity
- $78,000 in an IRA
- $45,000 in municipal bonds
- $37,000 in a stock fund
- $42,000 in CDs
- $3,500 in a checking account

To make this proposal visually appealing and more readily understandable to Sylvia, Bert will use visual aids.

Assignment

- Create a table to show how he wants to invest her money.
- Create a bar chart comparing her current allocations versus his proposed allocations.

ETHICAL CHALLENGE

According to ethical standards, visuals should be precise. Look at the following bar chart. It shows that during one quarter, the company lost over $50,000.

Question

Is the following visual aid ethical? Why or why not?

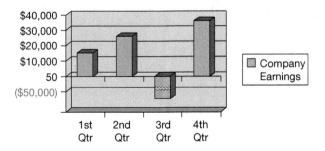

INDIVIDUAL AND TEAM PROJECTS

1. Present the following information in a pie chart, a bar chart, and a table.

   In 2015, Interstate Telephone Company bought and installed 100,000 relays. It used these for long-range testing programs that assessed failure rates. It purchased 40,000 Nestor 221s; 20,000 VanCourt 1200s; 20,000 Macro R40s; 10,000 Camrose Series 8s; and 10,000 Hardy SP6s.

2. Using the information presented in activity 1 and the following revised data, show the comparison between 2015 and 2016 purchases through two pie charts, a grouped bar chart, and a table.

   In 2016, after assessing the success and failure of the relays, Interstate Telephone Company made new purchases of 200,000 relays. It bought 90,000 VanCourt 1200s; 50,000 Macro R40s; 30,000 Camrose Series 8s; and 30,000 Hardy SP6s. No Nestors were purchased.

3. Analyze the following graphics and explain which ones succeed and which ones fail.

   a. Vertical bar chart   b. Pie chart

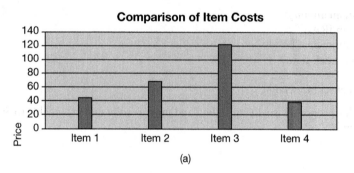

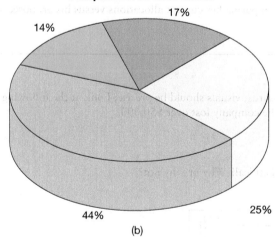

c. Horizontal bar chart

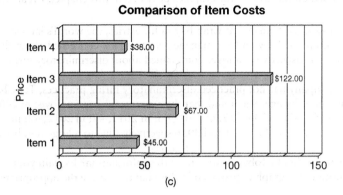

Comparison of Item Costs

(c)

d. Table

TABLE 8.4   Comparison of Item Costs

	Price	Percentage
Item 1	$ 45.00	17%
Item 2	$ 67.00	25%
Item 3	$122.00	44%
Item 4	$ 38.00	14%
Total	$272.00	100%

4. Based on the criteria provided in this chapter, revise any of the poor graphics from assignment 3. Add any additional information necessary to correct the graphics.

PROBLEM-SOLVING THINK PIECES

1. Angel Guerrero, Computer Information Systems technologist at HeartHome Insurance, was responsible for making an inventory of his company's hardware. He learned the following: The company had 75 laptops, 159 PCs, 27 printers, 10 scanners, and 59 smartphones.

   To write his inventory report, Angel needs to chart the above data. Which type of visual aid should he use? Explain your answer based on the information provided in this chapter. Create the appropriate visual aid.

2. Minh Tran works in the Marketing Department at Thrill-a-Minute Entertainment Theme Park (TET). Minh and her project team need to study entry prices, ride prices, food and beverage prices, and attendance of their park versus its primary competitor, Carnival Towne (CT).

   Minh and her team have found that TET charges $16.50 admission, while CT charges $24.95. Most of TET's rides are included in the entry price, but special rides (the Horror, the Bomber, the Avenger, and Peter Pan's Train) cost an additional $2.50 each. At CT, the entry fee covers many rides, excluding Alice's Teacup, Top-of-the-World Ferris Wheel, and the Zinger, which cost an additional $2.00 each. Food and beverages at TET cost $1.95 for a hot dog, $2.50 for a hamburger, $3.95 for nachos, and $1.50–2.50 for drinks. At CT, food and beverages cost $1.75 for hot dogs, $2.75 for hamburgers, $2.50 for nachos, and $1.50–2.50 for drinks. Attendance at TET last year was 250,000, while attendance at CT was 272,000.

What type of visual aid should Minh and her team use to convey this information? Explain your decision based on the criteria for graphics provided in this chapter. Create the appropriate visual aid.

3. Toby Hebert is Human Resource Manager at Crab Bayou Industries, the world's largest wholesaler of frozen Cajun food. Toby and her management team are concerned about the company's hiring trends. A prospective employee complained about discriminatory hiring practices at Crab Bayou.

   To prove that the company has not practiced discriminatory hiring practices, Toby has studied the last 11 years' hires by age. She found that the average age per employee in 2006 was 48; in 2007, 51; in 2008, 47; in 2009, 52; in 2010, 45; in 2011, 47; in 2012, 42; in 2013, 39; in 2014, 42; in 2015 and 2016, 29 and 30, respectively (due to a large number of early retirements).

   What type of visual aid should Toby use to convey this information? Explain your decision based on the criteria for graphics provided in this chapter. Create the appropriate visual aid.

4. Yasser El-Akiba is a member of his college's International Students' Club. Yasser is a native of Lebanon. Other members of the club are from other countries: three from Australia, two from Ecuador, eight from Mexico, five from Africa, two from England, three from Canada, four from the Dominican Republic, and nine from China.

   What kind of visual aid could Yasser create to show his club members' homelands? Defend your decision based on criteria for graphics in this chapter. Create the appropriate visual aid.

WEB WORKSHOP

Using an Internet search engine, type in phrases such as "automobile sales+line graph," "population distribution by age+pie chart," or "California+organizational chart." Create similar phrases for bar charts, pictographs, flowcharts, tables, and so forth. Open several links from your Web search and study the examples you have found. Which examples of graphics are successful and why? Which examples of graphics are unsuccessful and why? Explain your reasoning, based on this chapter's criteria.

CHAPTER NINE

Routine Correspondence

After completing this chapter, you will be able to

1. Understand the importance of memos, letters, e-mail, instant messages, and text messages

2. Determine which communication channel to use for routine correspondence

3. Distinguish among the different routine communication channels

4. Know why to write memos, letters, and e-mail

5. Use an all-purpose template for memos, letters, and e-mail

6. Recognize the essential components of memos

7. Recognize the essential components of letters

8. Identify three common letter formats

9. Recognize the essential components of e-mail

10. Follow criteria for writing routine correspondence

11. Understand the benefits of instant messages

12. Understand the reasons for using text messages

COMMUNICATION AT WORK

In this scenario, a biotechnology company frequently corresponds through letters, memos, e-mail, instant messages, and text messages.

COMRMed, a wholesale provider of biotechnology equipment, is home-based in Reno, Nevada. COMRMed's CEO, Jim Goodwin, plans to capitalize on emerging nanotechnology to manufacture and sell the following:

• Extremely lightweight and portable heart monitors and ventilators.

• Pacemakers and hearing aids, 1/10 the size of current products on the market.

• Microscopic biorobotics which can be injected in the body to manage, monitor, and/or destroy blood clots, metastatic activities, arterial blockages, alveoli damage due to carcinogens or pollutants, and scar tissue creating muscular or skeletal immobility.

COMRMed is a growing company with over 5,000 employees located in six cities and three states. To manage this business, supervisors and employees write on average over five letters, ten memos, fifty e-mail messages, and numerous instant messages and text messages a day.

The letters are written to many different audiences and serve various purposes. COMRMed must write letters for employee files, to customers, job applicants, outside auditors, governmental agencies involved in biotechnology regulation, insurance companies, and more. They write

- Letters of inquiry to retailers seeking product information (technical specifications, pricing, warranties, guarantees, credentials of staff, and so forth).
- Cover letters prefacing COMRMed's proposals.
- Complaint letters written to manufacturers if and when faulty equipment and materials are received in shipping and adjustment letters to compensate retailers when problems occur.

COMRMed's managers and employees also write memos to accomplish a variety of goals:

- Document work accomplished.
- Call meetings and establish meeting agendas.
- Request equipment from purchasing.
- Preface internal proposals.

To accomplish the majority of their routine correspondence, COMRMed's employees write many e-mail messages each day. These messages serve different purposes. Some e-mail messages are conversational. Other e-mail messages, however, must be professional in their style, organization, and content. This is especially true for e-mail messages sent to clients, vendors, and customers outside the company. These e-mail messages focus on timelines, deadlines, prices for service, meeting arrangements, cost breakdowns, procedural steps, and a host of other topics.

Finally, when COMRMed employees are working at distant locations, on the road, in hotels, or at the airport, they use instant messages or text messages to ask each other quick questions or to casually check up on the status of a project.

Routinely, COMRMed employees spend a great deal of their time writing memos, letters, e-mail messages, instant messages, and text messages.

The Importance of Routine Correspondence

1. Understand the importance of memos, letters, e-mail, instant messages, and text messages.

On a day-to-day basis, employees routinely write memos, letters, e-mail messages, instant messages, reports, and text messages. The National Commission on Writing, in their *Writing: A Ticket to Work ... Or a Ticket Out, A Survey of Business Leaders*, states that e-mail is common in the American workplace. Over half of the companies surveyed reported that communication channels "frequently" or "almost always" used included technical reports (59 percent), formal reports (62 percent), memos, and routine correspondence (70 percent). Figure 9.1 shows the significance of e-mail, memos, reports, and letters in the workplace.

FIGURE 9.1   Percentage of Employees Who Call E-mail, Memos, Letters, and Reports "Extremely Important"

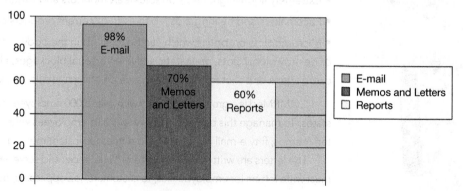

This survey of 120 American corporations with over 8 million employees states that you routinely can expect to write e-mail messages, letters, memos, and reports on the job.

With the growing importance of mobile communication, instant messages and text messages also are written routinely on the job. According to Daniel Sieberg, reporter for CBS News, the number one use of smartphones is e-mail and texting. Talking on the phone actually ranks as the third function.

Which Communication Channel Should You Use?

2. Determine which communication channel to use for routine correspondence.

Memos, letters, and e-mail messages are three common types of communication channels. Other communication channels include reports, Web sites, blogs, PowerPoint presentations, oral communication, instant messages, text messages, and more. When should you write an e-mail message instead of a memo? When should you write a memo instead of a letter? Is an instant message or a text message appropriate to the situation? You will make these decisions based on your audience (internal or external), the complexity of your topic, the speed with which your message can be delivered, and security concerns.

For example, e-mail is a convenient communication channel. It is easy to write a short e-mail message, which can be sent almost instantaneously to your audience at the click of a button. However, e-mail might not be the best communication channel to use. If you are discussing a highly sensitive topic such as a pending merger, corporate takeover, or layoffs, an e-mail message would be less secure than a letter sent in a sealed envelope. You might need to communicate with employees working in a manufacturing warehouse. Not all of these employees will necessarily have an office or access to a computer. If you sent an e-mail message, how would they access this correspondence? A memo posted in the break room or a text message sent to their handhelds would be better choices of communication channels.

Communication Channels
See Chapter 1, "An Introduction to Technical Communication," for more information about communication channels.

The Differences Among Routine Correspondence Channels

3. Distinguish among the different routine communication channels.

To clarify the distinctions among memos, letters, e-mail, instant messages (IM), and text messages (TM), review Table 9.1.

TABLE 9.1 Memos vs. Letters vs. E-mail vs. Instant Messages and Text Messages

Characteristics	Memos	Letters	E-mail	IM/TM
Destination	Internal: correspondence written to colleagues within a company.	External: correspondence written to people outside the business.	Internal *and* external: correspondence written to friends and acquaintances, coworkers within a company, and clients and vendors.	Internal *and* external: correspondence written to friends and acquaintances, coworkers within a company, and clients and vendors.
Format	Identification lines "Date," "To," From," and "Subject." The message follows.	Includes letterhead address, date, reader's address, salutation, text, complimentary close, and signatures.	This includes the following: "To" and "Subject." "Date" and "From" are computer generated. Options include complimentary copy (cc), forwarding the e-mail message to others, and replying to numerous readers.	No format.

(Continued)

TABLE 9.1   (*Continued*)

Characteristics	Memos	Letters	E-mail	IM/TM
Audience	Generally high tech or low tech, mostly business colleagues.	Generally low-tech and lay readers, such as vendors, clients, stakeholders, and stockholders.	E-mail messages are written to multiple audiences, both internal and external.	IM and TM messages are written to multiple audiences, both internal and external.
Topic	Generally topics related to internal corporate information; abbreviations and acronyms often allowed.	Generally topics related to vendor, client, stakeholder, and stockholder interests; abbreviations and acronyms usually defined.	A wide range of diverse topics determined by the audience.	A wide range of diverse topics determined by the audience.
Complexity and Length of Communication	Memos usually are limited to a page of text. If you need to write longer correspondence and develop a topic in more detail, you might consider using a different communication channel, such as a short report.	Letters usually are limited to a page of text, though you might write a two- or three-page report using a letter format. If you need to develop a topic in greater detail than can be conveyed in one to three pages, you might want to use a different communication channel, such as a longer, formal report.	An effective e-mail message usually is limited to one viewable screen (requiring no scrolling) or two screens. E-mail, generally, is not the best communication channel to use for complex information or long correspondence. If your topic demands more depth than can be conveyed in a screen or two, you might want to write a report instead.	IM and TM are very brief communication channels with a limited number of characters—under 200.
Tone	Informal due to peer audience.	More formal due to audience of vendors, clients, stakeholders, and stockholders.	A wide range of tones due to diverse audiences. Usually informal when written to friends, informal to coworkers, more formal to management or external readers.	Very informal due to the limits placed on size.
Attachments or Enclosures	Hard-copy attachments: can be stapled to the memo. Complimentary copies (cc) can be sent to other readers.	Additional information can be enclosed within the envelope. Complimentary copies (cc) can be sent to other readers.	Computer word processing files, HTML files and Web links, PDF files, RTF files, or downloadable graphics can be attached to e-mail. Complimentary copies can be sent to other readers.	Attachments can be sent using IM. You can send photo attachments with TM. However, sending file attachments with TM is limited.
Delivery Time	Determined by a company's in-house mail procedure.	Determined by the destination (within the city, state, or country). Letters could be delivered within three days but may take more than a week.	Often instantaneous, usually within seconds. Delays can be caused by system malfunctions or excessively large attachments.	Instantaneous.
Security	If a company's mail delivery system is reliable, the memo will be placed in the reader's mailbox. Then, what the reader sees on the hard-copy page will be exactly what the writer wrote. Security depends on the ethics of coworkers and whether the memo was sent in an envelope.	The U.S. Postal Service is very reliable. Once the reader opens the envelope, he or she sees exactly what was written. Privacy laws protect the letter's content.	E-mail systems are not secure. E-mail can be tampered with, read by others, and sent to many people. E-mail stays within a company's computer backup system and is the property of the company. Therefore, e-mail is not private.	The same problems with e-mail security exist with TM and IM.

Reasons for Writing Memos, Letters, and E-mail

4. Know why to write memos, letters, and e-mail.

You will write memos, letters, and e-mail to a wide range of readers. This includes your supervisors, coworkers, subordinates, vendors, stakeholders, customers, and multiple combinations of these audiences. Because of their frequency and widespread audiences, routine correspondence could represent a major component of your interpersonal communication skills within your work environment. Furthermore, memos, letters, and e-mail are flexible and can be written for many different purposes:

- **Documentation.** Expenses, incidents, accidents, problems encountered, projected costs, study findings, hiring, and reallocations of staff or equipment.
- **Confirmation.** A meeting agenda, date, time, and location; decisions to purchase or sell; topics for discussion at upcoming teleconferences; conclusions arrived at; and fees, costs, or expenditures.
- **Procedures.** How to set up accounts, operate new machinery, use new software, or solve a problem.
- **Recommendations.** Reasons to purchase new equipment, promote someone, award a contract to a vendor, or develop new software applications.
- **Feasibility.** Changes in the workplace (practices, procedures, locations, staffing, equipment, missions/visions).
- **Status.** Daily, weekly, monthly, quarterly, biannual, and yearly statements about where you, the department, or the company is regarding many topics (sales, staffing, travel, practices, procedures, or finances).
- **Directive (delegation of responsibilities).** Assignments to subordinates of their designated tasks.
- **Inquiry.** Questions about upcoming processes or procedures.
- **Cover.** The preface to a proposal, long report, or other attachments.

FAQs: Why write a letter?

Q: Haven't letters been replaced by e-mail?

A: Though e-mail is quick, it might not be the best communication channel for the following reasons:

1. E-mail might be too quick. In the workplace, you will write about topics that require detailed analysis. Because e-mail messages can be written and sent quickly, people too often write hurriedly and neglect to consider the impact of the message.

2. E-mail messages tend to be casual, conversational, and informal. Not all correspondence, however, lends itself to this level of informality. Formal correspondence related to contracts, for example, requires the more formal communication channel of a letter. The same applies to audience. You might want to write a casual e-mail to a coworker, but if you were writing to the president of a company, the mayor of a city, or a foreign dignitary, a letter would be a better, more formal choice of communication channel.

3. E-mail messages tend to be short. For content requiring more detail, a longer letter would be a better choice.

4. We get so many e-mail messages a day that they are easy to disregard—even easy to delete. Letters carry more significance. If you want to ensure that your correspondence is read and perceived as important, you might want to write a letter instead of an e-mail.

5. Letters allow for a paper trail. Most employees' e-mail inboxes fill up quickly. To clean up these inboxes, people tend to delete messages that they don't consider important. In contrast, hard-copy letters provide wonderful documentation.

Using an All-Purpose Template for Memos, Letters, and E-mail

Memos, letters, and e-mail contain the following key components:

- Introduction
- Discussion
- Conclusion

Figure 9.2 shows an ideal, all-purpose organizational template that works well for routine correspondence.

FIGURE 9.2 All-Purpose Template

> Introduction: Begin with a lead-in or overview stating *why* you are writing and *what* you are writing about.

> Discussion: Include detailed development, made accessible through highlighting techniques, explaining *exactly what* you want to say.
> *
> *
> *

> Conclusion: State *what* is next, *when* this will occur, and *why* the date is important.

Introduction

Get to the point in the introductory sentence(s). Write one or two clear introductory sentences which tell your readers *what* topic you are writing about and *why* you are writing. The following example invites the reader to a meeting, thereby communicating *what* the writer's intentions are. It also tells the reader that the meeting is one of a series of meetings, thus communicating *why* the meeting is being called.

EXAMPLE ▶

> In the third of our series of sales quota meetings this quarter, I'd like to review our productivity.

Discussion

The discussion section allows you to develop your content specifically. Readers might not read every line of your routine correspondence (tending instead to skip and skim). Thus, traditional blocks of data (paragraphing) are not necessarily effective. The longer the paragraph, the more likely your audience is to avoid reading. Make your text more reader-friendly by itemizing, using white space, boldfacing, creating headings, or inserting graphics.

BEFORE	AFTER
This year began with an increase, as we sold 4.5 million units in January. In February we continued to improve with 4.6. March was not quite so good, as we sold 4.3. April was about the same with 4.2. May's sales increased to 5.6. June was our best month at 6 million units sold.	Monthly Sales (in millions of units) • January: 4.5 • February: 4.6 • March: 4.3 • April: 4.2 • May: 5.6 • June: 6.0

Conclusion

Conclude your correspondence with "thanks" and/or directive action. A pleasant conclusion could motivate your readers, as in the following example. A directive close tells your readers exactly what you want them to do next or what your plans are (and provides dated action).

> Congratulations! If our quarterly sales continue to improve at the current rate, we will double our sales expectations by 2016. Next Wednesday (12/22/16), please provide next quarter's sales projections and a summary of your sales team's accomplishments.

◀ EXAMPLE

TECHNOLOGY TIPS

Using Memo and Letter Templates in Microsoft Word 2013

1. Click on **File** and **New**.

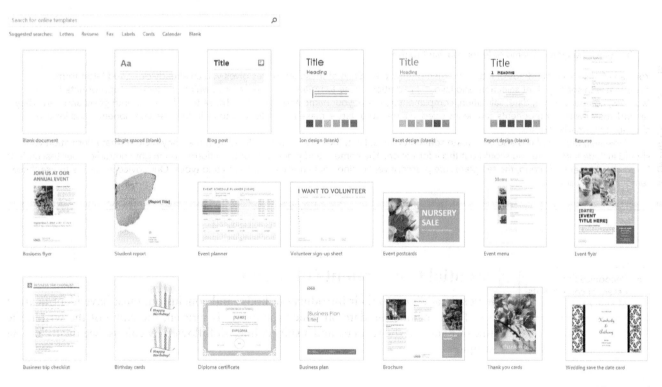

(Continued)

2. In the search field, type the kind of document you want to write, such as Letters or Memos.

When you choose the communication channel, the following window will pop up.

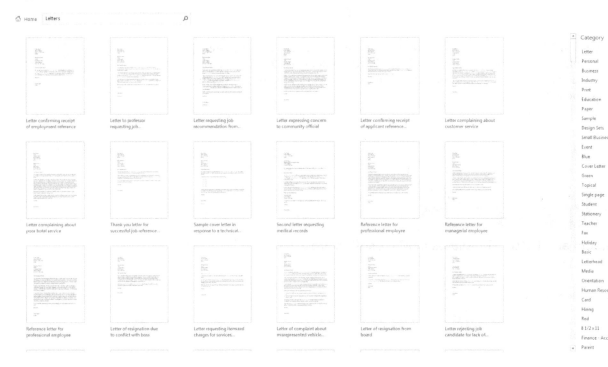

3. Select the type of memo or letter you want to use.

You can choose from hundreds of letter and memo templates. Each of these templates gives you an already designed letter format, complete with spacing, font selection, and layout. In addition, these templates provide fields in which you type the appropriate information (address, company name, date, salutation, complimentary close, your name and title, etc.). These templates are both good and bad. They remind you of which components can be included in a memo or letter, make it easy for you to include these components, and let you choose ready-made formats.

However, the templates also can create some problems. First, they are somewhat limiting in that they dictate what information you should include and where you should put this information. The content and placement of this information might contradict your teacher's or boss's requirements. Second, the templates are prescriptive, limiting your choice of font sizes and types. Our advice would be to use these templates with caution.

Microsoft Corporation, Inc.

| 6. Recognize the essential components of memos. |

Essential Components of Memos

Memos, in addition to their introduction, body, and conclusion, must have identification lines. These include the date on which the memo is written, the name of the writer, the name or names of the readers, and a subject line, all followed by colons. See the following example:

Date: March 15, 2016
To: John Staples
From: Marilyn Andrews
Subject: Sales Projections for Fiscal Year 2017

The subject line is especially important because it summarizes the memo's content. One-word subject lines do not communicate effectively, as in the following flawed subject line. The "Before" sample has a *topic* (a what) but is missing a *focus* (a what about the what).

BEFORE	AFTER
Subject: Sales	Subject: Report on Quarterly Sales

The Memo Checklist gives you the opportunity for self-assessment and peer evaluation of your writing.

MEMO CHECKLIST

_____ 1. Does the memo contain identification lines (Date, To, From, and Subject)?

_____ 2. Does the subject line contain a topic and a focus?

_____ 3. Does the introduction clearly state
  • Why this memo has been written?
  • What topic the memo is discussing?

_____ 4. Does the body explain exactly what you want to say?

_____ 5. Does the conclusion
  • Tell when you plan a follow-up or when you want a response?
  • Explain why this dated action is important?

_____ 6. Are highlighting techniques used effectively for document design?

_____ 7. Is the memo concise, limiting the length of words, sentences, and paragraphs?

_____ 8. Is the memo clear,
  • Achieving specificity of detail?
  • Answering reporter's questions?

_____ 9. Does the memo recognize audience,
  • Defining acronyms or abbreviations where necessary for various levels of readers (high tech, low tech, and lay)?

_____ 10. Did you avoid grammatical errors? Errors will hurt your professionalism. See Appendix A for grammar rules.

Essential Components of Letters

Your letter should be typed or printed on 8½" × 11" paper. Leave 1" to 1½" margins at the top and on both sides. Choose an appropriately businesslike font (size and style), such as Times New Roman or Arial (12 point). Though "designer fonts," such as Comic Sans and Shelley Volante, are interesting, they tend to be harder to read and less professional. Your letter should contain the essential components shown in Figure 9.3.

> 7. Recognize the essential components of letters.

> **Font and Readability**
> See Chapter 4, "Objectives in Technical Communication," for more discussion of font selection and readability.

FIGURE 9.3   Essential Letter Components

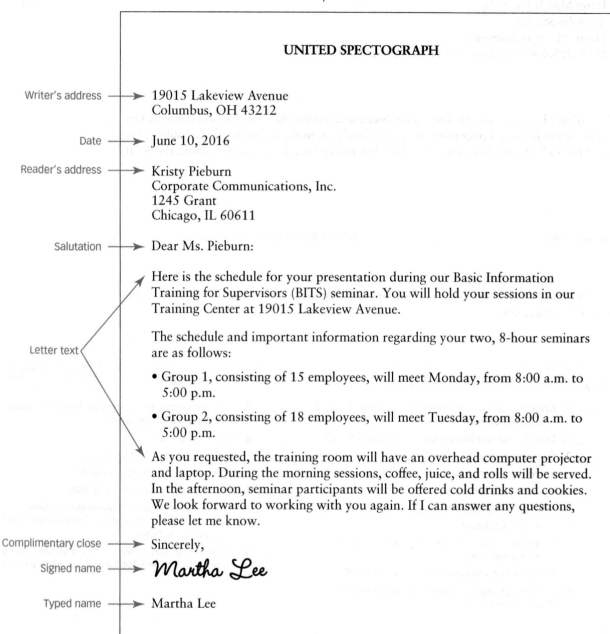

UNITED SPECTOGRAPH

Writer's address → 19015 Lakeview Avenue
Columbus, OH 43212

Date → June 10, 2016

Reader's address → Kristy Pieburn
Corporate Communications, Inc.
1245 Grant
Chicago, IL 60611

Salutation → Dear Ms. Pieburn:

Letter text →

Here is the schedule for your presentation during our Basic Information Training for Supervisors (BITS) seminar. You will hold your sessions in our Training Center at 19015 Lakeview Avenue.

The schedule and important information regarding your two, 8-hour seminars are as follows:

- Group 1, consisting of 15 employees, will meet Monday, from 8:00 a.m. to 5:00 p.m.

- Group 2, consisting of 18 employees, will meet Tuesday, from 8:00 a.m. to 5:00 p.m.

As you requested, the training room will have an overhead computer projector and laptop. During the morning sessions, coffee, juice, and rolls will be served. In the afternoon, seminar participants will be offered cold drinks and cookies. We look forward to working with you again. If I can answer any questions, please let me know.

Complimentary close → Sincerely,

Signed name → *Martha Lee*

Typed name → Martha Lee

Writer's Address

This section contains either your personal address or your company's address. If the heading consists of your address, include your street address, city, state, and zip code. The state may be abbreviated with the appropriate two-letter abbreviation. If the heading consists of your company's address, include the company's name, street address, city, state, and zip code.

Date

Document the month, day, and year when you write your letter. You can write your date in one of two ways: May 31, 2016, or 31 May 2016. Place the date one or two spaces below the writer's address.

Reader's Address

Place the reader's address two lines below the date.

- Reader's name (If you do not know the name of this person, begin the reader's address with a job title or the name of the department.)
- Reader's title (optional)
- Company name
- Street address
- City, state, and zip code

Salutation

The traditional salutation, placed two spaces beneath the inside address, is *Dear* and your reader's last name, followed by a colon (Dear Mr. Smith:). You can also address your reader by his or her first name if you are on a first-name basis with this person (Dear John:). If you are writing to a woman and are unfamiliar with her marital status, address the letter Dear Ms. Jones. However, if you know the woman's marital status, you can address the letter accordingly: Dear Miss Jones *or* Dear Mrs. Jones.

Letter Body

Begin the body of the letter two spaces below the salutation. The body includes your introductory paragraph, discussion paragraph(s), and concluding paragraph. The body should be single-spaced with double spacing between paragraphs. Whether you indent the beginning of paragraphs or leave them flush with the left margin is determined by the letter format you employ.

Complimentary Close

Place the complimentary close, followed by a comma, two spaces below the concluding paragraph. The typical complimentary close is "Sincerely."

Signed Name

Sign your name legibly beneath the complimentary close.

Typed Name

Type your name four spaces below the complimentary close. You can type your title one space beneath your typed name. You also can include your title on the same line as your typed name with a comma after your name.

Optional Components of Letters

In addition to the letter essentials, you can include the following optional components.

Subject Line. Place a subject line two spaces below the inside address and two spaces above the salutation.

Dr. Ron Schaefer
Linguistics Department
Southern Illinois University
Edwardsville, IL 66205

Subject: Linguistics Conference Registration Payment

Dear Dr. Schaefer:

You also could use a subject line instead of a salutation.

Linguistics Department
Southern Illinois University
Edwardsville, IL 66205

Subject: Linguistics Conference Registration Payment

A subject line not only helps readers understand the letter's intent but also (if you are uncertain of your reader's name) helps you avoid such awkward salutations as "To Whom It May Concern," "Dear Sirs," and "Ladies and Gentlemen." In the simplified format, both the salutation and the complimentary close are omitted, and a subject line is included.

New Page Notations.   If your letter is longer than one page, cite your name, the page number, and the date on all pages after page 1. Place this notation either flush with the left margin at the top of subsequent pages or across the top of subsequent pages. (You must have at least two lines of text on the next page to justify another page.)

Left, margin, subsequent page notation	Across top of subsequent pages
Mabel Tinjaca Page 2 May 31, 2016	Mabel Tinjaca Page 2 May 31, 2016

Writer's and Typist's Initials.   If the letter was typed by someone other than the writer, include both the writer's and the typist's initials two spaces below the typed signature. The writer's initials are capitalized, the typist's initials are typed in lowercase, and the two sets of initials are separated by a colon. If the typist and the writer are the same person, this notation is not necessary.

Sincerely,

W. T. Winnery

W. T. Winnery

WTW:mm

Enclosure Notation.   If your letter prefaces enclosed information, such as an invoice or report, mention this enclosure in the letter and then type an enclosure notation two spaces below the typed signature (or two spaces below the writer and typist initials). The enclosure notation can be abbreviated "Enc."; written out as "Enclosure"; show the number of enclosures, such as "Enclosures (2)"; or specify what has been enclosed—"Enclosure: January Invoice."

Copy Notation.   If you have sent a copy of your letter to other readers, show this in a copy notation. A complimentary copy is designated by a lowercase "cc." List the other readers' names following the copy notation. Type the copy notation two spaces below the typed signature or two spaces below either the writer's and typist's initials or the enclosure notation.

◀ EXAMPLE

> Sincerely,
> *Brian Altman*
> Brian Altman
>
> Enclosure: August Status Report
>
> cc: Marcia Rittmaster and Larry Rochelle

Letter Formats

Three common types of letter formats include *full block, full block with subject line,* and *simplified.* Two popular and professional formats used in business are full block and full block with subject line. With both formats, you type all information at the left margin without indenting paragraphs, the date, the complimentary close, or signature. The full block with subject line differs only with the inclusion of a subject line.

Another option is the simplified format. This type of letter layout is similar to the full block format in that all text is typed margin left. The two significant omissions include no salutation ("Dear _____:") and no complimentary close ("Sincerely,"). Omitting a salutation is useful in the following instances:

- You do not know your reader's name (NOTE: Avoid the trite salutation, "To Whom It May Concern:").
- You are writing to someone with a nongender-specific name (Jesse, Terry, Stacy, Chris, etc.) and you do not know whether to use "Mr.," "Mrs.," or "Ms."

The Administrative Management Society (AMS) suggests that if you omit the salutation, you also should omit the complimentary close. Some people feel that omitting the salutation and the complimentary close will make the letter cold and unfriendly. However, the AMS says that if your letter is warm and friendly, these omissions will not be missed. More importantly, if your letter's content is negative, beginning with "Dear" and ending with "Sincerely" will not improve the letter's tone or your reader's attitude toward your comments. The simplified format includes a subject line to aid the letter's clarity.

> 8. Identify three common letter formats.

EXAMPLE ▶

1–1½" margins on all sides of the letter

2 spaces above and below the date

2 spaces above and below the salutation

Single-space within the paragraphs.

Double-space between the paragraphs.

2 spaces before "Sincerely"

4 spaces between "Sincerely" and the typed signature

State Health Department
1890 Clark Road
Jefferson City, MO 67220

June 6, 2016

Dale McGraw, Manager
Elmwood Mobile Home Park
Elmwood, MO 64003

Dear Mr. McGraw:

On April 19, 2016, Ryan Duran and I, environmental specialists from the Health Department, conducted an inspection of the Elmwood Mobile Home Park Wastewater Treatment Facility. The purpose was to assess compliance with the following: the state's Clean Water Law, Clean Water Commission regulations, and your facility's plan for pollution control. The inspection also would allow the state to promote proper operation of Wastewater Facilities and to provide technical assistance where needed to the Elmwood Mobile Homes management.

Though the Elmwood Mobile Home pollution control plan had expired in 2015, a consent judgment was issued by the state's Attorney General's Office. The county court stipulated a timeline for correction by connection to an available sewer system. Your mobile home park's wastewater system has continually discharged to the Little Osage River. A copy of the abatement order, which requires that monthly discharge monitoring reports (DMRs) be submitted by the 28th of the month following the reporting periods, is attached. All DMRs for the previous twelve months have been received, and reported pollution parameters are not within limits. Due to the plant's performance, the stream was placed on the 2014 303 (d) stream for impairment by the Elmwood Mobile Home.

As part of the inspection, a review of the facility's DMR was conducted. Twenty-four-hour composite samples were collected using a composite sampler. Attached are the results of the 24-hour composite samples collected on April 20, 2016. Every one of the problems documented is an infraction that must be addressed.

Within 30 days of receipt of this letter, please submit to the Health Department written documentation describing steps taken to correct each of the concerns identified in the attachments. Also include engineering reports, and submit a timeframe to eliminate the problems. Thank you for your cooperation.

Sincerely,

Harvey Haddix

Harvey Haddix
Environmental Manager

Attachment

The Letters Checklist will give you the opportunity for self-assessment and peer evaluation of your writing.

LETTERS CHECKLIST

_____ 1. Does your letter include the eight essential components (writer's address, date, reader's address, salutation, text, complimentary close, writer's signed name, and writer's typed name)?

_____ 2. Does the introduction state _what_ you are writing about and _why_ you are writing?

_____ 3. Does your discussion clearly state the details of your topic depending on the type of letter?

_____ 4. Is your text accessible? To achieve reader-friendly ease of access, use headings, boldface, italics, bullets, numbers, underlining, or graphics (tables and figures). These add interest and help your readers navigate your letter.

_____ 5. Have you helped your readers follow your train of thought by using appropriate modes of organization? These include chronology, importance, problem/solution, or comparison/contrast.

_____ 6. Does your conclusion give directive action (tell what you want the reader to do next and when) and end positively?

_____ 7. Is your letter clear, answering reporter's questions and providing specific details that inform, instruct, or persuade?

_____ 8. Have you limited the length of your words, sentences, and paragraphs?

_____ 9. Have you written appropriately to your audience? This includes avoiding biased language, considering the multicultural/cross-cultural nature of your readers, and your audience's role (supervisors, subordinates, coworkers, customers, or vendors). Have you created a positive tone to build rapport?

_____ 10. Is your text grammatically correct? Errors will hurt your professionalism. See Appendix A for grammar rules.

FAQs: Professionalism in E-mail Messages

Q: Aren't e-mail messages just casual communication? Isn't it easy to write an e-mail message since you don't have to worry about grammar or correct style?

A: Nothing could be further from the truth. E-mail might be your major means of communication in the workplace. Therefore, you must pay special attention to correctness.

Listen to what managers at an engineering company say about e-mail messages:

- "Most workplace communication is now via e-mail. Business e-mail needs to be almost as formal and as carefully written as a letter because it is a formal and legal document. Never send an e-mail that you would not be comfortable seeing on the front page of a newspaper, because some day you may."

- "I see more and more new hires wanting to rely on e-mail. It is a totally ineffective way to resolve many issues on an engineering project. But they seem to feel it is OK for almost any communication. I suspect the general acceptance by their peers for this form of communication has led them to mistakenly assume the same is true for a business setting."

- "Many people tend to be very 'social' in e-mails. Your employer owns your e-mails written on your work computers. They are NOT private. They can be used not only against you, but against your firm in court. For example, if I send an e-mail to a coworker that states in it somewhere what a lousy job Frank is doing on the such-and-such project and that project goes bad, it is possible that e-mail could end up in court and be used against my employer. In my mind all I was doing was venting my frustrations to an understanding friend and coworker. But, in reality, I am creating a permanent record of anything I say."

(Gerson, et al. "Core Competencies.")

Essential Components of E-mail

To convey your messages effectively and to ensure that your e-mail messages reflect professionalism, follow these tips for writing e-mail.

Identify Yourself

Identify yourself by name, affiliation, or title. You can accomplish this either in the "From" line of your e-mail or by creating a signature file or .sig file. This .sig file acts like an online business card. Once this identification is complete, readers will be able to open your e-mail without fear of corrupting their computer systems.

Provide an Effective Subject Line

Readers are unwilling to open unsolicited or unknown e-mail, due to fear of spam and viruses. In addition, corporate employees receive approximately 50 e-mail messages each day. They might not want to read every message sent to them. To ensure that your e-mail messages are read, avoid uninformative subject lines, such as "Hi," "What's New," or "Important Message." Instead, include an effective subject line, such as "Subject: Meeting Dates for Tech Prep Conference."

Keep Your E-mail Message Brief

Readers skim and scan. To help them access information quickly, limit your message to one screen (if possible).

Organize Your E-mail Message

Successful writing usually contains an introductory paragraph, a discussion paragraph or paragraphs, and a conclusion. Although many e-mail messages are brief, only a few sentences, you can use the introductory sentences to tell the reader why you are writing and what you are writing about. In the discussion, clarify your points thoroughly. Use the concluding sentences to tell the reader what is next, possibly explaining when a follow-up is required and why that date is important.

Use Highlighting Techniques Sparingly

Many e-mail platforms will let you use highlighting techniques, such as boldface, italics, underlining, computer-generated bullets and numbers, centering, font color highlighting, and font color changes. Many other e-mail platforms will not display such visual enhancements. To avoid having parts of the message distorted, limit your highlighting to bullets, numbers, double spacing, and headings.

Be Careful When Sending Attachments

When you send attachments, tell your reader within the body of the e-mail message that you have attached a file. Specify the file name of your attachment and the software application that you have used (HTML, PowerPoint, PDF, RTF [rich text format], Word, or Works); and use compression (zipped) files to limit your attachment size.

Practice Netiquette

When you write your e-mail messages, observe the rules of "netiquette."

- **Be courteous.** Do not let the instantaneous quality of e-mail negate your need to be calm, cool, deliberate, and professional.
- **Be professional.** Occasionally, e-mail writers compose excessively casual e-mail messages. They will lowercase a pronoun like "i," use ellipses (. . .) or dashes

instead of more traditional punctuation, use instant messaging shorthand language or "textese" such as "LOL" or "BRB," and depend on emoticons (☺ ☻). These e-mail techniques might not be appropriate in all instances. Don't forget that your e-mail messages represent your company's professionalism. Write according to the audience and communication goal.

- **Avoid abusive, angry e-mail messages.** Because of its quick turnaround abilities, e-mail can lead to negative correspondence called *flaming*. Flaming is sending angry e-mail, often TYPED IN ALL CAPS. Readers can perceive the all caps as yelling, such as in the following example.

◀ EXAMPLE

Subject: EXCESSIVE PRINTER PAPER

    I HAVE TALKED WITH SEVERAL PEOPLE AND THIS SEEMS TO BE A PROBLEM IN YOUR DEPARTMENT. SOMEONE PRINTS INFORMATION AND WON'T PICK IT UP AT THE PRINTER. THEN THE NEXT PERSON HAS TO SORT THROUGH THE PRINTED MATERIAL TO FIND WHAT HE OR SHE WANTS. SOME PRINTOUTS ARE NEVER USED; THEY JUST SIT THERE FOR DAYS, GETTING IN OTHER PEOPLE'S WAY. PEOPLE SHOULD JUST PICK UP THEIR PRINTING AND GET IT OUT OF EVERYONE ELSE'S WAY. THAT'S ONLY COMMON COURTESY. BUT THE PEOPLE IN YOUR DEPARTMENT AREN'T EVEN REMOTELY CONSIDERATE OF OTHERS. IF YOU MANAGEMENT PEOPLE WOULD JUST DO YOUR JOBS, NONE OF THIS WOULD HAPPEN.

The following is an example of an unprofessional e-mail message.

◀ EXAMPLE

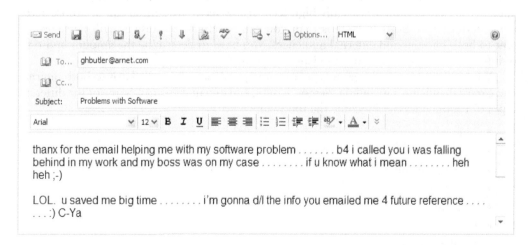

This e-mail is flawed due to its use of instant messaging abbreviations, lowercase text, ellipses (. . .), slang, and emoticons. Though these techniques might be appropriate for personal communication, avoid overly casual e-mail messages for job-related communication.

Use the e-mail checklist to evaluate your correspondence.

E-MAIL CHECKLIST

_____ 1. Does the e-mail use the correct address?

_____ 2. Have you identified yourself? Provide a sig (signature) line.

_____ 3. Did you provide an effective subject line? Include a *topic* and a *focus*.

_____ 4. Have you effectively organized your e-mail? Consider including the following:

- Opening sentence(s) telling *why* you are writing and *what* you are writing about.
- Discussion unit with itemized points telling *what exactly* the e-mail is discussing.
- Concluding sentence(s) *summing up* your e-mail message or telling your audience what to do next.

_____ 5. Have you used highlighting techniques sparingly?

- Avoid boldface, italics, color, or underlining.
- Use bullets, numbers, headings, and double spacing for access.

_____ 6. Did you practice netiquette?

- Be polite, courteous, and professional.
- Don't flame.
- Don't overuse textese.

_____ 7. Is the e-mail concise, limiting the length of words, sentences, and paragraphs? Have you limited your e-mail to one or two screens?

_____ 8. Did you identify and limit the size of attachments?

- Tell your reader(s) if you have attached files and what types of files are attached (PPT, PDF, RTF, Word, etc.).
- Zip large files.

_____ 9. Does the e-mail recognize audience?

- Define acronyms or abbreviations where necessary.
- Consider a diverse audience (factoring in multiculturalism or gender).

_____ 10. Did you avoid grammatical errors? Grammar errors will negate your professionalism. See Appendix A for grammar and mechanics rules.

---

10. Follow criteria for writing routine correspondence.

Audience

See Chapter 5, "Audience Recognition," for more discussion of audience types.

Criteria for Different Types of Routine Correspondence

Choosing whether or not to write a memo, letter, or e-mail message will be determined by your audience. Since letters go outside your company, the audience is usually a low-tech or lay reader, demanding that you define your terms specifically. In a memo, your in-house audience is easy to address (usually high tech or low tech). You often can use more acronyms and abbreviations in memos than you can in letters. Because e-mail messages are sent to both internal and external audiences, you have to determine the amount and type of information based on your knowledge of the audience.

You will write many different types of correspondence on a daily basis to both internal and external audiences including memos, letters, and e-mail messages. Some of the different types of routine messages include the following:

- Inquiry
- Response
- Cover (transmittal)
- Complaint
- Adjustment
- Order
- Confirmation
- Recommendation
- Thank-you

Inquiry

If you want information about degree requirements, equipment costs, performance records, turnaround time, employee credentials, or any other matter of interest to you or your company, you can request that information in a letter, memo, or e-mail. Inquiries demand specificity. For example, if you write, "Please send me any information you have on your computer systems," you are in trouble. You will either receive any information the reader chooses to give you or none at all.

Introduction.   Clarify your intent in the introduction. Until you tell your readers why you are writing, they do not know. It is your responsibility to clarify your intent and explain your rationale for writing. Also tell your reader immediately what you are writing about (the subject matter of your inquiry).

Discussion.   Specify your needs in the discussion. To ensure that you get the response you want, ask precise questions or list specific topics of inquiry. You must quantify. For example, rather than vaguely asking about machinery specifications, you should ask more precisely about "specifications for the 12R403B copier." Rather than asking, "Will the roofing material cover a large surface?" you need to quantify: "Will the roofing material cover 150" × 180"?"

Conclusion.   Conclude precisely. First, explain when you need a response. Do not write, "Please respond as soon as possible." Provide dated action and tell the reader exactly when you need your answers. Second, to sell your readers on the importance of this date, explain why you need answers by the date given. Figure 9.4 is an effective sample inquiry letter.

Response

In a response, you provide information, details, or answers to an inquiry. For example, you or your company might need to answer questions about quotes on equipment costs, maintenance fees, delivery options, and technical specifications on makes and models from various vendors. You might need to write a response with information about room availability at your hotel, food arrangements, technology and presentation equipment rental fees, presentation room setups, entertainment options, and transportation from the airport. Maybe a client needs to change its insurance carrier. Your insurance company responds to the inquiry with quotes for insurance premium costs, levels of coverage, deductibles, and accessibility of claims adjusters.

Introduction.   Begin with a pleasant reminder of when you spoke with a person or heard from the audience. This explains *why* you are writing. Then, specifically state *what* topic you are writing about.

Discussion.   Organize your discussion section into as many paragraphs as you need. When possible, remember to use bulleted or numbered lists for easier access. Include in this section the details or explanations needed. Consider including any of the following:

Times	Types of activities	Enrollment periods
Dates	Discounts	Enclosures
Amounts of people	Costs	Technology or equipment

FIGURE 9.4   Letter of Inquiry

COMRMed

8713 Hillview   Reno, NV 32901   1-800-551-9000   Fax: 1-816-555-0000
September 12, 2016

Sales Manager
OfficeToGo
7622 Raintree
St. Louis, MO 66772

Subject: Request for Product Pricing and Shipping Schedules

My medical technology company has worked well with OfficeToGo for the past five years. However, in August I received a letter informing me that OTG had been purchased by a larger corporation. I need to determine if OTG remains competitive with other major office equipment suppliers in the Reno area.

Please provide the following information:

1. What discounts will be offered for bulk purchases?
2. Which freight company will OTG now be using?
3. Who will pay to insure the items ordered?
4. What is the turnaround time from order placement to delivery?
5. Will OTG be able to deliver to all my satellite sites?
6. Will OTG technicians set up the equipment delivered, including desks, file cabinets, bookshelves, and chairs?
7. Will OTG be able to personalize office stationery onsite, or will it have to be outsourced?

Please respond to these questions by September 30 so I can prepare my quarterly orders in a timely manner. I continue to expand my insurance agency and want assurances that you can fill my growing office supply needs. You can contact me at the phone number provided above or by e-mail (jgood@COMRMed.com).

Jim Goodwin

Jim Goodwin
Owner and CEO

In the introduction, briefly explaining why you are writing establishes the context of the inquiry for the audience.

A detailed and itemized list informs the audience the exact questions you need answered. If you write precise questions, the audience will be able to provide the information you are requesting.

Pleasantly state when and why you need a response by a specific date to encourage your audience. In the conclusion, you can also include detailed contact information not provided in the letterhead.

Conclusion.   End your response in an upbeat and friendly tone. You can also include your contact information (e-mail, phone number, address). See Figure 9.5 for a sample response e-mail.

Cover (Transmittal)

Cover or transmittal correspondence precedes attached or enclosed documents, informing the reader by giving an overview of the material that follows. In business, you are often required to send information to a client, vendor, or colleague. You might send the following kinds of attachments or enclosures prefaced by a cover letter, memo, or e-mail:

Reports	Invoices	Drawings
Maps	Contracts	Specifications
Instructions	Questionnaires	Proposals

FIGURE 9.5 E-mail Response

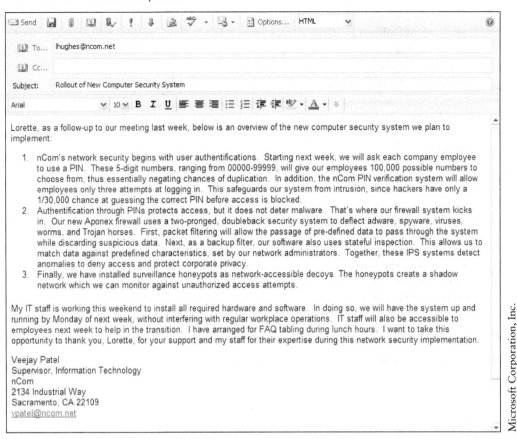

Anticipate questions the reader might have by including as much detail as possible. Using a list rather than sentences or paragraphs helps highlight key points.

Microsoft Corporation, Inc.

Cover correspondence accomplishes two goals: It tells readers up front what they are receiving and focuses your readers' attention on key points within the enclosures.

Introduction. What if the reader has asked you to send the documentation? Do you still need to explain why you are writing? The answer is yes. Although the reader requested the information, time has passed, other correspondence has been written, and your reader might have forgotten the initial request. Introductory sentences provide the reader information about why you are writing and what you are sending.

Discussion. In the body, accomplish two things. Either tell your reader exactly what you have enclosed or exactly what of value is within the enclosures. In both instances, you should provide an itemized list or easily accessible, short paragraphs. Page numbers are a friendly gesture toward your audience. You are helping the reader locate the important information. You are achieving audience recognition and involvement. However, including page numbers has a greater benefit than audience involvement. These page numbers also allow you to focus your reader's attention on what you want to emphasize.

Conclusion. Your conclusion should tell your readers what you want to happen next, when you want this to happen, and why the date is important. Figure 9.6 is an example of a successful cover memo.

FIGURE 9.6   Cover Memo Prefacing Attachments

A cover memo directs the reader to the content in an attachment, such as a proposal. You can summarize the most important parts of the proposal, providing page numbers. In addition, your memo can include dated action so that the reader knows what to do next.

Date: November 11, 2015
To: COMRMed Management
From: Bill Baker, Human Resources Director
Subject: Information about Proposed Changes to Employee Benefits Package

As of January 1, 2016, COMRMed will change insurance carriers. This will affect all 5,000 employees' benefits packages. I have attached a proposal, including the following:

1. Reasons for changing from our current carrier   page 2
2. Criteria for our selection of a new insurance company   pages 3–4
3. Monthly cost for each employee   pages 5–6
4. Overall cost to COMRMed   page 7
5. Benefits derived from the new healthcare plan   page 8

Please review the proposal, survey your employees' responses to our suggestions, and provide your feedback. We need your input by December 1, 2015. This will give the human resources department time to consider your suggestions and work with insurance companies to meet employee needs.

Enclosure: Proposal

Tone and Word Usage

See Chapter 4, "Objectives in Technical Communication," for discussion of tone and positive word usage.

Complaint Messages

To write a complaint message, include the following:

Introduction.   In the introduction, politely state the problem. To strengthen your assertions, include supporting details, such as the following: serial numbers, dates of purchase, invoice numbers, check numbers, names of salespeople involved in the purchase, and/or receipts. When possible, include copies documenting your claims.

Discussion.   In the discussion paragraph(s), explain in detail the problems experienced. This could include dates, contact names, information about shipping, breakage information, an itemized listing of defects, and poor service.

Conclusion.   End your letter positively. You want to ensure cooperation with the vendor or customer. You also want to be courteous, reflecting your company's professionalism. Your goal should be to achieve continued rapport with your reader. In this concluding paragraph, include your contact information and the times you can best be reached.

See Figure 9.7 for a sample complaint letter to an automotive supplies company.

Adjustment Messages

Responses to complaints, also called adjustment messages, can take three different forms.

- **100 percent yes.** You could agree 100 percent with the writer of the complaint.

- **100 percent no.** You could disagree 100 percent with the writer of the complaint.

FIGURE 9.7   Complaint Letter

1234 18th Street
Galveston, TX 77001
May 10, 2016

Mr. Holbert Lang
Customer Service Manager
Gulfstream Auto
1101 21st Street
Galveston, TX 77001

Dear Mr. Lang:

On February 12, I purchased two shock absorbers in your automotive department. Enclosed are copies of the receipt and the warranty for that purchase. One of those shocks has since proved defective.

The introduction includes the date of purchase (to substantiate the claim) and the problems encountered.

I attempted to exchange the defective shock at your store on May 2. The mechanic on duty, Vernon Blanton, informed me that the warranty was invalid because your service staff did not install the part. I believe that your company should honor the warranty agreement and replace the part for the following reasons:

1. The warranty states that the shock is covered for 48 months and 48,000 miles.
2. The warranty does not state that installation by someone other than the dealership will result in warranty invalidation.
3. The defective shock absorber is causing potentially expensive damage to the tire and suspension system.

The body explains what happened, states what the writer wants done, and justifies the request. This letter develops its claim with warranty information.

I can be reached between 1 p.m. and 6 p.m. on weekdays at 763-9280 or at 763-9821 anytime on weekends. You also can email me at cdelatorre12@hotmail.com. I look forward to hearing from you. Thank you for helping me with this misunderstanding.

The letter concludes by providing contact information and an upbeat, pleasant tone.

Sincerely,

Carlos De La Torre

Carlos De La Torre

Enclosures (2)

- **Partial adjustment.** You could agree with some of the writer's complaints but disagree with other aspects of the complaint.

Table 9.2 shows you the differences among these three types of adjustment messages. Writing a 100 percent yes response to a complaint is easy. You are telling your audience what they want to hear. The challenge, in contrast, is writing a 100 percent no response or a partial adjustment. Using a *buffer statement* delays bad news in written communication and gives you an opportunity to explain your position.

TABLE 9.2   Differences Among Adjustment Messages

	100% Yes	100% No	Partial Adjustment
Introduction	State the good news.	Begin with a buffer, a comment agreeable to both reader and writer.	State the good news.
Discussion	Explain what happened and what the reader should do and/or what the company plans to do next.	Explain what happened, state the bad news, and provide possible alternatives.	Explain what happened, state the bad news, and provide possible alternatives—what the reader and/or company should do next.
Conclusion	End upbeat and positive.	Resell (provide discounts, coupons, follow-up contact names and numbers, etc.) to maintain good will.	Resell (provide discounts, coupons, etc.) to maintain good will.

Buffers to Cushion the Blow.   Use the following techniques to buffer the bad news:

- **Establish rapport with the audience through positive words to create a pleasant tone.** Instead of writing, "We received your complaint," be positive and say, "We always appreciate hearing from customers."
- **Sway your reader to accept the bad news to come with persuasive facts.** "In the last quarter, our productivity has decreased by 16 percent, necessitating cost-cutting measures."
- **Provide information that both you and your audience can agree upon.** "With the recession, many jobs in business and industry have been lost."
- **Compliment your reader or show appreciation.** "Thank you for your June 9 letter commenting on fiscal year 2016."
- **Make your buffer concise, one to two sentences.** "Thank you for writing. Customer comments give us an opportunity to improve service."
- **Be sure your buffer leads logically to the explanation that follows. Consider mentioning the topic, as in the following example about billing practices.** "Several of our clients have noted changes in our corporate billing policies. Your letter was one that addressed this issue."
- **Avoid placing blame or offending the reader.** Rather than stating, "Your bookkeeping error cost us $9,890.00," write, "Mistakes happen in business. We are refining our bookkeeping policies to ensure accuracy."

See Figures 9.8, 9.9, and 9.10 for sample adjustment messages.

Order

In business, you will need to place orders. Order correspondence provides you documentation. When you write a memo, letter, or e-mail, be concise and clear to assure correctness of your order.

Introduction.   The introduction of order correspondence contains the following:

- **Reason for placing the order.** To meet holiday demands, upgrade office equipment, or maintain inventories, for example.
- **Authorization for placing the order.** Your position as purchasing manager, office manager, or supervisor gives you the authority to place orders for your

FIGURE 9.8 100 Percent Yes Adjustment

Thank you for your recent letter. Gulfstream will replace your defective shock absorber according to the warranty agreement.

The Trailhandler Performance XT shock absorber that you purchased was discontinued in April 2016. Mr. Blanton, the mechanic to whom you spoke, incorrectly assumed that Gulfstream was no longer honoring the warranty on that product. Because we no longer carry that product, we either will replace it with a comparable model or refund the purchase price. Ask for Mrs. Cottrell at the automotive desk on your next visit to our store. She is expecting you and will handle the exchange.

We appreciate your business, Mr. De La Torre. I'm glad you brought this problem to my attention. If I can help you in the future, please contact me at hlang@gulfstream.com.

Positive word usage ("Thank you") achieves audience rapport.

The introduction immediately states the good news.

The discussion explains what created the problem and provides an instruction telling the customer what to do next.

The conclusion ("We appreciate your business") resells to maintain customer satisfaction.

FIGURE 9.9 100 Percent No Adjustment Beginning with a Buffer Statement

Thank you for your May 10 letter. Gulfstream Auto always appreciates hearing from its customers.

The Trailhandler Performance XT shock absorber that you purchased was discontinued in April 2016. Mr. Blanton, the mechanic to whom you spoke, correctly stated that Gulfstream was no longer honoring the warranty on that product. Because we no longer carry that product, we cannot replace it with a comparable model or refund the purchase price. Although we cannot replace the shock absorber for free, we want to offer you a 10 percent discount off of a replacement.

We appreciate your business, Mr. De La Torre. I'm glad you brought this problem to my attention. If I can help you in the future, please contact me at hlang@gulfstream.com.

The introduction begins with a buffer. The writer establishes rapport with the audience through positive words to create a pleasant tone.

The discussion explains the company's position, states the bad news, and offers an alternative.

FIGURE 9.10 Partial Adjustment Stating the Good News in the Introduction

Thank you for your recent letter. Gulfstream will replace your defective shock absorber according to the warranty agreement.

The Trailhandler Performance XT shock absorber that you purchased was discontinued in April 2016. Mr. Blanton, the mechanic to whom you spoke, incorrectly assumed that Gulfstream was no longer honoring the warranty on that product. However, we no longer carry that product. We will replace the shock absorber with a comparable model, but you will have to pay for installation.

We appreciate your business, Mr. De La Torre. I'm glad you brought this problem to my attention. If I can help you in the future, please contact me at hlang@gulfstream.com.

Begin your letter with the good news.

Explain what happened, state the bad news, and provide a possible alternative.

corporation. On the other hand, your boss might have asked you to place an order.

- **Method of delivery.** FedEx, UPS, overnight, mail, express delivery, and so on are options.
- **Source of item information.** A catalog or sales brochure, for example.

Discussion. In the body, provide

- A sentence lead-in.
- An itemized list of the order.
- Precise details, including costs, sizes, shapes, colors, materials, descriptions, or titles. Though serial numbers are useful, not all catalogs (particularly those online) provide this information. The key is being as specific as you can to ensure receipt of correct merchandise. By writing precisely, you will avoid phone calls or e-mail messages asking for clarity.

Conclusion. You need to include the following in your conclusion:

- Date needed by.
- Method of payment.
- Contact information, such as telephone number, e-mail address, or fax number.
- Positive close, such as "Thank you for your help."

See Figure 9.11 for a sample of an e-mail order.

FIGURE 9.11   E-mail Order

In the body, include as many details as possible to ensure that the order is correctly filled. You should include exact dollar amounts and amounts of units being ordered. Also provide the serial number, catalog number, or name of the item.

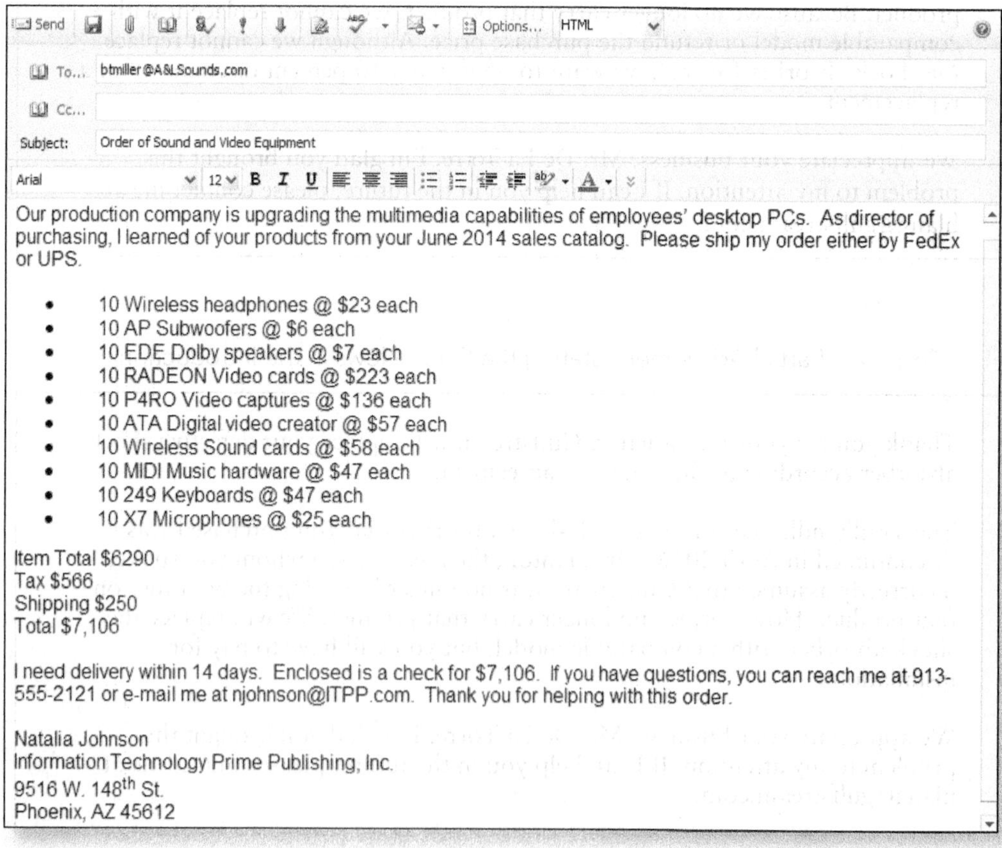

Send ⊟ 🖉 🕮 🔊 ! ⬇ 🖉 ⚡ ▾ 🖷 ▾ 🗐 Options... HTML ▾	⊘

To... btmiller@A&LSounds.com

Cc...

Subject: Order of Sound and Video Equipment

Arial   ▾ 12 ▾ **B** *I* U̲ ▤ ▤ ▤ ▤ ▤ 🔢 🔢 ▦ ▦ 🔤 ▾ **A** ▾ ⌄

Our production company is upgrading the multimedia capabilities of employees' desktop PCs. As director of purchasing, I learned of your products from your June 2014 sales catalog. Please ship my order either by FedEx or UPS.

- 10 Wireless headphones @ $23 each
- 10 AP Subwoofers @ $6 each
- 10 EDE Dolby speakers @ $7 each
- 10 RADEON Video cards @ $223 each
- 10 P4RO Video captures @ $136 each
- 10 ATA Digital video creator @ $57 each
- 10 Wireless Sound cards @ $58 each
- 10 MIDI Music hardware @ $47 each
- 10 249 Keyboards @ $47 each
- 10 X7 Microphones @ $25 each

Item Total $6290
Tax $566
Shipping $250
Total $7,106

I need delivery within 14 days. Enclosed is a check for $7,106. If you have questions, you can reach me at 913-555-2121 or e-mail me at njohnson@ITPP.com. Thank you for helping with this order.

Natalia Johnson
Information Technology Prime Publishing, Inc.
9516 W. 148th St.
Phoenix, AZ 45612

Microsoft Corporation, Inc.

Confirmation

In business, correspondence represents an official contract. Often, when clients and vendors make arrangements for the purchase of services or products, a confirmation letter, memo, or e-mail must be written to verify the details of the agreement. For example, as head of corporate training, you met with a consulting firm to discuss services they could provide your company. After the meeting, you must summarize the results of the discussion, confirming payment, dates, and training content.

Introduction. In your confirmation, remind the reader why the correspondence is being written and what topic is being discussed.

Discussion. The body clarifies the details of the agreement. Since this constitutes a legally binding document, you must specify anything agreed upon. Using highlighting techniques to make your content accessible, consider including any of the following:

Dates	Makes/models/serial numbers
Locations	Retainer fees
Times	Personnel and certification requirements
Audiovisual equipment	Length of agreement
Costs	Menus and decorations
Parking	Room setup

Conclusion. The conclusion tells the reader what to do next. You might include a request for signature, payment due dates, or method of payment. See Figure 9.12 for a sample confirmation letter.

Recommendation Letter

You might write a letter of recommendation for many reasons:

- An employee deserves a promotion
- An ex-employee asks for a reference for a new job
- A colleague is nominated for an award
- An acquaintance is applying for an education scholarship
- A governmental agency is checking references
- A consultant requests a reference for a new client

When someone asks you to write a letter of recommendation, consider the *Do's* and *Don'ts* in Table 9.3.

Introduction. In the introduction, include any of the following:

- Your position (or title)
- Your relationship with the person
- The length of your relationship
- The applicant's name
- The position, scholarship, or award

FIGURE 9.12 Confirmation Letter

Metro Consulting

600 Broadway Albuquerque, NM 23006
510-234-1818 www.metrocon.com

January 23, 2016

Mr. Carl Meyers
ProfCom
1999 Saguaro Dr.
Santa Fe, NM 23012

Dear Mr. Meyers:

The introduction explains why the letter is being written ("in response to your request"), specifies the type of letter (confirmation), and clarifies the topic ("training services" and "fees contracted").

In response to your request, this letter confirms our discussion from last week. Below I summarize the agreement we reached regarding the training services your company will provide and the fees contracted.

ProfCom is scheduled to offer the following workshops in 2016:

The letter's discussion provides specific details regarding the agreement costs, locations, and contact.

- Fifteen days of "Customer Service" training to Albuquerque municipal employees. The workshops will be held throughout the city at locations to be determined later. Each location will provide technology, per your specifications. ProfCom will be paid $500 for each half-day workshop and $700 for each full-day workshop.
- Thirty days of "Managing Diversity" training to Albuquerque municipal employees. These workshops will be held in the Albuquerque Civic Center, 1800 Mountainview Dr. ProfCom will be paid $900 a day for each of these workshops. Please contact Mr. Silvio Hernandez, 1-800-ALCIVIC, to request your technology needs.
- Ten days of "Supervisor/Management" training, leading to a "Supervisor/Management Certificate." These workshops will be held in the Albuquerque City Hall, Conference Room A. ProfCom will be paid $1,000 a day for each workshop. Please contact Mary O'Sullivan, 1-510-222-5150, to request your technology needs.
- Undecided additional training workshops, including "Ethical Decision-Making," "Accounting for Non-Accountants," and "Dynamic Presentation Skills." We will work with you to schedule these as enrollment figures are calculated. Metro Consulting will pay ProfCom a monthly retainer fee of $3,000 to ensure your availability for these workshops. The $3,000 retainer will be adjusted against complete payment for services rendered.

The conclusion ends positively and instructs the reader, stating what must be done next to confirm the agreement.

Thank you for agreeing to provide us these services, Mr. Meyers. Metro Consulting is excited about the prospect of working with your firm. We hope to continue offering these workshops annually if both parties agree upon the success of the training. By signing and dating this letter of confirmation below, you indicate your agreement with the assigned services and compensation.

Rob Harken

Rob Harken, Director

_____, Date: _____   _____, Date: _____

TABLE 9.3 Do's and Don'ts of Recommendation Letters

Do's	Don'ts
• Agree to write the letter only if you can be supportive • Request a current resume • Obtain information about the position, scholarship, or award the person is applying for • Write the letter with the specific position in mind • Study any information about the person to avoid omissions • Keep your letter to a reasonable length	• Avoid writing if you feel the candidate is weak • Avoid writing if you cannot be positive • Avoid writing letters longer than one page • Avoid writing these letters if you have only vague memories of the person's work • Avoid writing a reference letter if you cannot also talk with confidence about the person to an interviewer

Discussion. In the body, include the specific details of the applicant's skills. Consider including the following:

- Examples of the applicant's job performance
- Illustrations proving how and why the person will benefit the company
- Evaluations of the person's chances of success in the company or program
- Differences from other people
- Examples of projects the applicant worked on
- Projects supervised
- Team skills
- Communication abilities
- Names of classes attended or certifications acquired to enhance the applicant's skills
- Honors earned at work, school, or military

Conclusion. In the conclusion, sum up why this person is deserving of consideration for the job, award, or scholarship. Be sure to include contact information, such as your telephone number or e-mail address. This will help the recipient of the recommendation letter reach you for a follow-up discussion. See Figure 9.13 for a sample recommendation letter.

Thank-You Letter

When an employee, customer, vendor, supervisor, coworker, or any business professional does you a favor, you should write a follow-up thank-you letter. Doing so is not only courteous, but also it is good business. By writing a thank-you letter, you show your appreciation and build continued rapport. Create a pleasant tone and a positive attitude in your thank-you letter. Combined with a clear writing style and a well-designed letter, you can show your gratitude. In the introduction, remind the reader why you are writing. You can include the date of the reader's letter or kind words. You also should mention the topic of those comments. In the letter's body, explain how the reader's actions or words helped you. In the conclusion, thank the reader for his or her time and kindness.

See Figures 9.14, 9.15, and 9.16 (shown on pages 213 and 214) for the following series of letters: inquiry, response, and thank you.

FIGURE 9.13   Letter of Recommendation.

Midwest Technological College

15431 College Blvd.   Milwaukee, WI 32556   451-987-0101

March 15, 2016

Dr. Anne Cohen
University of Wisconsin Medical Center
1900 E. 39th Street
Madison, WI 35567

Subject: Letter of Recommendation for the University of Wisconsin Medical Center, Department of Health Information Management

Introduction giving the applicant's name, how the writer knows her, and the topic for this letter.

I am pleased to write this letter recommending Pekkahm Shoumavong for admission to the WU Department of Health Information Management. Pekkahm was a student in my Business Communication class last semester (Fall 2015). She impressed me not only with her scholarship, but also with her team skills and her conscientious desire to excel.

Discussion specifying ways in which the applicant excels. These include grades, team projects, interpersonal communication abilities, and personality traits.

Pekkahm earned an A in my class due to her excellent written and oral communication skills (letters, memos, reports, e-mail, instructions, and speeches accompanied by PowerPoint presentations). More importantly, I was impressed with her ability to work well with others and to take on leadership roles. My business communication class emphasizes collaboration. Pekkahm's team was one of my more successful groups due to their ability to work in concert toward a team goal and to communicate professionally with each other—valuable skills in any environment. This was an especially impressive achievement since Pekkahm's team was multicultural and comprised of students with vastly different abilities.

Note how the recommendation letter uses positive words throughout to highlight the applicant's abilities. Some of these words include "succeed," "happy," "deserving," "pleased," "impressed," and "excel."

In addition, I remember that her team encountered a problem with lost computer files. Rather than become distressed, Pekkahm remained calm and worked through the problem professionally. Her even-keeled and congenial personality, interpersonal communication talents, and ability to get along with others while demanding high quality work helped her team succeed. Finally, Pekkahm attended all classes and turned in all assignments on time, evincing her discipline, conscientiousness, and dedication to excellence.

Conclusion sums up the candidate's assets, focusing on ways in which this person meets the needs of a specific program. The conclusion also provides contact information for follow up.

Self-motivation, professionalism, problem-solving skills, and outstanding communication abilities suggest a high level of potential. I believe these traits distinguish Pekkahm and indicate her ability to succeed in the WU Department of Health Information Management. Pekkahm is deserving of your consideration for admission to your program. Thus, I am happy to be able to recommend her and serve as a reference. If you would like to discuss her attributes further, please call me at 451-987-0101, ext. 59, or e-mail me at cescobar@mtc.edu.

Dr. Carmen Escobar

Dr. Carmen Escobar
Communication Professor

FIGURE 9.14   Letter of Inquiry

1408 N. Hawker
Independence, MO 64050
March 24, 2016

R4 Technologies
1579 W. Pacific Highway
San Diego, CA 92447

Subject: Request for Game Designer Qualification

I am interested in becoming a video game designer and would like to know the requirements for the field.

Please send me the following information:

1. What type of education do I need?
2. Which universities offer this degree program?
3. How long does it take to acquire sufficient education or training for the field?
4. What is the starting pay for a new game designer?
5. What are the requirements for getting hired in your firm?

Your response by April 5, 2016, will help me plan for enrollment at a university. Thank you for your time.

Daniel Black

Daniel Black
Response to the Letter of Inquiry

FIGURE 9.15   Response to the Letter of Inquiry

R4 Technologies
1579 W. Pacific Highway
San Diego, CA 92447
April 2, 2016

Daniel Black
1408 N. Hawker
Independence, MO 64050

Dear Daniel:

I recently received your request for becoming a game designer. Unfortunately, I am unable to help you because R4 is a publishing company, and we outsource all design work.

However, I pulled some articles that I think you will find useful. They are enclosed. Also, several sources on the Internet can help you find what you are looking for (also enclosed).

Best wishes,

Susan Cardez

Susan Cardez
Director of Human Resources

FIGURE 9.16   Thank-You Letter

1408 N. Hawker
Independence, MO 64050
April 10, 2016

Susan Cardez
R4 Technologies
1579 W. Pacific Highway
San Diego, CA 92447

Dear Ms. Cardez:

The articles you sent on April 2, "How Do I Become a Game Designer?" from GameDesignX , and "I Really Like Games—How Do I Get a Job as a Game Designer?" from Obscure Productions are informative.

The information is very helpful for the following reasons:

1.   Both explain types of designers.
2.   The GameDesignX article tells about education needed.
3.   The Obscure Productions' article has an excellent list of references.

Thank you for your time and for sending me the articles. Your kindness in responding has helped provide me focus in my career search.

Sincerely,

Daniel Black

Daniel Black

Instant Messages

> 11. Understand the benefits of instant messages.

Memos and e-mail could be too slow for today's fast-paced workplace. Instant messages are already providing businesses many benefits. A professional writer we know works for an international provider of business solutions and services. Her office is in Kansas City; her boss lives in Orlando and telecommutes. How do they collaborate while working on team projects from their dispersed work sites? They communicate by e-mail, telephone, teleconference, and an occasional face-to-face meeting when the boss travels to the writer's home office. However, for their required daily office chats, they depend on instant messages.

Benefits of Instant Messages

Following are benefits of instant messages:

- Faster communication
- Improved efficiency for collaboration and dispersed workgroups
- Synchronous communication with coworkers and customers
- Cheaper than long-distance telephone rates
- More personal than e-mail; less intrusive than telephone calls
- Less intrusive than telephone calls

Challenges of Instant Messages

IM has potential problems in the workplace, including security, archiving, monitoring, and employee misuse (Hoffman; Shinder).

- **Security issues.** This is the biggest concern. IM users are vulnerable to hackers, identity theft, and uncontrolled transfer of documents. With unsecured IM, a company could lose confidential documents, internal users could download copyrighted software, or external users could send virus-infected files.

- **Lost productivity.** Use of IM on the job can lead to job downtime. First, we tend to type more slowly than we talk. Next, the conversational nature of IM leads to chattiness. If employees are not careful, or monitored, a brief IM conversation can lead to hours of lost productivity.

- **Employee abuse.** IM can lead to sending personal messages rather than job-related communication with coworkers or customers.

- **Distraction.** With IM, a bored colleague easily can distract you with personal messages, online chats, and unimportant updates.

- **Netiquette.** As with e-mail, due to the casual nature of IM, people could relax their professionalism and forget about the rules of business communication. IM can lead to rudeness.

- **Spim.** IM lends itself to spim, instant message spam—unwanted advertisements, pornography, pop-ups, and viruses.

Techniques for Successful Instant Messages

To solve potential problems, consider these five suggestions:

1. **Choose the correct communication channel.** Use IM for speed and convenience. If you need length and detail, other options—e-mail messages, memos, letters—are better choices. In addition, sensitive topics or bad news should never be handled through IM. These deserve the personal attention provided by telephone calls or face-to-face meetings.

2. **Summarize decisions.** IM is great for collaboration. However, all team members might not be online when decisions are made. Once conclusions have been reached that affect the entire team, the designated team leader should e-mail everyone involved. In this e-mail, the team leader can summarize the key points, editorial decisions, timetables, and responsibilities.

3. **Tune in, or turn off.** IM software tells everyone you are active online. Immediately, your IM buddies can start sending messages. IM pop-ups can be distracting. Sometimes, in order to get your work done, you might need to turn off your IM system. Your IM product might give you status options, such as "on the phone," "away from my desk," or "busy." Turning on IM could infringe upon your privacy and time. Turning off might be the answer.

4. **Limit personal use.** Your company owns the instant messages you write in the workplace. IM should be used for business purposes only.

5. **Never use IM for confidential communication.** Use another communication channel if your content requires security. As with e-mail, IM programs can let systems administrators log and review IM conversations.

See Figure 9.17 for an example of an instant message.

> **Teamwork**
>
> See Chapter 1, "An Introduction to Technical Communication," for additional information about working and writing in teams.

FIGURE 9.17 Instant Message

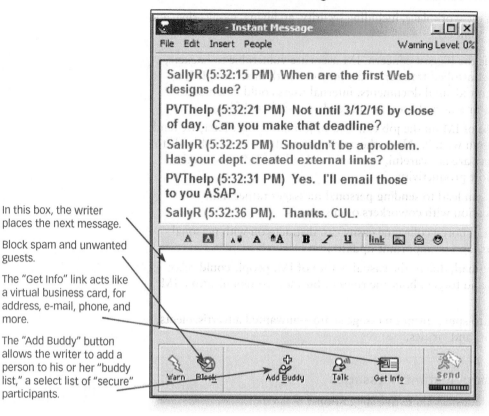

In this box, the writer places the next message.

Block spam and unwanted guests.

The "Get Info" link acts like a virtual business card, for address, e-mail, phone, and more.

The "Add Buddy" button allows the writer to add a person to his or her "buddy list," a select list of "secure" participants.

IM/TM Corporate Usage Policy

To clearly explain the role of IM and TM in the workplace, a company should establish a corporate IM and TM usage policy. Many industries already have such policies in place in relation to existing restrictions established by Sarbanes-Oxley or Health Insurance Portability and Accountability Act (HIPAA). Consider a policy that includes the following:

Ethical Considerations

See Chapter 6, "Ethical Considerations," for additional information about ethics and ethical strategies.

- Train employees to use IM and TM effectively for business.
- Explain which services are allowable in the workplace. For example, conversation between coworkers is acceptable, but chat between employees and outside individuals might not be advisable. File transfers of proprietary information need to be restricted and monitored.
- Do not allow employees to store IM or TM passwords on computer desktops or other sites easily accessed.
- Install appropriate security measures, including automated encryption, to protect against external threats, such as spam, spim, and viruses.
- Log and archive instant messages and text messages for compliance to company policy (Ollman; "How To"; Bradley).

12. Understand the reasons for using text messages.

Text Messages

Text messages, also known as *short message service* (SMS), have a growing presence in workplace communication. According to the United Nations International

Telecommunications Union, more than 1.6 trillion text messages, or 200,000 texts each second, are sent internationally (Toor). A unique characteristic of TM is its demand for conciseness. Text messages over 160 characters are delivered in multiple segments. Each segment is billed as a separate message. That's why users try to limit messages to 160 characters—to save expenses.

"Pretty soon, you're going to have to teach text messaging in your classes," Robert Clark said. "Why's that?" we asked. He responded, "30 percent of my interaction with staff is through text messaging" (Clark). Robert, a facilities manager for a real estate company, drives between eight apartment complexes to supervise his staff at each location. Robert says that he depends on text messages instead of e-mail for the following reasons:

Reasons for Using TM

- **Cost.** Though Robert's cell phone is supplied by his employer, the staff that he supervises pay for their own cell phones. It's cheaper for them to use text messaging if their cell phone plans allow for unlimited TM.
- **Technological access.** Staff members who work under Robert do not have computers at their work sites, nor does Robert have a computer in his car as he drives around town. However, all of Robert's employees have access to a cell phone.
- **Speed.** TM is a quick and easy way to communicate short messages like, "John will be late for work today. Is it OK if I stay overtime?" That's 63 characters (counting the letters, spaces, and punctuation marks). Robert and his staff use other means of communication for longer correspondence.
- **Multitasking.** TM is a great way for Robert to multitask. He might be in a meeting, for example. He says, "If three of my staff need to attend training on a certain day, I can text all three employees right then and there and not disrupt the meeting. They'll respond quickly, allowing me to tell the meeting facilitator which of my employees can or can't attend. Then we can reschedule accordingly."
- **Decrease the intimidation factor.** Many people don't like to write letters, memos, reports, or even e-mail messages. Many people don't like face-to-face communication, either. They're intimidated by writing or by bosses. Robert says that TM decreases this intimidation factor for his employees.
- **Documentation.** TM allows Robert to document his conversations, something that's not always possible with phone calls. TM is an instant record of a dialogue. A TM account saves incoming and outgoing calls for a few days until it's full. This allows an employee to clarify any later misconceptions.

Figure 9.18 shows how text messages are used at one company to determine an employee's availability to work a late shift.

The Writing Process at Work

The communication process is dynamic with the three steps frequently overlapping. To clarify the importance of the communication process, look at how Jim Goodwin, the CEO of COMRMed, used prewriting, writing, and rewriting to write a memo to his employees.

FIGURE 9.18   Text Message for Professional Communication

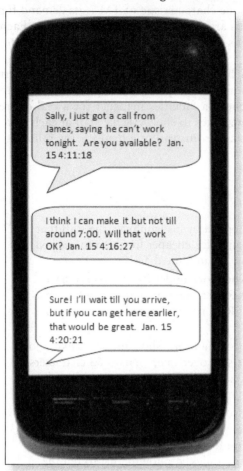

Prewriting

Prewriting

See Chapter 3, " The Communication Process," for additional prewriting techniques.

No single method of prewriting is more effective than another. Throughout this textbook, you will learn many different types of prewriting techniques geared uniquely for different types of communication. The goal of all prewriting is to help you overcome the blank page syndrome (writer's block). Prewriting will allow you to spend time before writing your document, gathering as much information as you can about your subject matter. In addition, prewriting lets you determine your objectives. Jim used mind mapping/clustering to gather data and determine objectives (Figure 9.19).

Writing

Once you have gathered your data and determined your objectives in prewriting, your next step is to draft your document. In doing so, you need to organize content.

Organize Your Ideas.   If your supporting details are presented randomly, your audience will be confused. As a writer, develop your content logically. When you draft your document, choose a method of organization that will help your readers understand your objectives. This could include comparison/contrast, problem/solution, chronology,

FIGURE 9.19   Mind Mapping/Clustering to Gather Data

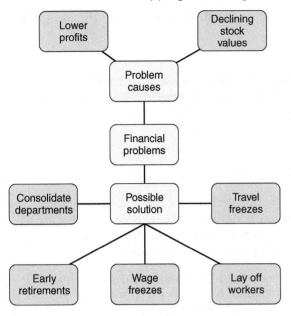

Clustering helps you see the different parts of any subject. Clustering helps you develop ideas and see the relationship between these concepts. Because clustering is less restrictive and less structured than outlining, this prewriting technique might allow you to gather data creatively.

cause/effect, and more. Jim drafted a memo focusing on the information he discovered in prewriting and then received revision suggestions from his administrative assistant (Figure 9.20).

FIGURE 9.20   Rough Draft with Revision Suggestions

Date: October 14, 2016
To: COMRMed Employees
From: Jim Goodwin
Subject: Problems

As you know, we are experiencing some problems at COMRMed. These include lower profits and stock value declines. We have alot of unhappy stockholders. Its up to me to help everyone figure out how to solve our problems.

I have some ideas I want to share with you. I'm happy to have you share your ideas with me too. Here are my ideas: we need to consider consolidating departments and laying off some employees. We also might need to freeze wages and certainly its time to freeze travel.

The best idea I have is for some of you to take early retirement. If all of you who have over twenty years vested in the company would retire, that would save us around 2.1 million dollars over the next fiscal year. And, you know, saving money is good for all of us in the long run.

Add a focus to the subject line, such as "Problems with . . . ."

I'd consider removing words like "as you know," "some," and "alot." Replace them with stronger words. Also, "alot" and "its" are spelled wrong.

List these problems and solutions to make them more accessible. Also, could you add more details?

Jim, I think you need to alter the tone of this memo. Is there some way to avoid talking about saving money by firing people?

Rewriting

Jim factored in his administrative assistant's suggestions and rewrote the memo. See Figure 9.21 for the finished product.

FIGURE 9.21 Problem-Solution Memo Incorporating Revision Suggestions

Date: October 14, 2016
To: COMRMed Employees
From: Jim Goodwin
Subject: Suggestions for Improving Company Finances

COMRMed is experiencing lower profits and declining stock value. Consequently, stockholders are displeased with company performance. I have been meeting with the Board of Directors and division managers to determine the best course of action. Here are ideas to improve our company's financial situation.

1. Consolidating departments—By merging our marketing and advertising departments, for example, we can reduce redundancies. This could save COMRMed approximately $275,000 over a six-month period.
2. Reducing staff—We need to reduce employees by 15 percent. This does not necessarily mean that layoffs are inevitable. One way, for instance, to reduce staff is through voluntary retirements. We will be encouraging employees with over 20 years vested in the company to take our generous early-retirement package.
3. Freezing wages—For the next fiscal quarter, no raise increases will go into effect. Internal auditors will review the possibility of reestablishing raises after the first quarter.
4. Freezing travel—Conference attendance will be stopped for six months.

I encourage you to visit with me and your division managers with questions or suggestions. COMRMed is a strong company and will bounce back with your help. Thank you for your patience and understanding.

CASE STUDY

After reading the following case studies, write the appropriate correspondence required for each assignment.

1. As Director of Human Resources at COMRMed biotechnology company, Andrew McWard helps employees create and implement their individual development plans (IDPs). Employees attend 360-Degree Assessment Workshops where they learn how to get feedback on their job performance from their supervisors, coworkers, and subordinates. They also provide self-evaluations.

Once the 360-Degree Assessments are complete, employees submit them to Andrew, who, with the help of his staff, develops the IDPs. Andrew sends the IDPs to the employees, prefaced by a cover letter. In this cover letter, he tells them why he is writing and what he is writing about. In the letter's body, he focuses their attention on the attachment's contents: supervisor's development profile, schedule of activities which helps employees implement their plans, courses designed to increase their productivity, costs of each program, and guides to long-term professional development.

In the cover letter's conclusion, Andrew ends upbeat by emphasizing how the employees' IDPs help them resolve conflicts and make better decisions.

Assignment

Based on the information provided, write this cover letter for Andrew McWard. He is sending the letter to Sharon Baker, Account Executive, 1092 Turtle Hill Road, Evening Star, GA 20091.

2. ITCom is committed to increasing the diversity of its workforce and its clientele. ITCom realizes that a diverse population of employees and customers (in terms of gender, ages, races, and religions) makes good business sense.

To ensure that ITCom achieves diversity awareness, the company plans the following:
- Develop a diversity committee
- Focus on ways the company can be a responsible member of the community's diverse constituency
- Hire a diverse workforce
- Train employees to respect diversity
- Write corporate communication (Web sites, e-mail, letters, corporate reports) that accommodate the unique needs of a diverse audience

As CEO of ITCom, you want buy-in throughout the company. You want the company to realize that diversity is good for society and good for business.

Assignment

Write an e-mail message to your employees. In this e-mail, explain the company's diversity goals, highlight what the company plans to do to accomplish these goals, and ask the audience to participate in upcoming diversity workshops. Follow the criteria for effective e-mail provided in the chapter.

3. You are the manager of WhiteOut, a store that sells snowboarding equipment and clothing. You have seven employees. Since you believed that today would be a light day for sales, you asked only one employee to work. However, suddenly, more customers than you had expected showed up, and you and your employee are overwhelmed. You need help—fast.

E-mailing your other employees might not work, since they aren't necessarily near a computer. In contrast, you know they all have smartphones. The fastest and most trustworthy way to communicate with them is through instant messaging or text messaging.

Assignment

Write an instant message or a text message to the employees. Tell them the circumstances at your store and ask them for assistance.

ETHICAL CHALLENGE

Carlos Delgado is a facilities maintenance employee for the city. His job requires that he travel throughout the city to various locations but to always be available for emergency calls. Therefore, the city issued him a smartphone for use during his 40-hour-a-week job.

As Carlos's manager, you are responsible for checking smartphone bills. When you review Carlos's bill, you find that he sent 2,248 text messages last month. Most facilities maintenance employees send an average of 500 text messages a month.

When the manager questioned Carlos about this excessive use of the smartphone, Carlos said that he had frequent TM conversations with coworkers at different job sites. However, Carlos also said that his son had been ill that month. Carlos had text messaged his home to check on his son's health.

Question

Are personal calls acceptable on company equipment? Is it ethical for Carlos to have made so many personal text messages on his corporate-issued smartphone? Why or why not?

INDIVIDUAL AND TEAM PROJECTS

1. Write an inquiry. You might want to write to a college or university requesting information about a degree program or to a manufacturer for information about a product or service. Whatever the subject matter, be specific in your request.

2. Write a cover message. Perhaps your cover letter, e-mail, or memo will preface a report you are working on in school, a report you are writing at work, or documentation you will need to send to a client.

3. Write an adjustment message. Envision that a client has complained about a problem he or she has encountered with your product or service. Write a 100 percent yes letter or e-mail in response to the complaint.

PROBLEM-SOLVING THINK PIECES

Northwest Regional Governmental Training Consortium (GTC) provides educational workshops for elected and appointed officials, as well as employees of city and state governmental offices.

One seminar participant, Mary Bloom, supervisor of the North Platte County Planning and Zoning Department, attended a GTC seminar entitled "Developing Leadership Skills," on February 12, 2016. Unfortunately, she was disappointed in the workshop and the facilitator. On February 16, Mary complained to GTC's director, Sue Randall, stating her dissatisfaction. Mary said that the training facilitator's presentation skills were poor. According to her, Doug Aaron, the trainer, exhibited the following problems:

- Late arrival at the workshop
- Too few handouts for the participants
- Incorrect cables for his computer, so he could not use his planned PowerPoint presentation

Mary also noted that the seminar did not meet the majority of the seminar participants' expectations. She and the other government employees had expected a hands-on workshop with breakout sessions. Instead, Mr. Aaron lectured the entire time. In addition, his information seemed dated and ignored the cross-cultural challenges facing today's supervisors.

Neither Sue nor her employees had ever attended this workshop. They offered the seminar based on the seemingly reliable recommendation of another state agency, the state's Data Collection Department. From Doug Aaron's course objectives and resume, he appeared to be qualified and current in his field.

However, Mary deserves consideration. Not only are her complaints justified by others' comments, but also she is a valued constituent. The GTC wants to ensure her continued involvement in its training program.

Assignments

1. Sue needs to write a 100 percent yes adjustment letter or e-mail. In this correspondence, Sue wants to recognize Mary's concern, explain what might have gone wrong, and offer satisfaction. Write the 100 percent yes adjustment letter or e-mail.

2. Sue needs to write an internal memo or e-mail to her staff. In this memo or e-mail, she will provide standards for hiring future trainers. Write the e-mail to her staff, explaining why the e-mail has been written, what standards are needed, and when the responses are needed.

WEB WORKSHOP

Research information about text messages by reading the following online articles. Then, write an e-mail message to your professor, summarizing your findings:

- Calvey, Mark. "RU Ready? Wells Initiates Banking by Text Message." *Kansas City Business Journal.*
- Kedrosky, Paul. "Why We Don't Get the (Text) Message." *CNN.Money*
- Noguchi, Yuki. "Life and Romance in 160 Characters or Less: Brevity Gains New Meaning as Popularity of Cell Phone Text Messaging Soars." *Washington Post.*
- Tedeschi, Bob. "Reaching More Customers with a Simple Text Message." *The New York Times.*
- "Text Messaging: Frequently Asked Questions." *AT&T.*

CHAPTER TEN

The Job Search

After completing this chapter, you will be able to

1. Appreciate the 3 Rs of searching for a job
2. Learn various techniques for finding jobs
3. Understand the benefits and drawbacks of social media in the job search
4. Use mobile apps in the job search
5. Use the Internet in the job search
6. Consider ethics when writing a résumé
7. Learn the criteria for effective résumés
8. Distinguish between different methods of delivery for sending résumés
9. Prepare electronic and hard-copy portfolios
10. Follow criteria for effective e-mail and letter applications
11. Review a sample job package, including job advertisement, e-mail, and résumé
12. Learn how to interview effectively
13. Follow criteria for effective follow-up correspondence
14. Write a successful job acceptance letter or e-mail

COMMUNICATION AT WORK

In this scenario, a business owner interviews potential job applicants.

The job search involves at least two people—the applicant and the individual making the hiring decision. Usually more than two people are involved, however, because companies typically hire based on a committee's decision. That is the case at **DiskServe**. This St. Louis-based company is hoping to hire a customer service representative for its computer technology department. DiskServe is eager to hire a new employee because one of its best workers has just advanced to a new position in the company. DiskServe asked applicants to apply using e-mail. Thus, the applicants submitted an application by way of a cover e-mail message, an attached résumé, and an attached portfolio.

DiskServe advertised this opening in the career placement centers at local colleges, through online search engines, and in its Web site: http://www.DiskServe.com.

In addition to DiskServe's chief executive officer (CEO), Sarah Beske, the hiring committee will consist of two managers from other DiskServe departments, the former employee whose job is being filled, and two coworkers in the computer technology department.

Ten candidates were considered for the position. All candidates first had teleconference interviews. While Sarah talked with the candidates, the other hiring committee members listened on a speakerphone. After the telephone interviews, four candidates were invited to DiskServe's work site for personal interviews—Macy Heart, Aaron Brown, Rosemary Lopez, and Robin Scott.

Sarah, who has worked hard to create a family-oriented environment at DiskServe, values three traits in her employees: technology know-how, an ability to work in teams, and a positive attitude toward customers and coworkers. When the candidates arrived at DiskServe, Sarah gave them a tour of the facilities, introducing them to many employees. Then the interviews began.

Each job candidate was asked a series of questions that included the following:

- What is your greatest strength? Give an example of how this reveals itself on the job.
- What did you like most and least about your previous jobs?
- How have you handled customer complaints in the past?
- Where do you see yourself in five years?

Then, each candidate was taken to the computer repair lab and confronted with an actual hardware or software problem. The candidates were asked to solve the problem, and their work was timed. Finally, the applicants were allowed to ask questions about DiskServe and their job responsibilities.

Sarah is a stickler for good manners and business protocol. She waited 48 hours after the final interview to make her hiring decision. The wait time allowed her to check references. More important, she wanted to see which of the candidates wrote follow-up thank-you notes, and she planned to assess the quality of their writing. Sarah takes the hiring process seriously. She wants to hire the best people because she hopes those employees will stay with the company a long time. Hiring well is a good corporate investment.

The 3 Rs of Searching for a Job

1. Appreciate the 3 Rs of searching for a job.

When beginning your job search, consider the 3 Rs: research yourself, research the company, and research the position. By performing this research, you will improve the focus of your job search. Learning more about your goals, the company's culture, and the specifics of the position will help you determine your goals, decide if the company and job meet your needs, and prepare you for an interview.

Research Yourself

You are not right for every job, and every job is not right for you. To ensure that you are applying for the correct position, ask yourself the following questions:

- What are your skills, attributes, and accomplishments?
- What can you bring to the company that is unique?
- How do your qualifications meet the position's requirements?
- What interests you about this company and the position?

Research the Company

By researching the company, you educate yourself about the company's culture, values, products, and services. To research the company, view the employer's Web site, read annual reports, speak with an employee, or go online to a site like Glassdoor.com to read employee reviews. Consider these questions:

- What is the employer's product or service?
- What is the company's vision or mission statement?
- What are the needs or problems of the employer?
- What role does the employer play in the community?
- Is the employer expanding?

Research the Position

When you consider the job opening, go online to the *Occupational Outlook Handbook* (http://www.bls.gov/oco/ooh_index.htm) to learn about a career field's salary range and job outlook. When you consider a specific company, review advertised job requirements and duties. This will help you determine if the job is right for you. Ask yourself these questions:

- What are the job responsibilities?
- What are the pay and benefits for this job?
- What skills and talents are needed for the position?
- What is the growth opportunity for this field?

2. Learn various techniques for finding jobs.

Searching for Jobs

Use a variety of techniques and sources in your job search. These can include social media, apps, the Internet, and other, more traditional ways to find jobs, as follows:

- Network with friends and past employers. A *Smart Money* magazine article reported that 62 percent of job searchers find employment through networking (Bloch 12).
- Visit your college or university job placement center.
- Attend a job fair.
- Talk to your instructors.
- Get involved in your community by volunteering for a community committee, pursuing religious affiliations, joining community clubs, or participating in fund-raising events.
- Access job postings from professional affiliations or business journals.
- Read the want ads.
- Take a temp job that will pay you while you look for a job, help you acquire new skills, allow you to network, and could lead to full-time employment.
- Get an internship to network, acquire additional skills, and improve your résumé.
- Shadow a mentor on a job to learn about job responsibilities in a certain field.
- Set up an informational interview with a professional in your field to learn about career opportunities.
- Use social media.
- Use the Internet.

3. Understand the benefits and drawbacks of social media in the job search.

Using Social Media to Find Jobs

You can network virtually through social media. Twitter, for example, allows you to connect with recruiters and employees at companies you are interested in. You also can use blogs to search for job openings. Technorati provides you access to approximately 130 million blogs which you can search to find information about job openings (Schawbel "7 Secrets"). Pinterest allows job seekers to advertise their creativity. In addition, search engines like Pipl, PeekYou, and Wink can help you find people to contact at companies you're interested in. LinkedIn posts jobs. In addition, if you follow a company and know any of its employees, you can ask them to aid in your networking.

Social media isn't just for the job seeker. The Society for Human Resource Management, the world's largest association of hiring managers, reports that a significant number of employers use social media to research job candidates. According to a SelectMinds survey, approximately 72 percent of companies use social media for finding job candidates.

According to Rachel Emma Silverman in *The Wall Street Journal*, "Twitter is becoming the new job board. It is also becoming the new résumé." Because of difficulties recruiting new employees, some recruiters are using social media for posting jobs and hunting for job candidates. People looking for jobs are summarizing their curriculum vitae (CVs) in 140 characters for Twitter or creating short videos to post online.

Social Media and the Job Search

See Chapter 12, "Social Media," for more information about searching for jobs using social media.

Problems with Digital Dirt in the Job Search

If you're like most people, you have a Facebook site. You post photographs of yourself having fun while on vacations or at parties. You list your favorite bands, restaurants, television shows, and movies. You chat with friends, state your relationship status, and post intimate information. Some of the information might be unflattering. That's called *digital dirt*. Do you want a prospective employer to see your life in such detail? Information that you consider harmless might, in fact, hurt your job chances. Therefore, be careful about what you post. Treat your social media sites as reflections of your professionalism.

LinkedIn and the Job Search

To enhance your job search profile, consider joining LinkedIn, a social media site for professionals. LinkedIn not only lets you post your own profile, but also through this social media site, you can find information about other professionals in your work environment. Fortune 500 companies, including Yahoo, Kraft, Microsoft, Lockheed Martin, eBay, EMC, Intuit, Netflix, Allstate, Target, and ConAgra, use LinkedIn to recruit, find, and hire new employees. The site has registered users in over 200 countries and is available in six languages—English, French, German, Italian, Portuguese, and Spanish. A mobile version of the site, created in 2008, provides limited access over a smartphone. The mobile service is available in six languages—Chinese, English, French, German, Japanese, and Spanish (*LinkedIn.com*).

Why Should a Prospective Employee or a Company Seeking Employees Create a LinkedIn Site Profile? LinkedIn Jobs lets members of LinkedIn search for or post jobs by key word, country, and postal code. In addition, human resource managers can access a network of over 65 million professionals to find candidates with specific skills and required experience. Job seekers can use their connections at potential employers to help them land a job within the company. In their LinkedIn sites, job seekers can post a profile of their experiences, talents, and professional references. Job seekers also can review a company's profile to learn more about its products, services, projects, clients, employees, ratings, community involvement, and corporate contacts. Finally, Google tends to rank LinkedIn profiles high, usually within the first five Google hits. This is dependent on the LinkedIn participant using key words and using his or her name instead of a URL or company name.

How to Create an Effective LinkedIn Profile Page. When building a profile, focus on the following:

- **Provide a headline.** List your full name, title, primary job responsibility, and company name. This headline is the first thing the audience sees, so make a clear statement.
- **Add a photo.** Include a headshot, making sure it's small (around 80 × 80 pixels).
- **Summarize your credentials.** In the summary, engage the audience quickly. To do so, pinpoint your primary industry (this aids another company's search), focus on your key areas of interest, highlight your experiences, and clarify your achievements.
- **List your work experience.** Include dates of employment, the names of the companies for which you have worked, and a detailed list of your job activities and achievements.
- **Provide keywords and skills.** Include keywords and skills to help others (recruiters, colleagues, clients, etc.) easily search for and find your profile.
- **Add links to your LinkedIn site.** Help your connections find out more about you. Link your LinkedIn site to your company Web site, blog, and alternative e-mail addresses, Facebook site, Twitter account, etc.
- **Get recommendations.** List others with whom you have worked and for whom you have worked. This is how connections are made and how your value is quantified (Brogan; Doyle).

FAQs: LinkedIn's Features

Q: Besides helping me with my job search, what other benefits does LinkedIn offer?

A: LinkedIn isn't just a social networking site that allows you to post a profile. Look at its other interesting features:

- LinkedIn Groups—allows users to form like-minded groups of peers within an organization or industry.
- LinkedIn DirectAds—lets you connect with a large audience by geography, job function, age, gender, industry, and company size.
- LinkedIn Blog—shares posts from professionals in many industries.

Q: Is LinkedIn secure?

A: LinkedIn assures the security of your personal data by participating in the United States and European Union's International Safe Harbor Privacy Principles. LinkedIn protects private and sensitive information and assures users that they will be able to correct errors and delete information.

> 4. Use mobile apps in the job search.

Using Mobile Apps in the Job Search

For the job search candidate on the go, use mobile apps to find job openings and to network with recruiters. Following are apps that allow you to locate jobs, create an online identity, and develop a database of contacts.

- Jobs by CareerBuilder
- JobCompass
- JobFinder
- ABContacts

- BeKnown
- Beyond.com Search Jobs

See Figure 10.1 for a screenshot of the Jobs by CareerBuilder's app.

FIGURE 10.1   Jobs by CareerBuilder App

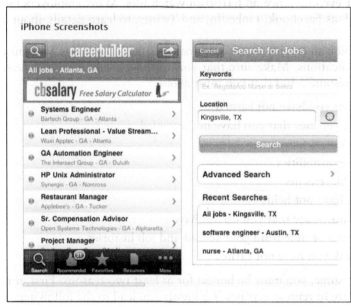

Source: Courtesy of CareerBuilder.

Using the Internet in Your Job Search

5. Use the Internet in the job search.

You also should make the Internet part of your job search strategy. Internet job search engines provide excellent job search resources, such as the following:

- Résumés—explaining the difference between résumés and CVs, addressing gaps in your career history, avoiding typical résumé mistakes, and writing winning résumés.
- Interviews—interviewing to get the job and handling illegal questions.
- Cover letters and thank-you letters—providing sample cover letter techniques and ways to write a better thank-you letter.
- Job search tips—employing the correct netiquette and job search techniques.

Search Engines for Jobs in the United States

- Indeed.com—lets you search job sites, newspapers, associations, and company job openings.
- Monster.com—lets you post résumés and search for jobs, and provides career advice.
- CareerBuilder.com—lets you search for jobs by company, industry, and job type.
- TwitJobsearch.com—lets you use Twitter for your job search.
- CollegeRecruiter.com—lists the latest job postings, "coolest career resources, and most helpful employment information."
- CareerJournal.com—the *Wall Street Journal's* career search site; provides salary and hiring information, a résumé database, and job hunting advice.
- Dice.com/—focuses on technology careers.
- Jobhunt.org/—called by *PC Magazine* and *Forbes* the Internet's best Web site for job hunting and resources.
- USAJOBS.gov/—a site dedicated to government job searches and advice.

Ethical Considerations When Writing a Résumé

If your résumé contains inaccurate information, your potential employer can find out. Prospective employers can perform follow-up reference checks, contacting your previous bosses to verify your résumé's assertions. In addition, it is very easy to perform Internet background checks related to a job candidate's credit issues, prior criminal offenses, driving records, military history, or even presence on terrorism watch lists. Many employers use social networking sites, such as Facebook, LinkedIn, and Twitter, to learn details about a job candidate.

Due to the ease with which an employer can verify your résumé's assertions, you must honestly present your qualifications. Make sure that your résumé is ethical and honest. Following are examples of résumé fraud that you must avoid:

- Including a degree that you have not earned
- Listing colleges or universities that you have not attended
- Inflating your job title
- Inflating your job responsibilities
- Inflating your job achievements
- Listing jobs that you have not held
- Claiming technical knowledge that you do not have
- Omitting large amounts of time from your school and job histories
- Including military rank you have not earned

When you write your résumé, you must be honest for at least two reasons. First, it is immoral to lie. Second, failure to represent yourself honestly can lead to the following:

- Misrepresentation by an employee during recruitment may result in termination.
- An employer can use your résumé misrepresentations as a defense in a wrongful dismissal case.
- An employer could sue an employee for damages due to a misrepresentation.

Phony degrees are easy to catch. All a recruiter needs to do is contact a college or university. Other résumé falsehoods are harder to catch. Recruiters look for the following red flags:

- Gaps and overlaps in work experience
- Unrealistic and vague qualifications
- Lack or misuse of industry terminology

Failing to follow ethical standards and being completely honest on your résumé could result in your not being offered a job. If you are offered a job based on false information, you could be terminated for this ethical lapse.

Criteria for Effective Résumés

Once you have found a job that interests you, it is time to apply. Your job application will start when you send the prospective employer your résumé. Résumés are usually the first impression you make on a prospective employer. If your résumé is effective, you have opened the door to possible employment—you have given yourself the opportunity to sell your skills during an interview. If, in contrast, you write an ineffective résumé, you have closed the door to opportunity.

Your résumé should present an objective, easily accessible, detailed biographical sketch. However, do not try to include your entire history. Because the primary goal of your résumé, together with your letter of application, is to get an interview, you can use your interview to explain in more detail any pertinent information that does not appear on your résumé. When writing a résumé, you have two optional approaches. You can write either a reverse chronological résumé or a functional résumé.

Reverse Chronological Résumé

Write a reverse chronological résumé if you

- Are a traditional job applicant (a recent high school or college graduate, aged 18 to 25)
- Hope to enter the profession in which you have received college training or certification
- Have made steady progress in one profession (promotions or salary increases)
- Plan to stay in your present profession

Functional Résumé

Write a functional résumé if you

- Are a nontraditional job applicant (returning to the workforce after a lengthy absence, older, not a recent high school or college graduate)
- Plan to enter a profession in which you have not received formal college training or certification
- Have changed jobs frequently
- Plan to enter a new profession

Key Résumé Components

Whether you write a reverse chronological or a functional résumé, include the following key components.

Identification.   Begin your résumé with the following:

- **Name (full first name, middle initial, and last name).**   Your name can be in boldface and printed in a larger type size (14-point, 16-point, etc.).
- **Contact information.**   Include your street address, city, state (use the correct two-letter abbreviation), and zip code. If you are attending college or serving in the armed forces, you might also want to include a permanent address. By including alternative addresses, you help your prospective employer contact you more easily.
- **Area code and phone numbers.**   Limit yourself to two phone numbers, and don't provide a work phone. Having prospective employers call you at your present job is not appropriate. First, your current employer will not appreciate your receiving this sort of personal call. Second, your future employer might believe that you often receive personal calls at work and will continue to do so if he or she hires you.
- **E-mail, Web site address, or fax number.**   Be sure that your e-mail address is professional sounding. An e-mail address such as "ILuvDaBears," "Hotrodder," or "HeavyMetalDude" is not likely to inspire a company to interview you.

Career Objectives.   The career objectives line is like a subject line in a memo, e-mail, or report. Your career objective informs the reader of your résumé's focus. If you decide to include a career objective, be sure it is precise and tailored to the job.

BEFORE

Flawed Career Objective

Career Objective: Seeking employment in a business environment offering an opportunity for professional growth.

AFTER

Improved Career Objective

Career Objective: To market financial planning programs and provide financial counseling to ensure positive client relations.

Too often, career objectives are so generic that their vagueness does more harm than good. In fact, because career objectives often focus more on what the writer wants instead of what the reader needs, many employers suggest omitting the career objectives from your résumé. Starting with a summary of qualifications might be a better strategy for creating a winning résumé.

Summary of Qualifications. After the career objectives, provide your audience a summary of qualifications. According to Monster.Com, people spend less than 15 seconds glancing at your résumé (Isaacs). "The Ladders," a job-matching service, even more dramatically states that "you have six seconds to make an impression" in a résumé. A summary of qualifications allows the employer an immediate opportunity to see how you can add value to the company.

A summary of qualifications should include the following:

- An overview of your skills, abilities, accomplishments, and attributes
- Your strengths in relation to the position for which you are applying
- How you will meet the employer's goals

To write an effective summary of qualifications, list your top three to seven most marketable credentials.

EXAMPLE ▶

Summary of Qualifications

- Over four years combined experience in marketing and business
- Developed a winning bid package for promotional brochures
- Promoted to manager in less than two years
- Maintained a database of over 1,000 customers, special ordered merchandise, and tracked inventory
- Managed over ten employees

Employment. The employment section lists the jobs you've held. This information must be presented in reverse chronological order (your current job listed first, previous jobs listed next). This section must include the following:

- Your job title (if you have or had one)
- The name of the company you worked for
- The location of this company (city and state)

- The time period during which you worked at this job
- Your job duties, responsibilities, and accomplishments

This last consideration is important. This is your chance to sell yourself. Merely stating where you worked and when you worked there will not get you a job. Instead, what did you achieve on the job? In this part of the resume, you should detail how you met deadlines, trained employees, cut expenses, exceeded sales expectations, decreased overage, managed money, and so forth. Plus, you want to quantify your accomplishments.

◀ EXAMPLE

Assistant Manager

McConnel Oil Change, Beauxdroit, LA 2014 to present

- Track and maintain over $25,000 in inventory
- Train a minimum of four new employees quarterly
- Achieved a 10 percent growth in service performed for three consecutive years
- Developed a user manual for hazardous waste disposal, earning a "Citizen's Recognition Award" from the Beauxdroit City Council

---

Listing your job title, company name, location, and dates of employment merely shows where you were in a given period of time. To prove your contributions to the company, provide specific details highlighting achievements.

Education. In addition to work experience, you must include your education. Document your educational experiences in reverse chronological order (most recent education first; previous schools, colleges, universities, military courses, and training seminars next). When listing your education, provide the following information:

- Degree. If you have not yet received your degree, you can write "Anticipated date of graduation June 2016" or "Degree expected in 2016."
- Area of specialization.
- School attended. Do not abbreviate. Although you might assume that everyone knows what *UT* means, your readers won't understand this abbreviation. Is UT the University of Texas, the University of Tennessee, the University of Tulsa, or the University of Toledo?
- Location. Include the city and state.
- Year of graduation or years attended.

As you can see, this information is just the facts and nothing else. Many people might have the same educational history as you. For instance, just imagine how many of your current classmates will graduate from your school, in the same year, with the same degree. Why are you more hirable than they are? The only way you can differentiate yourself from other job candidates with similar degrees is by highlighting your unique educational accomplishments. These might include any or all of the following:

Grade point average (generally if over 3.0 on a 4.0 scale)

Academic club memberships and leadership offices held

Unique coursework

Special class projects

Academic honors, scholarships, and awards

Fraternity or sorority leadership offices held

Number of hours you worked while attending school

Software and hardware certifications, or technical equipment you can operate

Please note a key concern regarding your work experience and education. You should have no chronological gaps when all of your work and education are listed. You can't omit a year without a very good explanation. (A missing month or so is not a problem.)

Professional Skills. If you are changing professions or re-entering the workforce after a long absence, you will write a functional résumé. Therefore, rather than beginning with education or work experience, which won't necessarily help you get a job, focus your reader's attention on your unique skills. These could include

Proficiency with computer hardware and software	New techniques you have invented or implemented
Procedures you can perform	Numbers of and types of people you have managed
Special accomplishments and awards you have earned	Machinery you can operate
On-the-job training you have received	Certifications you have earned
Training you have provided	Languages you speak, read, and write

These professional skills are important because they help show how you are different from all other applicants. In addition, they show that although you have not been trained in the job for which you are applying, you can still be a valuable employee.

EXAMPLE ▶

Highlight professional skills that will set you apart from other potential employees.

> **Professional Skills**
>
> - Proficient in Microsoft Word, Excel, Publisher, and PowerPoint
> - Knowledge of HTML, Java, Visual Basic, and C++
> - Certified OSHA Hazardous Management Safety Trainer
> - Fluent in Spanish and English
> - Completed Second Shift Administration Certificate

Military Experience. If you served for several years in the military, you might want to describe this service in a separate section. You would state the following:

Rank	Discharge status
Service branch	Special clearances
Location (city, state, country, ship, etc.)	Achievements and professional skills
Years in service	Training seminars attended and education received

Professional Affiliations. If you belong to regional, national, or international clubs or have professional affiliations, you might want to mention these. Such memberships might include the Rotarians, Lions Club, Big Brothers and Big Sisters, or Junior League. Maybe you belong to the Society for Technical Communication, the Institute of Electrical and Electronic Engineers, the National Office Machine Dealers Association, or the American Helicopter Society. Listing such associations emphasizes your social consciousness and your professional sincerity. Also include any offices you've held in these organizations.

References. Avoid a reference line that reads "Supplied on request," "Available on request," or "Furnished on request." Every employer knows that you will provide references if asked. Instead of wasting valuable space on your résumé with unnecessary text, use this space to develop your summary of qualifications, education, work experience, or professional skills more thoroughly. Create a second page for references, and bring this reference page to your interview. On the reference page, list three or four colleagues, supervisors, teachers, or community individuals who will recommend you for employment

(NOTE: Obtain their permission before listing them as references.). Provide their names, titles, addresses, and phone numbers. By bringing the reference page to your interview, you will show your prospective employer that you are proactive and organized.

Personal Data. Do not include any of the following information: birth date, race, gender, religion, height, weight, religious affiliations, marital status, or pictures of yourself. Equal opportunity laws disallow employers from making decisions based on these factors.

Effective Résumé Style

The preceding information suggests *what* you should include in your résumé. Your next consideration is *how* this information should be presented. As mentioned throughout this textbook, page layout is essential for effective technical communication. The same holds true for your résumé.

Choose Appropriate Font Types and Sizes. As with most technical communication, the best font types are Times New Roman, Calibri, and Arial. These are readable and professional looking. Avoid designer fonts, such as Comic Sans, and cursive fonts. In addition, use a 10- to 12-point font for your text. Smaller font sizes are hard to read; larger font sizes look unprofessional. Headings can be boldface and 14- to 16-point font size. Limit your résumé to no more than two font types: one for headings and one for text.

Avoid Sentences. Sentences create three problems in a résumé. First, if you use sentences, the majority of them will begin with the first-person pronoun *I*. You'll write, "I have . . . ," "I graduated . . . ," or "I worked. . . . " Such sentences are repetitious and egocentric. Second, if you choose to use sentences, you'll run the risk of committing grammatical errors: run-ons, dangling modifiers, agreement errors, and so forth. Third, sentences will take up room in your résumé, making it longer than necessary.

Format Your Résumé for Reader-Friendly Ease of Access. Instead of sentences, highlight your résumé with easily accessible lists. Set apart your achievements by bulletizing your accomplishments, awards, unique skills, and so on. In addition to bullets, make your résumé accessible by boldfacing headings and indenting subheadings to create white space. Avoid underlining headings or italicizing text. Most people find underlined text and italics hard to read.

Begin Your Lists with Verbs. To convey a positive, assertive tone, use verbs when describing your achievements. Use present tense verb for current jobs and education; use past tense verbs for past jobs and education. Following is a list of verbs you might use (Table 10.1).

TABLE 10.1 Active Verbs to Highlight Achievements

Accomplished	Designed	Initiated	Planned
Achieved	Developed	Installed	Prepared
Analyzed	Diagnosed	Led	Presented
Awarded	Directed	Made	Programmed
Built	Earned	Maintained	Reduced
Completed	Established	Managed	Resolved
Conducted	Expanded	Manufactured	Reviewed
Coordinated	Gained	Negotiated	Sold
Created	Implemented	Ordered	Supervised
Customized	Improved	Organized	Trained

Quantify Your Achievements.   Your résumé should not tell your readers how great you are; it should prove your worth. To do so, quantify by precisely explaining your achievements.

BEFORE	AFTER
Maintained positive customer relations with numerous clients.	Maintained positive customer relations with 5,000 retail and 90 wholesale clients.
Improved field representative efficiency through effective training.	Improved field representative efficiency by writing corporate manuals for policies and procedures.
Achieved production goals.	Achieved 95 percent production, surpassing the company's desired goal of 90 percent.
Trained employees.	Trained 20 employees annually in methods for safely removing hazardous waste from the workplace.

Make It Perfect.   You cannot afford to make an error in your résumé. Remember, your résumé is the first impression you'll make on your prospective employer. Errors in your résumé will make a poor first impression.

8. Distinguish between different methods of delivery for sending résumés.

Methods of Delivery

When writing either a reverse chronological or a functional résumé, you can deliver your document in several ways:

Mail Version

You can deliver a résumé by mailing it through the United States Postal Service. This résumé can be highly designed, using bullets, boldface, horizontal rules, indentations, and different font sizes. Because this document will be a hard copy, what the reader sees will be exactly what you mail. Do not be tempted to overdesign your résumé, however. For example, avoid decorative fonts, clip art, borders, or photos. Do not print your résumé on unusual colors, like salmon, baby blue, tangerine, or yellow. Instead, stick to heavy white paper and standard fonts.

TECHNOLOGY TIPS

Using Résumé Templates in Microsoft Word 2013

Microsoft Word 2013 provides a résumé template if you want help getting started.

1. Click on **File** and **New**.
2. Type Résumé in the search field to access résumé templates. You will be shown several options.

   These résumé templates provide benefits as well as create a few problems. On the positive side, the templates are great reminders of what to include in your résumé, such as objectives, work experience, education, and skills. In contrast, the templates also limit you and perhaps suggest that you include information that isn't needed. For example, the templates often mandate font sizes and page layout. Next, a few of the templates suggest that you include information about "Interests" and "References." Rarely should you include "Interests," such as hobbies. Furthermore, most experts suggest that you omit the "References" line, saving valuable résumé space for more important information. You can include references on a separate page, especially for interviews. More important, if you use the same templates that everyone else does, then how will your résumé stand out as unique? A good compromise is to review the templates for ideas and then create your own résumé with your unique layout.

New

Figures 10.2 and 10.3 are excellent examples of résumés, ready to be mailed. Figure 10.7, Letter of Application on page 246, is an example of a cover letter for Jody Seacrest's résumé.

E-mail Résumé

Delivering your résumé by mail can take several days. The quickest way to get your résumé to the prospective employer is as an e-mail attachment. Speed isn't the only issue. According to the Society for Human Resource Management, approximately one-third of recruiters and human resource managers prefer e-mailed résumés (Dixson).

Scannable Résumé

Many companies use computers to screen résumés with a technique called *electronic applicant tracking*. The company's computer program scans résumés as raster (or bitmap) images. Next, the software uses artificial intelligence to read the text, scanning for keywords. If your résumé contains a sufficient number of these keywords, the résumé will then be given to someone in the human resources department for follow-up.

FIGURE 10.2   Chronological Résumé

Sharon J. Barenblatt

1901 Rosebud Avenue
Boston, MA 12987
Cell phone: 202-555-2121
E-mail: sharonbb@juno.com

Summary of Qualifications

- Over five years customer service experience
- Experience in public relations, writing public service announcements
- Proven record of written and interpersonal communication abilities in technical writing
- Outstanding leadership skills, shown by my management background
- Fluent in Spanish

Education

BS, Business. Boston College. Boston, MA 2016

- 3.2 GPA
- Social Justice Chair, Sigma Delta Tau, 2016
- Study Abroad Program, Madrid, Spain 2015
- Internship, Ace Public Relations, Boston, MA 2014

Frederick Douglas High School. Newcastle, MA 2012

- 3.5 GPA
- Member, Honor Society
- Captain, Frederick Douglas High School tennis team

Work Experience

Salesperson/assistant department manager. Jessica McClintock Clothing Store. Boston, MA 2015 to present.

- Prepare nightly deposits, input daily receipts of up to $5,000
- Open and close the store
- Provide customer service to over 100 clients a day
- Trained six new employees

Salesperson. GAP Clothing. Newcastle, MA 2014 to 2015

- Assisted customers
- Stocked shelves

Professional Skills

- Made oral presentations to the Pan-Hellenic Council to advertise sorority philanthropic activities
- Helped plan community-wide "Paul Revere's Ride Day"
- Created advertising brochures and fliers for college-wide philanthropy netting over $25,000 donated to United Way

List your education and work experience in reverse chronological order.

List current jobs using present tense verbs and previous jobs using past tense verbs.

Do not only list where you worked and when you worked there. Also include your job responsibilities.

A scannable résumé can be e-mailed or sent through the mail. To create a scannable résumé, type your text using Notepad for Windows, Simpletext for Macintosh, or Note Tab, which is available as freeware. You also could type your résumé using Microsoft Word and save the document as a text file with a *.txt* extension.

FIGURE 10.3 Functional Résumé

JODY R. SEACREST
1944 W. 112th Street
Salem, OR 64925
(513) 451-4978
jseacrest12@hotmail.com

Professional Skills

- Operated a sporting goods/sportswear mail-order house. Business began as home-based but experienced 125% growth and was purchased by a national retail sporting goods chain.
- Managed a retail design studio producing over $500,000 annually.
- Hired, trained, and supervised an administrative staff of 15 employees for a financial planning institution.
- Provided purchaser training for office equipment, reducing labor costs by 25%.

Work Experience

Office manager, Simcoe Designs, Salem, OR 2015 to present.
Sales representative, Hi-Tech Office Systems, Salem, OR 2013 to 2015.
Office manager, Lueck Finances, Portland, OR 2011 to 2013.
President, Good Sports, Inc., Portland, OR 2009 to 2011.

Computer Proficiency

Microsoft Office XP, Visual Basic 6, C++, Oracle, Microsoft SQL Server, Network Administration

Military Experience

Corporal, U.S. Army, Fort Lewis, WA 2002-2008. Honorably discharged.
- Served as company network administrator.
- Planned and budgeted all IT purchases.

Education

BA, General Studies, Portland State University, Portland, OR 2001.

In a functional résumé, emphasize skills you have acquired which relate to the advertised position. Also quantify your accomplishments.

In a functional résumé, you list education and work experience in reverse chronological order.

A functional résumé is organized by importance. Begin with the skills or accomplishments that will get you the job. Place less important information lower in the résumé.

To create a successful scannable résumé, try these techniques:

- Use a Courier, Helvetica, or Arial typeface (10- to 14-point type).
- Place your name at the top of the page. Scanners read text at the top of a page as the writer's name. If your résumé goes to a second page, type your name and page 2 at the top.
- Avoid italics, underlining, colors, horizontal and vertical bars, and iconic bullets.
- White space is still important, but do not use your Tab key for spacing. Tabs will be interpreted differently in different computer environments. Use your space bar instead.
- Avoid organizing information in columns.
- Do not center text.
- Use headings and place your text below the headings, spacing for visual appeal.
- Create bullets using an asterisk (*) or a hyphen (-).
- Use keywords in your summary of qualifications, work experience, and professional skills.

Keywords are the most important feature of scannable résumés. Optical character recognition (OCR) searches focus on keywords and phrases specifically related to the job opening. The keywords include job titles, skills and responsibilities, corporate buzzwords, acronyms and abbreviations related to hardware and software, academic degrees, and certifications.

You can find which keywords to focus on by carefully reading the following:

- Job advertisements
- Your prospective employer's Web site
- Government job descriptions
- Industry-specific Web sites
- The *Occupational Outlook Handbook*
- Career-related discussion groups or blogs
- Sample résumés found online

When using keywords, be specific; avoid vague words and phrases.

BEFORE	AFTER
Knowledge of various software products	Can create online help using Author-iT and have expertise with PageMaker and Quark

BEFORE	AFTER
Familiar with computer technology	Proficient in multimedia, HTML, and Windows and Macintosh platforms

Figure 10.4 shows an excellent example of a scannable résumé.

Video Résumé

Union Square Ventures, a venture-capital firm, had an opening for an investment analyst. The company requested links to Twitter accounts or Tumblr blogs from applicants to show their Web presence. Applicants were also asked to submit vesumes. "Union Square says its process nets better-quality candidates" when asking for video profiles (Silverman).

To create an effective video résumé (or what some call a vesume), consider these five tips:

1. **Video quality.** The video must have clear resolution, excellent sound, and no background distractions (noises or images).
2. **Appearance and diction.** Just as you would strive to dress and behave professionally in an interview, you must also present a professional appearance in your video résumé. See our tips for professional appearance discussed later in this chapter. In addition to appearance, prospective employers will judge you on the content of your comments, what you say, and how you say it. Speak with confidence, clarity, and professionalism. Avoid mumbling, looking down, shuffling notes, or checking smartphones.
3. **Content.** Your video résumé, just like your hard-copy or electronic résumé, must focus on your objectives, education, work history, and special skills. Begin your video with an introductory overview introducing yourself and your goals; end the video résumé with a conclusion, summing up your value to the organization.

FIGURE 10.4   Scannable Résumé

Rochelle J. Kroft
1101 Ave. L
Tuscaloosa, AL 89403
Home: (313) 690-4530
Cell: (313) 900-6767
E-mail: rkroft90@aol.com

Place your name at the top of a scannable résumé and avoid centering text.

Summary of Qualifications

* Hazardous waste management with skills in teamwork, end-user support, OSHA quality assurance, and written documentation (minimum of ten reports weekly).
* Five years experience working with international and national businesses and regulatory agencies, including the Environmental Protection Agency and the Agency for Toxic Substances and Disease Registry.
* Skilled in assessing environmental needs and implementing hazardous waste improvement projects.
* Able to communicate effectively with multinational teams, consisting of clients, vendors, coworkers, and local and regional stakeholders.
* Excellent customer service (three-time winner of "Employee of the month").

Use key words to summarize your accomplishments.

Computer Proficiency

Microsoft Windows XP, PowerPoint, C++, Visual Basic, Java, CAD/CAM

Experience

Hazardous Waste Manager
Shallenberger Industries, Tuscaloosa, AL (2012 to present)

* Assess client needs for root cause analysis and recommend strategic actions.
* Oversee waste management improvements, using project management skills.
* Conduct and document follow-up quality assurance testing for over 25 clients monthly.
* Develop training manuals to ensure team and stakeholder safety. Shallenberger has had NO injuries throughout my management.
* Manage a staff of 25 employees.
* Achieved "Citizen's Recognition" Award from Tuscaloosa City Council for safety compliance record.

Type your scannable résumé in Courier, Arial, Verdana, or Helvetica. Avoid designer fonts like Comic Sans, Lucida, or Corsiva.

Hazardous Waste Technician
CleanAir, Montgomery, AL (2010–2012)

* Developed innovative solutions to improve community safety, including presentations at local K-12 public schools.
* Created new procedure manuals to ensure regulatory compliance.

Education

B.S., Biological Sciences, University of Alabama, Tuscaloosa, AL (2010)

* Biotechnology Honor Society, President (2009)
* Golden Key National Honor Society

Affiliations

Member, Hazardous Waste Society International

4. **Length.** A video résumé isn't a movie. Limit your video résumé to three to five minutes.

5. **Editing.** To achieve the desired length and to ensure that your video has a high quality, edit for errors, distractions, noises, and content. Review your video numerous times to make sure you haven't missed any problems. Then, if you need to, shoot the video again so you are an appealing job candidate.

Electronic and Hard-copy Portfolios

9. Prepare electronic and hard-copy portfolios.

As an enhanced component for your job search, consider using a portfolio. Résumés tell; portfolios show. A résumé tells an employer what you can do and how you'll benefit the company. Portfolios prove your résumé's assertions by showing examples of your skills, providing evidence of your accomplishments, and documenting your achievements.

If you are in technical writing, corporate communication, fashion merchandizing, heating/ventilation/air conditioning (HVAC), engineering, drafting, architecture, nursing, accounting, or graphic design, for example, you might want to provide the best examples of your work. These examples could include

Schematics	Published articles
Screenshots of PowerPoint presentations	Report samples
Outstanding performance reviews	Photographs
Outlines of presentations	Testimonials or letters of recommendation
CAD/CAM drawings	Brochures or fliers
Training and award certificates	Short video and audio files

Avoid sending an unsolicited portfolio to prospective employers. They will already be overwhelmed with résumés from job candidates and will not necessarily want to open additional documents. However, if an employer asks for a portfolio or when you go in for an interview, take your portfolio in hardcopy, on a CD, on a flash drive, or provide a URL link to an online site.

You can create a portfolio in at least two ways:

- Hardcopy
- Electronic

Hard-copy Portfolio

You can place your documents in a binder and bring it to the interview. Make the contents accessible by providing a title page, cover letter, table of contents, and tabs. Be sure your hard-copy portfolio examples are printed clearly and neatly on good quality paper—no blurred images, no wrinkled certificates, no smeared text from a printer low on ink.

Electronic Portfolio

Your portfolio also could be digital. Such portfolios are called *electronic* or *e-folio*s and can be delivered to your audience in PDF formats, PowerPoint formats, or Web-based, online formats.

- **PDF format.** Portable document formats (PDFs) start with a Word document. To create a PDF portfolio, include all items you want to highlight (publications, schematics, outlines, etc.) in your order of preference. Then, convert the document to a PDF format. A PDF ensures that your readers see your document exactly as you see it; formatting, layout, visuals, and fonts will not change dependent on your audience's software. Free PDF downloads include Adobe Reader, Cute PDF, PrimoPDF, PDFlite, Nitro PDF, and others.

Within your PDF portfolio, you can create links from portfolio tables of contents or résumé items. Then, if you e-mail a PDF portfolio to potential employers, they can click on the links to access individual pages or sections of the portfolio without having to scroll through the entire document. Figure 10.5 illustrates how PDF links are created by simply highlighting relevant text and right clicking.

FIGURE 10.5   Creating a PDF Link on a Résumé

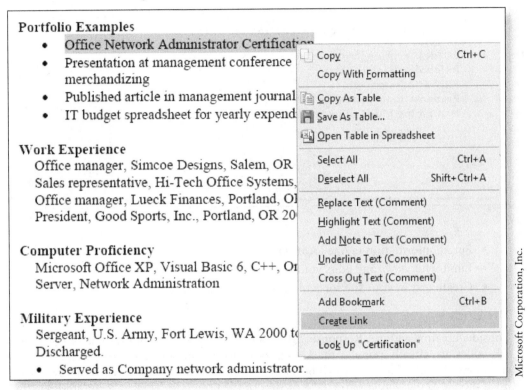

PowerPoint Format

Your PowerPoint (PPT) portfolio can be dynamic with video, audio, and motion. Save your PPT portfolio on either a CD or flash drive so the audience gets the full benefit of the PPT's capabilities. As with the PDF portfolio, the PPT portfolio can add hypertext links which allow your audience to access examples as they choose. See Figure 10.6 to learn how to create a hyperlink in a PPT portfolio by highlighting relevant text and right clicking.

PowerPoint Slides

See Chapter 20, "Oral Presentations," for more information about creating PowerPoint slides.

Web-based, Online Format

If you have a Web site or plan to create one, provide your audience a link to your online portfolio. With a Web-based portfolio (e-folio or webfolio), you can have hypertext links to PDF files, PowerPoint files, Word documents, video and audio files (vesumes), or Web pages.

An online portfolio proves to an employer that you have valuable technology skills. Google Sites provides an easy way to create a Web-based portfolio. In this free site, you can

Web Sites

See Chapter 13, "Web Sites and Online Help," for additional information about Web sites and hypertext links.

- Create and share Web pages
- Customize your site through templates

FIGURE 10.6 Hyperlink to Examples in a PowerPoint Portfolio

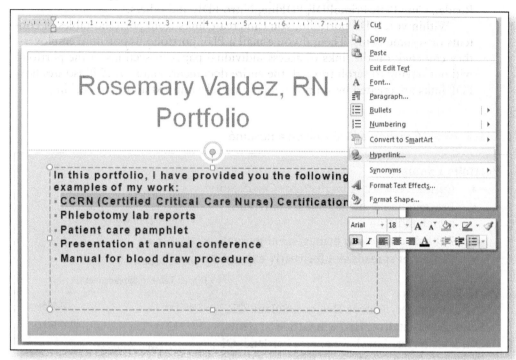

- Put all information in one location
- Limit who views and edits your content
- Create text in your site for mobile access (Google Sites)

QR Codes

Consider creating a Quick Response (QR) code and adding it to your résumé or portfolio. QR codes are already popular marketing tools, found on business cards, in magazines, and in retail stores. Now, they can be used in your job search. Prospective employers can use a mobile app or camera to take a picture of the QR code. Then, a scan interprets the code and provides access to your portfolio content, LinkedIn page, or telephone number directly from a smartphone or computer.

EXAMPLE ▶

To create your QR code, follow these steps:

1. Search for QR generators online.
2. Use a URL, phone number, LinkedIn profile, or e-mail address, for example, to generate the code.

3. Choose the output type, color, or size of your QR code.
4. Save and copy your code to your résumé or portfolio (Khare; Mantell).

Criteria for Effective E-mail Message or Letter of Application

10. Follow criteria for effective e-mail and letter applications.

Your résumé, whether hardcopy or electronic, will be prefaced by an e-mail or letter of application. The letter or e-mail serves different purposes than the résumé.

The résumé is fairly generic. You'll write a résumé, use it over and over again, modifying it for different positions. In contrast, the e-mail or letter of application is specific. Each e-mail or letter of application will be customized specifically for each job. Criteria for an effective e-mail or letter of application include the following.

Letter Essentials

Letters contain certain mandatory components: your address, the date, your reader's address, a salutation, the letter's body, a complimentary close, your signed name, your typed name, and an enclosure notation if applicable. If you are submitting an electronic résumé along with an e-mail cover message, you will not need these letter essentials. Both an e-mail message and letter of application include the following.

Formatting Letters and E-mail

See Chapter 9, "Routine Correspondence," for information about e-mail messages.

Introduction. In your introductory paragraph, include the following:

- Tell where you discovered the job opening. You might write, "In response to your advertisement in *CareerBuilder* . . . " or "Bob Ward, manager of human resources, informed me that. . . . "
- State which specific job you are applying for. Often, a company advertises several jobs. You must clarify which of those jobs you're interested in. For example, you could write, "Your advertisement for a computer maintenance technician is just what I have been looking for."
- Sum up your best credentials. "My BS in chemistry and five years of experience working in a hazardous materials lab qualify me for the position."

Discussion. In the discussion paragraph(s), sell your skills. To do so, describe your work experience, your education, and your professional skills. This section of your e-mail or letter of application, however, is not meant to be merely a replication of your résumé. In the discussion, follow these guidelines:

- State that you've attached a résumé for the e-mail message or enclosed a hard-copy résumé with the letter.
- Focus on your assets uniquely applicable to the advertised position. Select only those skills from your résumé that relate to the advertisement and which will benefit the prospective employer.
- Don't explain how the job will make you happy: "I will benefit from this job because it will teach me valuable skills." Instead, using the pronouns *you* and *your*, show reader benefit: "Working with governmental agencies has provided me a wide variety of skills from which your company will benefit."
- Quantify your abilities. Don't just say you're great ("I have outstanding customer service skills and communication abilities"). Instead, prove your assertions with quantifiable facts: "I won the 2016 Employee of the Year for providing solutions to customer concerns and working well with teammates."

Conclusion. Your final paragraph should be a call to action. You could say, "I am looking forward to discussing my application with you in greater detail. Then I can explain ways I could benefit your company." In addition, mention that you have enclosed a résumé. You can do this either in the introduction, discussion, or conclusion. Select the place that best lends itself to doing so. See Figure 10.7 for a letter of application example written to preface Jody Seacrest's functional résumé shown in Figure 10.3.

FIGURE 10.7 Letter of Application

1944 W. 112th Street
Salem, OR 64925
(513) 451-4978
jseacrest12@hotmail.com

February 11, 2016

Bill Baker
Human Resources Department
Eazi Marketing
10289 Ocean View
Portland, OR 67440

Subject: Application for Marketing Manager

On your Web site, I saw the posting for the position of Marketing Manager. As a dedicated business professional with expertise in marketing, I was excited to see this opening.

While I have enclosed my résumé, including education, work experience, and professional skills, allow me to elaborate on how I would be a positive addition to your organization.

- My marketing expertise is revealed through the 125% growth in revenues I helped achieve at my current job. In addition to face-to-face communication with customers and vendors, I also created and maintain the company's blog site and Facebook page.
- In my current position as manager at Simcoe Designs, I oversee 25 employees, a quarterly budget of $75,000, and a product line of over 1,000 different products. My job requires that I order, maintain stock, and troubleshoot delivery issues.
- My professional skill set includes proficiency with varied software languages, such as C++ and Visual Basic. I also have hardware management capabilities acquired through continuing education classes and workplace experience.
- At my current job, I recently won "Employee of the Month" for service exceeding the company's expectations.

I would like to have the opportunity to discuss the position and my applicable attributes further. Additionally, I would be happy to provide the selection committee with further information if needed. Thank you for your time and consideration.

Sincerely,

Jody Seacrest

Jody Seacrest
Enclosure: Seacrest Résumé

Jody highlights the skill set and positive attributes that will benefit the company in this letter of application. Note the quantification such as "125%," "$75,000," and "Employee of the Month."

Online Application Etiquette

If you send your résumé as an attachment to an e-mail message, be sure to follow online etiquette:

- **Do not use your current employer's e-mail system.** That clearly will tell your prospective employer that you misuse company equipment and company time.
- **Avoid unprofessional e-mail addresses.** Addresses such as Mustang65@aol .com, Hangglider@yahoo.com, or HotWheels@juno.com are inappropriate for business use. When you use e-mail to apply for a job, it is time to change your old e-mail address and become more professional. Use your initials or your name instead.
- **Send one e-mail at a time to one prospective employer.** Do not mass mail résumés. No employer wants to believe that he or she is just one of hundreds to whom you are writing.
- **Include a clear subject line.** Announce your intentions or the contents of the e-mail: "Résumé—Vanessa Diaz" or "Response to Accountant Job Opening."
- **Tell the reader how you have saved the attached résumé.** Specify whether the résumé is a Word, Works, RTF, or PDF file, for example.

Figure 10.8 shows an effective e-mail message prefacing an attached résumé.

FIGURE 10.8 Effective E-mail Message Prefacing an Attached Résumé

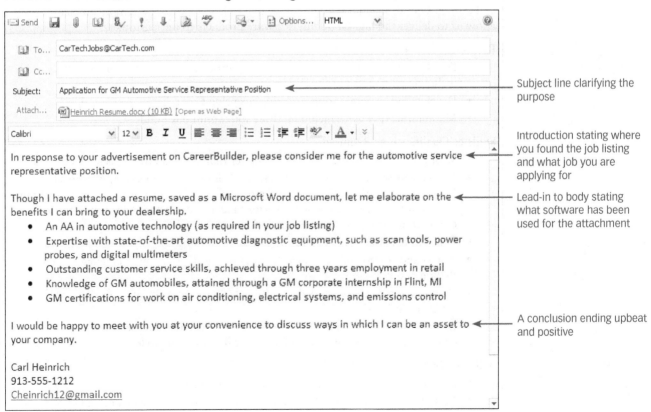

Microsoft Corporation, Inc.

11. Review a sample job package, including job advertisement, e-mail, and résumé.

An Example of a Job Package: Advertisement, E-mail, and Résumé

The following examples (Figures 10.9, 10.10, and 10.11) show how one individual responded to a job advertisement by submitting a cover e-mail and attached résumé. Note the ways in which the applicant customized his résumé and e-mail to meet the advertisement's requirements. The e-mail cover message takes the place of a letter of application because all of the communication is online.

FIGURE 10.9 Job Advertisement Found in an Online Site

Hi-Tech Industries

Employee Type: Full Time
Industry: Computer Software
Job Type: Entry Level

Description

Our company wants to hire a full-time Entry Level Software Engineer for the development of Aviation Engineering Tools.

Position Summary

Our company is looking for self-motivated, passionate, and creative software engineers to develop tools that will reduce manual overhead, improve processes, and automate where possible. As a member of this team, you will participate in planning, design, development, and testing. Our ideal candidate should be able to help us work even more efficiently and allow us to deliver the highest quality software products to our customers.

Responsibilities

- Develop and maintain tools to be used by product development engineers, including all phases of the SDLC
- Develop and maintain tools and plug-ins using C and C++
- Maintain servers used, including those for continuous builds

Skills & Qualifications Required

- Strong academics (cumulative GPA greater than or equal to 3.5 as a general rule)
- Bachelor of Science in Computer Science, Computer Engineering, Aerospace Engineering, Math, or Physics from a four-year college or university, or a minimum of four years of work experience performing a role substantially similar to the essential functions of this job description
- Relevant experience and/or training in programming languages such as C and C++
- Some exposure to and coursework in Java, as well as scripting languages such as Python and Perl
- Detail-oriented, able to manage multiple tasks proactively and effectively with minimal supervision
- Demonstrated strong and effective verbal, written, and interpersonal communication skills with a collaborative development style

Skills & Qualifications Preferred

- Previous experience working in a team environment
- More than two years of C/C++ and/or Java, and object oriented design experience
- Experience with software testing

FIGURE 10.10 E-mail Message Prefacing Attached Résumé in Response to a Job Advertisement

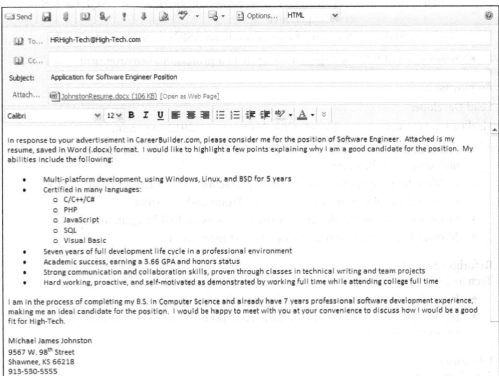

An applicant can rarely meet every job requirement listed in a job advertisement. However, Michael shows that he meets many of the advertised requirements.

Although Michael has not completed his B.S., a requirement for the job, he explains that his years of experience and other attributes outweigh the issue.

FIGURE 10.11 Reverse Chronological Résumé Responding to a Job Advertisement

Michael J. Johnston

9567 W. 98th St.   Shawnee, KS 66218
(Cell Ph.) 913-530-5555   michael.johns@gmail.com

Summary of Qualifications

- Experienced developer, total of seven years in professional software development
- Capable of multi-platform development with five years programming for Windows, Linux, and BSD platforms
- Four years experience with relational database programming and database table design
- Proficient in over ten programming languages, including the following:
  - C/C++   Knowledge of procedural and object oriented paradigms with three years of experience
  - Java   Developed Android based applications for one year
  - C# .Net   Well-versed with five years of professional experience making various applications
  - PHP   Experience in conjunction with Android application development for one year
  - JavaScript   Self-taught for five years

Michael's qualifications accent the ways in which he meets the company's advertised needs.

(Continued)

FIGURE 10.11 (Continued)

○ SQL — Over five years of professional experience of T-SQL for MS-SQL server experience

○ Visual Basic — One year of experience in an ASP.Net environment

- Seven years of full development life cycle in a professional environment

Employment

Lead Developer

Harken Law Group   Kansas City, MO   2012 to present

- Independently designed, developed, documented, and deployed applications including the following:
  ○ More than 30 applications for data extraction, transform, and load (ETL)
  ○ Over 20 applications for data manipulation and reporting
- Promoted to Lead Developer after three years as an EDI Programmer
- Managed two programmers during the last three years

Refurbish Specialist

Tech and U   Lenexa, KS   2008 to 2012

- Upgraded existing software for printer testing
- Built, repaired, and cleaned corporate class printers, including Hewlett Packard, Lexmark, and Epson

Education

A.S., General Science

Kingston Community College Overland Park, KS Anticipated Date of Graduation 2016

- GPA of 3.66
- Member, Dean's List
- Member, Phi Theta Kappa honor society
- Working 40 hours weekly while attending college full time (6 to 15 credit hours)

Professional Skills

- Proficient with MS Word, PowerPoint, Excel, Access, MS SQL Server Manager, and Visual Studio
- Capable of writing documentation due to acquired college education in professional writing
- Two years of experience teaching basic C# programming to three coworkers
- One year of working with network penetration testing

---

12. Learn how to interview effectively.

Techniques for Interviewing Effectively

The goal of writing an effective résumé and letter of application or cover e-mail message is to get an interview. The résumé and letter of application or e-mail may open the door; only a successful interview will win you the job. In fact, some sources suggest that the interview is the most important stage of your job search. The Society for Human Resource Management states that 95 percent of respondents ranked interviewing as a very influential factor when deciding to hire an employee. Your performance during an interview is more important than other criteria including education, references, test scores, and years of experience (Stafford).

Dress Professionally.   Professionalism starts with your appearance. The key to successful dressing is to wear clean, conservative clothing. No one expects you to spend money on high-fashion, stylish clothes, but everyone expects you to look neat and acceptable. Business suits are still best for both men and women.

Be on Time.   Plan to arrive at your interview at least 20 to 30 minutes ahead of schedule. That way, you won't have to worry about unexpected traffic, crowded parking lots, or finding the correct building.

Watch Your Body Language.   To make the best impression, don't slouch, chew your fingernails, play with your hair or jewelry, or check your watch. These actions will make you look edgy and impatient. Sit straight in your chair, even leaning forward a little to show your enthusiasm and energy. Look your interviewer in the eye. Smile and shake your interviewer's hand firmly.

Don't Chew Gum, Smoke, or Drink Beverages during the Interview.   The gum might distort your speech; the cigarette will probably offend the interviewer, particularly if he or she is a non-smoker; and you might spill the beverage.

Turn Off Your Smartphone.   Today, smartphones are commonplace. However, the interview room is one place where this device must be avoided. Taking a call while you are being interviewed is rude and will ensure that you will not be hired. Reading a text message or sending one during an interview is not appropriate. Your interviewer will not appreciate it if your smartphone rings during your meeting.

Watch What You Say and How You Say It.   Speak slowly, focus on the conversation, and don't ramble. Once you have answered the questions satisfactorily, stop.

Bring Supporting Documents to the Interview.   Supporting documents can include extra copies of your résumé, a list of references, letters of recommendation, employer performance appraisals, a portfolio (hard copies and electronic version), or transcripts.

Research the Company.   Show the interviewer that you are sincerely interested in and knowledgeable about the company. Most successful job candidates show interviewers that they know the company's product, service, history, mission statement, client base, and more.

Be Familiar with Typical Interview Questions.   You want to anticipate questions you will be asked and be ready with answers. Some typical questions include the following:

What are your strengths and weaknesses?	Can you travel for work-related activities?
Why do you want to work for this company?	Will you relocate?
Why are you leaving your present employment?	What do you want to be doing in 5 years, 10 years?
What did you like least about your last job?	How would you handle this (hypothetical) situation?
What computer hardware are you familiar with, and what computer languages do you know?	What was your biggest accomplishment in your last job or while in college?
What machines can you use?	What about this job appealed to you?
What special techniques do you know, or what special skills do you have?	What starting salary would you expect?
What did you like most about your last job?	How do you get along with colleagues and with management?

When Answering Questions, Focus on the Company's Specific Need.   For example, if the interviewer asks if you have experience using RoboHelp or C++, explain your expertise in that area, focusing on recent experiences or achievements. Be specific. In fact, you might want to tell a brief story to explain your knowledge. This is called "behavioral description interviewing." It allows an interviewer to learn about your speaking abilities, organization, and relevant job skills. To respond to a behavioral description interview question, answer as follows:

- Organize your story chronologically.
- Tell who did what, when, why and how.
- Explain what came of your actions (the result of the activity).
- Depict scenes, people, and actions.
- Make sure your story relates exactly to the interviewer's needs.
- Stop when you are through—do not ramble. Get to the point, develop it, and conclude. If, however, you do not have the knowledge required, then explain how you have applicable experience. You could say, "Although I've never used RoboHelp, I have created online help using HTML coding. Plus, I'm a quick learner. I was able to learn FlashHelp and XML well enough to create online help screens in only a week. Our customer was very happy with the results." This will show that you understand the job and can adapt to any task you might be given.

Criteria for Effective Follow-Up Correspondence

13. Understand criteria for effective follow-up correspondence.

Once you have interviewed, don't just sit back and wait, hoping that you will be offered the job. Write a follow-up letter or e-mail message. This follow-up accomplishes three primary goals: It thanks your interviewers for their time, keeps your name fresh in their memories, and gives you an opportunity to introduce new reasons for hiring you.

A follow-up letter or e-mail message contains an introduction, discussion, and conclusion.

- **Introduction.** Tell the readers how much you appreciated meeting them. Be sure to state the date on which you met and the job for which you applied.
- **Discussion.** In this paragraph, emphasize or add important information concerning your suitability for the job. Add details that you forgot to mention during the interview, clarify details that you covered insufficiently, and highlight your skills that match the job requirements. In any case, sell yourself one last time.
- **Conclusion.** Thank the readers for their consideration, or remind them how they can get in touch with you for further information. Don't, however, give them any deadlines for making a decision.

For an e-mail follow-up, you would include your reader's e-mail address and a subject line, such as "Thank You for the July 8 Interview" or "Follow-up to July 8 Interview."

EXAMPLE ▶

For a hard-copy follow-up letter, you would include all letter components: writer's address, date, reader's address, salutation, complimentary close, and signature.

> Thank you for allowing me to interview with Acme Corporation on July 8. I enjoyed meeting you and the other members of the team to discuss the position of account representative.
>
> You stated in the interview that Acme is planning to expand into international marketing. With my Spanish speaking ability and my study-abroad experience, I would welcome the opportunity to become involved in this exciting expansion.
>
> Again, thank you for your time and consideration. I look forward to hearing from you. Please e-mail me at gfiefer21@aol.com.

This correspondence succeeds for several reasons. First, it is short, merely reminding the reader of the writer's interest instead of overwhelming him or her with too much new information. Second, the message is positive, using words such as *enjoyed, ability, welcome, opportunity, exciting,* and *thank you.* Finally, the correspondence provides the reader an e-mail address for easy follow-up.

Job Acceptance Letter or E-mail Message

14. Write successful job acceptance letters or e-mail.

Great news! After working hard to find a job, your efforts have paid off. You've just received a job offer. Now what? Sometimes, accepting the offer over the phone isn't enough. Your new employer might want you to write and sign an official acceptance letter or submit an e-mail message. In this brief correspondence, you will want to accomplish the following goals:

- Thank the company for the job opportunity.
- Officially accept the job offer.
- Restate the terms of employment (salary, benefits, location, position, job responsibilities, and/or start date).

Address the letter or e-mail to the individual who offered you the position for a more personalized touch. Be sure to include your phone number, e-mail address, or mailing address, just in case the company needs to contact you. This acceptance correspondence actually could be seen as your first day on the job. Therefore, take as much care in writing this letter or e-mail as you did in applying for the job. Make sure your letter or e-mail is grammatically correct, well organized, and conveys a positive tone. Show your new boss or colleagues that you are a professional asset to the company.

Figure 10.12 provides a sample job acceptance letter.

FIGURE 10.12 Job Acceptance Letter

Amy Zhang
9103 Stonefield Rd.
Georgetown, TX 77829

May 15, 2016

George Smithson
Dell Computers
2134 Silicon Way
Round Rock, TX 77112

Dear Mr. Smithson:

In response to your phone call yesterday (5/14/16), I am happy to accept the position as a Technical Sales Representative II at Dell Computers. I know that with my education, experience, and energy, I will be an asset to your workforce.

As we discussed, my salary will be $35,000 with health and life insurance benefits provided after 90 days of employment. I will work in your Round Rock, Texas, facility, where my job will entail working with a team to meet the schedules, budgets, product costs, and production ramp rates for projects assigned to my team and me. In addition, I will manage small to medium platform/peripheral development programs.

I look forward to joining the Dell team on June 10, 2016. If you have any questions or need additional information, please let me know. Thank you, Mr. Smithson, for this outstanding opportunity.

Sincerely,

Amy Zhang

Amy Zhang
713-555-0112
azhang@gt.edu

Job Openings

_____ 1. Did you visit your college or university job placement center?

_____ 2. Did you talk to your professors about job openings?

_____ 3. Have you networked with friends or previous employers, and have you considered using social media for networking?

_____ 4. Have you checked with your professional affiliations or looked for job openings in trade journals?

_____ 5. Did you search the Internet for job openings?

Résumé

_____ 1. Are your name, address, and phone number correct?

_____ 2. If you included a career objective, is it specific and tailored to the job?

_____ 3. Have you included a summary of qualifications?

_____ 4. Is all information within your education, work experience, and military experience sections accurate?

_____ 5. Have you used lists beginning with verbs?

_____ 6. Have you quantified each of your achievements?

_____ 7. Have you avoided using sentences and the word "I"?

_____ 8. Does your résumé use bullets, headings, and no more than two different font types and sizes to make it reader-friendly?

_____ 9. Have you proofread your résumé to find grammatical and mechanical errors?

_____10. Have you decided whether you should write a reverse chronological résumé or a functional résumé?

Application Letter or E-mail Message

_____ 1. Have you included all of the letter essentials for an application letter?

_____ 2. Does your introductory paragraph state where you learned of the job, which job you are applying for, and your interest in the position?

_____ 3. Does your message's discussion unit pinpoint the ways in which you will benefit the company?

_____ 4. Does your letter or e-mail message's concluding paragraph end cordially and explain what you hope your reader will do next?

_____ 5. Is your application letter or e-mail free of all errors?

Interview

_____ 1. Will you dress appropriately?

_____ 2. Will you arrive ahead of time?

_____ 3. Have you practiced answering potential questions?

_____ 4. Have you researched the company so you can ask informed questions?

_____ 5. Will you bring to the interview additional examples of your work or copies of your résumé? Will you bring either an electronic or hard-copy portfolio?

Follow-up Letter or E-mail Message

_____ 1. Have you included all the letter essentials if you wrote a letter?

_____ 2. Does your introductory paragraph remind the readers when you interviewed and what position you interviewed for?

_____ 3. Does the discussion unit highlight additional ways in which you might benefit the company?

_____ 4. Does the concluding paragraph thank the readers for their time and consideration?

_____ 5. Does your letter or e-mail avoid all errors?

Job Acceptance Letter or E-mail Message

_____ 1. Have you included all the letter essentials if you wrote a letter?

_____ 2. Does your introductory paragraph explain why you are writing (in response to a job offer) and what you are writing about (accepting the job)?

_____ 3. Does the discussion unit confirm the particulars of the offer (salary, benefits, job duties, location, start date, and so on)?

_____ 4. Does the concluding paragraph thank the reader for the job opportunity?

_____ 5. Does your letter or e-mail message avoid errors and show your professionalism?

The Writing Process at Work

Effective communication in the job search follows a process of prewriting, writing, and rewriting. To illustrate the importance of the writing process in the job search, look at how LaShanda E. Brown used prewriting, writing, and rewriting to create an effective résumé so she could apply for a job in the field of human resources.

Prewriting

Before writing a résumé, use prewriting to help you begin a job search and determine your objectives. Beginning with a list of goals, audience, channels of communication, and material to include will get you started. LaShanda used reporter's questions to plan her job search (Figure 10.13).

FIGURE 10.13   LaShanda's Reporter's Questions

- **Who** should I send my resume to?
- **What** should I include about my experience in human resources? What should I include about education?
- **When** is the resume due?
- **Where** should I send the resume (what's the city/state address, e-mail address, or URL)?
- **Why** am I interested in this job posting?
- **What** have I learned about the company's culture?
- **How** should I send the resume—by e-mail, online form, or hard copy?
- **How** can I show benefit to the company hiring me?

Writing

After determining objectives by prewriting, you can begin drafting your résumé. LaShanda went to her husband's military career center for help writing her résumé. They told her to use a functional résumé because that's what they tended to suggest for people leaving the military. LaShanda drafted her résumé, creating a functional résumé that used information she discovered in prewriting (Figure 10.14).

Rewriting

Editing and revising a résumé will help you make it look professional and achieve the desired result of getting an interview. After getting no job interviews, LaShanda realized that her functional résumé wasn't working. She had to consider the following aspects of her résumé that needed revision:

- Her career objective's line did not match her new search for a job in human resources.
- The functional résumé listed jobs in one part of the résumé and then responsibilities in another part of the text.
- Directors of human resources could not match dates and duties.
- The functional résumé did not clarify her achievements and education.
- She needed to delete information from past employment that was not relevant for her current job search.
- She needed to include courses and professional development that proved she was up to date in the field of human resources.

She changed strategies and restructured the functional résumé into a reverse chronological résumé. With the revised résumé (Figure 10.15), LaShanda got an interview. Today she works in her desired field of human resources and gets to review job applicants' résumés and interview them.

FIGURE 10.14   Draft of a Functional Résumé

LaShanda E. Brown
833 Hampton Rd., Apt. 1
Virginia Beach, VA 23460
757-555-5555
labrown10@yahoo.com

Objective: Retail Management

HIGHLIGHTS OF QUALIFICATIONS

- 2 years retail management and recruiting experience.
- Expertise in oral and written communications. Able to present ideas and goals clearly.
- Motivated team player that thrives in a fast-paced, multi-faceted environment.
- Proficient in Microsoft Office (Word, Excel, PowerPoint, Outlook, Access), as well as basic office hardware (fax, copiers, cash registers)
- Familiar with most video game consoles (PS1, PS2, Xbox, NES, etc.)

EXPERIENCE

Management

- Supervised 5 management trainees and taught them all aspects of the rental car industry to include accounts receivable, marketing accounts as well as competition, managing a rental fleet, etc.
- Oversee 4 telemarketers and 2 sales representatives. Managed training and quality customer service when representing the company via telephone.
- Improved loss prevention programs within a retail setting by breaking the store up into teams and zones so that all areas were covered at all times. Loss prevention decreased 6% in 3 months.
- Earned All-stars sales award 4 months in a row. Average of 91% optional coverage sales on rentals within that four-month period.

Customer Service

- Implemented new strategies for customer satisfaction, such as interactive games to give customers an opportunity to win coupons on current and future rentals. Customer satisfaction increased 15%.
- Greeted customers enthusiastically and efficiently to provide a positive company image and impression.
- Researched and provided solutions to customer service issues, ensuring customer needs were met and they were satisfied.
- Coordinated customer follow-ups, ensuring they had a good rental experience with us and would return if they needed our rental services again.

EMPLOYMENT

2012–Present	Office Manager	Best Value Remodeling
2010–2012	Insurance Sales Agent	Geico Direct
2010	Management Assistant	Enterprise Rent-A-Car

EDUCATION AND TRAINING

- Bachelors in Communication, emphasis in Public Relations
  Missouri Western State College, St. Joseph, MO
- Elite Sales Training, Enterprise Rent-a-car
- Effective Management Training, Best Value Remodeling

FIGURE 10.15   Revised Reverse Chronological Résumé

LaShanda E. Brown
833 Hampton Rd., Apt. 1
Virginia Beach, VA 23460
757-555-5555
labrown10@yahoo.com

Career Objective

To work in the field of human resources management, using my communication skills to help an organization meet its goals

Summary of Qualifications

- Screened and hired over 24 applicants for employment
- Conducted annual performance reviews as an office manager
- Presented at college job fairs as a management assistant
- Made PowerPoint presentations
- Proficient at interpersonal communication and public speaking, as evident from my coursework in communication studies

LaShanda emphasizes her skills as they relate to a job search in human resources.

Education

Bachelor of Science in Communication Studies, Public Relations emphasis, May 2012 Missouri Western State College, St. Joseph, MO
- Related courses: small group communication, presentational communication, consumer marketing, persuasive speech, media in communications, public relations communication analysis, human resources management, nonverbal communication, advertising, desktop publishing.

The education section highlights human resource experience, allowing the city's human resources department to see that LaShanda meets the requirements for a position.

Employment
Office Manager
Best Value Remodeling   Virginia Beach, VA   2012–Present
- Make spreadsheets using MS Excel to track sales by outside sales teams and manage commissions
- Update Microsoft Access database to track results from sales calls
- Create PowerPoint presentations for general manager to use when going on corporate sales leads
- Oversee four telemarketers and two sales representatives

Insurance Sales Agent
Geico Direct   Virginia Beach, VA   2010–2012
- Provided excellent customer service on the phone
- Performed data entry by setting up insurance policies

Management Assistant
Enterprise Rent-a-Car   Oklahoma City, OK   2010
- Won award as "Best New Hire 2010"
- Supervised five management trainees
- Created advertisements locally for the branch
- Improved loss prevention programs by 6% in three months
- Earned All-stars sales award four months in a row. Average of 91% optional coverage sales on rentals within that four-month period

Software Knowledge
- Proficient with Microsoft Office programs, including Word, Excel, Access, PowerPoint, Outlook, and Project
- Proficient with PeopleSoft program (HR module)
- Proficient with Lotus programs, including Lotus Notes, Lotus 1-2-3

The reverse chronological résumé better suits her needs in that it matches job responsibilities with specific jobs and dates.

CASE STUDY

1. Rewrite the flawed resume on page 259. In doing so, revise the errors and create three different types of résumés for Macy G. Heart: a chronological résumé, a functional résumé, and a scannable résumé.

2. DiskServe, a St. Louis-based company, is hoping to hire a customer service representative for their computer technology department. In addition to DiskServe's CEO, Sarah Beske, the hiring committee will consist of two managers from other DiskServe departments and two coworkers in the computer technology department.

   The position requires a bachelor's degree in information technology (or a comparable degree) and/or four years' experience working with computer technology. Candidates must have knowledge of C++, Visual Basic (VB), SQL, Oracle, and Microsoft Office applications. In addition, customer service skills are mandatory. Four candidates were invited to DiskServe's work site for personal interviews—Macy Heart, Aaron Brown, Rosemary Lopez, and Robin Scott.

   Macy has a bachelor's degree in computer information systems. He has worked two years part time in his college's technology lab helping faculty and students with computer hardware and software applications, including Microsoft Office and VB. He worked for two years at a computer hardware/software store as a salesperson. His supervisor considers Macy to be an outstanding young man who works hard to please his supervisors and to meet customer needs. According to his supervisor, Macy's greatest strength is customer service, since Macy is patient, knowledgeable, and respectful.

   Aaron has an information technology certificate from Microsoft, where he has worked for five years. Aaron began his career at Microsoft as a temporary office support assistant, but progressed to a full-time salesperson. When asked where he saw himself in five years, Aaron stated, "The sky's the limit." References proved Aaron's lofty goals by calling him "a self-starter, very motivated, hardworking, and someone with excellent customer service skills." He is taking programming courses at night from the local community college, focusing on C++, Visual Basic, and SQL.

   Rosemary has an associate's degree in information technology. She has five years' experience as the supervisor of Oracle application. Prior to that, Rosemary worked with C++, VB, and SQL. She also has extensive knowledge of Microsoft Office. Rosemary was asked "How have you handled customer complaints in the past?" and responded "I rarely handle customer complaints. In my last job, I assigned that work to my subordinates."

   Robin has a bachelor's degree in information technology. To complete her degree, Robin took courses in C++, VB, Oracle, and SQL. She is very familiar with Microsoft Office. Since Robin just graduated from college, she has no full-time experience in the computer industry. However, she worked in various retail jobs (food services, clothing stores, and book stores) during high school, summers, and in her senior year. She excelled in customer service, winning the "Red Dragon Employee of the Month Award" from her last job as a server in a Chinese restaurant.

Assignment

Who would you hire? Give an oral presentation or write an e-mail or memo to Sarah Beske, DiskServe's CEO, explaining which of the candidates she should hire.

1890 Arrowhead Dr.
Utica, MO 51246
710-235-9999

Resume of Macy G. Heart

Work Experience

Jan. 2016 to now Aramco.net St. Louis, Missouri Tech Support Specialists
Primarily I provide customer support for customer problems with C++, Visual Basic, Java, networking, and databases (Access, Oracle, SQL). I provide solutions to software and hardware problems and respond to e-mail queries in a timely manner.

Oct. 2014 to Dec. 22, 2016 DocuHelp Chesterfield, MO Computer Consultant
Provide technical support for PCs and Macs. I also trained new PC and Mac users in hardware applications. When business was slow, I repaired computer problems, using my many technology skills.

May 2012 to Oct. 2014 Ram-on-the-Run East St. Louis, Illinois Computer Salesman
Sold laptops, PCs, printers, and other computer accessories to men and women. Answered customer questions. Won "Salesman of the Month Award" three months in a row due to exceeding sales quotas.

Jan. 2010 to May 2012 Carbondale High School Carbondale, IL Lab Tech
Worked in the school's computer lab, helping Mr. Jones with computer-related classwork. This included fixing computer problems and tutoring new students having trouble with assignments.

Education

Aug. 2016   Bachelor's Degree, Information Technology, Carbondale Institute of Technology, Carbondale, IL
Concentration: Database/Programming Applications
Relevant classes: Business Information Systems, Hardware Maintenance, Database Management, Visual Basic, Systems Analysis and Design, C++, Web Design

May 2014   Graduate Carbondale High School
Member of the Computer Technology Club
Member of FFA and DECA
Principal's Honor Roll, senior year

Computer Expertise

Cisco Certified, knowledge of C++, VB, Microsoft Office Suite XP, HTML, Java, SML

References

Mr. Oscar Jones, Computer Applications Teacher, Carbondale High School
Mr. Renaldo Gomez, Manager, Aramco.net
Mr. Ted Harriot, Technical Support Supervisor, DocuHelp

Additional Information

A good team player, who works well with others
Maintained 4.0 in my college major classes
Built my own computer from scratch in high school
Starting football player in high school, Junior Varsity tight end

ETHICAL CHALLENGE

Shelly Gurwell just graduated from college with a degree in business. She has begun interviewing for jobs at brokerage firms. During one interview, Shelly was asked the following questions:

1. Do you plan on getting married?
2. Do you want to have children?
3. When you have children, who would take care of them if they got sick?
4. Would you move to another city if your husband had to relocate?
5. What's your undergraduate GPA?
6. Do you require days off for religious observance?
7. What are your salary requirements?
8. Do you plan to participate in our company's wellness and weight loss program?
9. Do you have any tattoos, and if so, where are they?
10. Were you in a sorority?

Question

Which of these questions are ethical? Why or why not?

INDIVIDUAL AND TEAM PROJECTS

1. Practice a job search. To do so, find examples of job openings. Bring these job possibilities to class for group discussions. From this job search, you and your peers will get a better understanding of what employers want in new hires.

2. An informational interview can help you learn about the realities of a specific job or work environment. Interview a person currently working in your field of interest. Once you find an employee willing to help, visit with him or her and find out the following:

   - What job opportunities exist in your field?
   - What does a job in your field require, in terms of writing, education, interpersonal communication skills, teamwork, and so on, as well as the primary job responsibilities?

   After gathering this information, write a thank-you letter or e-mail to the employee who helped you. Then, write an e-mail to your professor documenting your findings and give an oral presentation to your classmates.

3. Write a résumé. To do so, follow the suggestions provided in this chapter. Once you have constructed this résumé, bring it to class for peer review. In small groups, discuss each résumé's successes and areas needing improvement.

4. Write a letter of application or e-mail message according to the suggestions provided in this chapter. Next, in small groups, review your correspondence for suggested improvements.

5. Practice a job interview in small groups, designating one student as the job applicant and other students as the interview committee. Ask the applicant the sample interview questions provided in this chapter or any others you consider valid. This will give you and your peers a feel for the interviewing process.

PROBLEM-SOLVING THINK PIECES

1. You need to submit a résumé for a job opening. However, you have problems with your work history, such as the following:

   - You have had no jobs.
   - You have worked as a baby sitter, or you have cut grass in your neighborhood.
   - You have been fired from a job.
   - You have been out of work for five years raising a family.
   - You have had three (or more) jobs in one year.

   Consider how you would meet the challenges of your job history.

2. You have found a job that you want to apply for. The job requires a bachelor's degree in a specific field. Though you had been enrolled in that specific degree program for three years, you never completed the degree. What should you say in your résumé and letter of application to apply for this position, even though you do not meet the degree requirement? Which type of résumé (reverse chronological or functional) should you write? Explain your answer.

3. During an interview, you are asked to "describe a problem you encountered at work and explain how you handled that challenge." How do you answer this typical question, but avoid giving an answer that paints a negative picture of a boss, coworker, or your work environment?

WEB WORKSHOP

Using an Internet search engine, find emerging careers, jobs that didn't exist until recently. Which companies are hiring, what skills do they want from prospective employees, and what keywords are used to describe preferred skills in this work field? Report your findings either to your instructor by writing a memo or e-mail, or give an oral presentation to your class about the job market in your field.

CHAPTER ELEVEN

Communicating to Persuade

After completing this chapter, you will be able to

1. Understand the importance of argument and persuasion in technical communication

2. Know the traditional methods of argument and persuasion

3. Use the *ARGU* technique for persuasive communication

4. Avoid unethical logical fallacies

5. Prepare different types of persuasive documents

COMMUNICATION AT WORK

In this scenario, Amanda Carroll at TechToolshop relies on persuasive writing to expand business.

TechToolshop provides sales, service, maintenance, installation, and data recovery for PC and Mac hardware and software tools. The company installs, repairs, and maintains workstations, servers, printers, and peripherals. Through an online catalog and storefront site, TechToolshop sells printers, desktops, mainframes, minitowers, software, accessories, and encryption devices. Located in Big Springs, Iowa, this business of 1,200 employees began as a small computer store in 2005. Now, however, the company wants to expand its business.

As Director of Marketing Communications at TechToolshop, Amanda Carroll's job is to communicate information to the public about product offerings, company product line changes, environmentally responsible products, and warranty changes. Many of Amanda's written documents and oral presentations depend on persuasion. To be persuasive, she relies on the traditional methods of argument by appealing to the emotions, being logical, and maintaining ethical standards. She wants to communicate effectively to ensure that information contributes to "product branding" and, in turn, enhances the profitability of the company.

Amanda uses e-mail messages, text messages, instant messages, sales letters, brochures, and fliers to promote TechToolshop's expansion plans. To persuade new customers to visit the store and online sites, Amanda is conducting a multifaceted viral marketing campaign. This will include

- **Full-color brochures.** These focus on the company's primary functions: computer hardware and software sales and maintenance, data recovery, installation, and repair.

- **Fliers.** Amanda uses these to advertise what's on sale this month, recent industry news that affects the computer user, innovative software information, repair and replacement steps for software updates, best prices for used computer parts, and so on. In contrast to the brochures, the company's single-sided, postcard-sized fliers are more cost effective. These fliers, which can be produced quickly, will focus on individual topics where quick turnaround is important.

- **A corporate Facebook site.** On this social media site, TechToolshop will encourage customer involvement. The company's Facebook site will post photographs of customers who have used TechToolshop's products. In addition, TechToolshop will advertise time-sensitive events, such as holiday specials or limited-time sales, to encourage return visits to the site. Customers who invite their friends to visit the TechToolshop Facebook site will receive discounts on future purchases.

- **Twitter updates.** Through 140-character tweets, TechToolshop will update customers on the arrival of new products, advertise daily specials, and notify followers of upcoming events. Through retweets, messages sent by TechToolshop customers to their friends, the company benefits from an electronic mass marketing plan.

- **Web site and company blog.** The Web site and blog will allow customers to interact with the company. In the blog, customers can suggest additional parts or service the company can offer as it expands.

One of Amanda's major goals is maintaining excellent customer relations. Hard work and client referrals obviously help contribute to this goal. However, Amanda has found that continuous communication is another way to build client rapport and persuade people to do business with TechToolshop. Though TechToolshop's business is increasing, it wants to keep adding customers. Outstanding technical communication through persuasive sales letters, social media, brochures, and fliers equals more business for TechToolshop.

The Importance of Argument and Persuasion in Technical Communication

> 1. Understand the importance of argument and persuasion in technical communication.

When communicating information, you will write and speak for many reasons. You might write a memo, letter, e-mail message, tweet, text message, or instant message to *inform* your readers about an upcoming meeting, a job opportunity, a new product release, or a facilities change. You might give an oral presentation to local businesses, government, or educational organizations in which you hope to *build rapport*. In writing a user manual, your goal will be to *instruct*. When you write a proposal, your goal is to *recommend* changes. If you write a technical specification, you *analyze* components of a piece of equipment.

Persuasive Writing Requesting Action

In addition to informing, building rapport, instructing, recommending, and analyzing, you also will need to communicate persuasively. Let's say that you are a customer who has purchased a faulty product. You might want to write a letter of complaint or an e-mail message to the manufacturer of this product. To make your case strongly, you will need to convince your audience, clarifying how the product failed. If your argument is effective, then you will persuade the company to give you a refund or new product.

Professionally, you will need to use argument and persuasion daily. As a manager, you might need to argue the merits of a company policy to an unhappy customer. If your colleagues have decided that the department should pursue a course of action, you might need to persuade your boss to accept it. Maybe you are asking your boss for a raise or promotion, for office improvements, or for changes to the work schedule. Your task is to persuade the boss to accept your suggestions. In these instances, you will communicate persuasively using any of the following communication channels: routine correspondence (memos, letters, or e-mail); social media (blogs); reports; proposals; or oral presentations.

Figure 11.1 is a persuasive e-mail from a subordinate to a boss, documenting a problem and suggesting a course of action.

Different Communication Channels

See Chapter 9, "Routine Correspondence," Chapter 12, "Social Media," Chapter 17, "Short, Informal Reports," Chapter 19, "Proposals and Business Plans," and Chapter 20, "Oral Presentations," for information about format and purpose.

FIGURE 11.1 Persuasive E-mail

The body develops the argument persuasively by providing facts and figures. In doing so, the writer reveals his knowledge of the subject matter.

The conclusion emphasizes persuasively the urgency of action.

> ☐ Send   ☐   ☐   ☐   ☐   !   ☐   ☐   ☐ ▾   ☐ ▾   ☐ Options...   HTML ▾   ⊙
>
> ☐ To...   Lynn Richards
>
> ☐ Cc...
>
> Subject: Need to Replace Current Cell Phones with Smartphones
>
> Calibri   ▾   12 ▾   **B**   *I*   U̲   ☰ ☰ ☰ ☰   ☷ ☷ ☷ ☷   ☷ ▾   **A** ▾   ⌄
>
> Lynn, our sales staff currently is using cell phones as their primary means of communication with customers, vendors, and management. Staff has complained about issues with their cell phones.
>
> Here are some facts that I've collected to help us decide on replacement phones.
>
> - Replacing the 30 phones will cost approximately $6,000. However, the initial cost will be offset by increased productivity.
>
> - Although 80 percent of the phones work well, twenty percent of the phones have reported problems with dropped calls.
>
> - Our current cell phones can make and take calls, send e-mail messages, and have limited photo abilities. Smartphones, in contrast, offer more capabilities which our sales staff and company can profit from. These improved abilities include mobile apps, such as GPS mapping, increased power for faster Internet search, QWERTY keyboards for improved typing, and syncing with the company's e-mail server. The smartphones have docking stations with external display and keyboards to create a laptop environment. Finally, our staff will benefit from at least 5 megapixel cameras with autofocus.
>
> Our sales staff relies on mobile phones to conduct business. We cannot depend on outdated equipment if we want to stay competitive.   Let's meet this week to discuss the best smartphones to purchase.

Persuasive Writing within Organizations

Frequently, you will write e-mail messages to supervisors and colleagues, persuading these readers to accept your point of view. Topics could range from requests for promotion, equipment needs, days off from work, assistance with projects, or financial assistance for job-related travel. The "Before" and "After" e-mail messages seek to persuade a colleague to attend a work-related conference instead of the e-mail writer. The "Before" sample is not effectively persuasive. It reads more like a command and fails to consider the audience's reaction. The "After" sample is more persuasive.

This e-mail message is not persuasive for a number of reasons: the verb "need" is too commanding. The e-mail fails to arouse the reader's interest or provide sufficient detail to convince the reader of the conference's worth. It also fails to urge the reader to action.

Ken, I need you to do me a favor. You need to attend the Annual Technology Conference. It's being held this year in San Antonio. Some of the topics that will be discussed include online help and RoboHelp workshops. Get in touch with me and I'll share the details.

Microsoft Corporation, Inc.

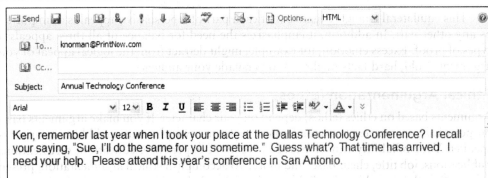

Ken, remember last year when I took your place at the Dallas Technology Conference? I recall your saying, "Sue, I'll do the same for you sometime." Guess what? That time has arrived. I need your help. Please attend this year's conference in San Antonio.

I know that you are busily working on the Art.com contract, but I've already spoken to Harry. He says your team is ahead of schedule and that he can finish the work on his own. In addition, I think that you would benefit more from the conference than I would. The keynote address is about online help, and that's one of your areas of interest. The agenda shows that the conference will provide hands-on RoboHelp workshops. This will help you learn more practical applications for our online help screens.

I'll need your answer by 5:00 p.m. tomorrow. After all, San Antonio is your favorite city.

Microsoft Corporation, Inc.

The introductory paragraph arouses reader interest with an anecdote and a question.

The e-mail body begins by refuting any objections. Then, the following sentences show audience benefit.

The conclusion urges action by giving a due date and highlighting reader enjoyment.

FAQs

Q: How does marketing fit in with technical communication?

A: Technical communication is more than hard, cold memos, letters, reports, and user manuals. Technical communication also has a soft side—*marketing*. The bottom line is that every company is in business to make money. Thus, every employee should perceive herself or himself as marketing personnel.

Examples:

George Butler Associates (GBA), an architectural and engineering company in Kansas, Missouri, and Illinois, asks every one of its employees to take marketing classes on site. The goal is to help GBA employees work well with clients, vendors, and partners.

The Society for Technical Communication also recognizes the importance of marketing for technical writers. STC's Web site (stc.org) provides links to "Special Interest Groups." One such group is Marketing Communication. In addition, two recent links for general news included articles about the importance of marketing for technical communicators: "Tech writers as sales reps? Interface Software's award-winning docs boost brand, revenues, and customer satisfaction" and "Technical writers turn to marketing to survive." Persuasive marketing materials are essential to technical communication.

Traditional Methods of Argument and Persuasion

To argue a point persuasively, you can use any of the traditional methods of argumentation: *ethos* (ethical), *pathos* (emotional), and *logos* (logical) appeals. These three appeals to an audience are called the *rhetorical triangle* as shown in Figure 11.2.

FIGURE 11.2   The Rhetorical Triangle

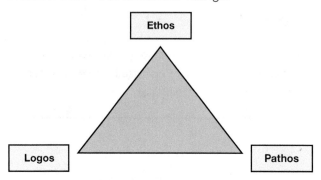

This equilateral triangle suggests that each part of a persuasive appeal is as important as any other part. In addition, it emphasizes the need for balance of all three appeals or types of proof. Excess emotion, for example, might detract from the logical appeal of your argument. Cold, hard facts might fail to persuade your audience.

Ethical Argumentation (Ethos)

Arguments based on ethics (ethos) depend on your character. If you make arguments based on personal experiences, you must appear to be trustworthy and credible as a writer or speaker. Your reliability can be based on years of experience, education, certifications, publications, job title, client base, and more. To accomplish ethical argumentation, present information that is unbiased, reliable, and evenhanded.

Emotional Argumentation (Pathos)

Arguments based on emotion (pathos) seek to change an audience's attitudes and actions by focusing on feelings. If you want to move an audience emotionally, you would appeal to passion. You can do this either positively or negatively. To sway an audience positively, you would focus on positive concepts like joy, hope, honor, pleasure, happiness, success, and achievement. You would use positive words to create an appealing message. In contrast, you also can appeal to emotions negatively. Fear, horror, anger, and unhappiness can be powerful tools in an argument.

Notice how Centers for Disease Control and Prevention warns about the dangers of carbon monoxide: "Carbon monoxide is a silent killer. This colorless, odorless, poisonous gas kills nearly 500 U.S. residents each year, five times as many as West Nile virus" ("CDC and CPSC Warn of Winter Home Heating Hazards"). To highlight the dangers associated with carbon monoxide, the Centers for Disease Control and Prevention uses emotional words, such as *killer* and *poisonous*, and compares this problem to the frightening "West Nile virus."

Logical Argumentation (Logos)

Argumentation based on logic (logos) depends on rationality, reason, and proof. You can persuade people logically when you provide them the following:

- Facts—statistics, evidence, data, and research
- Testimony—citing customer or colleague comments, expert authorities, and results of interviews
- Examples—anecdotes, instances, and personal experiences

- Strong, clear claims—including warranties and guarantees
- Acknowledgement of the opposing points of view to ensure that information is balanced

Figure 11.3 is an e-mail message that argues a case for a raise by focusing on ethics. The writer refers to his work-related achievements to support his credibility. He uses logic—facts and testimony. Also note how the e-mail factors in an opposing point of view. In addition, the e-mail uses emotion to sway the reader through positive words.

FIGURE 11.3 Persuasive E-mail Effectively Using Ethos, Pathos, and Logos

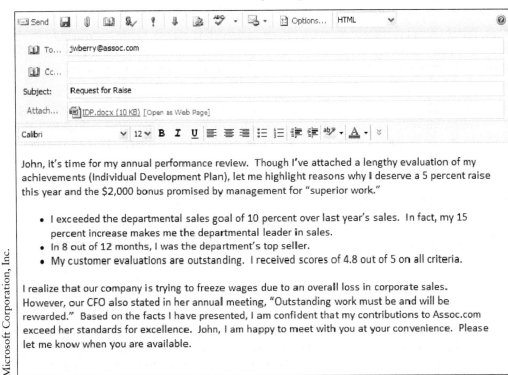

Ethics (*ethos*) is shown when the writer references his achievements at work.

Facts ("4.8 out of 5") and testimony from customer evaluations support the writer's case logically (*logos*).

Note the use of positive words, such as "achievements," "highlight," "exceeded," and "outstanding." These are examples of *pathos*—emotional words to sway the reader.

The e-mail addresses the opposing point of view ("freeze wages") to present a balanced argument.

ARGU to Organize Your Persuasion

Effective persuasive communication entails ethical, logical, and emotional appeals. Understanding the importance of this rhetorical triangle is only part of your challenge as a persuasive communicator. The next step is deciding how best to present your argument. Using our *ARGU* approach will help you organize your argument:

- *Arouse* audience involvement—grab the audience's attention in the introduction of your communication.
- *Refute* opposing points of view in the body of your communication.
- *Give* proof to develop your thoughts in the body of your communication.
- *Urge* action—motivate your audience in the conclusion.

> 3. Use the *ARGU* technique for persuasive communication.

Arouse Audience Involvement

You have only about five to eight seconds to grab your readers' attention in a sales letter, persuasive e-mail message, marketing brochure, speech, or any persuasive communication. You must arouse the audience's interest imaginatively in the first few sentences of your document or oral presentation. Try any of the following attention grabbers in the introduction of your persuasive message.

- **Use an anecdote—a brief, dramatic story relating to the topic.** Stories engage your audience. You can involve your readers or listeners by recounting an interesting story to which they can relate. The story should be specific in time, place, person, and action. The drama should highlight an event. However, this must be a short scenario, since your time is limited. If you do not capture their interest quickly, the audience might lose interest.

E-mail Message Persuading a Manager to Take Action

Sam, last week, a customer fell down in our parking lot, cutting her knee, tearing her slacks, and requiring medical attention. We can't let this happen again. Please consider hiring a maintenance crew to salt and sand our lot during icy weather.

- **Start with a question to interest your audience.** By asking a question, you involve your audience. Questions imply the need for an answer. A question can make readers or listeners ask themselves, "How would I answer that?" This encourages their participation.

Sales Letter from a Financial Planner

Where will I get money for my kid's college education? How can I afford to retire? Will my insurance cover all medical bills? You have asked yourself these questions. Our estate planning video has the answers.

- **Begin with a quotation to give your communication the credibility of authority.** By quoting specialists in a field or famous people, you enhance your credibility and support your assertions. A quote from Warren Buffett, world-renowned investor and businessman, about economics, for example, gives credence to a financial topic. Similarly, quotes from Bill Gates, founder of Microsoft, about the computer industry are trustworthy.

An Oral Presentation about Stock Performance to Shareholders

As Warren Buffett says, "Our favorite holding period is forever." Though our stock prices are down now, don't panic and sell. The company will rebound.

- **Let facts and figures enhance your credibility.** Factual information (percentages, sales figures, amounts, or dates) catches the audiences' attention and lays the groundwork for your persuasion.

Sales Letter about Computer Technology

Eighty-seven percent of all college students own a computer. Don't go off to college unprepared. Buy your laptop or PC at CompuRam today.

- **Appeal to the senses.** You can involve your audience by letting them hear, taste, smell, feel, or visualize a product.

Marketing Flier Advertising a New Restaurant

Hickory-smoked goodness and fire-flamed Grade A beef—let Roscoe's BBQ bring the taste of the South to your neighborhood.

- **Use comparison or contrast to highlight your message.** Comparison/contrast lets you make your point persuasively. For example, you might want to show a client that your computer software is superior to the competition. You might want to show a boss that following one plan for corporate restructuring is better than an alternative plan. You could justify why an employee has not received a raise by comparing or contrasting his or her performance to the departmental norm.

E-mail from Management to Subordinates Proposing Changes in Procedures

Last year our department fell short of corporate sales goals. This year we must surpass expectations. To do so, I propose a seven-step procedure.

◄ EXAMPLE

Comparing one year to another provides a benchmark for the writer's argument.

- **Begin with poetic devices.** Advertising has long used alliteration (the repetition of sounds), similes (comparisons), and metaphors to create audience interest. Poetic devices are memorable, clever, fun, and catchy.

Sales Letter from a Bank

Looking for a low-loan lease? Let US Bank take care of your car leases. Our 2 percent loans beat the best.

◄ EXAMPLE

Use alliteration, the repetition of sounds, to appeal to an audience.

- **Create a feeling of comfort, ease, or well-being.** To welcome your audience, make them feel calm and peaceful or invoke nostalgia or good times.

Slogan from a Mortgage Company Brochure

Come home again to Countryside Mortgage. We treat our customers like neighbors.

◄ EXAMPLE

Interest your readers by making them feel comfortable and welcomed.

- **Create a feeling of discomfort, fear, or anxiety.** Another way to involve your audience is through stress. Beginning communication by highlighting a problem allows you to persuade the audience that you offer the solution.

Sales Letter from an Apartment Complex

Why pay too much? If you paid over $300 a month for your apartment last semester, you got robbed. We'll charge 20 percent less—guaranteed.

Refute Opposing Points of View

You can strengthen your argument by considering opposing opinions. Doing so shows your audience that you have considered your topic thoroughly. Rather than looking at the subject from only one perspective, you have considered alternatives and discarded them as lacking in merit. In addition, by refuting opposing points of view, you anticipate negative comments an audience might make and defuse their argument.

To refute opposing points of view in the discussion (body) of your communication, follow these steps:

- Recognize and admit conflicting views.
- Let the audience know that you understand their concerns.
- Provide evidence.
- Allow for alternatives.

Figure 11.4 shows a letter of application that successfully uses refutation as part of its persuasion.

FIGURE 11.4 Letter of Application Using Refutation

The writer's refutation anticipates negative comments the audience might have and defuses the argument through proof and alternatives.

The writer refutes the job requirement for a BA.

The first five bullets provide evidence to show that the writer has skills appropriate for the job versus a degree.

The current enrollment provides an alternative to the degree requirement.

1901 West King's Highway
San Antonio, TX 77910

February 27, 2016

Marissa Lee
Manager, Human Resources
PMBR Marketing
20944 Wildrose Dr.
San Antonio, TX 78213

Dear Ms. Lee:

In response to your advertisement in the *San Antonio Daily Register,* please consider me for the position of marketing accounts associate. I have enclosed a résumé elaborating on ways I can benefit your company.

Your advertisement requires someone with a BA in marketing. Though I do not have this degree, my experience and skills prepare me for this position:

- Worked for ten years in marketing.
- Prepared press releases and created public service announcements.
- Created 30-second spots for local radio and television stations.
- Participated in press conferences providing corporate information about declining stock values and company layoffs.
- Created PowerPoint presentations for local governmental agencies.
- Enrolled in a Marketing class at the University of Texas.

My experience and abilities will make me an asset in your marketing department. I would be happy to meet with you to discuss ways in which I can benefit PMBR. Please contact me at arosa22@hotmail.com.

Sincerely,

Armondo Rosa
Armondo Rosa

Attachment: Resume

Give Proof to Develop Your Thoughts

In the body of your communication, develop your argument with proof. Arousing interest and refuting opposing points of view will not necessarily persuade the audience. Most people require details and supporting evidence before making decisions. You can provide specific details to support your argument using any of the techniques in Table 11.1.

TABLE 11.1 Techniques for Supporting an Argument

Provide facts and figures to document your assertions.	Eighty-five percent of the homeowners contend that . . . or Seven of 10 buyers said they would . . .
Persuade through graphics.	[Pie chart: 15 — Homeowners without Computers; 85 — Homeowners with Computers] This pie chart not only conveys data, but also it makes a visual impact. It graphically highlights how many people own computers.
Give testimony from satisfied customers, vendors, or coworkers.	The Job Corps of Blue Valley, Titan Co., and Amex Inc. have used our product for over 10 years. or Julie Jones, sales associate, attests to my effective interaction with customers.
Document your credentials—years in business or certification of employees.	In business since 2000 . . . or All of our staff members are certified public accountants (CPAs).
Give examples.	For example, I attended three continuing education seminars this year. These included training workshops on Microsoft Word applications, Effective Business Writing, and Listening Skills. Because of my continuing education, I have the credentials needed to be an effective project manager. Please consider me for promotion.
Cite rules and regulations.	Section II.a of the State Waste Management Act specifies that wood debris, railroad ties, and used tires must be disposed of.
Attach an expert's name.	Linda Freeman, certified public accountant, suggests that allocating $12,500 to a self-employment pension (SEP) account will lower your federal taxes.

Urge Action—Motivate Your Audience

Throughout your communication, you have worked to persuade the audience to accept your point of view. In the conclusion, you need to motivate the audience to action. This could include any of the following: attend a meeting, purchase new equipment, invite you to interview for a position, vote on a proposition, promote you, give you a raise, allow you to work a flexible schedule, or change a company policy.

To urge the audience to action, consider the techniques in Table 11.2.

The paragraph example concludes a persuasive letter from a governmental agency to a business owner.

TABLE 11.2 Techniques for Motivating an Audience to Action

Give due dates.	Please respond by January 15.
Explain why a date is important.	Your response by January 15 will give me time to prepare a quarterly review and meet with you if I have additional questions.
Provide contact information for follow-up.	Please submit your proposal to Hank Green, project director. You can e-mail him at hgreen@modernco.com.
Suggest the next course of action.	We need to plan our presentation before the next City Council meeting, so please attend Tuesday's meeting at 9:00 a.m.
Show negative consequences.	You must repair your sidewalk within 30 days to comply with city laws regarding pedestrian safety. Failure to do so will result in a $150 fine.
Reward people for following through.	Following these ten simple steps will help you load the software easily and effectively.

EXAMPLE ▶

This conclusion provides a due date, explains why the date is important, shows negative consequences, and tells the reader what to do next. This information helps to motivate the reader.

> Your business has not been in compliance for over 16 months. We have written three letters to give you guidance and information necessary to get the facility into compliance. You have not responded to our efforts, with the exception of last week's phone call. The deadline for the pending abatement order is 9/7/16. Please reply to this letter in a timely manner to avoid fines. We are open to having a meeting at our regional office to discuss the corrective measures you can take to address the pending order. Thank you for considering your options.

4. Avoid unethical logical fallacies.

Avoiding Unethical Logical Fallacies

In a corporate environment, you must persuade your audience not only logically but also ethically. Your persuasive communication must be honest and reasonable. Honesty demands that you avoid the following logical fallacies.

Inaccurate Information

Visual Aids

See Chapter 8, "Visual Aids," for information about the ethical use of visuals.

Facts and figures must be accurate. One way in which dishonest appeals are made in communication is through inaccurate graphics. Look at Figures 11.5 and 11.6. Figure 11.5 shows a $25,000 deficit in the third quarter. So does Figure 11.6. However, the size of the bar in Figure 11.6 is misleading and imprecise, visually suggesting that the loss is less. This is an inaccurate depiction of information.

Unreliable Sources

Being an expert in one field does not mean you are an expert in all fields. For example, quoting a certified public accountant to support an important health care issue is illogical. Not all specialists are reliable in all situations.

Sweeping Generalizations

Avoid exaggerating. Allow for exceptions. For instance, it is illogical to say that "All marketing experts believe that newsletters are effective." Either qualify this with a word like *some* or quantify with specific percentages.

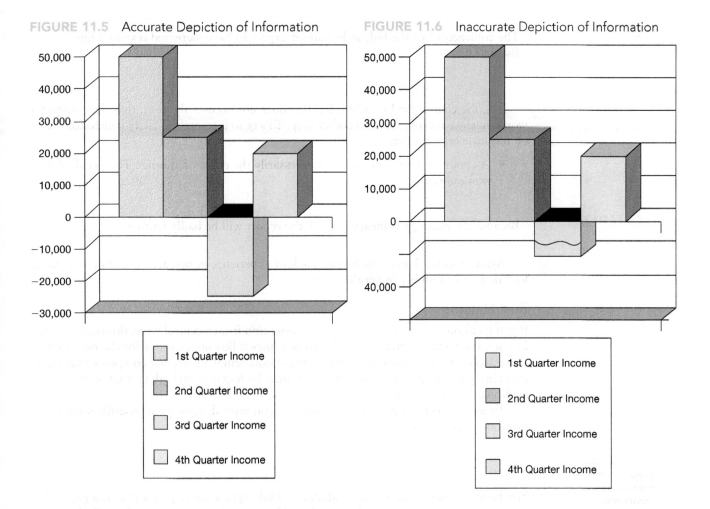

FIGURE 11.5 Accurate Depiction of Information

FIGURE 11.6 Inaccurate Depiction of Information

Either . . . Or

Suggesting that a reader has only two options is deceitful if other options exist. Allow for other possibilities. It is wrong to write, "Either all employees must come to work on time, or they will be fired." This blanket statement excludes alternatives or exceptions. Is the "either . . . or" statement true if an employee has a car accident, if an employee's child is sick, or if the employee is caught in heavy traffic due to a snowstorm?

Circular Reasoning (Begging the Question)

"Some advertising executives are ambitious because they wish to succeed." This statement is illogical and uses circular reasoning because it states the same thing twice. *Ambitious* and *succeed* are essentially synonyms. The writer fails to prove the assertion.

Inaccurate Conclusions

When communicating persuasively, consider all possible causes and effects. Exact causes of events often are difficult to determine.

- A condition that *precedes* another is not necessarily the *cause* of it. This error is called *post hoc, ergo propter hoc.*

> The contractor lost the bid, so he cannot expect to have increased revenues this fiscal year.

Yes, the contractor lost a bid, but this does not necessarily mean that the contractor won't increase revenues in some other way. To say so is to make a hasty conclusion based on too little information.

- A condition that *follows* is not necessarily the *effect* of another. This is called a *non sequitur*.

> Because the manager is inexperienced, the report will be badly written.

Again, this is a hasty conclusion. Lack of experience in one area does not necessarily lead to a lack of ability in another area.

Red Herrings

If you focus on an irrelevant issue to draw attention from a central issue, this is called a *red herring*. For instance, you have failed to pay fines following citations for the mishandling of hazardous wastes. You contact the state environmental agency and complain about state taxes being too high. This is an irrelevant issue. By focusing on high taxes, you are merely avoiding the central issue.

To write effective persuasive documents, you must develop your assertions correctly, avoiding logical fallacies.

Types of Persuasive Documents

> 5. Prepare different types of persuasive documents.

You have manufactured a new product (a mobile app, a minidigital camera, a paper-thin plasma computer monitor, a fiber optic cable, or a microfiber rain jacket). Perhaps you have just created a new service (home visitation health care, a mobile accounting business, a Web design consulting firm, or a gourmet food preparation and delivery service). Congratulations!

However, if your product sits in your basement gathering dust or your service exists only in your imagination, what have you accomplished? To benefit from your labors, you must market your product or service.

To convince potential customers to purchase your merchandise, you could write any of the following persuasive documents:

- Sales letters
- Fliers
- Brochures
- Press releases

Sales Letters

> **Letter Formats**
>
> See Chapter 9, "Routine Correspondence," for additional information.

To write your sales letter, follow the format for letters discussed in Chapter 9. Include the letter essentials (letterhead address, date, reader's address, salutation, text, complimentary close, and signature). Your sales letter should accomplish the following objectives relating to effective persuasion.

Arouse Reader Interest. The introductory paragraph of your sales letter tells your readers why you are writing (you want to increase their happiness or reduce their anxieties, for example). Your introduction should highlight a reader problem, need, or desire. If the readers do not need your services, then they will not be motivated to purchase your merchandise. The introductory sentences also should mention the product or service you are marketing, stating that this is the solution to their problems. Arouse your readers' interest with anecdotes, questions, quotations, or facts.

Refute Opposing Points of View. Your audience will not always agree with your assertions. To persuade the reader to accept your point of view, anticipate disagreements or alternative points of view. Consider opposing comments about your new product or service. Think about what your competitors offer as alternatives to your company. By mentioning competitors or alternative ideas and refuting them, you emphasize your product's value.

Give Proof to Develop Your Thoughts. In the discussion paragraph(s), specify exactly what you offer to benefit your audience or how you will solve your readers' problems. You can do this in a traditional paragraph. In contrast, you might want to itemize your suggestions in a numbered or bulleted list. Whichever option you choose, the discussion should provide data to document your assertions, give testimony from satisfied customers, or emphasize your credentials.

Urge Action. Make readers act. If your conclusion says, "We hope to hear from you soon," you have made a mistake. The concluding paragraph of a sales letter should motivate the reader to act.

Conclude your sales letter in any of the following ways:

- Provide your Web site URL; online help desk e-mail address; and social media links to the company blog, Twitter account, or Facebook site.
- Give directions (with a map) to your business location.
- Provide a tear-out to send to you for further information.
- Supply a self-addressed, stamped envelope for customer response.
- Offer a discount if the customer responds within a given period of time.
- Give your name or a contact name and a phone number (toll-free if possible).

Figure 11.7 provides a sample sales letters using the *ARGU* method of persuasion.

Fliers

Corporations and companies; educational institutions; religious organizations; museums, zoos, and amusement parks; cities and states—all need to communicate with their constituencies (clients, citizens, members). An effective way for these organizations to communicate persuasively is with a one- or two-page flier, electronic or hard-copy.

Fliers provide the following benefits:

- **Cost effective.** A flier costs less than an expensive advertising campaign and can be produced in-house.
- **Time efficient.** Creating a flier can take only a few hours of work or less by the company's employees.
- **Responsive to immediate needs.** Different fliers can be created for different audiences and purposes to meet unique, emerging needs.
- **Personalized.** Fliers can be created with a specific market or client in mind. Then, these fliers either can be mailed or e-mailed to that client or hand-delivered for more personalization.
- **Persuasive.** In a compact format, fliers concisely communicate audience benefit.

FIGURE 11.7 Sales Letter Using *ARGU* Method of Persuasion

The introduction arouses reader interest by asking questions and states that *ConnectionNet* can solve the problems. The questions highlight a reader problem: the need to increase business.

The letter refutes prior attitudes toward social media, showing how it impacts search engine rankings.

The letter gives specific proof to sway the reader and build credibility: "77 percent," "IBM-trained," and "63 percent." The letter also uses alliteration ("power to propel profits").

The conclusion urges action by giving a seasonal discount.

ConnectionNet
9516 Pecan St.
Santa Clara, CA 66321

November 10, 2016

GamesterLine
8821 Wilkes Way
Houston, TX 77300

Subject: Leveraging Social Media to Increase Your Customer Base

Is your business growing as fast as you'd like? Are you only depending on word of mouth and repeat customers to increase company sales? It's time to reach a new and larger clientele. Let *ConnectionNet* help you reach the world's largest consumer base. Join the 77 percent of businesses who use social media to increase their market share.

ConnectionNet's IBM-trained information technologists will help you with the following options:

- Create Twitter and Facebook accounts to introduce new products and services.
- Start "blogrolling"—create a corporate blog to initiate topics of discussion and provide answers to customer questions.
- Use Twitter as an online help line for quick and personalized responses to customer concerns.
- Allow customers to communicate with each other through your blog site.
- Provide customer support with YouTube video instructional manuals.
- Use our "Trackback" options to monitor what customers are saying about your products and services.

According to statistics, three quarters of global customers who use the Internet also visit social network sites. Studies show that social media links improve your search engine rankings by 63 percent. With our help, you can harness this power to propel profits. Act today! Cyber Monday is coming soon, when online shoppers spend billions of dollars. If you contact us this week, we'll build your social media sites for a one-time only 20 percent discount.

Sam Greenberg

Sam Greenberg

CEO, ConnectionNet

sgreen@connectionnet.com   www.connectionnet.com   800-555-CONN

Flier Criteria. When writing your flier, follow these criteria:

Determine the length of the flier. Though one page might be preferable, you could create a two-page flier, using the front and back of an 8½" × 11" piece of paper. If you keep your flier to one page (front only), then you can save money by folding the flier in thirds, stapling it, and using the blank side for mailing purposes (addresses and stamp). A flier even could be smaller, the size of a postcard, for example. Many companies create electronic fliers, transmitted via their Web sites or through e-mail. This is more cost effective than a hard-copy flier, since no postage is needed.

Focus on one idea, topic, or theme per flier. A flier should make one key point. This is how you make the flier's content relevant to your audience, fulfilling that audience's unique needs. For example, if your company's focus is automotive parts, avoid writing a flier covering every car accessory. Write the flier with one accessory in mind, such as windshield wipers, batteries, or custom rims.

Use a title at the top of your flier to identify its theme. The title can be one or two words long, you could use a phrase, or you might want to write an entire sentence at the top of the flier. An effective persuasive approach is to begin your flier with a question to immediately arouse reader interest: "Is your software giving you a headache?" or "How usable is your Web site?"

Limit your text. Using few words, provide reader benefit, involve the audience, and motivate them to act. The action could be to purchase a product, attend an event, or contact you for additional information. By limiting your text, you avoid overwhelming either the flier's appearance or the audience's attention span, or both. Getting to the point in a flier is a key concern. Limiting your text helps you achieve this goal.

Increase font size. In a flier, you can use a 16-point font and up for text, and a 20-point font and up for titles. This will make the text more readable and dramatic. Your heading must be eye catching. To accomplish this goal, make sure the heading's font size and style are emphatic—at a glance, even from across a room.

Use graphics. One graphic, at least, will emphasize your theme and visually make your point memorable. Another graphic could include your company logo (for corporate identity and namesake recognition). In addition, the logo should be accompanied by a street address, e-mail address, Web site URL, fax number, or phone number so clients can contact you or visit your site.

Use color for audience appeal. Pick one dominant color to emphasize key points. Use a color in your company logo to remind your reader of your company's identity. However, don't overuse color. Excess will distract your reader.

Use highlighting techniques. Bullets, white space, tables, boldface, italics, headings, text boxes, or subheadings will help your reader access information. A little highlighting goes a long way, especially on a one-page flier. Too much makes a jumble of your text and distracts from the message.

Find the phrase. Select a catchy phrase, which you can use in all your fliers, to personify your company's primary focus. For example, McDonald's repeats the line "We love to make you smile"; Pepsi uses "The Joy of Cola"; Nike says "Just Do It"; and Ford states "Quality is Job #1." What phrase captures your company's personality?

Recognize your audience. You want to show the readers how your product or service will benefit them. Understand your audience's needs and direct the flier to meet those concerns. In addition, you want to engage the reader. To do this, use pronouns which speak to the reader on a personal level and positive words which motivate the reader to action. Remember to speak at the reader's level of understanding, defining terms as necessary. Figure 11.8 provides a sample persuasive flier.

FIGURE 11.8 Sample Persuasive Flier

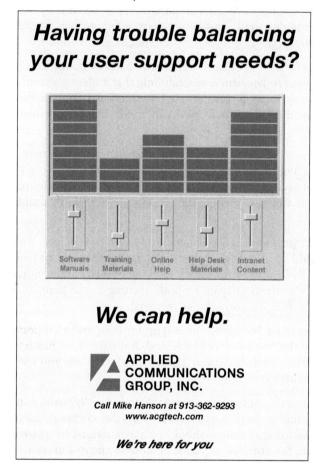

An introductory question appeals to the reader's needs.

A graphic adds visual appeal to make the flier more persuasive. The graphic also develops how the company can meet the audience's need.

This sentence answers the introductory question.

The contact name and number are provided to motivate the reader.

The last sentence uses empathy to persuade the audience.

TECHNOLOGY TIPS

Using Word 2013 for Flier and Brochure Design Templates

You can access many design templates for flier and brochures using your word processing software. For example, in Microsoft Word 2013, try this approach:

1. Click on **File** and then **New.**
2. Type Flier or Brochure in the search field to find hundreds of templates.

New

3. Choose from different brochure options for events, travel, recruiting, professional services, real estate offerings, and financial information. These include templates for 8½" × 11" landscape with three folds, 8½" × 14" landscape with three and four folds, and 8½" × 11" portrait letter folds.
4. Choose from different flier options including "Events" fliers, "Marketing" options, "Real Estate" choices, and "Other Fliers."

NOTE: The templates will not tell you what to write in each panel. Instead, Word 2013 provides a design layout, a shell for you to fill with your content.

Creating Your Own Brochure Layout

Instead of using a predetermined template, you can create your own as follows:

1. Change from portrait to landscape by clicking on the **Page Layout** tab and then **Orientation**.

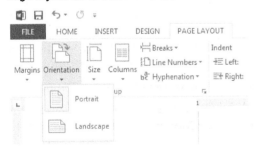

2. Create three panels for your brochure by clicking on the **Columns** icon.

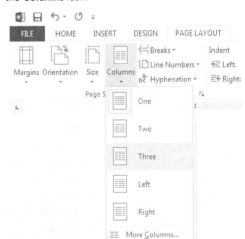

Brochures

A flier *must* be short—one or two pages. If you have more information to convey than can fit on a one-page flier, then a brochure might be a good option. Brochures offer a detailed overview of products, services, options, and opportunities, complete with photographs, maps, or charts. Brochures are persuasive for the following reasons:

- Create awareness of your company, product, or service
- Increase understanding of a product, service, or your company's mission
- Advertise new aspects about your company, product, or service
- Change negative attitudes
- Show ways in which your company, product, or service surpasses that of your competition
- Increase frequency of use, visit, or purchase
- Increase market share

Criteria for Writing Brochures. Brochures come in many shapes and sizes. They can range from a simple front and back, four panel 8½" × 5½" brochure (one landscape 8½" × 11" page folded in half vertically) to six-, eight-, or even 12-panel brochures printed on any size paper you choose. See Figure 11.9 for two ways to fold paper to create panels in a brochure. Brochures, like fliers, also can be transmitted electronically. Your topic and the amount of information you are delivering will determine your brochure's size and means of transmission.

FIGURE 11.9 Four- and Six-Panel Brochures

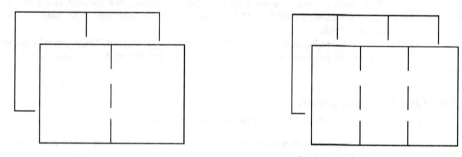

To determine what you will write in each panel of your brochure and how the brochure should look, follow these criteria for writing an effective, persuasive brochure.

Title Page (front panel). Usually, the title page includes at least three components:

- **Topic.** In the top third of the title panel, name the topic. This includes a product name, a service, a location (such as "Rocky Mountain Park" or "San Diego Zoo"), or the subject of your brochure.
- **Graphic.** In the middle third of the panel, include a graphic to appeal to your reader's need for a visual representation of your topic. The graphic will sell the value of your subject (its beauty, its usefulness, its location, or its significance) or visually represent the focus of your brochure.
- **Contact information.** In the bottom third of the panel, place contact information. Include your name, your company's name, street address, city, state, zip code, telephone number, Web site, and/or e-mail address.

Back Panel.   The back panel could include the following:

- **Conclusion.** Summarize your brochure's content. Restate the highlights of your topic or suggest a next step for your readers to pursue.
- **Mailing information.** The back panel could be used like the face of an envelope. On this panel, when left blank, you could provide your address, a place for a stamp or paid postage, and your reader's address.
- **Coupons.** As a tear-out, this panel could be an incentive for your readers to visit your site or use your service. Here you could urge action by providing discounts or complimentary tickets.
- **Location.** A final consideration would be to provide your readers with your address, hours of operation, phone numbers, e-mail, Web site, and a map to help them locate you.

Body Panels (fold-in and inside).   The following are some suggestions for creating the brochure's text:

- **Provide headings and subheadings.** These act as navigational tools to guide your readers, direct their attention, and help them find the information they need. The headings and subheadings should follow a consistent pattern of font type and size. First-level headings should be larger and more emphatic than your second-level subheadings. The headings must be parallel grammatically to each other.

  For example, if your first heading is entitled "Introduction," a noun, all subsequent headings must be nouns, like "Location," "Times," "Payment Options," and "Technical Specifications." If your first heading is a complete sentence, like "This is where it all began," then your subsequent headings must also be complete sentences: "It's still beautiful," "Here's how to find us," and "Prices are affordable."
- **Use graphics.** Use photographs, maps, line drawings, tables, or figures to vary the page layout for visual appeal and to enhance your text.
- **Develop your ideas.** Consider including locations, options, prices, credentials, company history, personnel biographies, employment opportunities, testimonials from satisfied customers, specifications, features, uses of the product or service, payment schedules, or payment plans.
- **Persuade your audience.** Review the tips provided in this chapter for persuasive arguments. Use ethics, logic, and emotion to sway your reader.

Document Design.   Visual appeal helps to interest and persuade an audience. Compelling graphics, for example, can help to convince an audience. Use pie charts, bar charts, tables, or photographs to highlight key concerns. In addition to graphics, make your brochure visually appealing by doing the following:

- Limit sentence length to 10–12 words and paragraph length to 4–6 lines. When you divide paper into panels, text can become cramped very easily. Long sentences and long paragraphs then become difficult to read. By limiting the length of your text, you will help your readers access the information.
- Use white space instead of wall-to-wall words. Indent and itemize information so readers won't have to wade through too much detail.
- Use color for interest, variety, and emphasis. For example, you can use a consistent color for your headings and subheadings.
- Bulletize key points.
- Boldface or underline key ideas.
- Do not trap yourself within one panel. For variety and visual appeal, let text and graphics overlap two or more of the panels.

- Place graphics at angles (occasionally) or alternate their placement at either the center, right, or left margin of a panel. Panels can become very rigid if all text and graphics are square. Find creative ways to achieve variety.

See Figure 11.10 to get a better idea of what a typical brochure looks like.

FIGURE 11.10 Sample of a Brochure

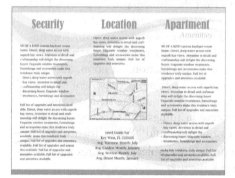

Press Releases

Press releases are written to persuade an audience that a company has new information to offer. Press releases use many channels of communication, including hard-copy newspapers, social media, Web sites, blogs, and e-mail. A company will use press releases to announce new products, services, locations, product options, management changes, promotions, or any topic of interest to the public.

When writing your press release, include the following:

- A headline to concisely summarize the topic, focusing on one key point. Use techniques discussed earlier in this chapter to grab the reader's attention.
- A subhead for clarity. While your heading arouses reader interest, the subhead provides more specific details about the topic.
- An introductory lead-in answering who, what, when, why, where, and how. Present the most important information early in the press release where the audience's attention is greatest.
- Development in the press release's body. Create interest in the topic by giving supporting evidence.
- A conclusion urging follow-up action.
- Persuasive tone and word usage.
- An effective technical communication style, focusing on objectivity and conciseness.
- A visually appealing format (headings, subheadings, bullets, font changes, and more).

See Figure 11.11 for a successfully written press release.

FIGURE 11.11   Press Release

Press and Media

TweetaleeD: Hear Twitter Sing at the App Store

Social app delivers innovative Twitter experience to iPhone and iPod touch users

The headline uses a clever pun "Hear Twitter Sing" to arouse reader interest.

Greenbay, AR - (March 3, 2016) – Our worldwide leading mobile application developers today announced a new Twitter App, **TweetaleeD**, for iPhone and iPod touch, now available at your local App Store.

TweetaleeD is an intuitive App, featuring unique, fast, and easy-to-use Twitter functionality. With **TweetaleeD**, you get support from our experts and social media integration. Look at what else we offer:

- Zipper speeds your ability to hashtag a keyword or phrase, follow, and unfollow.
- Social integration helps your link to Facebook, LinkedIn, Yelp, Foursquare, . . . and, of course, Twitter.
- PaperIt lets you customize your App with backgrounds and designs.
- Filester saves your key files, addresses, phone numbers, and URLs.

The press release uses positive words like "fast," "intuitive," "features," and "unique" to persuade the reader.

"This App delivers the best, most up-to-date interface for iPhone and iPod touch on the market today. Using **TweetaleeD** will help you leverage every feature you'll want to bring a one-of-kind social experience," said Eric Halloday, Social Media IT VP.

Quoting a noted social media expert provides persuasive proof.

The **TweetaleeD** App is *free* from your neighborhood, online App Store on iPhone and iPod touch. Why wouldn't you want this NOW?

For more information about **TweetaleeD**, visit www.TweetaleeD.com.

PERSUASIVE COMMUNICATION CHECKLIST

_____ 1. **Use ethical argumentation.** Have you made an ethical argument based on character? You must be trustworthy and credible as a writer or speaker.

_____ 2. **Present emotional argumentation.** Have you used emotion to change an audience's attitudes? You can appeal to an audience's emotions either positively or negatively.

_____ 3. **Use logical argumentation.** Have you developed your persuasion by depending on rationality, reason, and proof? You can persuade people logically by providing facts, testimony, and examples.

_____ 4. **Arouse audience interest.** Have you used questions, quotes, anecdotes, comparison/contrast, poetic language, or an appeal to senses to interest your audience?

_____ 5. **Refute opposing points of view.** Have you presented a balanced argument? To do so, recognize and admit conflicting views, let the audience know that you understand their concerns, and allow for alternatives.

_____ 6. **Give proof.** Have you provided evidence to prove your point?

_____ 7. **Urge to action.** Have you motivated your audience to act? To do so, provide incentives, give discounts, mention warranties, provide contact information, or suggest a follow-up action.

_____ 8. **Use highlighting and engaging page layout.** Is your text accessible? To achieve reader friendly ease of access, use headings, boldface, italics, bullets, numbers, underlining, or graphics (tables and figures). These add interest and help your readers navigate your text.

_____ 9. **Be concise.** Have you limited the length of your sentences, words, and paragraphs?

_____10. **Identify your audience.** Have you written appropriately to your audience? This includes avoiding biased language, considering the multicultural/cross-cultural nature of your readers, and analyzing your audience's role (supervisors, subordinates, coworkers, customers, or vendors).

_____11. **Read for errors.** Is your text grammatically correct? Errors will hurt your professionalism. Have you avoided logical fallacies?

The Writing Process at Work

Gerri Norton is Director of Training at MidContinent Community Service's Municipal Training Institute (MTI). MidContinent is the metropolitan planning organization for local governments. MTI offers a comprehensive selection of professional development opportunities to municipal and county officials (elected and appointed) and their employees. She relies on the writing process of prewriting, writing, and rewriting to create brochures.

Prewriting

To create an effective, persuasive document, first prewrite to determine goals, audience, channel of communication, and data. To plan her communication, Gerri considered the following:

- Goal—help communities meet their infrastructure needs
- Audience—address small-town city councils
- Channels—create an informative and persuasive flier, mailed as hard-copy text and posted on MidContinent's Web site
- Data—provide specific workshop topics of discussion

Gerri says, "My challenge at MidContinent is communicating to a diverse audience in different sized cities. To prewrite, I brainstormed and created a list" (Figure 11.12).

FIGURE 11.12   Gerri's Brainstorming List

1. Brainstorm topics with internal staff and advisory committee (made up of planners and community development representatives).
2. Narrow the topics to three or four workshops.
3. Draft course descriptions.
4. Recruit instructors.
5. Determine dates and times.
6. Secure locations.
7. Provide draft information to public affairs to create the flier.
8. Proofread and finalize marketing materials.
9. Send e-mail announcements with flier attachment to targeted audiences.
10. Hand out fliers at MidContinent-sponsored events.
11. Post the flier on our Web site.
12. Send fliers electronically to MidContinent's customer list serves.

After listing ideas for the workshop, Gerri sought help from Jane, a colleague. To get quick input on the topic, she text-messaged Jane for her thoughts about workshop speakers, locations, topics, and more. See Figure 11.13 for this text message exchange.

FIGURE 11.13   Text Message Exchange

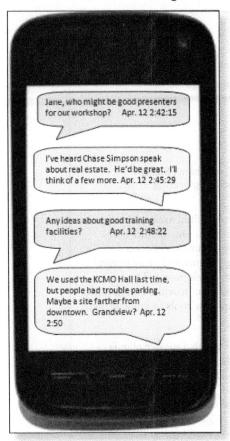

Writing

According to Gerri, "After prewriting, I drafted the document. My rough draft included all of the information I needed to share with my intended audience. One goal was to provide answers to reporter's questions: who, what, when, where, why, and how. I know that I'll revise the draft later, adding visuals for graphic appeal. For the draft, I just wanted to get as much information down as possible." Figure 11.14 shows Gerri's rough draft.

Rewriting

After reviewing Gerri's draft of the persuasive document, her colleagues suggested the following revisions: "Rewrite by considering goals, audience, document design, and completeness of the content. You should add highlighting techniques to emphasize key points in the flier, such as bullets, headings, textboxes, and color. Because a flier is short, we'll want to offer the audience a visually appealing document filled with relevant content. Please consider cutting some of the content; we can use it later in another flier. That way we can better emphasize one key topic." See Gerri's rewritten flier in Figure 11.15.

FIGURE 11.14   Gerri's Rough Draft for the Flier

Master Plans for Smaller Communities
Thursday, May 17, 2016
6:00–8:30 p.m.
Grandview Community Center—The View
13500 Byars Road
Grandview, MO

Moderator: Bob Linder, Planning Commissioner

Panelists: Chase Simpson, Real Estate Attorney; William Shane, Shane
Appraisal, Inc.; Scott Mitchell, Vice President, Mitchell Engineering

Smaller communities throughout the Kansas City metropolitan area are facing
growth challenges—residential, commercial, and industrial. Our smaller cities
are eager to accommodate new residents and businesses, but are concerned
about losing the small-town quality of life, impacting streams and other natural
features, and paying for infrastructure to support new development. The
preparation and adoption of a master plan or land-use plan provides communities
with an important citizen engagement opportunity and new tools to guide growth
and development.

The two presenters will describe the conditions under which a master plan is
useful or necessary; the elements that might be included in a master plan; the
various ways that master plans might be developed, including ways to involve
residents and other stakeholders; the adoption process; and how the planning
commission uses the adopted plan in making decisions on development
applications.

FIGURE 11.15   Gerri's Revised Flier

CASE STUDY

Creative Seminars sends persuasive sales letters to prospective customers. Read the sales letter and complete the following assignments.

Assignments

- Determine where ethos, logos, and pathos are used persuasively. Give examples to prove your point.
- Based on the criteria provided in this chapter, decide how the sales letter is successful and where it could be improved. Rewrite the letter according to your suggestions for improvement.

May 22, 2016

Tom Johnson
Director of Human Resources
ABC Trucking
1234 S. Main St.
Kansas City, MO 64111

Dear Tom:

I am pleased to have made contact with you. As promised, I am sending the packet of information about Creative Seminars. Since 2010, we have been making it easier for many Kansas Citians and others in the region to do their jobs, ultimately making them more productive and their work more consistent.

Our specialty is customized training programs and material that focus on what workers need to *do*, not just what they need to *know*. Our tried and true process is applicable to any industry, any content area, and any delivery medium. A few of our projects include

- Performance management training for trucking company supervisors.
- Orientation training for over-the-road drivers.
- Operation training for tire retreading equipment.
- Safety and facilities management training for federal prison personnel.

We are a local company with strong ties to the community. Our clients include FTI Insurance, Dave's Diamonds, Shoetime, Nature's Neuroscience, and Truckers Freight. We are a valuable resource right in your backyard. I would appreciate the opportunity to visit with you about the match between your needs and our capabilities. I will call next week to set up an appointment.

Sincerely,

Carolyn Johnson
President

Enclosure

2782 State Line, Suite 2   Kansas City, MO 64182   816-555-5555   Fax: 816-555-5511
www.creativeseminars.com

ETHICAL CHALLENGE

Alyssa Adams is a sales representative for a biomedical software company. She frequently travels throughout the country to meet with potential clients. In the past year, her company created a new policy for dealing with clients: Sales representatives are limited to a maximum entertainment expenditure reimbursement of $25 monthly per client.

For the past ten years, Alyssa has been a successful sales representative for the company with a high level of productivity. She likes to do the following for her clients and potential clients: buy them tickets to baseball games, take them out to play golf, buy them lunches, and give them fruit baskets at Christmas.

She abides by the company's new policy limiting her to $25 monthly per client, but her productivity has decreased by 22 percent.

Question

Is it ethical for Alyssa to supplement company funds with her own money to do favors for her clients? Why or why not?

INDIVIDUAL OR TEAM PROJECTS

ARGU

For the *ARGU* method of organization, you *A*rouse reader interest, *R*efute opposing arguments, *G*ive proof to support your argument, and *U*rge reader action. Read the following situations and complete the assignments.

- You plan to sell a mobile app. For a sales letter, write five different introductions to arouse reader interest using any of the options provided in the chapter.
- Write a body that refutes opposing arguments (too expensive, easily damaged, easily lost, etc.) and that gives proof to support your product claims.
- In the sales letter for the mobile app, conclude by urging reader action. Use at least two of the methods discussed in this chapter.

Analysis of Persuasive Writing

Find examples of persuasive writing (e-mail messages, sales letters, fliers, brochures, or press releases). Bring these to class, and in small groups or individually, accomplish the following:

1. Decide which methods of persuasion have been used. Where in the documents do the writers appeal to logic, emotion, and ethics? Give examples and explain your reasoning, either in writing or orally.
2. Have the writers aroused reader interest? Give examples and explain your reasoning, either in writing or orally. If the writers have not aroused reader interest, should they have? Explain why. Rewrite the introductions using any of the techniques discussed in this chapter to arouse reader interest.
3. Have the writers refuted opposing points of view? Give examples and explain your reasoning, either in writing or orally. If the writers have not negated opposing points of view, should they have? Explain why. Rewrite the text using any of the techniques discussed in this chapter to negate opposing points of view.
4. Have the writers developed their arguments persuasively? Give examples and explain your reasoning, either in writing or orally. If the writers have not provided persuasive proof, rewrite the text using any of the techniques discussed in this chapter to improve the arguments.
5. Do any of the examples you have found use logical fallacies to persuade the readers? Give examples and explain your reasoning, either in writing or orally. If the writers have used logical fallacies, rewrite the text using any of the techniques discussed in this chapter to improve the arguments.

PROBLEM-SOLVING THINK PIECES

Logical, Emotional, and Ethical Appeals

Read the following situations and determine whether the persuasive arguments appeal to logic, emotion, or ethics. These argumentation techniques can overlap. Explain your answers.

1. If you purchase this product, you can benefit from a healthier, happier, and longer life!
2. Seventy-two percent of SUV owners say that high gas prices will influence their next car purchases.
3. CEO Jim Snyder, an expert in the field of sports management, says, "Building a downtown sports arena enhances a city's image."
4. Style-tone Hair Gel improves your hair quality by preventing split ends, generating new hair growth, and inhibiting "frizzies."
5. Failure to recycle will cause 52 percent more dangerous hydrofluoric carbons to be released into the atmosphere, leading to harmful decreases in the ozone layer.

Logical Fallacies

Read the following logical fallacies and revise them, ensuring that the sentences provide logical, ethical, and correct argumentation.

1. All marketing experts believe that social media is an effective way to communicate persuasively.
2. Either all employees must come to work on time, or they will be fired.
3. Some accountants are ambitious because they wish to succeed.
4. The contractor lost the bid, so he cannot expect to have increased revenues this fiscal year.
5. Because the manager is inexperienced, the report will be badly written.

WEB WORKSHOP

You can find electronic brochures and fliers on the Internet. Using a search engine of your choice, type in phrases like "online brochure," "e-brochure," "online flier," "e-flier," "electronic brochure," or "electronic flier." Once you find examples of these online persuasive documents, do the following:

- Compare and contrast the electronic documents with hard-copy brochures and fliers.
- Compare and contrast the electronic documents with the criteria provided in this chapter.
- Decide how the online communication is similar and different from the criteria and from hardcopy versions.
- If you decide that the online versions can be improved, print them out and revise them.

CHAPTER TWELVE

Social Media

After completing this chapter, you will be able to

1. Learn why social media is important in the workplace
2. Recognize different types of social media
3. Understand guidelines of social media on the job
4. Learn how to use different types of social media
5. Know how to use a blog in the workplace
6. Use YouTube in the workplace
7. Use Twitter for technical communication
8. Rely on Facebook for corporate communication
9. Appreciate the diverse Google applications for business
10. Use micro-videos in the workplace

COMMUNICATION AT WORK

In the LandBeyond scenario, Pasha Urtag explains how he uses social media in his position as marketing and communications manager.

LandBeyond is a software and services company that scours the Internet to bring back the most relevant information on any topic so that businesses can engage strategically online and make smart decisions based on real data. Clients can do the following:

- Monitor corporate brand/identity
- Manage crises/customer service
- Measure campaigns, product launches, and online sentiment
- Analyze the competition

LandBeyond helps companies make sense of the Web with a comprehensive analysis and engagement toolset. This allows businesses to put their entire brand's Web presence in context and engage strategically. The company monitors the following:

- Blogs
- News sites
- Social networks like Twitter, Facebook, LinkedIn, Google+, etc.
- Video sites like YouTube and Vimeo
- Reference and review sites, such as Wikipedia and Yelp
- Public forums and microblogs

Pasha Urtag, marketing and communications manager at LandBeyond, has an interesting and varied job with numerous responsibilities. He often works 55 hours or more each week and rarely turns off his smartphone. "I even watch TV with the remote in one hand, my smartphone in the other, and a tablet in front of me." For Pasha, a typical workday consists of

- Researching LandBeyond every morning when he gets to work using the company's Web-monitoring software. Pasha performs this research to find out what is being said about LandBeyond and who is talking about the company. He responds to mentions of the company and thanks the people who have communicated about LandBeyond.
- Monitoring the brand at all times. Pasha says, "I have all of our company's platforms on my smartphone, and I will respond as the company during off hours."
- Creating case studies and reports for clients.
- Writing entries on the company's blog or working on current projects. One recent project was to organize content for a presentation on "social media measurement" at The 360° Pivot Conference held in New York City.
- Attending social networking events. These events include social media club meetings and events with business professionals whom he has met through his social media connections.
- Lending LandBeyond's expertise and services to community-related and other high-profile events.

According to Pasha, organization is a critical part of performing his job. He finds that the following tools are essential to his success:

- Smartphone allows 24/7 access.
- Company software gives him the capability to study the information found on the Web.
- Laptop/tablet is essential, especially when he is traveling on business.
- TweetDeck is an application designed around Twitter that filters tweets for easy review/ response.

Why Is Social Media Important in the Workplace?

Social media has a growing impact in today's workplace. According to Marius Ciortea, senior manager of project management and office Web architecture at Oracle, using social media in the workplace makes employees approachable, allowing for transparent conversations with customers and quick response to customer concerns. Michael Brito, social media strategist for Intel, says that with social media, companies can humanize their brand and achieve a sense of community and participation with customers.

Sun Microsystems is one example of a company that is using social media channels to connect its customers and employees. According to Sumaya Kazi, senior social media manager at Sun Microsystems, employees at Sun are expected to participate in social networking (wikis, blogs, tweets, and Facebook) to build community, listen, share, learn, and communicate with each other and with their customers. Social media is such a large part of life at Sun that the company has created social media ecosystems—multiple avenues of connectivity. Recently, Sun employees participated in social media as follows:

- 5,426 Sun bloggers
- 4,437 Sun blogs (including group blogs)
- 71,000 contributors to wikis.sun.com
- 946,000 contributors to forums.sun.com
- 200 communities on Sun's Facebook site
- 5,240 followers on Sun's Twitter site

Uses of Social Media in the Workplace

Today's companies are using social media to interact with clients, colleagues, and corporate partners as follows:

- Learn what clients are saying about a company's products and services.
- Manage, monitor, and influence these conversations.
- Improve customer satisfaction through increased interaction with customers.
- Personalize customer contact.
- Respond more quickly to customer comments and concerns.
- Meet customer needs through the development of new products and services, as suggested by comments overheard through social media.

Crisis Communication

Another important use of social media is crisis communication. According to the FBI, social media is outpacing services such as 911, police, emergency medical technicians, firefighters, and even journalists in reporting crisis situations.

Today's workplace communicators must be prepared to communicate information about crises that occur in the workplace. A crisis could occur for any number of reasons, including an armed intruder on the premises, severe weather, bomb threat, receipt of potentially hazardous material through the mail system, the threat of terrorism, and more.

One of the most important ways in which corporations can protect their assets is by establishing a clear crisis communication plan. Such a plan must include a variety of communication channels, including telephone calls, e-mail messages, text messages, or loudspeaker announcements. Additional communication channels mandatory in a 21st-century digital workplace must include Twitter, Facebook, YouTube, LinkedIn, and blogs. Such social media channels can convey crisis information related to large groups of people quickly and often directly to their smartphones and tablets (Braud; Tinker).

What Is Social Media?

2. Recognize different types of social media.

Social media can take many different forms, including Internet discussion forums, blogs, vlogs, podcasts, online videos, and wikis. A company can use a blog to transmit information about the company, its personnel, and its products and services. This communication can be delivered quickly and efficiently to an individual's smartphone, iPod, or laptop computer—anytime, anywhere. Social media, in combination with mobile communication, allows for real-time video instructions to be delivered through YouTube and information delivered through Twitter, Facebook, or LinkedIn, among others.

Through a Facebook wall, customers can post directly to the company and communicate with many different levels of management. Bill Brelsford, owner of Rebar Business Builders, says, "I believe the 'cool' factor lies not in the technology but in the fact that now I can have a conversation with someone like the president of a large corporation, something I would probably not be able to do without social media tools" (Goforth). Whitney Mathews, social media expert at Digital Evolution Group, says that social media is the "new customer help line" (Mathews). See Table 12.1 for examples of social media.

TABLE 12.1   Examples of Social Media

Blogger—a blog publishing system	**Meetup**—an online social networking site that facilitates group meetings and allows members to find and join groups unified by common interests	**TypePad**—a blog publishing service
Delicious—a social bookmarking or social tagging Web service for storing and sharing Web bookmarks		**Ustream.tv**—a network of channels for live video streaming of online events
Digg—a social news Web site allowing users to submit Internet links and stories, vote, and comment on submitted stories	**MySpace**—a social networking Web site	**Bulletin**—an Internet site for forums and community blogs
	Open Diary—a blog publishing system	**Vimeo**—a social networking site that supports video sharing, video storage, and user comments
Facebook—a social networking Web site	**Photobucket**—an image and video hosting Web site	
Flickr—an image and video hosting Web site	**Pinterest**—a social photo and information sharing Web site	**Wikis**—community Web sites, such as Wetpaint, PBWiki, and Wikispaces, that allow for collaborative creation and editing of Web pages
Foursquare—a site for mobile connectivity to find the location of friends	**Plaxo**—an online address book and social networking service	

(Continued)

TABLE 12.1 *(Continued)*

Google Groups—a service from Google allowing discussions about common interests	**Plurk**—a social networking, microblogging service	**WordPress**—an open source blog publishing application
Instagram—an online photo-sharing service	**Revver**—a video sharing Web site	**Yelp**—a social networking online opinion site for user reviews
Justin.tv—a network of channels for online live video streaming	**Skype**—a software application that lets users make telephone calls over the Internet	**Yahoo! Groups**—an electronic mailing list and Internet forum
LinkedIn—a business-oriented social networking site	**Slideshare**—for uploading and sharing PowerPoint presentations, Word documents, and Adobe PDF Portfolios	**YouTube**—a video sharing Web site allowing users to upload and share video clips
LiveJournal—a blog publishing system	**Twitter**—a social networking, microblogging service	**Zooomr**—a Web site for sharing digital photos and text messages

3. Understand guidelines of social media on the job.

Guidelines of Social Media Usage on the Job

Companies might hope that their employees' use of social media will be ethical and professional. However, this assumption could lead to problems. Without precise social media guidelines, employees could disclose confidential information or trade secrets, create public relations problems, or damage a company's image.

To avoid such situations, companies should create guidelines for employee use of social media, including the following:

- Use a personal e-mail address versus your employer's e-mail address to register for social media sites.
- Never reveal confidential company information.
- Avoid using company logos on personal social media sites.
- Avoid using social media during business hours unless doing so is part of your job.
- Never complain about an employer on a social media site.
- Write professionally in any social media postings.
- Never harass a colleague, customer, or employer on a social media site.
- Avoid expressing any disparaging views about an employer or coworker on a social media site.

An excellent policy database for many company social media policies is available at Social Media Governance's "Policy Database."

4. Learn how to use different types of social media.

Types of Social Media

Although social media involves many different channels of delivery, this chapter focuses on the following:

- Blogging
- YouTube

- Twitter
- Facebook
- Google+

Understanding these types of social media will allow you to participate in online communities and incorporate social media into your workplace.

Blogging for Business

5. Know how to use a blog in the workplace.

Jonathan Schwartz, president and chief operating officer of Sun Microsystems, says that blogging is a "must-have tool for every executive. 'It'll be no more mandatory that they have blogs than that they have a phone and an e-mail account'" (Kharif). Bill Gates, Microsoft's CEO, says that blogs could be a better way for firms to communicate with customers, staff, and partners than e-mail and Web sites. Microsoft employees use blogs to update clients and colleagues about projects ("Gates"). Marc Cuban, owner of the Dallas Mavericks basketball team, has a blog (Ray). IBM created a corporate blogging initiative by encouraging its employees to become active bloggers with the goal of achieving "thought leadership" in the global information technology market (Foremski). Nike launched an adverblog to market its products. What is blogging, and why are so many influential companies and business leaders becoming involved in this social media channel?

Blogging—a Definition

A blog, the shortened version of the words *Web log,* is created online and can include text, graphics, links to other Web sites, and video. Blogs not only allow but also encourage input from many readers. Daily, millions of people use blogs to publish their thoughts. Though many blogs feature text, they take different forms, including the following:

- **vlogs.** The posting of videos through Really Simple Syndication or Rich Site Summary feeds (RSS) to create miniprograms such as those seen on YouTube.
- **Podcasting.** News, market information, product announcements, public service announcements, how-to videos, photographs, audio files, and more can be conveyed through podcast blogs. The term *podcast* is a combination of the words *iPod* and *broadcast.*
- **Microblogging.** Brief text, usually less than 200 characters. Facebook and MySpace, two popular blog sites, include microblogging features.
- **Twitter.** A free microblogging service. Users can send short tweets or updates limited to 140 characters to cell phones or other blog sites.

Reasons for Blogging

Following are reasons why companies blog:

Communicate with Colleagues. Many companies encourage their employees to use blogging for project updates, issue resolutions, and company announcements. The engineering department at Disney ABC Cable Networks Group uses blogs for documentation of requests to the help desk (Li).

Communicate with Customers. In contrast to private, intranet-based blogs used for internal corporate communication, companies also have public blogs. Through public blogs, a company can initiate question/answer forums, respond to customer concerns, allow customers to communicate with each other, create interactive newsletters, introduce new products and services, and build rapport with customers, vendors, and stakeholders.

Improve Search Engine Rankings.   Marketing is a key attraction of corporate blogs. A blog post using keywords, allowing for comments and responses, and providing references and links to other sites, tends to rank in the top 10 to 20 listings in Internet search engines (Ray).

Network through Syndication.   To access a Web site, a reader must know the URL or use a search engine, such as Google. In contrast, blogs can be distributed directly to the end users through a Web feed. The writers of a blog can syndicate their content. Then, through the Web feed, readers can subscribe to the blog. Syndication makes Web feeds available for people to access. By using feed programs, such as RSS or Atom, bloggers can syndicate their blogs or be notified when topics of interest are published. Thus, blogs can be very personalized, essentially delivered to your door on a moment's notice.

Track Public Opinion.   Trackback features, available from many blog services, let companies track blog usage. Tracking lets companies monitor their brand impact and learn what customers are saying (good news or bad news) about their products or services.

Guidelines for Effective Corporate Blogging

If you and your company decide to enter the blogosphere, the world of blogging, follow these guidelines:

Identify Your Audience.   As with all professional communication, audience recognition and involvement are crucial. Before blogging, decide what topics you want to focus on, what your unique spin will be, what your goals are in using a blog, and who your blog might appeal to.

Achieve Customer Contact.   Blogs are innately personal. Take advantage of this feature. Make your blogs fun and informal. You can give your blog personality and encourage customer outreach by including corporate news, personnel biographies, question/answer forums, updates, an opportunity to add comments, as well as information about products and services. In addition, make sure the blog is interactive, allowing for readers to comment, check out new links, or add links.

Start Blogrolling.   Once you have determined audience and blog location, it's time to start blogrolling. Start talking. Not only do you need to start the dialogue by adding content to your blog, but also you want to link your blog to other sites.

Keep It Fresh.   Avoid a stale blog. By posting frequently (daily or weekly) and responding quickly to comments and criticism, you encourage bloggers to access your site and return to it often.

Ethics and Social Media

See Chapter 6, "Ethical Considerations," for information about ethics and social media.

Develop Guidelines for Corporate Blog Usage.   Because blogs encourage openness from customers as well as employees, a company must establish guidelines for corporate blog usage.

6. Use YouTube in the workplace.

YouTube and Business

YouTube is increasingly important in business to communicate with a technologically adept consumer.

Why Should Businesses Have a YouTube Presence?

Another social media opportunity for a business is YouTube. YouTube is a way for a company to connect with the customer. Because television, radio, and print advertising have become very expensive, companies can benefit from a more affordable means of advertising and marketing products and services. YouTube videos can be posted for free and allow businesses to reach millions of potential customers.

What Should Businesses Post on YouTube?

A business can benefit from posting the following on YouTube:

- Presentations from executives to discuss finances, corporate activities, company performance, and so forth
- Announcements of important corporate news
- News releases of company activities
- Training for purchasers of products
- Videos of consumer questions and help desk answers with procedural steps
- Marketing of new products with pictures
- Events such as corporate celebrations, meetings, and other activities
- Instructional video for use of products

Screencasting

See Chapter 15, "Instructions, User Manuals, and Standard Operating Procedures," for more information about instructional video.

Guidelines for Shooting Video to Post on YouTube

You have to shoot a video that people actually want to watch. A picture might be worth a thousand words, but if no one sees the picture, then you've wasted time. To shoot a successful video, make sure your production values are professional. This includes the following:

- Title your video and use appropriate subtitles periodically in the video.
- Include the company name and contact information such as a telephone number and the Web site URL.
- Use onscreen graphics to enliven the video.
- Edit a longer video into short segments. YouTube allows videos 10 minutes long.
- Focus on one message; avoid attempting to cover all your products or all your services.
- Avoid plagiarism by creating your own content. Only use music, television, or movie clips with permission from the copyright holder (Brewer).

Tweeting for Business

7. Use Twitter for technical communication.

Businesses rely increasingly on Twitter for corporate communication and to reach the consumer.

What's Twitter? What's a Tweet?

To use Twitter, you tweet. Twitter, a micro-blogging service, lets you send and receive messages limited to 140 characters. These messages are called tweets. Twitter is a free service that lets you subscribe to, share, or follow as many twitter messages as you like.

Jessica Vascellaro of the *Wall Street Journal* writes, "Email has had a good run as king of communications. But its reign is over" (R1). In Vascellaro's opinion, Twitter is taking over and affecting how people communicate. The primary reason for this change is access. Twitter is a distributed communication system, and you don't need to open a Web browser to access its service. Instead, you and your readers (called *followers*) can receive tweets in a variety of ways. From your desktop or laptop, you can use programs like Twinkle, Twitterific, or Feedalizr. From your smartphone, you can use programs like PocketTweets, Tweetie, or iTweet.

Tweets don't have to be intrusive, either. Not everyone needs to read what you tweet. You can be more discerning by sending a Direct Message, which only one follower sees versus responding with a public tweet. Tweets can be sent directly to your RSS reader, and you can republish (re-tweet) a stream of content on your Web site or blog. Table 12.2 defines commonly used Twitter terms.

TABLE 12.2 Definitions of Commonly Used Twitter Terms

Follower	People who read your tweets on Twitter
Re-tweet	Copy someone else's tweet or republish a stream of tweets
Tweet	A single Twitter message
Tweetdeck	Software that runs on your desktop, allowing you to post tweets, respond to tweets, set up searches, and get updates in real time
Tweetie	The Twitter application for iPhone and iPod Touch
Tweet scan	Twitter's search tool
Tweeple	People who tweet
Twellow	A Twitter directory (something like the Yellow Pages)
Twhirl	A desktop application that updates incoming tweets
Twitturly	A service for tracking URLs that people are talking about

Why Do Companies Use Twitter?

Restaurants can tweet their daily specials, realtors can tweet new homes on the market, and a chamber of commerce could tweet about local city events and promotions. Companies can accomplish the following with Twitter:

- Direct people's attention to corporate news, products, services, and events.
- Give customers and coworkers a forum, allowing a company to study concerns, manage issues, initiate discussions, answer questions, and respond to complaints.
- Assist with business development—promotional and marketing collateral.
- Add to and improve customer service. Through Twitter, you can send service updates and respond to customer concerns. Best Buy, for example, created Twelpforce, a customer service team to answer customer questions about products and services.
- Help you show the human side of your company (tell what employees are doing and show how your company is involved in charitable activities).

- Allow dispersed work teams to communicate quickly with each other. For example, Handmark is a leading developer of mobile apps. Its employees work in Kansas City, Dallas, and Palo Alto. Due to their dispersed work locales, employees can't just walk down the hall to visit about projects. Paul Reddick, CEO of Handmark, says he and his coworkers use Twitter for "virtual cube talk."

Which Companies Have Twitter Sites?

Table 12.3 shows companies with Twitter sites, their uses of these sites, and a few examples of tweets.

TABLE 12.3   Companies Using Twitter

Companies Using Twitter	Purposes	Tweet Examples
JetBlue	Respond to customers' questions	Q: "Which destination would you pick for a short vacation?: Las Vegas (1) or New Orleans (2)?" A: "2 to 1 in favor of Vegas for a short trip after the initial volley of replies."
TheHomeDepot	Respond to customer concerns	"Sorry you didn't get the attention you deserved. We'd still like to help with your project. Can I help?"
Dell	Highlight individual Dell employee activities	"RT@LionelatDell donates $5 to Susan G. Komen for the Cure for each Promise Pink laptop we sell. Lots of options (http://twitter.com/richardatdell)."
SouthwestAir	Provide links to online articles for more information	"Did you know Southwest Pilots mentor thousands of students in science, math, geography, and writing? Read more: http://tinyurl.com/bpf6wb"
GM	Share customer comments	"Re-tweeting @WarLordwrites: I've been using the GM OnStar Navigation system. I'm pleased. GM got this very right!"
M&Ms	Update company promotional activities	"Need a little va-va-voom in your love life? Tell me why over at www.mms.com/green and you could win a romantic Paris getaway!"
HRBlock	Build rapport	"Happy to hear it went well! It's a relief to have your taxes filed isn't it? You're ahead of me;)"

Rules for Twittiquette

The bottom line to etiquette of any sort is treating people with respect. After that essential consideration, the following are rules for etiquette—*twittiquette*—when using Twitter.

1. Use services like Twitter Search to find who might be writing about you or your company.
2. Don't tweet all day while at work. That's not an effective use of your time or corporate time.
3. Don't use Twitter just as another channel for advertising. It should have a more personal, social media component—people talking to people.
4. Separate business from fun. Have two Twitter accounts: one that is strictly business and one that is for friends, family, and fun.
5. Don't tweet in anger. A tweet sent in anger can be read by thousands of people. They might not appreciate your haste and negativity. In contrast, positive interaction with a company representative could have a more positive impact.

<table>
<tr><td>8. Rely on Facebook for corporate communication.</td></tr>
</table>

Facebook

In addition to other social media sites, businesses use Facebook for corporate communication.

Business Uses for Facebook

Why would a business want to have a Facebook presence? Perhaps the greatest reason for companies to use Facebook is the enormous growth of this social media site, as shown in Table 12.4. The information is drawn from Facebook's newsroom.

TABLE 12.4 Facebook Statistics

People on Facebook	Facebook has more than 1.23 billion monthly users.
Activity on Facebook	945 million mobile users 757 million daily users
Global Reach	More than 70 translations are available on the site. About 80 percent of Facebook users are outside the United States.

With such large numbers of potential customers, a company can use Facebook as a one-stop location to market products or services, create brand recognition, recruit new employees, and connect with the customer on a personal basis. On their Facebook sites, companies can host their corporate blog, provide access to their Twitter entries, post company YouTube videos, show photographs of corporate employees, list upcoming events and product releases, and provide access to discussion forums. Businesses use Facebook in a variety of ways, including the following:

- Collaborating in teams for sharing documents and tasks and collaborating on projects
- Uploading PowerPoint slides to share presentations, tutorials, and demonstrations
- Conducting polls to find out end-user needs and preferences
- Managing a crisis by providing updates and responding to customer concerns

Guidelines for Businesses Participating on Facebook

For you and your company to participate in Facebook, follow these guidelines:

Create a Clear Business Goal.   What do you hope to accomplish with your company's Facebook site? Do you want to sell, promote your charitable activities, share information about your employees, show videos of how-to tutorials, allow for give and take with customers through blog entries, and so forth? Determine a plan before going online in Facebook.

Keep It Fresh.   You should update your Facebook profile consistently. Add new videos, new blog postings, new tweets, new information about company products and services, and news about company activities.

Add Links to Social Media Tools.   Your company's Facebook site will be enhanced by links to other social media tools such as Twitter, YouTube videos, your company's blog, LinkedIn, and so on.

Create Reasonable Conduct Policies for Employees.   Publish corporate policies for Facebook use regarding security, confidentiality, and personal conduct online. That way, both supervisors and staff will be aware of corporate expectations. No surprises should occur that could lead to terminations or loss of sensitive data.

When using social media sites such as Facebook, you must be ethical, honest, and represent your company respectfully. Consider the following guidelines before you post on Facebook:

- Always write truthfully about yourself or your company.
- Do not disclose confidential information about the company.
- Post photographs that represent employees in a professional manner.
- Avoid links to unprofessional, inappropriate, or personal sites.
- Post videos that are short and professional.
- Respond to public inquiries on the site immediately (within 24 hours of the posting).
- Update the work-related information on the Facebook site frequently.
- Avoid wasting work time with frequent postings of a personal nature.
- Avoid endorsing other people or companies unless your company approves.
- Do not post other people's writing without getting permission.

Google+

Google+ is becoming a major competitor to Facebook in the social media arena. A key predictor of Google+ success is its massive ecosystem of other popular services: Google Search, Google Maps, Google Translate, and Google Docs for wiki collaborations, YouTube, Google Sites for Web design, cloud service, Gmail, apps, Chrome, and more.

In addition, Google+ might have an additional advantage over Facebook. More people access Google+ through their mobile devices (smartphones and tablets) than from their desktops. This asset has been made possible by the fact that Google+ comes preinstalled on the almost one million Android devices sold every day.

> 9. Appreciate the diverse Google applications for business.

How else might Google+ be a social media option to Facebook for businesses?

- **Privacy.** People in your Google+ circle can see all of your posts while others might only see your public postings.
- **Search Engine Optimization (SEO).** Google+ provides "DoFollow" links to improve blog search engine rankings.
- **Group video chats.** Google+ lets 10 people participate in video chats in one "hangout" time for free, unlike Skype, which charges a fee for multiple users.
- **Instant uploads.** From an Android phone, you can automatically post photographs (Agarwal; Ostrander; Brown).

10. Use micro-videos in the workplace.

Micro-Video—the Newest Social Media Trend

Micro-video is short bursts of audio/video shot from and conveyed to mobile devices. These short videos range in length from just 6-second long videos from Vine, 15 seconds on Instagram, and 16 seconds on MixBit.

Currently, the major companies providing this hot social media channel include the following:

- Twitter's Vine, the originator of this social media channel
- Instagram
- MixBit
- Tumblr
- Viddy
- Qwiki
- Tout
- Klip
- Socialcam

Why has micro-video become a hot trend? As YouTube proved, people like video content. As Twitter proved, people like short content. Micro-video combines the two in short videos. Micro-videos offer an abbreviated view of people (selfies with sound and action), places, ideas, products, and objects. Companies are happy to use micro-video for short narratives, animations, and how-to instructions because micro-video is cost effective. The channel depends on viral sharing versus the more traditional paid promotional spot, and micro-video is easy to create.

Who is using micro-video to promote products and services? The illustrious list includes Coca-Cola, Oreo, Red Bull, Tide, Samsung, and Audi (Kelly; Payne).

SOCIAL MEDIA CHECKLIST

_____ 1. Have you considered your audience's needs in your social media communication?

_____ 2. If you are using social media, how will you engage your audience and encourage participation in the blog, tweet, Google+, or Facebook site?

_____ 3. If you are tweeting, have you limited your tweet to 140 characters?

_____ 4. If you are using Facebook or Google+, have you developed a corporate plan?

_____ 5. If you have created a video to post to YouTube, have you followed the guidelines, such as limited length, appropriate subject matter, titles and subtitles, contact information, and so forth?

_____ 6. Have you considered ethical standards when using social media?

_____ 7. Did you review your communication for grammar, punctuation, or spelling errors?

8. If you have used material such as pictures, video, or words created by someone other than yourself, have you linked to the source or provided copyright information?

9. Are you updating your blog posts frequently to keep the information fresh and current?

10. Do you avoid spending excessive amounts of work time communicating via social media?

The Writing Process at Work

To communicate effectively, Pasha Urtag, marketing and communications manager at LandBeyond, used prewriting, writing, and rewriting.

Prewriting

Figure 12.1 illustrates Pasha Urtag's use of a mind map to gather and organize data for his blog posting for LandBeyond.

FIGURE 12.1 Mind Map for a Blog Post

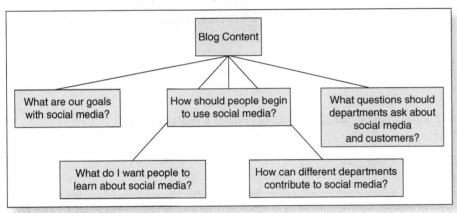

Writing

Figure 12.2 is Pasha's draft of the blog entry. He then thought about redesign issues to revise the text.

FIGURE 12.2 Pasha's Draft of the Blog Entry with His Thoughts on Revision

I'll break the text into smaller chunks, smaller paragraphs for easier access. I'll italicize the quoted section for ease of reading.

I'll add headings and subheadings to help our readers navigate the text more easily.

I'll link to related articles on our blog to drive traffic and position our company as an expert.

I'll illustrate the point of the article with a screenshot of our product and use strategic keywords in the caption for better SEO (search engine optimization).

I'll tie in our company's value proposition so that the research I'm writing about and the narrative I'm weaving in supports the idea that using our product will help solve this problem.

> There are still plenty of CEOs who are unsure about the value of social media. A new report says that a shocking 43% of B_2B companies admit that their CEOs never consider their social media reputation. 74% of CEOs think that marketers focus too much on the "latest marketing trends such as social media," and say they can rarely demonstrate its business value, according to a U^2 Marketing Group.
>
> Companies need to be able to prove that their social media efforts are paying off in a business sense. They need to be able to go beyond surface-value metrics and correlate their social media and web presence with tangible business metrics. The problem is that ROI is a financial metric and social media metrics are decidedly non-financial.
>
> Nicholas Keilty from Social Media Research approaches it simply. Here are three specific questions to ask your CEO that will get them thinking about the business value (or ROI) of social media marketing:
>
> 1. Do I want to get to potential buyers before my competition?
> 2. Do I want the opportunity to build brand loyalty with these potential buyers so they select my company over the competition?
> 3. If I could build that loyalty for comparatively low cost, would it make good business sense?

Rewriting

See Figure 12.3 for Pasha's successful blog post from LandBeyond.

FIGURE 12.3 Pasha's Blog Post for LandBeyond

> # How Social Media Can Improve Your ROI
>
> by Pasha Urtag on March 15, 2016
>
> Many CEOs are still skeptical about the value of social media.
>
> According to a U^2 Marketing Group report, CEOs at 43% of B2B companies never focus on their social media reputation. An even larger 74% of CEOs think that marketers emphasize the value of "hot and hip marketing trends such as social media" which "never demonstrate business value."
>
> I understand why these CEOs are skeptical. If my marketing department reported that our company's social media sites were getting new likes and follows, I would say "Huh? What are 'likes and follows'?"

(Continued)

FIGURE 12.3 *(Continued)*

Bridging the Gap between Social Media and ROI

To satisfy stakeholders, companies must prove that social media efforts earn money. The problem is that ROI is a profit/loss spreadsheet, but social media benefits are less tangible.

Nicholas Keilty from Social Media Research approaches it simply. Here are three questions to ask your CEO about the ROI of social media:

1. *Do I want to reach a potential client before my competitor?*
2. *Do I want to build a brand loyalty with my potential clients so they'll buy from my company before they visit the competition?*
3. *If I could achieve customer loyalty for low cost, would that help my ROI?*

Every CEO will answer yes to these questions. Then they'll say, "Prove it." Here's how to track the ways in which social media impacts ROI:

Let LandBeyond's metric software show how a social media rollout

- Increases sales
- Reduces traditional marketing expenses
- Acquires new customers
- Reduces the cost of transactions
- Increases customers and sales leads
- Increases customer satisfaction

Showing how the work you do in social media impacts your company's financial goals will let you speak the language of the CEO.

Leave a Comment

Name*

E-mail*

Web site

Submit

CASE STUDY

S&L Engineering has been working on the construction of a multi-purpose site in downtown Los Angeles. The proposed site includes a boutique hotel, office space, condos, and retail shops. However, financial and legal problems have negatively impacted the construction schedule. Although 60 percent complete, the construction company that S&L hired has filed for bankruptcy. S&L has hired a new construction company that promises to complete the project in 18 months.

S&L Engineering needs to update the community on the project's status. This includes damage control, because S&L wants to be a good neighbor to other businesses and residents near the construction site.

To accomplish their communication goals, S&L plans the following:

a. Post a YouTube video of S&L's CEO speaking to the public about the project's challenges.

b. Write a tweet announcing the status of the project.

c. Post a blog entry from the construction manager of the building site, detailing plans for the project's completion.

d. Post a Facebook wall entry about the new construction company (its local reputation and its plans for hiring from the local labor force).

Assignment

What content should be included in each of the above communication channels? Write the scripts for a, b, c, and d.

ETHICAL CHALLENGE

All corporations adhere to a non-disclosure policy about stock valuation and company profits and losses. Employees are always reminded not to talk about changes in the corporate environment that might affect stock values.

Two colleagues, Rebecca Bacero and Tom Suttles, are eating lunch in a restaurant. During lunch, they discuss a new software product that has not met the company's expectations. Problems have occurred during usability testing which could negatively impact company revenues. In their disappointment, Rebecca tells Tom that she just received a tweet about how the company's stock is expected to decrease by at least 35 percent as a result of this new product's failure.

At a nearby table, a fellow employee, Daniel Solomon, eating with some friends, overhears Rebecca and Tom's conversation. One of Daniel's friends says, "Isn't that your company they're talking about? I have money invested in that stock. Should I sell all of your company's stock?"

Question

What should Daniel do? Is it ethical for Daniel to respond to his friend's question? Is Daniel required to report the conversation to his boss? Why or why not?

INDIVIDUAL AND TEAM PROJECTS

Facebook

1. Create a Facebook site for your technical communication class.

2. Review the Facebook sites of various businesses, governmental organizations, or industries. What do these sites have in common? What is unique to each site that you visited? What is each company trying to accomplish in its Facebook site? How would you improve a site if you were the social media manager of this organization or company? For a few samples, check out the following:

   - The Missouri Department of Transportation
   - Sun Microsystems
   - Sprint
   - Harley-Davidson
   - Nike
   - Gap
   - H&R Block

Twitter

After reading the following scenarios, rewrite the text in 140-character tweets.

1. Your bank, Hearth and Home Savings and Loan, has new, low rates for mortgages (3.5 percent for 30-year loans—this is the lowest rate for home loans in over 30 years) and high rates for certificates of deposit (2.5 percent for three-year CDs—CDs have been higher in the past, but recent rates have been much lower). Market these benefits to your bank's customers by sending a tweet.

2. At a recent X-Games motorcycle jumps and tricks competition, hosted by the EnVi Hotel in Las Vegas, Nevada, a rider, Jason Whitworth, performed a first-ever trick, which he called the "double loop superman kick out." Jason rides your company's top-of-the-line motorcycle, the Turbo ZZ210, and wears your company's gear (helmet, boots, gloves, knee pads, and jumpsuit). To inform your company's audience of this new trick and to market your brand, send a tweet.

3. Four employees of your internationally marketed fast-food restaurant, Hot Wings & Cool Things, created a video of themselves acting inappropriately in one of your restaurants. Posted on YouTube, they can be seen eating food from customers' trays and using cooking utensils in an unhealthy manner. Yes, you fired the employees and created new rules of appropriate behavior, incorporated into your training manuals, but still, you must manage this publicity crisis. Write a tweet to your international audience in which you confront bad publicity quickly and show how you manage poor employee behavior.

4. Your software company is rolling out a new product, a software program that will help potential college students apply for college loans, write their letters of application, and gather all application materials in a single portfolio file. To help customers use this new software, your company is planning a free-to-the-public informational webinar on March 11. The webinar will provide hands-on tutorials for using the new software and provide a "virtual" help desk for questions and answers. Share this update with potential customers by writing a tweet.

YouTube

1. Following are YouTube sites providing step-by-step instructional videos:

   - Installing Mac OS X onto a RunCore and Dell Mini
   - Battery Operated Holiday LED Lights
   - How to Replace RAM in Your DELL Inspiron Laptop

- Sun Microsystems t2000 Server
- How to Adjust Honda 50/70 cc Valves

   Access any of these YouTube instructional videos and analyze ways in which they succeed or could be improved. Base your comments on the guidelines provided in this chapter. Write an e-mail message to your professor with a summary of your findings.

2. Create a company, product, or service. Write a script for a YouTube video that introduces or markets the new company, product, or service.

Google+

1. Create a Google+ profile. Here you can provide a profile photograph, tagline, biography, and tabs for YouTube videos and additional photographs.

2. Create a post from Google+. Include text, links, hashtags, and photos. You can use any of the Twitter Assignments in this chapter for your text, or you can create your own content.

3. Create a community in Google+. Then use this for collaborative work to ask questions, post content, and provide comments.

PROBLEM-SOLVING THINK PIECES

1. Check out Stonyfield Farm's blog site at http://www.stonyfield.com/blog/. This site is named "The Yogurt Dish."

Assignment

Read entries from the blog site. Then answer these questions:

- What are Stonyfield's goals in creating this blog?
- How does this blog relate to corporate business?
- If you were hired by a company of your choice, what storylines would your blogs feature and why?

2. Sanya Guptel is attending a social media conference in San Francisco. She is presenting a speech about her company's use of YouTube for customer service training. In addition, she has attended workshops on ethics regarding the use of social media in the workplace, how to target audience demographics through social media, and the development of new social media channels of communication. To bring her colleagues up to date on her experiences at the conference, Sanya can write on her company's Facebook wall, send a tweet, post a blog entry, or make a YouTube video.

Assignment

Explain the benefits and drawbacks for each of Sanya's options. Then, write the text for Sanya, using the social media option that you consider best. Create any additional text you would like to complete this task.

WEB WORKSHOP

1. Bloggers provide up-to-date information on newsbreaking events and ideas.

   - For e-commerce news, visit ClickZ Network—Solution for Marketers.
   - For business blogging, visit Business Week Online to learn where the "worlds of business, media and blogs collide."
   - For technology news (information technology, computer information systems, biomedical informatics, and more), visit ZDNet. Once in this site, use its search engine to find a technology topic that interests you.
   - Every news agency, such as ABC, NBC, CBS, CNBC, has a news blog.

       To see what's new in your field of interest, check out a blog. Then, report key findings to your instructor in an e-mail message.

2. There are many published guidelines for accepted use of social media. Research the following topics to learn what these guidelines entail. Then write an e-mail message to your professor reporting your findings.

   - "Twitter Tips"
   - "IBM Social Computing Guidelines: Blogs, wikis, social networks, virtual worlds and social media"
   - "Sun Guidelines on Public Discourse"
   - "Intel Social Media Guidelines"

3. Not everyone believes that social media is appropriate for the workplace. Read the following articles about potential problems using social media, Twitter, and Facebook in the workplace.

   - Juliet Barbara's "Is Social Media Bad For Business?"
   - Edmund Lee's "Why Facebook Can't Succeed"

       Based on your readings, write an e-mail message to your instructor or give an oral presentation about the pros and cons of social media in the workplace.

4. Research the uses of social media for crisis communication. Then, write an e-mail message or memo to your professor reporting your findings:

   - "Crisis Communications Steps for the Social Media Era"
   - "Best Practices for Crisis Communications over Social Media"

CHAPTER THIRTEEN

Web Sites and Online Help

After completing this chapter, you will be able to

1. Recognize the importance of the Internet for corporate communication
2. Appreciate the need for Web site accessibility
3. Distinguish among the unique characteristics of online communication
4. Establish credibility and security in a Web site
5. Consider ethics when creating a Web site
6. Follow criteria for creating a successful Web site
7. Understand the purpose of online help

COMMUNICATION AT WORK

In the following scenario, Future Promise is creating a team to develop a Web site.

Future Promise is a not-for-profit organization created to help at-risk high school students. This agency realizes that to reach its target audience (teens ages 15–18), it needs an Internet presence. Future Promise's CEO, Brent Searing, has decided to form a team to create the agency's Web site. He will encourage the team to work collaboratively to determine the Web site's content, its level of interactivity, and its design features. Brent wants the Web site to include

- College scholarship opportunities
- After-school intramural sports programs
- Job-training skills (résumé building and interviewing)
- Service learning programs to encourage civic responsibility
- An FAQ page

- Future Promise's 800-hotline (for suicide prevention, STD information, depression, substance abuse, and peer counseling)
- Social media links to Future Promise's Facebook, Twitter, and blog sites
- Additional links (for donors, sponsors, educational options, job opportunities, etc.)
- Easy access from mobile devices

To accommodate these Web components, the Future Promise Web team will consist of the agency's accountant, sports and recreation director, public relations manager, counselor, training facilitator, graphic artist, and computer and information systems director. In addition to these Future Promise employees, Brent also has asked two local high school principals, two local high school students, and a representative from the mayor's office to serve on the committee. Future Promise's public relations manager Jeannie Kort will chair the committee.

Jeannie has a big job ahead of her. First, she must coordinate everyone's schedules. The two principals and high school students, for example, can attend meetings only after school hours. Next, she must meet Brent's deadline; he wants the Web site up and running within three months. Jeannie also must manage this diverse team (with varying ages, levels of responsibility, and levels of knowledge).

The task is daunting, but the end product will be invaluable for the city and the city's youth. Jeannie and Brent know that by conveying information about jobs, training, scholarships, and counseling to their end users (at-risk teens), Future Promise can improve the quality of many people's lives. Jeannie's Web design team has an exciting project on its hands.

Why the Web Is Important for Corporate Communication

1. Recognize the importance of the Internet for corporate communication.

Due to its speed, affordability (for end users as well as companies), and international access, the Web is a key component for corporate communication. Like no other communication medium, the Web has changed the way companies do business by providing companies an international channel for

- Selling their products and services through e-commerce
- Updating employees and customers about corporate changes through wikis, forums, intranets, FAQs, blogs, and more
- Providing employment opportunities
- Offering online forms and instructions for internal and external communication
- Creating a point of contact for customers through online chat, order entry, social media sites, and customer service systems

With very low investments of time, money, and personnel, practically any business can access a huge market, quickly and economically, regardless of the type of product or service being sold, the size of the company, or the location of the business.

International Growth of the Internet

The Internet has flooded the world market. In 1995, only 16 million people had access to the Internet, representing just 0.4 percent of the world's population. By 2014, this number had grown to over three billion people ("Internet World Stats"). Figure 13.1 compares today's Internet usage breakdown in the world.

FIGURE 13.1   Internet Usage in the World by Geographic Region

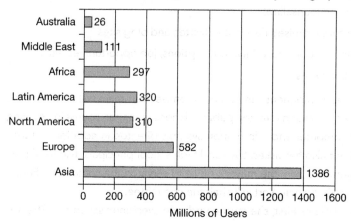

Asia has over a billion Internet users; Europe is second in usage with over 580 million users ("Internet World Stats"). These numbers represent an enormous and growing market for companies. No other communication channel, such as e-mail, memos, letters, reports, newsletters, or brochures, can reach so many people in so many distant locations, 24 hours a day, seven days a weeks, as can the Internet.

Web Accessibility

> 2. Appreciate the need for Web site accessibility.

An international goal is Internet accessibility for persons with disabilities. The Web Accessibility Initiative (WAI) in coordination with organizations around the world pursues accessibility of the Web. The following are Web accessibility problems ("Internet Accessibility").

Cognitive

These include learning disabilities, reading disorders, and attention-deficit hyperactivity disorder (ADHD). Web access can be enhanced with illustrations, graphics, and headings, which provide visual cues for easier Web understanding.

Hearing

Hearing-impaired Web users may need assistive technology to read Web audio or captioned text for multimedia content.

Visual

Color blindness causes problems on the Internet. To combat this challenge, Web writers need to choose colors correctly, perhaps avoiding green and red. The U.S. government recognizes the importance of the Internet and the need for inclusiveness for both employees and external Web readers. In 1998, Congress amended the Rehabilitation Act, requiring federal agencies to make electronic and information technology accessible to people with disabilities. Section 508 for federal employees and the public eliminates barriers in information technology ("508 Law"). For information about Web accessibility, see Table 13.1.

TABLE 13.1 Sites for Web Accessibility

Web Site Name	URL
Web Accessibility Initiative (WAI)	http://www.w3.org/WAI/
The Society for Technical Communication Accessibility Special Interest Group	http://www.stcsig.org/sn/internet.shtml
National Center for Accessible Media	http://ncam.wgbh.org/
International Web Accessibility Concerns	http://www.stcsig.org/sn/international.shtml

The Characteristics of Online Communication

> 3. Distinguish among the unique characteristics of online communication.

Writing for a Web site is different than writing hard-copy text, such as proposals, reports, letters, or user manuals. This means changing your mind-set as a writer. When you design a Web site, you must reconsider the following:

- **Skimming versus linear reading.** We read books linearly, line by line. In contrast, Web sites are skimmed and scanned. Speed is essential online. "Research shows that you have just three seconds to communicate your message on a web page to visitors before they'll click away," says LogoGarden, a Web design company (Mielach).
- **Linking versus chronological reading.** We read books from beginning to end, sequentially. Web sites, however, allow us, even encourage us, to leap randomly from screen to screen, site to site through hypertext links.
- **Reading on varying screen sizes versus 8½" × 11" hard-copy text.** For traditional, 8½" × 11" hard-copy text, we easily can read from margin to margin. In contrast, reading from margin to margin online is not as easy, due to glare, varying screen sizes, and a desire to skim and scan. To help the user access your Web content, pay attention to the Web site's page length, margins, and font selection. In addition, for mobile devices, remember that users must pan by moving their fingers down and across the screen.
- **Choosing the best font.** Your computer's word processing software probably defaults to Times New Roman, 12 point, or Calibri, 11 point. That is considered the best type and size font for readable hard-copy text. In contrast, sans serif fonts, like Arial or Verdana, seem to work best for online reading. Cursive, italicized, and decorative fonts are harder to read online.
- **Minimizing noise.** Because hard-copy text is motionless and printed on dull paper that absorbs light, it can be easier to read than online text. For online text, however, the screen can reflect light, creating visual glare. In addition, Web sites often contain lots of colors, blinking text, animated graphics, frames of layered text, and sound and video. Extended viewing of a computer screen is more demanding than continued reading of paper text. Help online readers by limiting computer noise—sound and visual distractions. In such a busy communication vehicle as a Web site, the less we give the reader on one screen in terms of words and images, the better.

Designing Web Sites for Mobile, Multi-browser, Multi-platform Devices

Additional considerations when you design Web sites relate to today's mobile, multi-browser, multi-platform devices. Web designers realize that their audiences will access a site from a variety of devices with a variety of screen sizes, including desktops, laptops, tablets, e-readers, televisions, and handheld mobile devices. Pew Research tells us that close to 60 percent of users access the Internet using smartphones (Smith). Due to the grow use of mobile devices, Web designers need to be aware of Wireless Application Protocol (WAP), a standard for accessing information over mobile wireless networks. According to Marziah Karch (author of *Android for Work, Droids Made Simple,* and *Android Tablets Made Simple*) and Shannon Conner (freelance Web site developer), consider the following when you design Web sites for mobile devices:

- Remember that the screen of a handheld mobile device is much smaller than desktop PCs.
- Connection time is slower for devices that rely on WiFi or mobile phone networks for connectivity (like 3G and 4G).
- Use XML or XHTML for coding.
- Avoid background images that could reduce readability.
- Don't expect the mobile device user to have the same plug-ins for tools, like PDF, Flash, Javascript, videos, or ActiveX.
- Smaller graphics will aid download time.
- Less text, fewer paragraphs, and more links save room.
- Increased font size improves legibility.
- Design the site for fingers, so make sure that links are about 30–40 pixels in size and can be tapped easily.

Designing Web Sites for Different Screen Sizes. To manage the challenges of multi-browser, multi-platform devices, Web designers are taking the following approaches:

- **Build for minor enhancements.** The easiest way to accommodate multiple devices is to write the Web site's hypertext markup language (html) for the lowest common denominator—a desktop—and then add more advanced features, such as XML and Java scripting.
- **Build for responsive design.** In the past, a Web developer could code different sites for different devices (desktop, mobile, television, etc.). Responsive Web design (also called *scalable design*), in contrast, tries to provide a "consistent viewing experience across a wide range of devices (from desktop to laptops to tablets to smart phones)" (Badminton). Responsive design detects the size of the screen and applies a different layout and style according to the device. A Web site viewed on a desktop browser might look one way, but when the site is viewed on a tablet or smartphone browser, it will render differently. This is accomplished by a cascading style sheet called CSS3. Doing so tailors the site's height, width, and color to a specific output device but does not change the Web site's content. The caveat is that responsive design isn't right for everyone. Not all mobile apps can be scaled up to the size of a desktop; not all Web sites can be scaled down to the size of a cell phone. See Figure 13.2 for an example of responsive Web site design.
- **Implement HTML5.** The Worldwide Web Consortium (W3C) is working to make HTML5 the new standard for Web design. W3C CEO Jeff Jaffe says that when HTML5 is fully operational, developers will be able to "reach smartphones, cars, televisions, e-books, digital signs, and devices not yet known" (Goldman).

FIGURE 13.2 Example of Responsive Web Site Designs with Image and Text Scaled to Various Devices

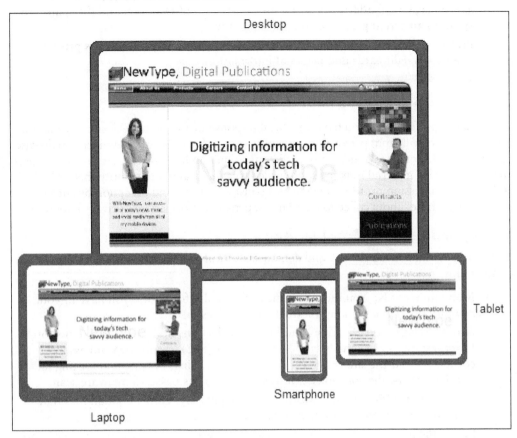

Establishment of Credibility and Security in a Web Site

4. Establish credibility and security in a Web site.

As the creator of a Web site, you must be concerned about your site's credibility. When the user accesses your site, he or she must be assured that the content within the site is credible, safe, and secure. For example, if a consumer enters credit card information when purchasing a product or service from your site, that consumer demands a sense of security. When people buy products from your site, they need to know that their purchase is warranted and that your customer service is accountable.

To determine how the public views a Web site's credibility, Stanford University studied how 2,440 people evaluated the credibility of sites. Almost half of all respondents to the Stanford study (46.1 percent) ranked visual design most important for establishing credibility, focusing on layout, font type, font size, and color. However, Stanford also concluded that a Web site's visual appeal had limited value for long-term credibility (Fogg). Though a Web site's design might initially attract an audience, the consumers will go to other Web sites if they are not satisfied with other aspects of the site, including the following key credibility concerns:

- **Identity of the site or its operator.** Prove your credentials by including your company's years in business, your staff's education and certifications, lists of satisfied customers, and customer testimonials.

- **The site's customer service.** Provide easy access for consumers when contacting people within the company for help. This includes 800 numbers, hotlines, 24/7 assistance, e-mail addresses, social media sites, rapid response, and the chance to speak to an actual person within the company.
- **Privacy policies.** Clearly and briefly state how your company ensures privacy related to credit cards and personal information.

Ethical Considerations in a Web Site

5. Consider ethics when creating a Web site.

As a business professional, you have an ethical responsibility on your Web site to your audience. Customers will input personal information, such as Social Security numbers, credit card numbers, expiration dates, and personal identification numbers, when purchasing products and services. This material must be safeguarded to avoid fraud and identity theft. The Federal Trade Commission protects consumers against unfairness and deception by enforcing a company's promises about collections and use of personal information ("Privacy Initiatives").

Privacy and Security Considerations

To ensure your audience's privacy and demonstrate your company's ethical practices, focus on the following:

- **Maintain records.** You must document personal information you have in your computer files and back up these files in archives.
- **Acquire security certifications and keep certifications updated.** With Transport Layer Security (TLS) and Secure Sockets Layer (SSL), certificates will assure consumers that the site is safe for searching, browsing, buying, and logging in. In other words, the consumer can trust your site has passed a malware scan.
- **Delete inactive files.** Don't stockpile files indefinitely. Only maintain files that your company needs to conduct business.
- **Secure the information.** Frequently update firewalls, anti-virus, and anti-spyware software.
- **Protect passwords.** Avoid sharing customer passwords online, over the phone, or in e-mail messages. Ask customers to change passwords frequently and to avoid using the same password for multiple accounts ("Protecting Personal Information"; *OnGuard Online*).

Criteria for a Successful Web Site

6. Follow criteria for creating a successful Web site.

Because Web sites are a different type of correspondence than traditional, paper-bound text, you need to employ different criteria when creating your site.

Home Page

The home page on a Web site is your welcome mat. Because it is the first thing your reader sees, the home page sets the tone for your site. A successful home page should consist of the following components.

Identification Information. Who are you? What is the name of your company, product, or service? How can viewers get in touch with you if they want to purchase your product? A good home page should provide your audience the following:

- The name of your company, service, or product.
- Contact information (or a link to corporate contacts), including phone numbers, e-mail addresses, fax numbers, street address, city/state/zip, customer service contacts, and social media links.

Amazingly, many Web sites omit this information. They provide the reader information about a product or service, but they prevent the reader from making a follow-up purchase or inquiry. That's not good business.

Graphics. Don't just tell the reader who or what you are. Show the reader and establish brand recognition. An informative, attractive, and appealing logo depicting your product or service could convey more about your company and be more memorable than words. Check out various Web sites to prove this point. Be careful, however, when you choose the size of your home page graphic. If the graphic is too large, it will take a long time to load, especially on mobile devices. This delay might cause Web users to lose patience, stop the HTTP transfer, and visit another site. Then you've lost their business or interest.

Lead-in Introduction. In addition to a graphic, you might want to provide a one- or two-sentence lead-in, or a three- to eight-word tagline, explaining the purpose of the Web site. Few Web visitors will read a paragraph-long mission or vision statement.

For example, if your company is named NovaTech, what do you market? The name alone does not explain the company's focus. A company named NovaTech could market telecommunications equipment, intergalactic armaments, or computer repairs. To clarify your company's purpose, provide a short lead in. Quickly welcome readers to the site and tell them what they can find in the following screens.

At this textbook's writing, Dell's home page introduced itself with this sentence: "Shop Dell PCs, Laptops, Servers and More." CSC knows that its three-letter company name isn't informative, so they lead into their Web site by stating, "We are a global leader in providing technology enabled business solutions and services."

See Figure 13.3 for a sample Web site home page.

FIGURE 13.3   Web Site Home Page with Lead-in Information

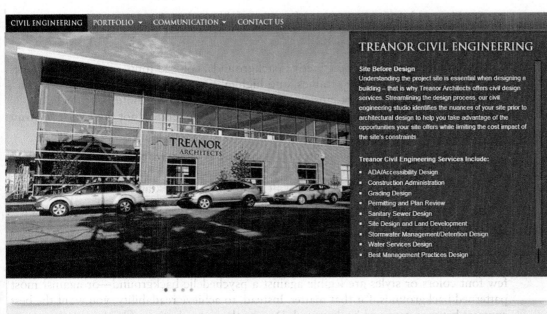

To explain their company's focus, Treanor Architects' home page lists the specific services they offer.

Source: Courtesy of Treanor Architects.

Navigation Links/Buttons. The navigation bar is what distinguishes a Web site from all other types of written communication: the ability to click on a hot button and link to another screen. By providing the reader with hypertext links, a home page acts as an interactive table of contents or index. The reader selects the topic he or she wishes to pursue, clicks on that link, and jumps to a new screen. Instead of being forced by the constraints of paper-bound text to read sequentially, line by line, page by page, the reader can follow a more intuitive approach. With hypertext links, you read what you want to read, when you want to read it.

Linked Pages

Once your reader clicks on the hypertext links from the home page, he or she will jump to the designated linked pages. These linked pages should contain the following.

Headings/Subheadings. To ensure that readers know where they are in the context of the Web site, you need to use headings. These give the readers visual reminders of their location. Effective linked-page headings should be consistent. They should be the same font size, font type, and in the same location on the screen. For example, if one heading is centered and typed in 16-point Arial font, all headings should follow this format.

Navigation. Help the reader navigate online. You can do this in two ways:

- **Home buttons.** The reader needs to be able to return to the home page easily from any page of a Web site. The home page acts as a table of contents or index for all pages within the site. By returning to the home page, the reader can access any of the other pages. To ensure this easy navigation, provide a hypertext-linked "Home" button or icon on each page.
- **Links between Web pages.** Why make the reader return to the home page each time he or she wants to access other pages within a site? If each page has hypertext links to all pages within the site, then the reader can access any page he or she chooses in any order of discovery.

See Figure 13.4 for an example of a successful linked page.

Document Design

On a Web site, you can add distinctive backgrounds, colored fonts, different font faces and sizes, animated graphics, frames, and highlighting techniques (such as lines, icons, and bullets). However, just because you can doesn't mean you should. Document design should enhance your text and promote your product or service, not distract from your message.

Background. You can add backgrounds running the gamut from plain white, to various colors, to an array of patterns: fabrics, marbled textures, simulated paper, wood grains, grass and stone, psychedelic patterns, water and cloud images, and waves. They all look exciting, but some are not as effective as others.

When choosing a background, consider your corporate image. If you are a child care center, do you want a black background? This could convey a negative image. A white background with toys as a watermark might more successfully convey your center's mission statement.

In addition, remember that someone will attempt to read your Web site's text. Very few font colors or styles are legible against a psychedelic background—or against most patterned backgrounds, for that matter. Instead, to achieve readability, you want the best contrast between text and background. Despite the vast selection of backgrounds at your disposal, the best contrast is still black text on a white background.

FIGURE 13.4 Linked Page

Excellent document design including bullets, headings and subheadings, and symmetrically placed graphics for visual appeal.

Source: Courtesy of George Butler Associates.

Color. When you create a Web site, you can use any font color in the spectrum, but do you want to? What corporate image do you have in mind? The font color should be suitable for your Web site, as well as readable. A yellow font on a light blue background is hard to read. A red font on a black background is nightmarish.

The issue, however, isn't just esthetics. As noted, a primary concern is contrast. Red, blue, and black font colors on a white background are legible because their contrast is optimum. Other combinations of color do not offer this contrast, making readability difficult. In the following example, notice how yellow font on a blue background "strobes." The colors bleed into each other and become unreadable. Though black font on a white background might seem uninteresting, this color combination is the most readable choice.

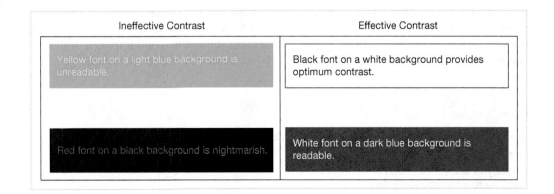

◀ EXAMPLE

Graphics. Another problem is image size. Your Web site will be affected by the size of your graphics and the number of graphics you have included. For example, a small graphic (approximately 90 × 80 pixels, 1½ inches × 1½ inches) consumes about 2.5 kilobytes. A medium-sized graphic (approximately 300 × 200 pixels, 5 inches × 4 inches) consumes about 25 kilobytes. A large graphic (approximately 640 × 480 pixels, 10 inches × 8 inches) consumes about 900 kilobytes. When you use graphics with varying amounts of color depth, your file size increases. The larger your file, the longer it will take for the images to load. The longer it takes for your file to load, the less interested your reader might be in your Web site. Limit the size and number of your graphics.

Highlighting Techniques. Font sizes and styles, lines, icons, bullets, frames, Java applets, Flash animation, video, audio! You can do it all on your Web page, but should you? Excessive highlighting techniques can be distracting.

Effective Online Highlighting	Ineffective Online Highlighting
1. *Lines* (horizontal rules) can separate headings and subheadings from the text.	1. *Frames* are considered to be one of the worst highlighting techniques. Too many frames can produce a patchwork effect. If you want to achieve the frame look, but without the hassle, use tables instead. They are easier to create and revise, they load faster, and they are less distracting to your readers.
2. *Bullets and icons* enliven your text and break up the monotony of wall-to-wall words.	2. *Italics* and *underlining*. Both are hard to read online. Plus underlining looks like a hypertext link.
3. *First- and second-level headings,* achieved by changing your font size and style, separate key ideas.	3. *Java applets* take a long time to load.
4. *Boldface* also emphasizes important points.	4. *Video* requires an add-on users must download before they can enjoy your creation.
5. *White space,* created by indenting, makes text more readable.	

Grammar

If your Web site represents your company, you don't want prospective clients to perceive your company as lacking in quality control. In fact, that's what poor grammar online denotes—poor attention to detail. It's easy to see why many sites are grammatically flawed. First, anyone can go online almost instantaneously. All you need to do is save a Word document as an HTML file and transfer the file to your Web server. Speed is great for business, but it's bad for proofreading. Good proofreading takes time. When building your Web site, you must be vigilant. The integrity of your Web site demands correct grammar.

WEB SITE USABILITY CHECKLIST

_____ **1.** Audience recognition and involvement

- Does the Web site meet your reader's needs?
- Does the Web site give your audience a reason to return (tutorials, tips, links to other interesting sites, regular updates, social media links, and so on)?

- Does the Web site involve the audience by asking for user feedback, displaying customer comments in a blog, providing an FAQ page, or providing links to the company's Twitter or Facebook site?
- Does the Web site make it easy for your audience to purchase online?

- Does the Web site use pronouns to engage the reader?

_____ 2. Home page
  - Does the home page provide identification information (name of service or product, company name, e-mail, fax, telephone, address, etc.)?
  - Does the home page provide an informative and appealing graphic?
  - Does the home page provide a welcoming, informative, and concise introduction?
  - Does the home page provide hypertext links connecting the reader to subsequent screens?

_____ 3. Linked pages
  - Do the linked pages provide headings clearly indicating to the reader which screen he or she is viewing?
  - Are the headings consistent in location and font size/style?

_____ 4. Navigation
  - Does the Web site allow for easy return from linked pages to the home page?
  - Does the Web site allow for easy movement between linked pages?

_____ 5. Document design
  - Does the Web site provide an effective background suitable to the content and creating effective contrast for reading the text?

- Does the Web site use color effectively, in a way suitable to the content and creating effective contrast for reading the text?
- Does the Web site use graphics effectively, so that they are suitable to the content, not distracting, and not causing delays for the site to load?
- Does the Web site use highlighting techniques effectively (lines, bullets, icons, audio, video, font size and style, etc.) in a way that is suitable to the content and not distracting?
- Does the Web site's design allow for accessibility, meeting Web Accessibility Initiative and Section 508 mandates?

_____ 6. Style
  - Is the Web site concise?
    - Short words (one to two syllables)
    - Short sentences (10–12 words per sentence)
    - Short paragraphs (four typed lines maximum)
    - Text per Web page minimizing the need to scroll excessively
  - Is the Web site personalized, using positive words and pronouns?

_____ 7. Grammar
  - Does the Web site avoid grammatical errors?

TECHNOLOGY TIPS

Using Microsoft Word 2013 to Create a Web Site

You can use Word 2013 to create your Web site. Follow these steps to do so:

1. From the Insert tab, insert a 3 × 2 table.

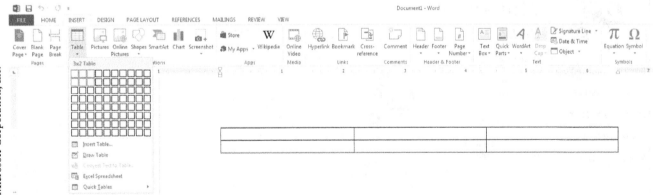

(Continued)

2. Merge the top right two columns.

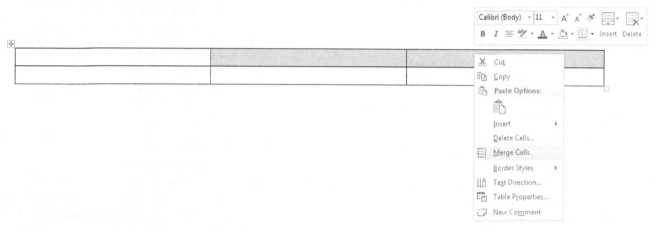

3. Place your cursor in the right rows and press the **Enter** key to add space. At this point, you have the beginning design for a Web page.
4. Add graphics, text, color, and font types just as you would with any Word document.

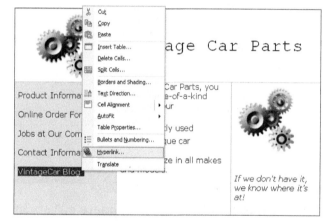

5. To create hypertext links, highlight selected text, right click, and scroll to **Hyperlink.**

6. To create your Web site, click on the Word 2013 **File** tab and scroll to **Save As**.

(Continued)

7. In the pop-up **Save As** window, click on the **Save as type** down arrow and click on **Web Page**. This saves your work as an HTML (hypertext markup language) file which you can open in a Web browser.

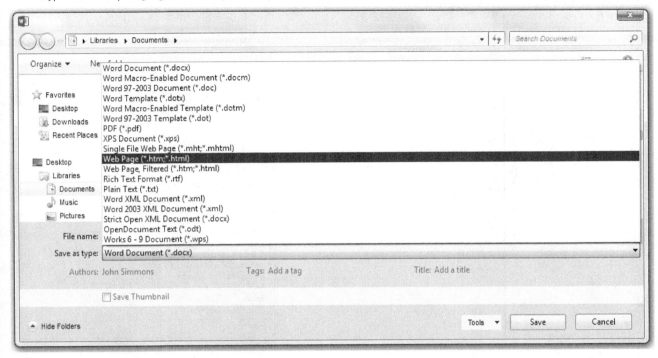

Online Help

> 7. Understand the purpose of online help.

Imagine this possibility. Your boss calls you into a meeting and says, "We're changing policies. No more paper documents. From today, our company is going digital." It's not as farfetched as it sounds. Online Web help is a major part of the technical communicator's everyday job for several reasons, including

- The increased use of computers in business, industry, education, and the home
- The reduced dependence on hard-copy manuals by consumers
- The need for readily available online assistance
- Proof that people learn more effectively from online tutorials than from printed manuals

In addition, online help screens provide the practical value of portability. Rather than take up computer archival space with thousands of pages of PDF files or print out hundreds of pages of instructional help, online help screens allow companies to save space and provide readers ease of use.

What is an online help system? Online help systems, which employ computer software to help users complete a task, include procedures, reference information, indexes, and definitions (see Figure 13.5). Typical online help navigation provides readers hypertext links, tables of contents, and full-text search mechanisms.

Help menus on your computer are excellent examples of online help systems. As the computer user, you pull down a help menu, search a help list, and click on the topic of your choice revealed as hypertext links. When you click on a link, you could get a *pop-up*

Online help screen with hypertext link to definition.

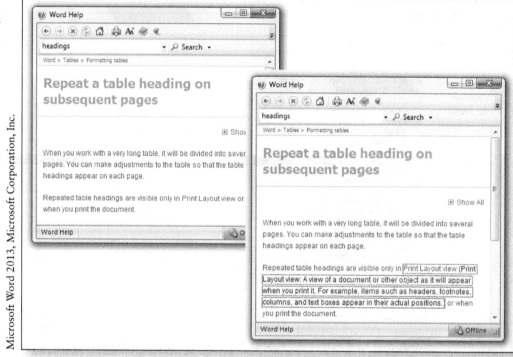

Microsoft Word 2013, Microsoft Corporation, Inc.

(a small window superimposed on your text), or your computer might *link* to another full-sized screen layered over your text. In either instance, the pop-up or hyperlink gives the following information about your topic:

- **Overviews.** Explanations of why a procedure is required and what outcomes are expected.
- **Processes.** Discussions of how something works.
- **Definitions.** An online glossary of terms.
- **Procedures.** Step-by-step instructions for completing a task.
- **Examples.** Feedback verifying the completion of a task or graphic depictions of a completed task. These could include a screen capture or a description with callouts (labels identifying key components of the mechanism).
- **Cross-references.** Hypertext links to additional information.
- **Tutorials.** Opportunities to practice online.

Online help systems allow the technical communicator to create interactive training tools and informational booths within a document. These online systems can be created using a wide variety of help authoring tools (HATs). Some popular HATs include Adobe RoboHelp, Author-it, Doc-To-Help, DocArchitector, HelpScribble, HyperText Studio, MadCap Flare, and Sandcastle. The value of online help is immense. Because online help offers "just in time" learning or "as needed" information, readers can progress at their own pace while learning a program or performing a task.

Techniques for Writing Effective Online Help

To create effective online help screens, consider the following suggestions:

Organize Your Information for Easy Navigation. Poorly organized screens lead to readers who are lost in cyberspace. Either they cannot find the information they need, or they have accessed so many hypertext links that they are five or six screens deep into text.

You can avoid such problems and help your readers access information in various ways:

- Allow users to record a history of the screens they've accessed through a bookmark or a help topics pull-down menu.
- Provide an online contents menu, allowing users to access other cross-referenced help screens within the system.
- Provide a back button or a home button to allow the readers to return to a previous screen.
- Provide links such as "In this section" and "Related topics" on each screen for immediate access to topic-specific content.

A good test is the three clicks rule. Readers should not have to access more than three screens to find the answer they need. Similarly, readers should not have to backtrack more than three screens to return to their place in the original text.

Recognize Your Audience. Online help must be user oriented. After all, the only goal of online help is to help the user complete a task. Thus, a successful online help system must be designed at the user's level. This means that technical communicators have to determine a user's level of knowledge.

Audience
See Chapter 5, "Audience Recognition," for more information.

If the system is transmitted by way of an intranet or extranet, you might be tempted to write at a high- or low-tech level. Your readers, you assume, work within a defined industry and possess a certain level of knowledge. That, of course, is probably a false assumption. Even within a specific industry or within a specific company, you will have coworkers with widely diverse backgrounds: accountants, engineers, data processors, salespeople, human resource employees, managers, technicians, and so on.

Don't assume. Find out what information your readers need. You can accomplish this goal through usability testing, focus groups, brainstorming sessions, surveys, and your company's hotline help desk logs. Then don't scrimp on the information you provide. Don't just provide the basic or the obvious. Provide more detailed information and numerous pop-ups or links. Pop-up definitions are an especially effective tool for helping a diverse audience. Remember, with online help screens, readers who don't need the information can skip it. Those readers who do need the additional information will appreciate your efforts. They will more successfully complete the tasks, and your help desk will receive fewer calls.

Achieve a Positive, Personalized Tone. Users want to be encouraged, especially if they are trying to accomplish a difficult task. Thus, your help screens should be constructive, not critical. Your text should be affirmative, a concept supported by human-computer interaction (HCI) concerns. HCI-driven online help systems coach their users rather than command them. The messages also should be personalized, including pronouns to involve the reader.

Design Your Document. How will your help screens look? Document design is important because it helps your readers access the information they need.

Document Design
See Chapter 7, "Document Design," for more information.

To achieve an effective document design, consider the following points:

- **Use color carefully.** Bright colors and too many colors strain your reader's eyes. Furthermore, a color that looks good on a high-resolution monitor might be difficult to read on a monitor with poor resolution. Your primary goal is contrast. To help your audience read your text, maximize the contrast between the text color and the background color, as seen in Figure 13.5.
- **Be consistent.** Pick a color scheme and stick with it. Your headings should be consistent, along with your word usage, tone, placement of help screen links and pop-ups, graphics, wizards, and icons. Readers expect to find things in the same

place each time they look. If your help screens are inconsistent, readers will be confused.

- **Use an easy-to-read font.** A 12-point type size is standard for most printed documents, but 10-point type will save you valuable space online. Serif fonts are the standard for most technical writing. A serif font, with small, horizontal "feet" at the bottom of each letter, helps guide the eye while reading printed text. However, on lower resolution monitors, serif text is more difficult to read. Many online help screens, such as those provided by Microsoft Word, use a sans serif font, such as Arial or Calibri. Avoid designer fonts, such as Algerian, Bauhaus 93, or Brush Script MT, which are hard to read.
- **Use white space.** Don't clutter your help screens. Avoid excessive emphasis techniques. For example, typical PowerPoint "fly-in" effects not only are distracting, but also they negatively impact the online screen's performance, slowing load time down dramatically. Minimize your reader's overload by adding ample horizontal and vertical white space. Online, less is best.

Conciseness and Clarity

See Chapter 4, "Objectives in Technical Communication," for more information.

Be Concise.   It takes longer to read on-screen than in print, so minimize text. In addition to limiting word and sentence length, a help screen, just like any Web page, should limit horizontal and vertical scrolling. Each screen should include one self-contained message. If readers need additional information, use "In this section" and "Related topics" links.

Be Clear.   Your audience reads the help screen only to learn how to perform a task. Thus, your only job is to meet the reader's needs—clearly. To accomplish this goal, be specific. In addition, clarity online could include the following:

- Tutorials to guide the reader through a task. Microsoft provides clear, step-by-step instructions with embedded screen captures to help its audience complete activities.
- Pop-up definitions with expanding blocks for longer definitions.

Provide Access on Multiple Platforms.   You will want every reader to be able to access your online help, not just those using a PC or a Mac or UNIX. After all, if readers can't access online help, how have you helped them? To ensure that all readers can access your system,

- Do not use platform-specific technologies, such as ActiveX, which only runs on 32-bit Windows.
- Create content that can be viewed on multiple browsers, including Internet Explorer and Firefox.
- Create online help for the oldest version that you will support—the lowest common denominator. Remember, all users won't have the latest version browser, software, or hardware.
- Test your online help on at least three different browsers and platforms. Though you might be creating your online help screens on a 30" widescreen, your users might be accessing the text on a 15" panel or smaller smartphone. Make sure they can see what you see, even if they are viewing text on smaller screens.

Correct Your Grammar.   As in all technical communication, incorrect grammar online leads to two negative results: a lack of clarity and a lack of professionalism. Don't embarrass your company or confuse your reader with grammatical errors. Proofread.

ONLINE HELP CHECKLIST

_____ **1.** Have you used your online help screens to explain why a procedure is required, show how something works, define terms, give examples, allow opportunities to practice tasks, and/or provide access to additional information?

_____ 2. Have you helped readers access information with pull-down menus, tables of content, back and/or home buttons, and links to related topics?

_____ 3. Have you tried to ensure that readers do not need more than three screens to find answers to their questions?

_____ 4. Have you met your audience's needs by designing content at the level of their knowledge and expertise?

_____ 5. Have you written your content using a positive tone, striving to "coach" rather than "command"?

_____ 6. Did you design your help screens effectively by limiting color, using a consistent color scheme, choosing an easy-to-read font, and incorporating white space for easy access?

_____ 7. Is your online help concise, limiting word length, sentence length, paragraph length, and horizontal and vertical scrolling?

_____ 8. Have you develped your text thoroughly?

_____ 9. Can your audience access the online help using different platforms and browsers?

_____ 10. Is your online help grammatically correct?

The Writing Process at Work

Molly Seacrest, freelance Web designer, used the communication process to create an effective and dynamic Web site for a client's photography business, S5Photography.

Prewriting

See how Molly used prewriting to begin her Web site. To plan the Web site, Molly considered the following:

- Goal—provide policy and procedures for the photography studio's Web site
- Audience—existing and potential photography studio clients
- Channels—statement of purpose for the Web site
- Data—policy and procedures including fees, ordering, pricing structure, and so on

Molly used a questionnaire to define the audience, communication channel, and shelf life of the content for the Web site. Figure 13.6 shows how Molly used a questionnaire to plan the Web site.

FIGURE 13.6   Questionnaire Completed by Client used for Prewriting

What are session fees?
Session fees are $125. This covers my time, talent, and equipment and does not include any prints. Session fees are due at time of booking. Due to my busy schedule, session dates are only held once the session fee has been received.

What are print prices?
Prices range from $20 for wallet sized to $400 for large, wall-hanging photos.

What is available in the boutique?
Boutique items are available after a minimum purchase of $300, unless otherwise noted. Session Album—$100
A small 5 × 7, spiral bound album that includes watermarked proofs from your session. It's perfect for carrying around to show to friends and family.

When are final prints available?
Prints will be delivered in 2–3 weeks. Larger images, canvas prints, and specialty items may take 4–6 weeks. S5Photography will contact you for the best time to deliver the prints. All sales are final.

Writing

You next have to write a draft of every Web page. Figure 13.7 shows Molly's rough draft of a Web page.

FIGURE 13.7   Web Page Mock-up with Input from the Client

The client said to Molly, "I see why you wanted me to pick the most important ideas for the potential customer to take away from the Web site. Let's make the Web site simpler, with far fewer words, and use lots more linked pages so that the customer can focus. I want the Web site to be as elegant as possible and illustrate some of my most treasured photographs so the customers can see what they will get when they hire me."

Thank you for choosing S5Photography! Please take a moment to look through our policies and procedures. If you have any questions, please contact our company at S5Photography@gmail.com.

Session Fee

Session fees are $125. This covers our photographer's time, talent, and equipment and does not include any prints. Session fees are due at time of booking. Due to her busy schedule, session dates are only held once the session fee has been received.

Your Session

Your session is individualized to fit your needs. I do not rush you through the session. We will take our time to get the best image possible. For small children and newborns, this may mean taking short breaks for feeding and snacks.

Ordering

A minimum of 15 proofs will be provided in a private, password-protected online gallery within one week of your session date. We will email you a link to your gallery. You're welcome to forward your gallery to family and friends. The ordering process is secure. Cash and credit/debit payments are made through the PayPal system.

Print Prices

Wall Portraits (mounted and coated)	Gift Prints (coated)
30 × 40 $400	8 × 10 $40
20 × 24 $275	5 × 7 $25
16 × 20 $170	8 Wallets $20
Add $150 for gallery-wrapped canvas.	

Boutique

Boutique items are available after a minimum purchase of $300, unless otherwise noted. S5Photography also offers custom sterling silver photo jewelry and photo purses and handbags. Due to the handmade nature of these items, ask about current prices, styles and availability.

Rewriting

Rewriting is the final part of the writing process. Using input from the client, Molly revised and reformatted the Web page considering how the design would be effective for the audience.

Figure 13.8 is Molly's revised Web page.

FIGURE 13.8   Revised Linked Page

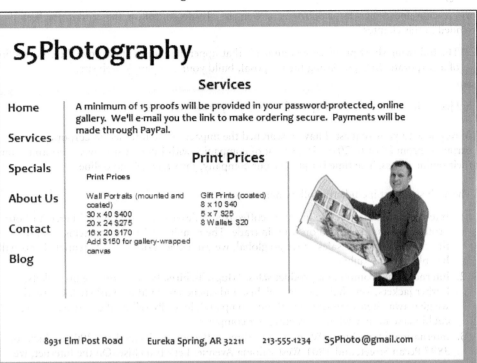

S5Photography

Services

Home

Services

Specials

About Us

Contact

Blog

A minimum of 15 proofs will be provided in your password-protected, online gallery. We'll e-mail you the link to make ordering secure. Payments will be made through PayPal.

Print Prices

Print Prices

Wall Portraits (mounted and coated)
30 x 40 $400
20 x 24 $275
16 x 20 $170
Add $150 for gallery-wrapped canvas

Gift Prints (coated)
8 x 10 $40
5 x 7 $25
8 Wallets $20

8931 Elm Post Road   Eureka Spring, AR 32211   213-555-1234   S5Photo@gmail.com

APPLY YOUR KNOWLEDGE

CASE STUDIES

1. Future Promise is a not-for-profit organization geared toward helping at-risk high school students. This agency realizes that it needs an Internet and social media presence to reach its target audience (teens age 15–18). To do so, it has formed a team, consisting of the agency's accountant, public relations manager, social media manager, computer and information systems director, two local high school principals, local high school students, and a representative from the mayor's office. Jeannie Kort, the PR manager, is acting as team leader.

   The team needs to determine the Web site's content, design, and levels of interactivity. Jeannie's boss, Brent Searing, has given the team a deadline and a few components that must be included in the site:
   - College scholarship opportunities
   - Job-training skills (résumé building and interviewing)
   - Social media sites (Facebook, Twitter, YouTube, Future Promise blog, and Foursquare)
   - Future Promise's 800-Hotline (for suicide prevention, STD information, depression, substance abuse, and peer counseling)

Assignment

Form a team and design Future Promise's Web site. To do so, follow the criteria for Web design provided in this chapter.

2. The following short proposal recommends that upper management approve the construction of a corporate Web site. Using the proposal, build your company's Web site.

Subject: Recommendation Report for New Corporate Web Site

In response to your request, I have researched the impact of the Internet on corporate earnings. From 2014 to 2016, 78 percent of companies with Internet sites were profiting from their online access. The time is right for our company, Java Lava, to go online.

The Web site must include the following:

1. **International bean sales.** Currently, coffee bean sales account for only 27 percent of our profit. These sales depend on walk-in trade. The remaining 73 percent stems from over-the-counter beverage sales. If we go global, we can expand our sales. Potential clients will be able to order online.
2. **International promotional product sales.** Mugs, T-shirts, boxer shorts, jeans jackets, leather jackets, key chains, paperweights, and calendars could be marketed. Currently we give away items imprinted with our company's logo. By selling these items online, we could make money while marketing our company.
3. **International franchises.** We now have three coffeehouses, located at 1200 San Jacinto, 3897 Pecan Street, and 1801 West Paloma Avenue. Let's franchise. On the Internet, we could offer franchise options internationally.
4. **Online employment opportunities**. Once we begin to franchise, we'll want to control hiring practices to ensure that Java Lava's standards are met. Through a Web site, we could post job openings internationally and list the job requirements. Then potential employees could apply online.

In addition to the above information, used to increase our income, we could provide the following:

- A map showing our three current sites.
- Our company's history—founded in 1992 by Hiram and Miriam Coenenburg, with money earned from their import/export bean business.
- Sources of our coffee beans—Guatemala, Costa Rica, Columbia, Brazil, Sumatra, France, and the Ivory Coast.
- Corporate contacts (addresses, phone numbers, e-mail, fax numbers, etc.).
- A corporate blog for news, events, sales, and promotional discounts.

Coffee is a "hot" commodity now. The international market is ours for the taking. We can maximize our profits and open new venues for expansion. I'm awaiting your approval.

ETHICAL CHALLENGE

Sam Henry, a salesperson for a children's toy store, loves the Internet. He wrote a proposal to his company's CEO suggesting that the company's Web site could be improved by adding a corporate Web blog. This blog would allow for better communication between the company and its client base. The CEO agreed and allowed Sam to create the blog.

As part of the corporate blog, Sam included many links to related Web sites that he thought would appeal to customers. One of the links was to Sam's personal Web site. He believed that this link would allow him to achieve a more personal contact with clients.

Sam's Web site provides personal information about his hobbies, his children, and his weekend activities, which include tailgating at sporting events. One photograph on the site shows Sam sitting on his motorcycle and drinking beer.

Question

Is Sam's Web site an appropriate link for a corporate blog? Why or why not?

INDIVIDUAL AND TEAM PROJECTS

Web Sites

1. Create a corporate Web site. To do so, make up your own company and its product or service. Your company's service could focus on dog training, computer repair, basement refinishing, vent cleaning, Web site construction, child care, auto repair, personalized aerobic training, or online haute cuisine. Your company's product could be interactive computer games, graphics software packages, custom-built engines, flooring tiles, or duck decoys. The choice is yours. To create this Web site, follow our writing process.

2. Create a Web site for your technical communication class. To create this Web site, follow our writing process.

3. Create a personal Web site for yourself or for your family.

4. Research several Web sites, either corporate or personal. Use our Web Site Usability Checklist to determine which sites excel and which sites need improvement. Then write a report justifying your assessment. In this report, clarify exactly what makes the sites successful. To do so, you could use a table, listing effective traits and giving examples from the Web sites to prove your point. Next, explain why the unsuccessful sites fail. Finally, suggest ways in which the unsuccessful sites could be improved.

Online Help

1. In a small group, have each individual ask a software-related question regarding a word processing application. These could include questions such as "How do I print?" or "How do I set margins?" Then, using a help menu, find the answers to these questions. Are the answers provided by way of pop-up windows or hyperlinks? Rewrite any of these examples to improve their conciseness or visual layout.

2. Take an existing document (one you have already written in your technical writing class or writing from your work environment) and rewrite it as online help. To do so, find key words, phrases, or concepts that need to be defined. Then, expand on them as either pop-up windows or hyperlinks.

3. Using our suggestions in this chapter, rewrite the following text as an effective online help screen. To do so, reformat the information for access on a smaller screen. Next, determine which of the words, phrases, or concepts can be expanded with a short pop-up window and which require a longer hyperlink.

Because your company is located in the East Side Commercial Park, the outside power is subject to surges and brownouts. These are caused when industrial motors and environmental equipment frequently start and stop. The large copiers in your office also cause power surges that can damage electronic equipment.

As a result of these power surges and outages, your three network servers have become damaged. Power supplies, hard drives, memory, and monitors have all been damaged. The repair costs for damages totaled $55,000.

(Continued)

Power outages at night also have caused lost data and failed backups and file transfers. Not being able to shut down the servers during power failures is leading to system crashes. Data loss requires time-consuming tape restorations. If the failures happen before a scheduled backup, the data are lost, which requires new data entry time. Ten percent your data loss can not be restored. This will lead to the loss of customer confidence in the reliability of your services.

Finally, staff time losses of $40,000 are directly related to network downtime. Data-entry staff are not able to work while the servers are being repaired. These staff also must spend extra time reentering lost data. Fifteen percent of the data management staff time is now spent recovering from power-related problems.

To solve your problems, you need to install a BACK-UPS 9000 system. It has the configuration and cost-effectiveness appropriate for your current needs.

PROBLEM-SOLVING THINK PIECES

1. FlyHigh Travel does not book trips to Washington, DC, for tours of national monuments; to Las Vegas to see shows; or to Boston to visit historical sites. Instead, FlyHigh specializes in "adventure trips . . . in the air, sea, and land." If it's shark watching off the Barrier Reef, paragliding from a cliff in Acapulco, or mountain climbing in Nepal, FlyHigh is the traveler's answer. FlyHigh books adventures like feeding stingrays in the Caribbean, animal photo shoots in Kenya, and kayaking down the Colorado River. FlyHigh also makes travel arrangements for scuba diving in Hawaii, spelunking in French caves, and skydiving anywhere in the world.

   To reach as wide an audience as possible, FlyHigh is building its Web site (www.WeFlyHigh.com). FlyHigh's CEO wants this Web site to include trip information, testimonials from satisfied travelers, pricing, and photos that highlight each trip's excitement.

Assignment

Build the Web site for FlyHigh including the following content:

- Specific trip information travelers need to plan their trips (costs, accommodations, options, and so on)
- Information international customers will need (passports, visas, vaccinations, or innoculations)
- Information for FlyHigh's clients who travel with families (alternative, age-specific activities)
- Information for FlyHigh's "fit and fearless" clients whose travels will involve extreme sports

2. Web site design is challenging. Some Web sites are outstanding; others do not fare as well. Check out http://www.webpagesthatsuck.com/ and http://www.webpractices.com/samplesites.htm, two sites that assess flawed Web pages. Do you agree with their assessments?

Assignment

Write a memo to your instructor or give an oral presentation explaining your decisions.

WEB WORKSHOP

1. Access any company's Web site and study the site's content, layout (color, graphics, headings, use of varying font sizes and types, etc.), links (internal and external), ease of navigation, tone, and any other considerations you think are important. Then, determine how the Web site could be improved if you were the site's webmaster. Once you have made this determination, write a memo or e-mail message recommending the changes that you believe will improve the site. In this memo or e-mail,

   - Analyze the Web site's current content and design, focusing on what is successful and what could be improved.
   - Recommend changes to improve the site.

2. Audience recognition involves a reader's level of knowledge as well as his or her gender, religion, age, culture, and language. No communication channel allows for a more diverse audience than the Internet where anyone can click on a link and enter a site.

   Still, effective Web sites recognize and appeal to specific audiences. To better understand how successful technical communication recognizes and involves audiences, click on the following sites:

   - McDonalds (food)
   - Gap (clothing)
   - Harley-Davidson (motorcycles)

   Who are the intended audiences within each Web site, and what techniques do the Web sites use to involve and recognize their unique audience types?

3. The current challenge for Web designers is mobile devices. Research mobile Web trends. Then, report your findings to your professor in an e-mail message or memo.

CHAPTER FOURTEEN

Technical Descriptions and Process Analyses

After completing this chapter, you will be able to

1. Define technical descriptions

2. Understand the uses of technical descriptions

3. Define process analysis

4. Understand the uses of process analyses

5. Follow criteria for writing effective technical descriptions and process analyses

COMMUNICATION AT WORK

The information technology scenario shows the importance of technical descriptions and process analyses in the engineering profession.

Cottrell and Light Engineering Associates, Inc. (C&L) specializes in creating software for public service engineering purposes, such as storm water, water distribution, transportation and traffic control, parks and recreation maintenance, facility and building management, and inventory of fleets and equipment.

The city of Maple Valley, Texas, is interested in purchasing one of C&L's software suites related to storm water maintenance. The suite focuses on accurate and easy-to-use inventory and inspection tools for maintaining and improving storm water systems. C&L's software will help Maple Valley inspect and test all components of the city's storm water system including conduits, structures, detention basins, pump stations, and pumps.

Before C&L can sell the software to Maple Valley, the company's Information Technology director, Neeha Patel, must do the following:

- **Determine the city's unique needs.** Neeha will interview city residents, city council members, and city government officials. To accomplish this goal, Neeha will create a questionnaire focusing on infrastructure needs. Then, through a combination of hard-copy mailers and online social media (Maple Valley's Facebook site and Twitter account), Neeha will collect constituent responses.

- **Research software options.** Based on the interviews, Neeha will determine if the city's current software is failing to meet the city's needs and what C&L software options exist.

- **Create or adapt software to meet the city's technical specifications.** If new software is needed, Neeha will work with her colleagues in IT to customize software for Maple Valley.

- **Write technical descriptions of the hardware.** Based on her analysis of hardware requirements, Neeha will write the technical descriptions of hardware needed to run the software. These descriptions will accompany the operations manual C&L will give Maple Valley employees.

- **Provide process analyses.** To clarify how the software will solve Maple Valley's infrastructure problems, Neeha will write a process analysis. This process analysis will show how the software will improve storm water, transportation, and traffic flow.

Once Neeha creates this documentation, she can provide it to Maple Valley's Planning Commission, which will decide whether to purchase the software and work with C&L.

Define Technical Descriptions

1. Define technical descriptions.

A *technical description* is a part-by-part depiction of the components of a mechanism, tool, or piece of equipment. Technical descriptions are important features in several types of technical communication.

Types of Technical Descriptions

2. Understand the uses of technical descriptions.

Operations Manuals

Manufacturers often include an operations manual in the packaging of a mechanism, tool, or piece of equipment. This manual helps the end user construct, install, operate, and service the equipment. Operations manuals also can contain technical descriptions.

Technical descriptions provide the end user with information about the mechanism's features or capabilities. For example, this information may tell the user which components are enclosed in the shipping package, clarify the quality of these components, specify what function these components serve in the mechanism, or allow the user to reorder any missing or flawed components. Following is a brief technical description found in an operations manual:

◀ **EXAMPLE**

> The Modern Electronics Tone Test Tracer, Model 77A, is housed in a yellow, high-impact plastic case that measures 1¼ inch × 2 inch × 2¼ inch, weighs 4 ounces, and is powered by a 1604 battery. Red and black test leads are provided. The 77A has a standard four-conductor cord, a three-position toggle switch, and an LED for line polarity testing. A tone selector switch located inside the test set provides either solid tone or dual alternating tone. The Tracer is compatible with the EXX, SetUp, and Crossbow models.

Product Demand Specifications

Sometimes a company needs a piece of equipment that does not currently exist. To acquire this equipment, the company writes a product demand specifying its exact needs.

The following product demand specification is written for a high-tech audience. It assumes knowledge on the part of the reader, such as definitions for "high-speed" and "deep-hole," which a lay reader would not understand. In addition, the high-tech abbreviation "AISI" is not defined.

EXAMPLE ▶

> Subject: Pricing for EDM Microdrills
>
> Please provide us with pricing information for the construction of 50 EDM Microdrills capable of meeting the following specifications:
>
> - Designed for high-speed, deep-hole drilling
> - Capable of drilling to depths of 100 times the diameter using 0.012-inch to 0.030-inch diameter electrodes
> - Able to produce a hole through a 1.000-inch-thick piece of AISI D2 or A6 tool steel in 1.5 minutes, using a 0.020-inch diameter electrode
>
> We need your response by January 13, 2016.

Study Reports Provided by Consulting Firms

Companies hire a consulting engineering firm to study a problem and provide a descriptive analysis. The resulting study report is used as the basis for a product demand specification requesting a solution to the problem. One firm, when asked to study crumbling cement walkways, provided the following technical description in its study report.

EXAMPLE ▶

> The slab construction consists of a wearing slab over a ½-inch-thick waterproofing membrane. The wearing slab ranges in thickness from 3½ inches to 8½ inches, and several sections have been patched and replaced repeatedly in the past. The structural slab varies in thickness from 5½ inches to 9 inches with as little as 2 inches over the top of the steel beams. The removable slab section, which has been replaced since original construction, is badly deteriorated and should be replaced. Refer to Appendix A, Photo 9, and Appendix C for shoring installed to support the framing prior to replacement.

Construction Design

Prior to building any structure, architectural companies must clarify the construction design for the city and the clients. This requires a legal technical description of the property limits (location and dimensions). Once this description is written and drawn, the architectural company will use the text and graphics for a variety of purposes. These can include presentations of the design to a city council or homeowner's association, requests for a change of zoning, requests for special use permits, and re-platting of the site if a property line must be relocated. Finally, based on the legal technical description, the architectural firm can obtain city approval. See Figure 14.1 for a layout and landscaping plan and the legal technical description that accompanies it.

Sales Literature

Companies want to make money. One way to market equipment or services is to describe the product. Such descriptions are common in sales letters, proposals, and on Web sites. Figure 14.2 is a technical description in sales literature for a smartphone.

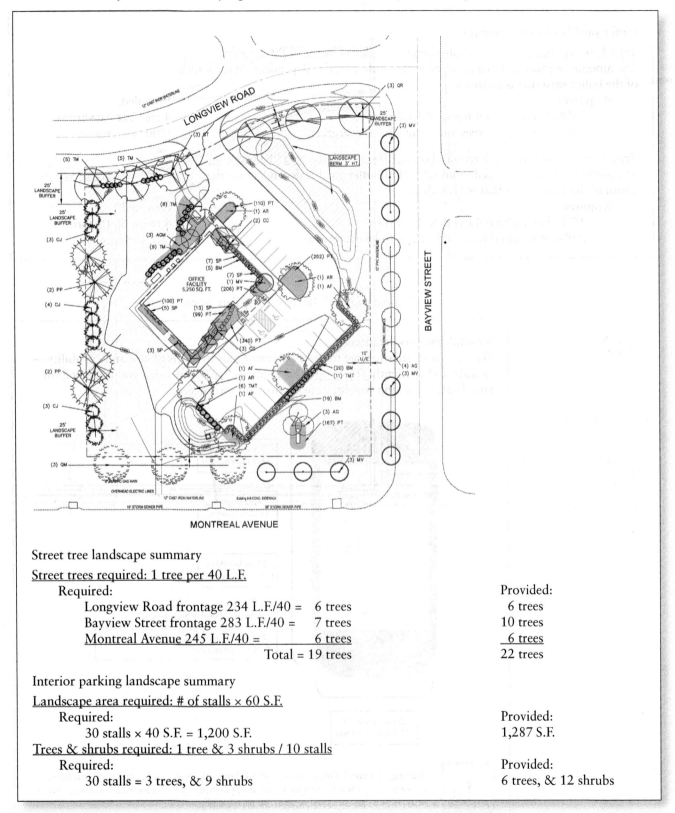

Street tree landscape summary

<u>Street trees required: 1 tree per 40 L.F.</u>

Required:		Provided:
Longview Road frontage 234 L.F./40 =	6 trees	6 trees
Bayview Street frontage 283 L.F./40 =	7 trees	10 trees
<u>Montreal Avenue 245 L.F./40 =</u>	<u>6 trees</u>	<u>6 trees</u>
	Total = 19 trees	22 trees

Interior parking landscape summary

<u>Landscape area required: # of stalls × 60 S.F.</u>

Required: Provided:

    30 stalls × 40 S.F. = 1,200 S.F.   1,287 S.F.

<u>Trees & shrubs required: 1 tree & 3 shrubs / 10 stalls</u>

Required: Provided:

    30 stalls = 3 trees, & 9 shrubs   6 trees, & 12 shrubs

(Continued)

FIGURE 14.1 *(Continued)*

Buffer yard landscape summary

<u>Type 1 buffer yard required: required on west side (adjacent to PRD zoning)</u>
The amount of plant material required within buffer yard is dependent on the width of the buffer yard that is provided.

  Required: Provided:
    25' buffer width = 4 trees, & 10 shrubs   4 trees, & 10 shrubs
    (50% of required trees and shrubs to be evergreen) (100% evergreen)

<u>Type 1 buffer yard required: required on north side (adjacent to PRD zoning)</u>
The amount of plant material required within buffer yard is dependent on the width of the buffer yard that is provided.

  Required: Provided:
    25' buffer width = 4 trees, & 10 shrubs   4 trees, & 10 shrubs
    (50% of required trees and shrubs to be evergreen) (50% evergreen)

FIGURE 14.2 Technical Description for a Lay Audience

The title identifies the equipment to be described.

Impressionistic words such as "quickly and efficiently," "helpful," "easy troubleshooting," and "user-friendly advice" are used for sales purposes to the lay audience. The pronoun "your" personalizes the text.

Smartphone Specifications
We want your smartphone to meet all your needs in terms of size and capability—quickly and efficiently. To ensure this happens, we've provided you these helpful specifications, listing weight, dimensions, and screen size.

3.5 inches tall
(88.9 millimeters)

4.32 ounces
(122.47 grams)

.23 inches thick
(5.54 millimeters)

Retina HD Display Screen Size
3-inch (diagonal) LED widescreen display
1277-by-750-pixel resolution
1400:1 contrast ratio

2.2 inches wide
(55.88 millimeters)

Warning: To avoid damaging your smartphone, please read all instructions carefully, or contact us at www.smartphone.com/support/specifications for additional help. We'll provide you easy troubleshooting and user-friendly advice.

Definition of Process Analysis

3. Define process analysis.

What is a process analysis? A *process analysis* is comparable to an instruction. Instructions provide a step-by-step explanation of how to do something. For example, in an instruction, you explain how to change the oil in a car, how to connect a printer to a computer, how to make polenta, or how to put together a child's toy. In an instruction, the audience wants to know how to do a job. A process analysis, in contrast, focuses not on how to do something but on how something works. For instance, a process analysis might explain how viruses attack our bodies, how airbags save lives, how metal detectors work, or how e-mail messages are transmitted.

Examples of Process Analyses

4. Understand the uses of process analyses.

A discussion of process is a common part of many technical descriptions:

- **Engineering.** Dubai, second largest emirate of the United Arab Emirates, plans to develop several islands off its coast. To increase tourism, these islands will have both private residences and hotels. Your engineering company has been hired to build bridges connecting the islands to the coastline. Your options include beam bridges, arch bridges, and suspension bridges. To clarify which type of bridge will best meet your client's needs, you provide a PowerPoint presentation explaining how each bridge works.
- **Automotive sales.** A potential customer wants to buy a new car. This customer is concerned about the environment and is interested in buying a hybrid, knowing that these cars are energy efficient and ecologically friendly. However, the customer doesn't understand how hybrids work. Your job is to explain the process of how gasoline-electric hybrid cars work.
- **Biomedical technology.** You and a team of scientists have been working on new treatments for diabetes, a health problem in the United States that affects approximately 20.8 million Americans. One possible treatment is pancreatic islet transplantation. Islets are injected through a catheter into a diabetic's liver. In time, the islets begin releasing insulin. This treatment has great potential for diabetics, but research is ongoing. You are writing a proposal requesting additional funding for research. In this proposal, you provide a process analysis of how diabetes affects a patient and how the pancreatic islet transplantation will work to treat this disease.

Figure 14.3 is a process analysis of why, when, and how automobile air bags inflate.

Criteria for Writing Technical Descriptions and Process Analyses

5. Follow criteria for writing effective technical descriptions and process analyses.

As with any type of technical communication, there are certain criteria for writing technical descriptions and process analyses.

Title

Preface your text with a title precisely stating the topic. This could be the name of the mechanism, tool, piece of equipment, landform, product, or service you are describing or analyzing.

FIGURE 14.3   Why, When, and How Does an Air Bag Inflate?

The introduction explains what the topic is, why it is important, and where the mechanism is located.

This technical description discusses the parts of the mechanism, the materials used, the location of these components, and the chemical compounds required for activation.

Here, providing the mechanism's process, the text explains how an air bag works. Note the specificity of detail: 150–250 mph and 1/20 of a second.

This illustration graphically depicts the process in action. When the crash sensor is triggered, it activates the inflator. Nitrogen gas explodes, inflating the air bag.

The conclusion sums up the process analysis by quantifying the significance of air bags as a means of saving lives.

Driver and passenger air bags save lives. Although steering wheel, right front instrument panel, or side-panel air bags will not inflate if a change in automobile motion is minor (such as hitting a bump), they will inflate in major incidents (such as collisions), minimizing injury and saving lives.

Parts of an Air Bag System
Air bag systems consist of the following:
- Air bag holding compartments: Compartments are located in the steering wheel, dashboard, seat, or door.
- Sensor: This device triggers bag inflation.
- Inflator canister: The canister contains sodium azide (NaN_3), potassium nitrate (KNO_3), and silicon dioxide (SiO_2).
- Air bag: The thin nylon air bag contains a nontoxic powder to keep the bag flexible and dry. It also features minute vents to allow the bag to deflate after a crash.

How an Air Bag Inflates
Within the sensor, a steel ball inside a smooth bored cylinder is secured by a magnet or stiff spring that inhibits the ball's motion during minor motion changes. However, in a dramatic stop or severe impact, the air bag sensing system will deploy immediately. The following occurs:
1. The steel ball within the sensor moves quickly forward, triggering an electrical circuit, sending electricity into a heating element, and igniting the sodium azide within the inflator canister to produce nitrogen gas (N_2).
2. The gas inflates the air bag at a speed of 150–250 mph. In about 40 milliseconds (1/20 of a second), the inflation is complete.
3. Although sodium azide normally produces sodium metal, which is potentially explosive and harmful to the eyes, nose, or mouth, in an air bag it reacts with the potassium nitrate and silicon dioxide to produce silicate glass, a harmless and stable compound.
4. The air bag vents allow the deployed air bags to deflate immediately after impact, ensuring that the bags do not smother the car's inhabitants.

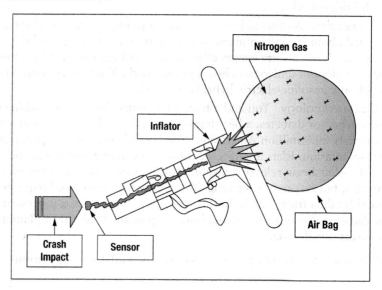

The Proven Benefits of Air Bags
Air bags were first introduced in 1973 and were only provided by some car manufacturers. However, they have been mandatory in all cars since 1998 and have made a major difference. The National Highway Traffic Safety Administration estimated that air bags saved 1,043 lives in the year they became mandatory, and that they have saved over 28,000 lives in the United States alone since their introduction.

Overall Organization

In the *introduction,* specify and define your topic and explain the topic's functions, capabilities, or processes. You can also include a list of components, parts, or equipment.

Technical Description

The Apex Latch (#12004), a mechanism used to secure core sample containers, is composed of three parts: the hinge, the swing arm, and the fastener.

Technical Description and Process Analysis

The DX 56 DME (Distance Measuring Equipment) is a vital piece of aeronautical equipment. Designed for use at altitudes up to 30,000 feet, the DX 56 electronically converts elapsed time to distance by measuring the length of time between your transmission and the reply signal. The DX 56 DME contains the transmitter, receiver, power supply, and range and speed circuitry.

In the *discussion*, describe each of the mechanism's components and how the mechanism works—its process. Focus on details, word usage, and an organizational pattern that helps your audience easily understand the text.

- **Details.** To develop your discussion, include the following details:

Weight	Density	Make/model
Size (dimensions)	Parts	Texture
Color	Materials (composition)	Capacity
Shape	Identifying numbers	Procedural steps

- **Word usage.** Your word usage in the technical description and process analysis depends on your purpose. For factual, objective technical descriptions or process analyses, use photographic words. For subjective, sales-oriented descriptions or process analyses, use impressionistic words. Photographic words are denotative, quantifiable, and specific. Impressionistic words are vague and connotative. Table 14.1 shows the difference between photographic and impressionistic words.
- **Internal organization.** When describing your topic in the discussion portion of the technical description or process analysis, itemize the topic in some logical sequence. For a technical description, use *spatial* organization. When a topic is

> **Organization**
> See Chapter 3, "The Communication Process," for more information on organization.

TABLE 14.1 Photographic versus Impressionistic Word Usage

Photographic	Impressionistic
6'9"	Tall
350 lb	Heavy
Gold	Precious metal
6,000 shares of United Can	Major holdings
700 lumens	Bright
0.030 mm	Thin
1966 XKE Jaguar	Impressive classic car

spatially organized, you literally lay out the components as they exist in space. You describe the components as they are seen either from left to right, right to left, top to bottom, bottom to top, inside to outside, or outside to inside. In contrast, when writing your process analysis, you will use chronological organization to show how the product or service works in sequence.

Your *conclusion* depends on your purpose in describing the topic or analyzing how the process works. Some options are as follows:

Sales—"Implementation of this product will provide you and your company . . . "	**Guarantees**—"The product carries a 15-year warranty against parts and labor."	**Comparison/contrast**—"Compared to our largest competitor, this product has sold three times more . . . "
Uses—"After implementation, you will be able to use this product to . . . "	**Testimony**—"Our satisfied customers include . . . "	**Reiteration of introductory comments**—"Thus, the product is composed of the above interchangeable parts and works according to the process explained."

Visual Aids

See Chapter 8, "Visual Aids," for more information.

Highlighting Techniques

To aid understanding and allow for easy access, use highlighting techniques, including headings, itemization, and graphics.

You can use line drawings, schematics, CAD/CAM drawings, photographs, architectural renderings, clip art, exploded views, or sectional cutaway views of your topic, each accompanied by *callouts* (labels identifying key components of the mechanism). An example of an exploded view with callouts can be seen in Figure 14.4.

FIGURE 14.4 Labeled Callouts

Valve Parts List		
Item	Description	Product Number
1	Nut	054859
2	Cap	098997
3	O–Ring	066584
4	Diaphragm	023337
5	Body Valve	V45665
6	Connector Elbow	C33678

How to Make Callouts Using Microsoft Word 2013

Callouts are a great way to label mechanism components in your technical description or process analyses. Follow these steps to create callouts.

1. Click on **Insert** in your Menu bar.

2. Click on **Shapes**. You will see a dropdown box.

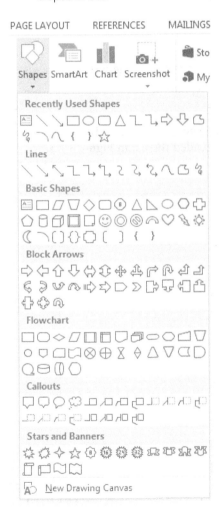

3. Select the type of callout you want to use, such as the **Rounded Rectangular Callout**.
4. When you choose the callout, a crosshair will appear. Use it to draw your callout, as shown.

NOTE: When you draw the callout, Word 2013 opens up the **Format** tab, as shown.

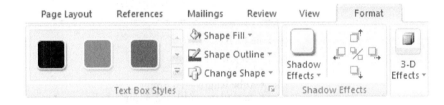

From the **Format** tab, you can change the color of the text box (from **Text Box Styles**), change its line type and density (from **Shape Outline**), create shadow effects, 3-D effects, and add content.

TECHNICAL DESCRIPTION AND PROCESS ANALYSIS CHECKLIST

_____ **1.** Does the technical description or process analysis have a title noting your topic's name (and any identifying numbers)?

_____ **2.** Does the technical description or process analysis's introduction (a) state the topic, (b) mention its functions or the purpose, and (c) list the components?

_____ **3.** Does the technical description or process analysis's discussion use headings to itemize the components for reader-friendly ease of access?

_____ **4.** Do you need to define the mechanism and its main parts?

_____ **5.** Does the discussion portion of the description or process analysis specify the following?

Colors	Capacities
Sizes	Textures
Materials	Identifying numbers
Shapes	Weight
Density	Make/Model
Parts	Process

_____ **6.** Are all of the calculations, measurements, or process steps correct?

_____ **7.** Do you sum up your discussion using any of the optional conclusions discussed in this chapter?

_____ **8.** Does your technical description or process analysis provide graphics that are correctly labeled, appropriately placed, and appropriately sized?

_____ **9.** Do you write clearly and concisely, using a personalized tone?

_____ **10.** Have you avoided biased language and grammatical and mechanical errors?

The Writing Process at Work

The following technical description shows how Hector Cruz used the writing process to create a document.

Prewriting

As part of his job, Hector needed to describe a cash register pole display. To do so, he first had to gather data using brainstorming/listing. He provided the list in Figure 14.5.

FIGURE 14.5 Prewriting List

Pole PCB

- length—15 mm
- width—5.1 mm
- tube length—10.8 mm
- face plate width—2.3 mm
- thick—1.7 mm
- PCB thickness—0.2 mm
- 10 inch stranded wire with female connectors
- fiberglass and copper construction

Pole Case Assembly

- long—15.5 mm
- bottom width—2.5 mm
- top width—0.9 mm
- mounting pole—5 mm high × 3.2 mm diameter
- tongue for mounting—3.1 mm
- lower mounting tongue—1.5 mm
- side mounting tongue—0.8 mm high
- high—6.1 mm
- almond-colored plastic

Filter

- long—15.6 mm
- high—6.2 mm
- thick—0.7 mm
- plastic, blue

Writing

After the prewriting, Hector drafted a technical description. Focusing on overall organization, highlighting, detail, and a hand-drawn graphic, he wrote a rough draft. He had his colleagues suggest revisions (Figure 14.6).

FIGURE 14.6   Rough Draft of a Pole Display with Suggested Revisions from Colleagues

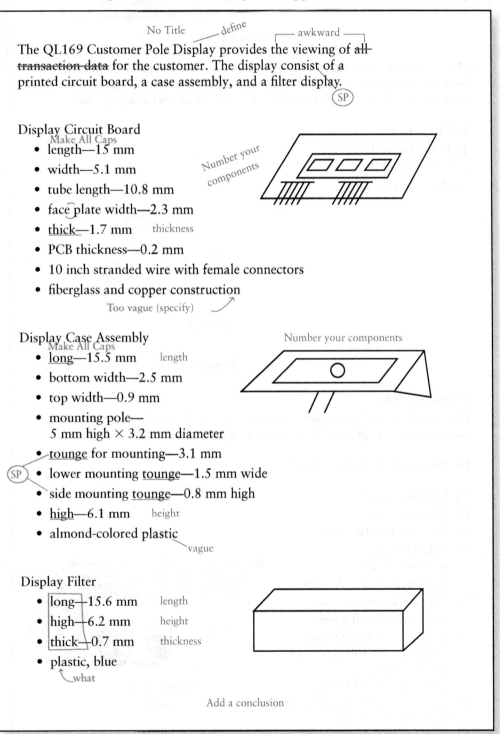

No Title   define   — awkward —

The QL169 Customer Pole Display provides the viewing of all transaction data for the customer. The display consist of a printed circuit board, a case assembly, and a filter display.   (SP)

Display Circuit Board
Make All Caps
- length—15 mm
- width—5.1 mm
- tube length—10.8 mm
- face plate width—2.3 mm
- thick—1.7 mm   thickness
- PCB thickness—0.2 mm
- 10 inch stranded wire with female connectors
- fiberglass and copper construction
  Too vague (specify)

Number your components

Display Case Assembly
Make All Caps
- long—15.5 mm   length
- bottom width—2.5 mm
- top width—0.9 mm
- mounting pole—
  5 mm high × 3.2 mm diameter
- tounge for mounting—3.1 mm
(SP) - lower mounting tounge—1.5 mm wide
- side mounting tounge—0.8 mm high
- high—6.1 mm   height
- almond-colored plastic
  vague

Number your components

Display Filter
- long—15.6 mm   length
- high—6.2 mm   height
- thick—0.7 mm   thickness
- plastic, blue
  what

Add a conclusion

Rewriting

Hector incorporated these suggestions and prepared the finished copy (Figure 14.7).

FIGURE 14.7   Hector's Revised Technical Description

QL169 Pole Display

Including a definition of the QL169 pole helps the reader understand the topic.

The QL169 pole is an electronic mechanism that provides an alphanumeric display for customer viewing of cash register sales. The display consists of a printed circuit board (PCB) assembly, a case assembly, and a display filter.

Item Description

1. PCB ASSEMBLY
2. CASE ASSEMBLY
3. DISPLAY FILTER

Exploded graphics help readers visualize the topic.

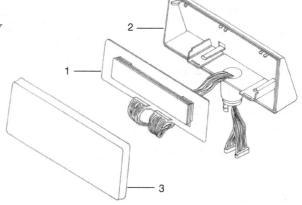

Figure 1. QL169

1.0 PCB Assembly

The printed circuit board, containing the display's electrical circuitry, is constructed of fiberglass with copper etchings.

The board consists of the following features:

1.1 Length—15 mm
1.2 Width—5.1 mm
1.3 Tube length—10.8 mm
1.4 Tube faceplate width—2.3 mm
1.5 Tube total width—2.8 mm
1.6 Tube thickness—1.7 mm
1.7 PCB thickness—0.2 mm

Specificity of detail adds clarity.

1.8 20 conductor 10" 22-gauge stranded wire with two AECC female connectors (AECC part #7214-001)

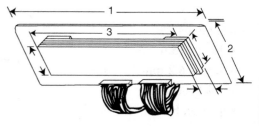

Figure 2. PCB

1.9 American Display Company blue phosphor display tube (ADC part #1172177)

FIGURE 14.8 *(Continued)*

2.0 Case Assembly

Almond-colored ABS plastic, used to construct the case assembly, protects the PCB.

2.1 Length—15.5 mm
2.2 Bottom width—2.5 mm
2.3 Top width—0.9 mm
2.4 Mounting pole—5 mm high and 3.2 mm in diameter
2.5 Mounting tongue inside width—3.1 mm from side of assembly
2.6 Lower mounting tongue— 1.5 mm wide
2.7 Side mounting tongue— 0.8 mm high
2.8 Tongue thickness—0.2 mm
2.9 Height—6.1 mm

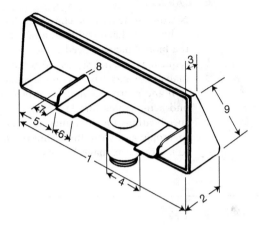

Figure 3. Case Assembly

3.0 Display Filter
3.1 Length—15.6 mm
3.2 Height—6.2 mm
3.3 Thickness—0.7 mm

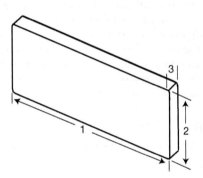

Figure 4. Display Filter

Transparent blue plastic, used to construct the display filter, allows the customer to view the readings. The QL169 pole provides easy viewing of clerk transactions and ensures cashier accuracy.

The conclusion focuses on the end-user benefits.

CASE STUDY

Nitrous Systems Biotechnology, Inc. plans to expand by hiring 50 new bio-technicians. Each will need a tablet PC. But what kind of tablet PC best meets the biotechnicians' needs: one with a touch screen or a pen-enabled interface requiring a digital stylus? You have been asked to research tablet PCs. Your manager wants a brief comparison/contrast, complete with a product description and a process analysis explaining the differences between touch screens and pen-enabled interfaces requiring a digital stylus. What are the technical specifications for each interface, and how does each interface work? Research this topic and write the technical description, complete with process analysis, to meet your manager's request.

ETHICAL CHALLENGE

Gary Porter is the human resource manager at a biomedical technology company. The company has an anti-nepotism policy for full-time employees. Gary's wife Deb is a self-employed consultant who makes presentations to corporations about changes in security modules. She has an MBA from the state university, over 20 years of experience in security, and a long list of well-known clients. Each participant in her training seminars receives an instructional manual with technical descriptions and process analyses on how to use new hardware security modules. As a consultant, Deb receives no benefits or regular pay from her clients. She is hired on an as-need, contract basis.

Gary's company needs to review its current security procedures versus the benefits of hardware security modules. He wants to bring in an expert in the field. Deb meets Gary's company's requirements for experience and knowledge.

Question

Is it ethical for Gary to hire his wife as a consultant? Why or why not?

INDIVIDUAL AND TEAM PROJECTS

1. Write a technical description, either individually or as a team. To do so, first select a topic. You can describe any tool, mechanism, or piece of equipment. However, don't choose a topic too large to describe accurately. To provide a thorough and precise description, you will need to be exact and minutely detailed. A large topic, such as a computer, an oscilloscope, a respirator, or a Boeing airliner, would be too demanding for a two- to four-page description. On the other hand, do not choose a topic that is too small, such as a paper clip, a nail, or a shoestring. Choose a topic that provides you with a challenge but that is manageable. You might write about any of the following topics:

USB flash drive	Backpack	DVD remote control
Wrench	Computer mouse	Mechanical pencil
Screwdriver	Light bulb	Ballpoint pen
Pliers	Calculator	Computer monitor
Wall outlet	Automobile tire	Smartphone

2. Write a process analysis, either individually or as a team, on one of the following topics:
   - How does blood coagulate?
   - How does an x-ray machine work?
   - How do WiFi connections work?
   - How do rotary engines work?
   - How do fuel gauges work?
   - How is metal welded?
   - How does a metal detector work?
   - How do browsers work?
   - How do computer viruses work?
   - How does a wireless mouse work?

3. Select a simple topic for description, such as a pencil, coffee cup, toothbrush, or textbook (you can use brainstorming/listing to come up with additional topics). Describe this item without mentioning what it is or providing any graphics. Then, read your description to a group of students/peers and ask them to draw what you have described. If your verbal description is good, their drawings will resemble your topic. If their drawings are off base, you haven't succeeded in providing an effective description.

4. In degree-specific teams, choose a topic from your area of interest and expertise to describe (including a process analysis). For example, students majoring in HVAC could describe a humidifier and explain how it works. Biotech and nursing students could describe and provide a process analysis for a nebulizer, blood pressure monitor, or glucometer. EMT students could write about defibrillators. Automotive technicians could describe and include a process analysis for jump starters, battery chargers, or hydraulic jacks. Welding students could write about MIG, TIG, or stick welding.

PROBLEM-SOLVING THINK PIECE

Read the following process analysis and reorganize the numbered sentences to achieve a clear, chronological order. If necessary, research this process to learn more about how blood clots.

How Does Blood Clot?

1. In our bodies, blood can clot due to platelets and the thrombin system.
2. When bleeding occurs, chemical reactions make the surface of the platelet "sticky." These sticky platelets adhere to the wall of blood vessels where bleeding has occurred.
3. Platelets, tiny cells created in our bone marrow, travel in the bloodstream and wait for a bleeding problem.
4. Soon, a "white clot" is formed, so called because the clotted platelets look white.
5. Blood clots consist of both platelets and fibrin. The fibrin strands bind the platelets and make the clot stable.
6. In the thrombin system, several blood proteins become active when bleeding occurs. Clotting reactions produce fibrin, long, sticky strings. These sticky strands catch red blood cells and form a "red clot."
7. Primarily, arteries clot due to platelets, while veins clot due to our thrombin system.

WEB WORKSHOP

1. You are ready to purchase a product. This could include printers, monitors, smartphones, scanners, PCs, tablet PCs, speakers, cables, adapters, automotive engine hoists, generators, battery chargers, jacks, power tools, truck boxes, screws, bolts, nuts, rivets, hand tools, and more. A great place to shop is online. By going to an online search engine, you can find not only prices for your products but also technical descriptions or technical specifications. These will help you determine if the product has the size, shape, materials, and capacity you are looking for.

   Go online to search for a product of your choice and review the technical description or specifications provided. Using the criteria in this chapter and your knowledge of effective technical communication techniques, analyze your findings.

   - Are graphics used to help you visualize the product?
   - Are callouts used to help you identify parts of the product?
   - Are high-tech terms defined?
   - Is the use of the product explained?

     a. Report your findings, either in an oral presentation or in writing (e-mail message, memo, letter, or report).
     b. Rewrite any of the technical descriptions that need improvement according to the criteria provided in this textbook.

2. HowStuffWorks is an extensive online library of process analyses. Revise a process analysis in this Web site.

CHAPTER FIFTEEN

Instructions, User Manuals, and Standard Operating Procedures

After completing this chapter, you will be able to

1. Learn why writing instructions is important in the workplace

2. Follow criteria for writing effective instructions

3. Collaborate to create effective user manuals

4. Use instructional videos to depict steps visually

5. Follow criteria for writing standard operating procedures (SOPs)

6. Test manuals or standard operating procedures for usability

This scenario shows the importance of effective instructions in the biomedical industry and various communication channels used to convey step-by-step procedures.

As the baby-boom generation ages, medical needs are expanding—sometimes faster than medical care facilities and medical professionals can manage. One area in which this has been felt most acutely is in medical laboratories. Thousands of medical technicians need to be trained to accommodate increased demand.

PhlebotomyDR is a medical consulting firm seeking to solve this problem. Its primary area of concern is training newly hired technicians responsible for performing blood collection. PhlebotomyDR facilitates training workshops to teach venipuncture standards and venipuncture procedures.

PhlebotomyDR focuses on the following venipuncture instructions:

- Proper patient identification procedures

- Proper equipment selection, sterilization, use, and cleaning

- Proper labeling procedures

- Order of phlebotomy draw

- Patient care before, during, and following venipuncture
- Safety and infection control procedures
- Procedures to follow when meeting quality assurance regulations

Each of the instructions mentioned requires numerous steps complete with visual aids.

PhlebotomyDR offers its audience various communication channels. Hospitals, labs, and treatment centers can access PhlebotomyDR's instructions as follows:

- Hard-copy instructional manuals
- Online instructions, accessible at http://www. phlebotomydr.com
- Videos showing step-by-step performances of blood collection, complete with case studies enacted by technicians, patients, and supervisors
- Standard operating procedures used to achieve uniformity and efficiency
- Smartphone apps for individual tutorials
- One-on-one tutorials with trained phlebotomists
- Instructional workshops designed for groups of seminar participants

PhlebotomyDR's outstanding staff realizes that trained technicians make an enormous difference. Training, achieved through instructional manuals, electronic aids, and individual facilitation, ensures the health and safety of patients. Of equal importance, excellent training also benefits many stakeholders. Untrained technicians make errors that cost us all. Medical errors create insurance problems, the need to redo procedures, increased medical bills, the potential involvement of regulators and legislators, and dangerous repercussions for patients.

In contrast, effective communication, achieved through successful instruction, saves lives, time, and money.

---

1. Learn why writing instructions is important in the workplace.

Why Write Instructions?

Almost every manufactured product comes complete with instructions. You will receive instructions for baking brownies; making pancakes; assembling children's toys; or changing a tire, the oil in your car, or the coolant in your engine. Instructions help people set up iPod and MP3 systems, construct electronic equipment, maintain computers, and operate fighter planes.

We frequently see short instructions about computer-related problems. If you are at home or in the office, for example, and you can't figure out how to operate one of your computer's applications, you can access online help, e-mail a help desk technician, or contact a company's Twitter account. Figure 15.1 is an e-mail message giving a customer the steps to follow for computer technical support.

Include instructions or user manuals whenever your audience needs to know how to

Operate a mechanism	Collect lab specimens
Install equipment	Service equipment
Manufacture a product	Repair a system
Package a product	Troubleshoot a system
Perform lab experiments	Use software
Test components	Set up a product
Maintain equipment	Implement a procedure
Clean a product	Assemble a product

FIGURE 15.1   Instructions for Computer Technical Support

To	AN	aneesha.colt@aol.com ✕	+
Cc			

Technical Support for Creating a New Profile

🖼 ↻ | **B** *I* U̲ Aa A⁺ ✿ A̲ ⋮≡ ≔ ⋯ ⋯ ⌄

Thank you for contacting us about your computer difficulties accessing @Scrapbook services. Recently, we have encountered cases where an online @Scrapbook update modified existing customer profiles. This change can disallow our customers from accessing their accounts.

To solve this problem, please follow these steps to create your new customer profile:

1. Go to www.@Scrapbook.com.
2. From the home page, click on "Configure New Profile."
3. When the new screen pops up, click on "Add a New Account."
4. Type in your first name, middle initial, and last name in the "Name" field.
5. Type in your e-mail address and PIN (Personal Identification Number). This must be more than 5 characters and no more than 10. Use a combination of letters, numbers, and typographical symbols: Ex: bluebird3*
6. Retype your PIN to confirm.
7. Click "OK."
8. Click "Finish."

Once you complete these steps, you'll be able to create new scrapbook entries to share with your friends and family. If you need further help, please let us know. You can reach us at @Scrapbook.help.com.

Microsoft Corporation, Inc.

Criteria for Writing Instructions

2. Follow criteria for writing effective instructions.

Odds are good that you've read badly written instructions. You've probably encountered instructions that failed to define unfamiliar terminology, didn't tell you what equipment you needed to perform the task, didn't warn you about dangers, didn't clarify how to perform certain steps, and didn't provide you enough graphics or a video to complete the job successfully. Usually, such poorly written instructions occur because writers fail to consider their audience's needs. Successful instructions, in contrast, start with audience recognition.

Audience Recognition

The instructions tell you to "place the belt on the motor pulley," but you don't know what a "motor pulley" is nor do you know how to "place the belt" correctly. You are told to "discard the used liquid in a safe container," but you don't know what is safe for this specific type of liquid. You are told to "size the cutting according to regular use," but you have never regularly performed this activity before.

Recognizing Audience

See Chapter 5, "Audience Recognition," for more information.

Here are typical instructions written without considering the audience:

◀ EXAMPLE

> To overhaul the manual starter, proceed as follows: Remove the engine's top cover. Untie the starter rope at the anchor and allow the starter rope to slowly wind onto the pulley. Tie a knot on the end of the starter rope to prevent it from being pulled into the housing. Remove the pivot bolt and lift the manual starter assembly from the power head.

Although many high-tech readers might be able to follow these instructions, many more readers will be confused. How do you remove the engine's top cover? Where is the anchor? Where is the pivot bolt, and how do you remove it? What is the power head?

The problem is caused by writers who assume that their readers have high-tech knowledge. This is a mistake for several reasons. First, even high-tech readers often need detailed information because technology changes daily. You cannot assume that every high-tech reader is up-to-date on these technical changes. Thus, you must clarify. Second, low-tech and lay readers—and that's most of us—carefully read each and every step, desperate for clear and thorough assistance.

As the writer, you should provide your readers with the clarity and thoroughness they require. To do this, recognize accurately who your readers are and give them what they want, whether that amounts to technical updates for high-tech readers or precise, even simple, information for low-tech or lay readers. The key to success as a writer of instructions is the following: Don't assume anything. Spell it all out—clearly and thoroughly.

Ethical Instructions

Avoiding Plagiarism
See Chapter 16, "Research," for more information.

Your job as a professional writer is to meet your audience's needs. This not only entails clarity in instructions, but also it demands ethical behavior. In Chapter 6 we discuss legalities and practicalities in professional communication. User manuals and instructions are perfect examples of the importance of ethical communication. Clearly and correctly citing sources of information will help you avoid plagiarism. Only boilerplate content does not necessarily need to be cited.

Legalities in User Manuals.   As a corporate employee, you never want your writing to lead to customer injuries or to cause a customer to damage equipment. Instead, to avoid issues of liability, your communication must identify the potential for harm. Clearly stated hazard notations, warranties, and disclaimers, discussed later in this chapter, allow you to warn your audience of potential dangers and to set limits and exceptions for product guarantees. By doing so, you adhere to your legal responsibilities to the company and to the client.

Practicalities in User Manuals.   Ethical writing doesn't just keep customers safe. It also satisfies your customer's need for up-to-date information, and that's just good business. For example, let's say that a decade ago, your company wrote a technical specification for a product. The product has changed over the years, improving through numerous iterations. However, your company's customer support (training manuals, online help, user manuals, and troubleshooting guides) have not been revised to reflect these changes. Management says that the company doesn't have the time or budget to update the customer support. This can lead to many problems in the manuals, including inconsistencies, inaccuracies, irregularities, and customer complaints. If a client stops buying your products because the supporting documents are flawed, the company loses money. The practical solution, then, is to revise the manuals, bringing them up to date. Doing so is a company's ethical responsibility to its customers.

Components of Instructions

Not every set of instructions will contain the same components. Some very short instructions will consist of nothing more than a few, numbered steps. Other instructions, however, will consist of the components shown in Figure 15.2.

Title Page

Preface your instructions with a title page that consists of the *topic* about which you are writing, the *purpose* of the instructions, and a *graphic* depicting your product or service. For example, to merely title your instructions "iPod" would be uninformative. This title names the product, but it does not explain why the instructions are being written. Will the text discuss operating instructions, troubleshooting, service, or maintenance? A better title would be "Operating Instructions for the iPod: Steps for Downloading Music, Taking

FIGURE 15.2 Key Components of Instructional Manuals

Title Page	Hazards	Table of Contents	Introduction	List of Required Tools / Equipment
Topic				
Graphic				
Purpose	⚠			

Glossary of Terms	Steps	Steps	Additional Components	Corporate Contact Information
	1.	3.		
	2.	4.		
		Etc.		

Photos, and Watching Movies." Adding a graphical representation lets your audience see what the finished product will look like and helps you market your product or service.

Safety Requirements

You can place safety requirements anywhere throughout your text. If a particular step presents a danger to the reader, call attention to this hazard just before asking the reader to perform the step. Similarly, to help your audience complete an action, place a note before the step, suggesting the importance of using the correct tool, not overtightening a bolt, or wearing protective equipment.

In addition to placing safety requirements before a step, consider prefacing your entire instructions with hazard notations. By doing so, you make the audience aware of possible dangers, warnings, cautions, or notes in advance of performing the instructions. This is important to avoid potentially harming an individual, damaging equipment, and avoiding costly lawsuits. Correctly using hazard notations will avoid expensive liability issues and adhere to ethical communication standards. Include the following in your safety requirements.

Access. Make the hazard notations obvious. To do so, vary your typeface and type size, use white space to separate the warning or caution from surrounding text, box the warning or caution, and call attention to the hazards through graphics.

Definitions. What does *caution* mean? How does it differ from *warning, danger*, or *note*?

Four primary organizations that seek to provide a standardized definition of terms are the American National Standards Institute (ANSI), the U.S. military (MILSPEC), the Occupational Safety and Health Administration (OSHA), and the MSDS (material safety data sheets) search national repository. To avoid confusion, we suggest the following hierarchy of definitions, which clarifies the degree of hazard:

1. **Danger.** The potential for death
2. **Warning.** The potential for serious personal injury
3. **Caution.** The potential for damage or destruction of equipment
4. **Note.** Important information necessary to perform a task effectively or to avoid loss of data or inconvenience

Colors. Another way to emphasize your hazard message is through a colored window or text box around the word. Usually, *Danger* is printed in red, *Warning* in orange, *Caution* in yellow, and *Note* in blue, green, or black.

Text. To further clarify your terminology, provide the readers text to accompany your hazard alert. Your text should have the following three parts:

1. **A one- or two-word identification alerting the reader.** Words such as "High Voltage," "Hot Equipment," "Sharp Objects," or "Magnetic Parts," for example, will warn your reader of potential dangers, warnings, or cautions.
2. **The consequences of the hazards, in three to five words.** Phrases like "Electrocution can kill," "Can cause burns," "Cuts can occur," or "Can lead to data loss," for example, will tell your readers the results stemming from the dangers, warnings, or cautions.
3. **Avoidance steps.** In three to five words, tell the readers how to avoid the consequences noted: "Wear rubber shoes," "Don't touch until cool," "Wear protective gloves," or "Keep disks away."

Icons. Equipment is manufactured and sold internationally; people speak different languages. Your hazard alert should contain an icon—a picture of the potential consequence—to help everyone understand the caution, warning, or danger.

Figure 15.3 shows an effective page layout and the necessary information to communicate hazard alerts.

FIGURE 15.3 Hazard Alert

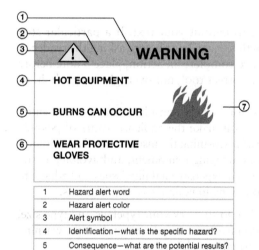

1	Hazard alert word
2	Hazard alert color
3	Alert symbol
4	Identification—what is the specific hazard?
5	Consequence—what are the potential results?
6	Avoidance steps—what should you do?
7	Hazard icon

Table of Contents

Table of Contents

See Chapter 18, "Long, Formal Reports," for a table of contents sample.

Your instructions might have several sections. In addition to the actual steps, the instructional manual could include technical specifications, warranties, guarantees, frequently asked questions (FAQs), troubleshooting tips, and customer service contact numbers. An effective table of contents will allow your readers to access any of these sections individually on an as-needed basis.

Introduction

Companies need customers. A user manual might be the only contact a company has with its customer. Therefore, instructions often are reader-friendly and seek to achieve audience recognition and audience involvement. The manuals try to reach customers in a personalized way. Look at the following introduction from a user manual:

Thank you for your purchase. Installation and operating procedures are contained in this manual about your new 3-D TV. The minutes you spend reading these instructions will contribute to hours of viewing pleasure.

This introduction uses pronouns (*you* and *your*) to personalize the manual. The introduction also uses positive words, such as *Thank you* and *pleasure* to achieve positive customer contact. An effective introduction promotes good customer–company relationships.

Glossary

If your instructions use the abbreviations *BDC, CCW,* or *CPR,* will the readers know that you are referring to "bottom dead center," "counterclockwise," or "continuing property records"? If your audience is not familiar with your terminology, they might miss important information and perform an operation incorrectly.

To avoid this problem, define your abbreviations, acronyms, or technical terms. You can define your terms early in the instructions, throughout the manual, or in a glossary located at the end of your manual. The following example defines terms alphabetically in a glossary:

GLOSSARY

BDC	Bottom dead center
CCW	Counterclockwise
Danger	This hazard alert designates the possibility of death.
	Be extremely careful when performing an operation.
RMS	Root mean square

Required Tools or Equipment

What tools or equipment will the audience need to perform the procedures? You don't want your audience to be in the middle of performing a step and suddenly realize that they need a missing piece of equipment. Provide this important information either through a list or graphics depicting the tools or equipment necessary to complete the tasks.

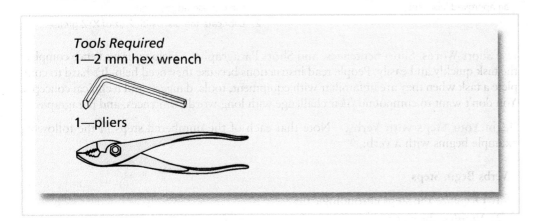

Tools Required
1—2 mm hex wrench

1—pliers

Instructional Steps

The instructional steps are the most important part of your manual—the actual actions required of the audience to complete a task. To successfully write your instructions, follow these steps:

Organize the Steps Chronologically.   You cannot tell your readers to do step 6, then go back to step 2, then accomplish step 12, and then do step 4. Such a distorted sequence would fail to accomplish the task. To operate machinery, monitor a system, troubleshoot equipment, or perform a standard operating procedure, your readers must follow a chronological sequence. Be sure that your instructions are chronologically accurate.

Number Your Steps.   Do not use bullets or the alphabet. Numbers, which you can never run out of, help your readers refer to the correct step. In contrast, if you used bullets, your readers would have to count to locate steps—seven bullets for step 7, and so on. If you used the alphabet, you'd be in trouble when you reached step 27.

Use Highlighting Techniques.   You can use boldface, different font sizes and styles, text boxes, emphatic warning words, color, or italics to call attention to special concerns. A danger, caution, warning, or specially required technique must be evident to your reader. If this special concern is buried in a block of unappealing text, it will not be read. This could be dangerous to your reader or costly to you and your company. To avoid lawsuits or to help your readers see what is important, call it out through formatting.

Limit the Information within Each Step.   Don't overload your reader by writing lengthy steps:

BEFORE	AFTER
Overloaded Steps Start the engine and run it to idling speed while opening the radiator cap and inserting the measuring gauge until the red ball within the glass tube floats either to the acceptable green range or to the dangerous red line.	**Separated Steps** 1. Start the engine. 2. Run the engine to idling speed. 3. Open the radiator cap. 4. Insert the measuring gauge. 5. Determine whether the red ball within the glass tube floats to the acceptable green range or up to the dangerous red line.

Develop Your Points Thoroughly.   Avoid vague content by clarifying directions, needed equipment, and cautions.

The "After" example clarifies what is meant by correct rotating and approved lubricant. The steps are also separated for enhanced readability.

BEFORE	AFTER
After rotating the discs correctly, grease each with an approved lubricant.	1. Rotate the disks clockwise so that the tabs on the outside edges align. 2. Lubricate the discs with 2 oz of XYZ grease.

Use Short Words, Short Sentences, and Short Paragraphs.   Help your audience complete the task quickly and easily. People read instructions because they need help. It's hard to complete a task when they are unfamiliar with equipment, tools, dangers, and technical concepts. You don't want to compound their challenge with long words, sentences, and paragraphs.

Begin Your Steps with Verbs.   Note that each of the numbered steps in the following example begins with a verb:

EXAMPLE ▶

> **Verbs Begin Steps**
>
> 1. *Organize* the steps chronologically.
> 2. *Number* the steps.
> 3. *Use* highlighting techniques.
> 4. *Limit* the information within each step.
> 5. *Develop* the points thoroughly.
> 6. *Use* short words and phrases.
> 7. *Begin* the steps with verbs.

Do Not Omit Articles. Articles, such as *a*, *an*, and *the*, are part of our language. Although you might see instructions that omit these articles, please don't do so yourself. Articles do not take up much room in your text, but they make your sentences read more fluidly.

BEFORE	AFTER
1. Press right arrow button to scroll through list of programs.	1. Press **the** right arrow button to scroll through **a** list of programs.
2. Select program to scan.	2. Select **the** program you want to scan.
3. Place item to scan face down on scanner glass in upper left corner.	3. Place **an** item to scan face down on **the** scanner glass in **the** upper left corner.

Additional Components

Your instructions might include the following additional components.

Technical Descriptions. In addition to the step-by-step instructions, many manuals contain technical descriptions of the product or system. A description could be a part-by-part explanation or labeling of a product or system's components. Such a description helps readers recognize parts when they are referred to in the instruction. For example, if the user manual tells the reader to lay shingles with the tabs pointing up, but the reader doesn't know what a "tab" is, the step cannot be performed. In contrast, if a description with appropriate callouts is provided, then the reader's job has been simplified.

Perhaps the user manual will contain a list of the product's specifications, such as size, shape, capacity, capability, and materials of construction. Specifications allow the user to decide whether the product meets the reader's needs. Does it have the desired resolution? Is it the preferred weight and size? The specifications for a smartphone answer these questions.

Smartphone Specifications	
Height:	114 mm (4.48 inches)
Width:	66 mm (2.6 inches)
Depth:	15 mm (0.59 inches)
Weight:	136 grams (4.8 ounces)
Monitor Resolution:	480 × 320 pixel color display

◀ EXAMPLE

Finally, a user manual might include a schematic depicting the product or system's electrical layout. This would help the readers troubleshoot the mechanism.

Warranties. Warranties protect the customer and the manufacturer. Many warranties tell the customer, "This warranty gives you specific legal rights, and you may also have other rights that vary from state to state." A warranty protects the customer if a product malfunctions sooner than the manufacturer suggests it might: "This warrants your product against defects due to faulty material or installation." In such a case, the customer usually has a right to free repairs or a replacement of the product.

However, the warranty also protects the manufacturer. No product lasts forever, under all conditions. If the product malfunctions after a period of time designated by the manufacturer, then the customer is responsible for the cost of repairs or a replacement product. The designated period of time differs from product to product. Furthermore, many warranties tell the customers, "This warranty does not include damage to the product resulting from normal wear, accident, or misuse."

Disclaimers are another common part of warranties that protect the manufacturer. For example, some warranties include the following disclaimers:

- Note: Any changes or modifications to this system not expressly approved in this manual could void your warranty.

- Proof of purchase with a receipt clearly noting that this unit is under warranty must be presented.
- The warranty is only valid if the serial number appears on the product.
- The manufacturer of this product will not be liable for damages caused by failure to comply with the installation instructions enclosed within this manual.

Accessories. A company always tries to increase its income. One way to do so is by selling the customer additional equipment. A user manual promotes such equipment in an accessories list. This equipment isn't mandatory, required for the product's operation. Instead, an accessories list offers customers equipment such as extra-long cables or cords, carrying cases, long-life rechargeable batteries, belt clips and cases, wireless headsets, automotive chargers, and memory cards. Often the specifications are also provided for these accessories.

Frequently Asked Questions. Why take up your customer support employees' valuable time by having them answer the same questions over and over? By including an FAQs page in the user manual, common consumer concerns can be addressed immediately. This will save your company time and money while improving customer relations.

Corporate Contact Information. Conclude your manual by providing your company's street, city, and state address; help desk contact information; Web site URL; e-mail address; fax number; social media sites; or other ways your audience could contact the company with questions or requests for more information.

Graphics

Clarify your points graphically. Use drawings, photographs, and screen captures that are big, simple, clear, keyed to the text, and labeled accurately. Not only do these graphics make your instructions more visually appealing, but also they help your readers and you. What the reader has difficulty understanding, or you have difficulty writing clearly, your graphic can help explain pictorially.

EXAMPLE ▶

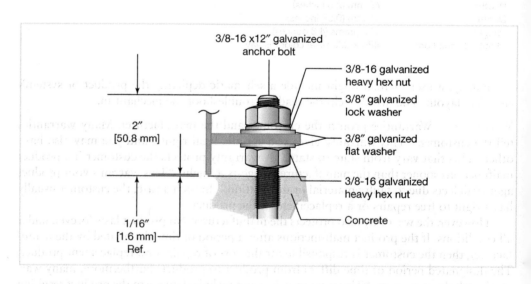

A company's use of graphics often depends on the audience. With a high-tech reader, the graphic might not be needed. However, with a low-tech reader, the graphic is used to help clarify. Look at the examples of instructions for the same procedure. Table 15.1 uses text without graphics for the high-tech audience. Figure 15.4 uses text with graphics to help the low-tech audience better understand the required steps.

TABLE 15.1 Text without Graphics for a High-Tech Reader

Location Item	Action	Remarks
Note: For additional operating instructions, refer to The All Seasons Room Humidifier 2000 operator's manual.		
1. Operation Dial	Set the dial to "Power On."	
2. Rear Panel	Ensure that Warning and Caution instructions on all labels are observed.	
3. Settings	Turn the Operation Dial to your desired setting.	See Table 2, step 3.
Note: If your home thermostat is set between 70 degrees and 75 degrees, the best humidifier setting is at "Winter." If your home thermostat is set between 60 degrees and 70 degrees, the best humidifier setting is "Summer."		
4. Power Cord	Connect the Power Cord to a 120/220 V AC grounded outlet.	If the power source is an external battery, connect the external battery cable to the unit following Table 2, step 4.

FIGURE 15.4 Instructions with Large, Bold Graphics for a Low-Tech Audience

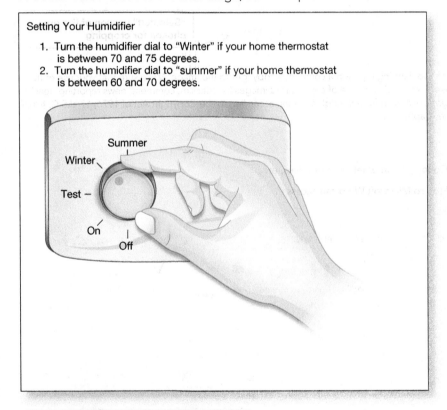

Setting Your Humidifier

1. Turn the humidifier dial to "Winter" if your home thermostat is between 70 and 75 degrees.
2. Turn the humidifier dial to "summer" if your home thermostat is between 60 and 70 degrees.

TECHNOLOGY TIPS

Using Screen Captures to Visually Depict Steps

Here's a good, simple way to visually depict steps in your instructions—screen captures. A screen capture lets you copy any image present on your computer monitor. To make screen captures, follow these steps:

Capturing the Image

1. Find the graphic you want to include in your instructions.
2. Press the **Print Screen** key on your computer keyboard.

(Continued)

Cropping the Image

Pressing the print screen key captures the entire image seen on the monitor. This might include images that you do not want to include in your instruction. To crop your image, follow these steps:

1. Paste the **Print Screen** image into a graphics program like MS Paint.
2. Click on the **Select** icon.

3. Copy and paste this cropped image into your user manual.

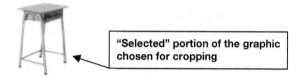

"Selected" portion of the graphic chosen for cropping

NOTE: When capturing screenshots, be careful to avoid infringing upon a company's copyright to the image. Make sure that your use of the image meets the principles of "fair use." "Fair use" permits *limited* use of copyrighted images for criticism, comment, news reporting, teaching, scholarship, or research, without having to get permission from the rights holder, according to the Copyright Act of 1976, 17 U.S.C. § 107.

In addition, you can use MS Word for screen capturing.

Microsoft Word 2013 Screenshots

1. Find the image you want to copy (from Google, the Internet, your own files, etc.).

    NOTE: This image must be on your desktop so Microsoft Word can access it.

2. In Microsoft Word, click on the **Insert** tab.

3. Click on the Screenshot down arrow and then Screen Clipping.

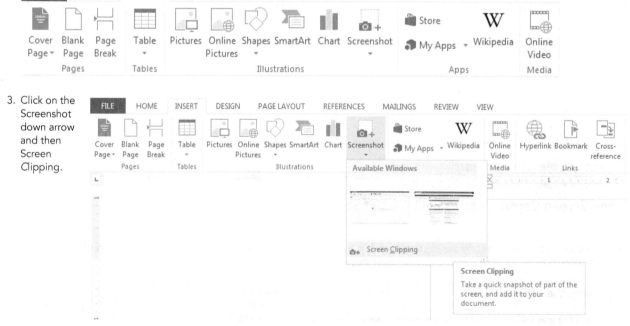

(Continued)

NOTE: The image you've selected in step 1 will pop up and gray out. You'll see a crosshair.

4. Use the crosshair to click, drag, and select your image.

NOTE: Once selected, the image will automatically be inserted into your Word document.

Collaboration to Create User Manuals

> 3. Collaborate to create effective user manuals.

Collaboration is essential for team members who write user manuals. User manuals are often long and complex, requiring input from a variety of individuals, each with a different job function. Sloane Harken, documentation specialist, works collaboratively with multiple team members when writing her user manuals. Her team consists of three separate cohorts: other writers, clients, and a project team of subject matter experts (SMEs).

The SMEs are headed by a project manager. This person is in charge of the budget and schedule for the user manual creation. The manager oversees the following:

- Account specialists, who liaison with the clients to determine customer needs
- Business analysts, who determine what software changes need to be made to meet the clients' expectations
- Software developers, who code the software programs
- Quality assurance personnel, who make sure the coding is correct

Sloane takes the information from these team members and writes the manual. If the manual is very long, she might work with a team of writers, most of whom are dispersed in various cities. Team members collaborate virtually through online chats, text messages, e-mail, and telephone calls. They discuss how the manual will look, mock up parts of the project, decide how to delegate responsibilities, work on their individual parts, proofread each other's work, and then merge the components for the finished project.

Finally, Sloane meets with a representative group of clients who complete the steps within the manual for usability testing. By gauging the challenges her test group might encounter, Sloane can correct glitches and fine-tune the manual. Once this is completed, the manual is ready for publication, either in hard-copy or electronic form.

Instructional Videos

> 4. Use instructional videos to depict steps visually.

Another way to depict your instructional steps graphically is through videos. Screencasting, a mixture of animated screen visuals, voiceovers, and captioned text, will benefit your audience in the following ways:

> **Social Media**
>
> See Chapter 12, "Social Media," for information about online videos.

- Animated videos show the end-user how to perform a step more successfully than static text might.
- Videos are an excellent communication channel for visual learners.
- Through podcasting and smartphone apps, instructional videos can be downloaded anywhere, anytime by mobile users.
- A company can enhance its marketing strategy by advertising its instructional videos as an alternative communication channel to hard-copy text. YouTube is an excellent social networking tool for sharing instructional videos and for marketing a company's product.
- Instructional videos allow for end-user interaction. The audience can fast forward, slow down, go backward, or freeze screens to learn how to perform tasks at their own pace.

- A video can be more informative than static text and photography. A video screencast shows how an end user interacts with machinery, equipment, tools, and more.
- Video instructions help end users see the consequences of hazard alerts. For example, the user can see a short animation showing the results of pushing the wrong button, using the incorrect equipment, or putting the wrong chemical in a test tube.
- You can make instructional videos using a variety of tools that are supported by Windows, Macs, or Linux. These include WebEx, Camtasia, Captivate, CamStudio, iShowU, SnapzPro, Wink, and Encorder (Tietjen; Sharp).

Standard Operating Procedure (SOP)

5. Follow criteria for writing standard operating procedures (SOPs).

A *standard operating procedure* (SOP) is a set of written instructions that documents routine or repetitive technical or administrative activities followed by business and industry.

Reasons for Writing an SOP

SOPs ensure accurate job performance and consistent quality and integrity of the end product. SOPs also ensure that governmental regulations are followed. SOPs must be reviewed and enforced by management. Employees need to reference SOPs for accuracy of procedures, so SOPs must be available as hardcopy documents and in electronic formats. Annual reviews of SOPs by both management and employees are necessary to ensure that the SOPs are being followed.

SOPs are used in many fields, including science, healthcare, biomedical technology, government, military, computer industry, and others. These industries use SOPs for the following reasons:

- Calibrate and standardize instruments
- Collect lab samples
- Handle and preserve food
- Analyze test data
- Troubleshoot equipment, machinery, and procedures
- List mathematical steps to follow for acquiring data and making calculations
- Assess hardware and software analytical data

Components of SOPs

SOPs contain many of the same components as user manuals already discussed in this chapter, such as procedural steps. However, several of the parts of an SOP are different. Figure 15.5 shows the key components of an SOP.

FIGURE 15.5   Key Components of SOPs

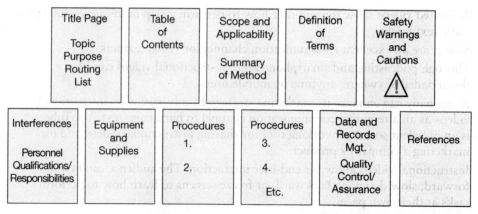

Title Page. Preface your SOP with a title page that consists of the topic about which you are writing, the date of the SOP, the purpose of the SOP, and the routing list of people who must sign off on this procedure.

◀ EXAMPLE

Standard Operating Procedure

January 14, 2016
Preparing and Processing Algae Samples to Ensure Proper pH

Author: _____
Date Approved: _____
Manager: _____
Date Approved: _____
Quality Assurance Manager: _____
Date Approved: _____

Scope and Applicability. This section of the SOP can be limited to approximately three to five sentences. However, some SOPs, due to the complexity of the subject matter, can have scope and applicability sections that are several paragraphs long. This section of the SOP provides an overview for the audience. To do so, explain why the SOP has been written, show how it meets regulatory requirements, specify any limits for the use of the procedure, and state the applicability of the procedure.

◀ EXAMPLE

Scope

The purpose of this SOP is to establish uniform procedures for water compliance inspections (WCIs) performed by the Georgia Science and Assessment Division (GSAD). The WCIs evaluate the effectiveness and reliability of the state agency's inspection procedures to meet regulations of the Clean Water Act (CWA). The SOP will ensure thoroughness of all water compliance inspections and reports. The inspector may alter SOPs due to unexpected or unique problems in the field. Deviations from the SOP must be reported.

Applicability

The policies and procedures of the SOP apply to all personnel who take part in the WCIs.

Summary of Method. In this brief summary of the procedure (three to five sentences or short paragraphs), you might focus on any of the following, determined by the topic of the SOP:

- Title of people involved
- Their roles and duties
- Sequence of their involvement
- Sequence of activities performed

◀ EXAMPLE

Summary of the Method

During the WCI, the state inspector is the lead person supervising all tests. The state inspector manages a staff of scientists who gather water samples either manually or automatically with a portable compositor. Once the samples are collected, the state scientists divide the collection into containers, preserve the samples in ice, and ship the collection to a state laboratory for analysis. The laboratory technicians compare

---

the test results to acceptable limits to determine compliance. The state inspector must sign off on all parts of the procedure.

Once these tests are completed, the WCI inspector and the United States EPA inspector discuss procedures, findings, variances, and corrective actions. The EPA inspector fills out the Water Compliance Overview Inspection Checklist. Both inspectors are required to sign this form, which is kept on file in the state and national archives.

Interferences.   Interferences include any components of the process that may interfere with the accuracy of the final product. This can include but is not limited to the following:

- Humidity
- Temperature
- Depth
- Altitude
- Weather
- Cleanliness
- Carelessness
- Sample size
- Contamination

EXAMPLE ▶

Interferences

If the water has high concentrations of chlorine, this can interfere with test results. Therefore, if the inspector determines that chlorine is present, add sodium sulfide to the sample bottle before autoclaving.

Personnel Qualifications/Responsibilities.   This section lists the required experience and certifications of the individuals performing the SOP. These qualifications can include education, years of experience, courses taken, continuing education hours completed, and more.

EXAMPLE ▶

Personnel Qualifications/Responsibilities

This SOP is written specifically for water compliance inspections (WCIs) performed by the Georgia Science and Assessment Division (GSAD). All personnel who perform this activity must have the Basic Inspector Training for Water Compliance certification and have taken the state-required, 8-hour continuing education class. Additional training qualifications include GSAD regulation certification as well as on-the-job training.
  Personnel performing this SOP have the following responsibilities:

1. WCI inspector—this person must have a minimum of a Bachelor of Science in chemistry (or related discipline), five years field work experience, three years laboratory work, and must be certified in state water compliance regulations. This inspector is responsible for coordinating staff and conducting the WCI test to meet deadlines. The inspector must be able to enforce the policies and procedures outlined in this SOP.
2. Scientists and laboratory technicians—these employees must have a minimum of an Associate of Science in chemistry, biotechnology, or a related field. These employees also must have passed state-level training in water compliance regulations. Scientists and laboratory technicians must meet deadlines for the WCI, submit all samples in accordance with regulatory policies, and adhere to the SOP.

Equipment and Supplies. This section can include the following items necessary to perform the SOP:

- Tools
- Equipment
- Reagents needed for the procedure
- Standards for the chemicals
- Biological specimens

If the steps in the process break down into multiple sections, you can divide the equipment and supplies into the units in a chronological pattern of organization. Remember to include quantities of items if needed to perform different tests in the process.

◀ EXAMPLE

Equipment and Supplies
1.1 Sample Collection
    1.1.1 Sterile sample jars
    1.1.2 Syringes with carbon filters
    1.1.3 Sterile screw-top lids
1.2 Lab Analysis
    1.2.1 Air incubator
    1.2.2 Water bath
    1.2.3 Disposable sterile pipettes
    1.2.4 Sterile applicator sticks
    1.2.5 Sterile culture tubes
    1.2.6 Sterile screw-cap tops
    1.2.7 Pyrex graduated cylinders
    1.2.8 Culture tube racks

Data and Records Management. Include the following information in this unit of the SOP:

- Calculations to be performed during the procedure
- Forms for the reports
- Required reports
- Reporting intervals
- Report recipients
- Process to follow for recording and storing data and information generated by the SOP

◀ EXAMPLE

Records Management
1. Appropriate SOPs will be kept in the laboratory library in black binders. All SOPs will be available to management and employees. When an SOP is removed from the library, sign your name and add the checkout date on the posted SOP inventory sheet.
2. No SOPs can leave the laboratory area.
3. The laboratory director will update the SOP binders as procedures or regulations change.
4. Staff will read all revised SOPs within five working days and sign and date the updated SOP list.

Quality Control and Quality Assurance. Quality control activities help you ensure that you are checking for the highest quality and consistency of the procedure.

- Explain how you will verify your work through quality control. What self-checks will you implement? Will you retest, recount, or recalibrate findings?
- Include the specific process to follow for retesting the procedure, recounting to determine the performance level of the procedure, or recalibrating the machinery.
- Include the self-check intervals for retesting, recounting, or recalibrating. Detail the procedure for dealing with divergent results found through the self-checks.
- State how and to whom you will report the results of the quality control activities.

EXAMPLE ▶

> *Quality Control*
>
> To achieve quality assurance, the laboratory will follow this procedure:
>
> - Sterility—each sample is incubated for 24 hours at 0.5°C and examined for contaminants. If a sample tests positive for contaminants, the entire lot is discarded.
> - pH—if the pH does not meet manufacturer's specifications prior to autoclaving, pH can be adjusted according to EPA specifications. If the pH does not meet specifications after autoclaving, the entire sample must be discarded.
> - Testing—the laboratory will test samples twice a year (January and June).

References. In this unit, to avoid plagiarism and communicate ethically, cite the documents or procedures used in or referred to in your SOP. You should include documentation of additional SOPs referenced, literature used for research, and any additional instruction manuals used for reference. Attach copies of the documents if they are not readily available to your potential audience.

---

6. Test manuals or standard operating procedures for usability.

Test for Usability

You've written the instructions or SOP, but is it usable? If the instructions *don't* help your audience complete the task, what have you accomplished? To determine the success of your manual, test the usability of the instructions or SOP as follows:

1. **Select a test audience.** The best audience for usability testing would include a representative sampling of individuals with differing levels of expertise. A review team composed only of high-tech readers would skew your findings. In contrast, a review team consisting of high-tech, low-tech, and lay readers would give you more reliable feedback.
2. **Ask the audience to test the instructions.** The audience members would attempt to complete the instructions, following the procedure step by step.
3. **Monitor the audience.** What challenges do the instructions seem to present? For example, has the audience completed all steps easily? Are the correct tools and equipment listed? Are terms defined as needed and presented where they are necessary? Has the test audience abided by the hazard notations? Are any steps overloaded and, thus, too complicated? Has each step been explained specifically?
4. **Time the team members.** How long does it take each member to complete the procedure? More important, why has it taken some team members longer to complete the task?
5. **Quantify the audience's responses.** Once your test audience has completed the procedure, debrief these individuals to determine what problems they encountered.

Use the Instructions, User Manual, and SOP Usability Checklist to help gather quantifiable information about the instruction's usability.

Sample Instructions

See Figure 15.6 for a sample of instructions.

FIGURE 15.6   Viewing Photos on Your Smartphone

How to View Your Photos on a Smartphone
You want to send a photo to a friend or show a photo to an acquaintance. How can you accomplish these goals with your smartphone? You easily can view photos that you've taken with your smartphone's built-in camera or saved from an e-mail or text message as follows:

 Caution: To avoid damaging your smartphone or your photos, please read all instructions in this guide carefully or contact us at www.smartphone.com/support/instruction for additional help.

(Continued)

FIGURE 15.6 *(Continued)*

1. From your smartphone's homepage, tap on the Gallery icon to access your saved photos.

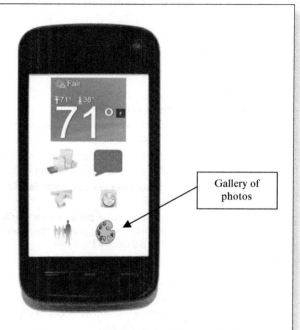

Gallery of photos

2. Once you've accessed your photo gallery, tap on the photo of your choice to see the photo full screen.

(Continued)

FIGURE 15.6 *(Continued)*

3. Zoom in or out of the photo by pinching your finger on the screen.

4. View the photo in either landscape or portrait orientation by rotating your smartphone. The photo will reorient automatically to fit the screen.

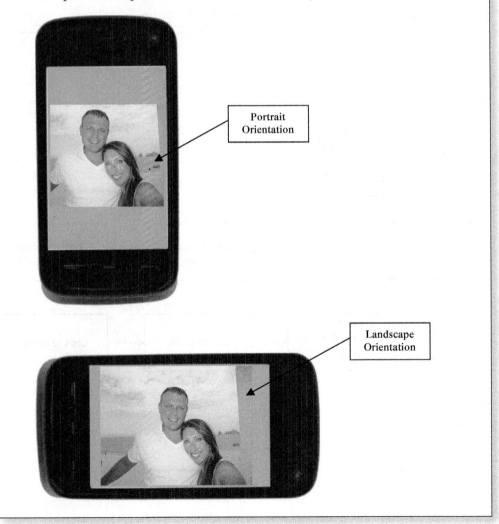

The Writing Process at Work

Sloane Harken, documentation specialist, is participating in an in-house training conference for her company's employees. Sloane says, "My role in this conference is to create 15 training manuals." To write an instruction manual, Sloane gathered data, determined objectives and sequenced the instruction through prewriting, drafted an instruction in writing, and finally revised the draft.

Prewriting

Flowcharts chronologically trace the stages of an instruction, visually revealing the flow of action, decision, authority, responsibility, input/output, preparation, and termination of process. Flowcharting is not just a graphic way to help you gather data and sequence your instruction, however. It is two-dimensional writing. It provides your reader content as well as a panoramic view of an entire sequence.

Figure 15.7 illustrates Sloane's flowcharting to gather data and to arrange it chronologically.

FIGURE 15.7 Flowchart Showing How to Share Google Docs

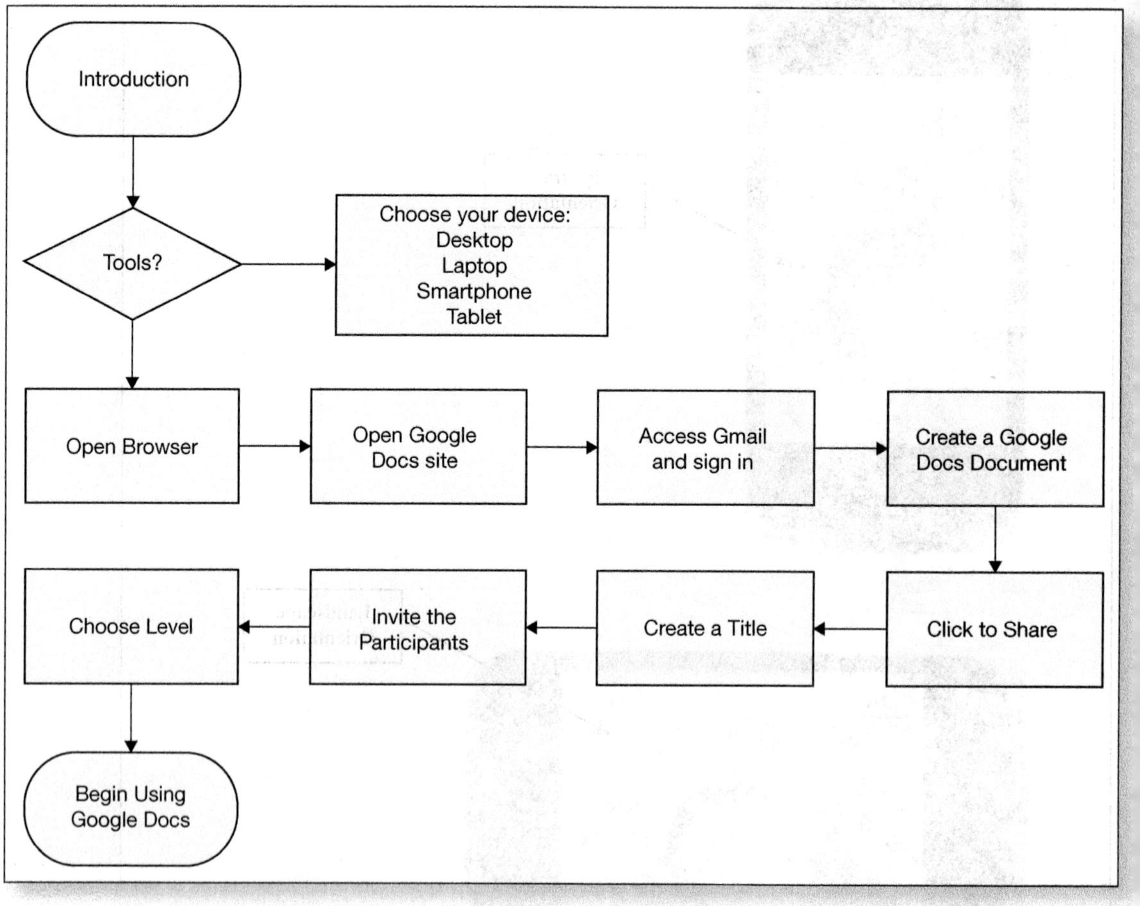

Writing

After completing the prewriting, Sloane composed and reviewed her rough draft (Figure 15.8).

FIGURE 15.8   Sloane's Instructional Rough Draft with Commentary

Google Docs

Hazard
Be careful about viruses.
Only give permission to share with individuals you know.

Introduction
I'll tell the employees why using Google Docs is good.

Required Tools and Equipment
They can work from their desktops, laptops, smartphones, tablets, anywhere.

Steps for Sharing on Google Docs

- Open browser and type docs.google.com search bar/press enter.
- Then access Gmail account.
- Sign in.
- Next click to move forward.
- Create a document.
- Click the share button.
- Now create a title/click Save.
- Invite team members.
- Choose level of access.
- "Notify people via e-mail."
- Finally click Send.

This will let them successfully share document and collaborate.

After I reviewed my rough draft, I knew I needed to add a title page and a welcoming introduction.

I added unnecessary transitional words, such as "then," "next," "now," and "finally." I'll remove these words since the numbered steps will act as transition.

My first step in the rough draft is actually three, separate steps. The use of the bullets confused my test audience, so I'll revise that setup.

Last, I'll add screen captures to help my audience understand the procedure. Plus, I'll make this user manual in PowerPoint for easier presentation.

Rewriting

Sloane revised the rough draft. The revision is shown in Figure 15.9.

FIGURE 15.9   How to Share Google Docs

(Continued)

FIGURE 15.9 *(Continued)*

Introduction

Thank you for using Google Docs. With Google Docs, you can

- create text
- share your writing
- allow your writing to be edited by others
- ensure that original versions of your text are saved

Required Tools/Equipment

To use Google Docs, all you'll need is Internet-enabled devices like a

- Desktop computer
- Laptop
- Smartphone
- Tablet

You'll also need an Internet connection via Ethernet or WiFi and a Gmail account.

Steps for Sharing on Google Docs

1. Open your browser.

2. Type docs.google.com in your search bar and press enter.

3. Access your Gmail account by typing in your Gmail address and password.

Steps for Sharing Google Docs

4. Click Sign in.

Steps for Sharing Google Docs

5. Click the X to move forward *or* take the tour for further details.
6. To create a document, click the plus symbol (+) at the bottom right corner of the page.

Steps for Sharing Google Docs

7. Click the share button, located on the top right corner of the page.

Steps for Sharing Google Docs

8. Create a title for your document and click Save.

9. Type the e-mail addresses of your team members who will share your document.

Steps for Sharing Google Docs

10. Click on the down arrow by "Can edit" to choose the level of access for your user(s). These include "Can edit," "Can comment," and "Can view."

Sloane says, "I knew I had succeeded when my boss said, 'Thanks so much for your hard work, Sloane. We've gotten great reviews of your manuals.' He also said they had been hosting these annual conferences since the company began five years ago, and he had never heard such positive comments about any of the manuals.

Because I have only been with the company for a few months, his comments were particularly important to me. They let me know that I was doing my job and that prewriting, writing, and rewriting all the documents was really paying off."

(Continued)

FIGURE 15.9   (Continued)

Steps for Sharing Google Docs	Contact
11. Click the blue Send button. Congratulations! You have successfully shared your Document and can now collaborate on Google Docs . . . in real time!	**Google Inc.** 1600 Amphitheatre Parkway Mountain View, CA 94043 USA Telephone: +1 650 253 0000 Fax: +1 650 253 0001

APPLY YOUR KNOWLEDGE

CASE STUDIES

1. PhlebotomyDR provides instructions for its staff to ensure cleanliness and employee protection. These include hand washing procedures and the correct use of sterile equipment, such as gloves, masks, aprons, and shoe coverings.

   Following is a rough draft of one set of instructions for personnel safety. Revise the rough draft according to the criteria provided in this chapter. Correct the order of information, the grammar, and the instruction's content. In addition, improve the instruction by including appropriate graphics.

Hand washing
- Lather hands to cover all surfaces of hands and wrists.
- Wet hands with water.
- Rub hands together to cover all surfaces of hands and fingers. Pay special attention to areas around nails and fingers. Lather for at least 15 seconds.
- Dry thoroughly.
- Rinse well with running warm water.
- Avoid using hot water. Repeated exposure to hot water can lead to dermatitis.
- Use paper towels to turn off faucet.

Gloves
- Replace damaged gloves as soon as patient safety permits.
- Don gloves immediately prior to task.
- Remove and discard gloves after each use.

Masks
Wear masks and eye protection devices (goggles or eye shields) to avoid droplets, spray, or splashes and to prevent exposure to mucous substances. Masks are also worn to protect nurses, doctors, and technicians from infectious elements during close contact with patients.

Aprons and Other Protective Clothing
- Wear aprons or gowns to avoid contact with body substances during patient care procedures
- Remove and discard aprons and other protective clothing before leaving work area.
- Some work areas might require additional protective clothing such as surgical caps and shoe covers or boots.

2. BurgerNet Online Order System is a software product in development. As a professional communicator, write the instruction to accompany this online order system.

BurgerNet Online Order System

1. **Description.**
   The BurgerNet application will be used to order burgers for delivery to a customer's home or work.

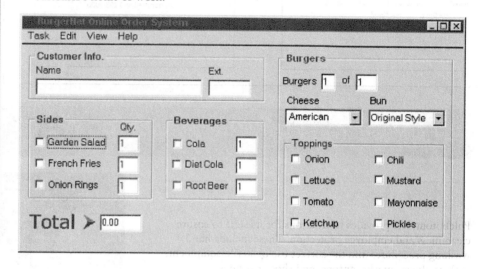

2. **Dropdown Menu Bar.** The dropdown menu bar includes a Task, Edit, View, and Help menu.
   - Task menu. On the "Task" menu customers will find a "Send Order" command, used to electronically send the completed order to the restaurant. The "Print Order" command on the "Task" menu lets the customer print a copy of the order. The "Exit" command terminates the order session.
   - Edit menu. On the "Edit" menu, an "Add Favorites" command allows customers to add a frequently ordered combination of food items to make future online ordering quick and easy. The "Edit Favorites" command lets customers modify an existing favorite order.
   - View menu. On the "View" menu, the "Preferences" command is used to modify default customer. In addition, when favorites have been defined, they are listed under "View." Customers can select a favorite for display before sending the order.
   - Help menu. On the "Help" menu, customers can access FAQs.

3. **Key Components of BurgerNet.**
   - Customer Info. frame: Customers type in their name and telephone number in the "Name" and "Extension" fields.
   - Burgers frame. In the "Burgers" fields customers specify the number of burgers desired. The "Cheese" dropdown list lets the customer select either American cheese, Pepperjack, Mozzarella, or Swiss cheese to top their burgers. "Bun" lets customers choose either plain or onion buns from the dropdown list.
   - Toppings. Here, customers check what to put on their burgers.
   - Sides: Select which side dish or dishes and specify the quantity for each.
   - Beverages: Select drink(s) and specify the quantity for each.
   - Total field. The total cost of the customer's order is automatically displayed.

ETHICAL CHALLENGE

Your company is releasing a new version of a software application. Your supervisor says, "There aren't that many changes from the prior version. I don't think we really need to write a new user manual. Let's just use our old manual." You agree that the new version only has minor changes, but the changes are significant. They include new hazards, new hardware requirements, and steps that must be completed in the specified order to initialize the software. You have concerns about the lack of clarity in the online user manual if a new version is not written. However, to revise the manual would delay release of the application, plus you have to consider your supervisor's opinion and concerns about budget.

Question

Which ethical considerations would you reference to discuss your concerns with your supervisor?

INDIVIDUAL AND TEAM PROJECTS

1. Write an instruction or SOP. To do so, first select a topic. You can write instructions telling how to monitor, repair, test, package, download music or videos, play online games, clean, operate, maintain, troubleshoot, install, use software, and so on. Choose a topic from your field of expertise or one that interests you. Follow the writing process techniques to complete your instructions. Prewrite (using a flowchart), write a draft (abiding by the criteria for instructions or SOPs presented in this chapter), and rewrite to perfect your text.

2. Find examples of instructions, user manuals, or SOPs written in the work environment. Bring these to class. Using the criteria for instructions and SOPs presented in this chapter, decide whether the instructions and SOPs are successful or unsuccessful. If the instructions and SOPs are good, show how and why. If they are flawed, explain the problem(s). Then rewrite the instructions or SOPs to improve them.

3. Good writing demands revision. Following is a flawed instruction. To improve it, rewrite the text, abiding by the criteria for instructions and the rewriting techniques included in this chapter.

Date: November 1, 2016
To: Maintenance Technicians
From: Second Shift Supervisor
Subject: Oven Cleaning

The convection ovens in kiln room 33 need extensive cleaning. This would consist of vacuuming and wiping all walls, doors, roofs, and floors. All vents and dampers need to be removed and a tack cloth used to remove loose dust and dirt. Also, all filters need replacing. I am requesting this because when wet parts are placed in the ovens to cure the paint, loose particles of dust and dirt are blown onto the parts, which causes extensive rework. I would like this done twice a week to ensure cleanliness of product.

PROBLEM-SOLVING THINK PIECE

Read the following instructional steps. Are they in the correct chronological order? How would you reorder these steps to make the instruction more effective?

> **Changing Oil in Your Car**
>
> Run the car's engine for approximately ten minutes and then drain the old oil.
> Park the car on a level surface, set the parking brake, and turn off the car's engine.
> Gather all of the necessary tools and materials you might need.
> Open the hood.
> Jack up and support the car securely.
> Place the funnel in the opening and pour in the new oil.
> Replace the cap when you have finished pouring in new oil.
> Locate the oil cap on top of the engine and remove the cap.
> Tighten the plug if you find leakage.
> Run the engine for a minute, then check the dipstick. Add more oil if necessary.
> Pour the used oil into a plastic container and dispose of it safely and legally.

WEB WORKSHOP

1. Review any of the following Web sites' online instructions. Based on the criteria provided in this chapter, are the instructions successful or not?
   - If the answer is yes, explain why and how the instructions succeed.
   - If the answer is no, explain why the instructions fail.
   - Rewrite any of the flawed instructions to improve them.

Web Sites	Topics
Home Tips World.com	Cleaning, home improvement, patio, landscaping, and garden
Hammerzone.com	Kitchen projects, tubs, sinks, toilets, showers, and water heaters
Dmoz.org	Links to step-by-step procedures for painting, welding, soldering, plumbing, walls, windows, and door repair and installation
Quakerstate.com/	Instructions for car care

2. The U.S. Department of Labor article "Hazard Communication: A Review of the Science Underpinning the Art of Communication for Health and Safety" focuses on many important aspects of writing instructions, including the use of icons, readability, and audience variables. Access the article and report your findings either orally or in a memo, letter, or e-mail message.

3. YouTube provides thousands of instructional videos for topics including creating Origami shapes, playing World of Warcraft, playing pool and billiards, snowboarding and paragliding, playing guitars, tossing pizzas, using Macs and PCs, installing software and hardware, and downloading music for iPods. Research any of these videos (or others). Based on the criteria provided in this chapter, assess your findings. Were the videos successful? How could they be improved? Share your findings with your teacher in an e-mail message or present your findings to your class in a brief oral presentation.

CHAPTER SIXTEEN

Research

After completing this chapter, you will be able to

1. Learn reasons for writing research reports in business

2. Distinguish between primary and secondary research sources

3. Perform secondary research using the communication process

4. Prewrite using a variety of research techniques

5. Research using traditional sources

6. Use the Internet to research source material

7. Take notes by quoting, paraphrasing, and summarizing

8. Write the research document by organizing effectively

9. Document sources to avoid plagiarism

10. Rewrite the researched document for correctness and style

The Oak Springs, Iowa, scenario illustrates the use of primary and secondary research before writing construction plans for a new road.

The **City of Oak Springs, Iowa,** needs to improve a 10-mile stretch of road, Ironhorse Avenue, that runs east and west through the town. The winding, two-lane road was built in 1976. Since then, the city has grown from a population of 45,000 to 86,000. In 1976, the road primarily served farmers whose houses were separated by miles of privately owned land. Now, over 40 years later, many of the landowners have sold their land for development. Ironhorse Avenue is occupied by numerous housing subdivisions, apartment complexes, and strip malls. Due to this development, the road no longer is sufficient for the increased traffic load.

The city also added a middle school with access directly off Ironhorse Avenue. This middle school draws students from a two-mile radius including a new subdivision to the west of Ironhorse Avenue. The arrival at school of students in the morning and their departure in the afternoon necessitated the addition of pedestrian crosswalks to Ironhorse Avenue in two locations. Each crosswalk is patrolled morning and afternoon by a crossing guard. Traffic on the street slows measurably during both times of the day. Because Ironhorse Avenue is the main artery to the downtown shopping and business areas, the city has to make certain that traffic flows steadily and safely.

Ironhorse Avenue must be expanded to four lanes and straightened. An expanded and straightened street will allow the city to provide better snow removal; quicker response time to emergencies by fire, police, and medical personnel; easier travel time for commuters; and safer travel for pedestrians taking into consideration middle school students. To accomplish these goals, the city must add new sewer lines, sidewalks, easements, esplanades, pedestrian crosswalks, bicycle lanes, and lighting. However, this construction will impact current homeowners (whose land, through eminent domain, can be expropriated without the owner's consent). In addition, part of the envisioned road construction will impact a wildlife refuge for waterfowl. This wildlife refuge is important to the ecosystem and is also used for field research by Oak Spring's schools' science classes.

Before construction plans can be made, city engineers must conduct research in order to determine needs and considerations of the public. First, the engineers will perform primary research by interviewing city residents, as well as employees from the city's parks and recreation, police, fire, and transportation departments. Then, the engineers will conduct secondary research. They will read statutes regarding eminent domain, real estate, environmental considerations, state wildlife refuges, zoning, planning, and land use.

Only after conducting primary and secondary research will the engineers be able to produce a construction plan for Oak Springs' city management for their consideration and approval.

---

1. Learn reasons for writing research reports in business.

Why Conduct Research?

Research skills are important in your school or work environment. You may want to perform research to better understand a technical term or concept; locate a magazine, journal, or newspaper article for your supervisor; or find data on a subject to prepare an oral or written report. Technology is changing so rapidly that you must know how to do research to stay up to date.

You can research information using online catalogs; electronic indexes and databases; reference books in print and online; and by using Internet search engines and directories. Reference sources vary and are numerous, so this chapter discusses only research techniques.

Research, a major component of long, formal reports, helps you develop your report's content. Often, your own comments drawn from personal experience will lack sufficient detail, development, and authority to be sufficiently persuasive. You need research for the following reasons:

- Create content
- Support commentary and content with details, quotes, paraphrases, and summaries
- Emphasize the importance of an idea
- Enhance the reliability of an opinion
- Show the importance of a subject to the larger business community
- Address the audience's need for documentation and substantiation

Research Including Primary and Secondary Sources

2. Distinguish between primary and secondary research sources.

You can use researched material to support and develop content in your formal reports. Researched material generally breaks down into two categories: primary and secondary research. All primary and secondary research requires parenthetical source citations (discussed in Appendix B).

Primary Research and Field Research

Primary research is research performed or generated by you. You do not rely on books or periodicals for this type of research. With primary research, you will be generating the information based on data or information from a variety of sources that might include observations, tests of equipment, and laboratory experiments. Another type of primary research is called *field research*. This can entail the following:

Proposals and Researched Reports

See Chapter 18, "Long, Formal Reports," and Chapter 19, "Proposals and Business Plans," for sample research papers and reports.

- Preparing a survey or a questionnaire targeting a group of respondents
- Conducting face-to-face interviews
- Conducting participant observations
- Interviewing individuals through telephone calls and online surveys
- Using social media, such as Twitter, Facebook, and LinkedIn, to collect information
- Visiting job sites

You may perform this field research to determine for your company the direction a new marketing campaign should take, how to use social media in the workplace, the importance of diversity in the workplace, the economic impact of relocating the company to a new office site, the usefulness of a new product, or the status of a project.

You also may need to interview people for their input about a particular topic. For example, your company might be considering a new approach to take for increased security on employee computers. You could ask employees for a record of their logs which would highlight the problems they have encountered with their computer security.

Surveying for Primary Research. If you choose to use a survey for your primary research, consider doing the following to get other people's assistance:

- Schedule a convenient time for the interview.
- Explain why you need the interview and information.
- Explain how you will use the information.
- Consider performing a test survey with one or two people so you can determine if the questions are written clearly.
- Tell the survey participants approximately how long the survey should take.
- Come prepared. Research the subject matter so you will be prepared to ask appropriate questions. Write your interview questions or the survey before you meet with the person. Take the necessary paper, forms, pencils, pens, laptop, electronic notepads and tablets, smartphones, or recording devices you will need for the meeting.
- Create survey questions that allow for quantifiable answers. In other words, try to avoid simple yes or no questions. Give survey participants a choice from a value-based scale, such as 4 = absolutely critical; 3 = important; 2 = somewhat important; 1 = not important.
- Ask the survey participant(s) for permission to use their responses in your research. You may need to ask the participant(s) to sign a release form.
- Thank the participant(s) for the interview and taking time to fill in the questionnaire.

Questionnaires

See Chapter 19, "Proposals and Business Plans," for a sample survey.

Figure 16.1 is a sample survey. An information technology manager used this primary research to substantiate his long, formal report recommending that the company switch from a Microsoft Windows operating system to a Linux operating system.

FIGURE 16.1   Survey Used to Conduct Primary Research

User Survey on Computer Operating Systems

Our company is considering migrating from a Windows-based computing system to a Linux-based system. To help us make this decision, please answer the following questions, using a four point scale.

1. Would you like to customize your desktop, choosing your own GUI links to computer operations (graphical images that you choose as iconic buttons to click on)?

   _____ 4 = absolutely critical

   _____ 3 = important

   _____ 2 = somewhat important

   _____ 1 = not important

2. Your DOS commands in Windows are a "text mode interface." Are you comfortable with our current system that relies on limited commands, or would you prefer a text mode interface that allows for multiple commands and is customizable?

   _____ 4 = absolutely critical

   _____ 3 = important

   _____ 2 = somewhat important

   _____ 1 = not important

3. Windows impacts your departmental budget because Microsoft allows only a single copy of Windows to be used on only one computer. In contrast, if we purchased a Linux system, you can run it on any number of computers for no additional charge. Such a purchase would require training to help computer users learn the new system. Realizing that a change in systems would necessitate training, how important is this budgetary issue to your department?

   _____ 4 = absolutely critical

   _____ 3 = important

   _____ 2 = somewhat important

   _____ 1 = not important

4. Windows comes with an operating system. Linux-based machines require that consumers install the operating system. Our company would provide you the software and hardware; an individual or individuals within your department would be required to perform this installation. Would this requirement make you vote *against* a purchase of a Linux system?

   _____ 4 = absolutely critical in a vote of "no"

   _____ 3 = important in a vote of "no"

   _____ 2 = somewhat important in a vote of "no"

   _____ 1 = not important—our department would not vote "no" due to this requirement

---

(Continued)

FIGURE 16.1 *(Continued)*

5. Has spyware or malware become a problem in your work environment?

_____ 4 = absolutely critical

_____ 3 = important

_____ 2 = somewhat important

_____ 1 = not important

Technology has made it easy to create and conduct surveys. Many online sites, such as SurveyMonkey, Surveydaddy, Zoomerang, Free Online Surveys, and Kwik Surveys, help you design and customize e-mail or Web site forms, polls, and surveys.

Using Social Media for Research. If you can't perform fieldwork, try to network as a source of research. Social media can supplement your research by telling you what the large base of Facebook, Twitter, LinkedIn, and other social media users are thinking and saying. Social media provides an almost instantaneous focus group in the following ways:

1. **Monitor customer comments.** Sites like SocialMention, uberVu, and Spiral16 can help companies monitor what the public is saying about products and services. If you don't want to use a secondary source for this information, you can access blog sites' comments sections to learn what people are thinking.

2. **Pose research questions.** Use social media for crowd sourcing, getting information from the general public. Companies, using corporate social media sites like Facebook or Twitter, can ask their customers precise questions about products and services. Then, when customers respond, these answers can be collected, collated, and analyzed. Social media allows you to create almost synchronous surveys.

3. **Develop infographics.** Infographics are tables and figures used to help audiences visualize data. You can use a social media site like Delicious.com to search for content. Then, after drilling down into topics of your choice, you can access statistics, usage, visuals, and more to help you support research contentions.

4. **Research the latest news.** Access breaking news by using Twitter's hashtags. By clicking on a hashtagged (#) word in a tweet, you can access other tweets in that category or topic. The hashtag, community-driven metadata, provides you additional, immediate content for your research project (Messina).

5. **Get quotes to support your ideas.** You can access Twitter Search. Enter the hashtag #quotes or #quote and a keyword to find pertinent quotes.

6. **Access academic and research blogs.** Social media doesn't have to depend solely on Facebook or Twitter. Many academic research blogs exist related to specific disciplines, such as science, archeology, and history (Gray). Following are a few to look at:

Blog Site
Finds and Features
Love of History
MicrobiologyBytes
Research Blogging
Science in the Open
Science of the Invisible
Stanford Blog Directory
Starting Out in Science

Secondary Research

When you conduct secondary research, you rely on already printed and published information taken from sources including books, periodicals, newspapers, encyclopedias, reports, proposals, or other business documents. You might also rely on information taken from a Web site, blog, or YouTube instructional video. For a detailed discussion of how to conduct research, see the next unit on researching secondary sources using the communication process.

3. Perform secondary research using the communication process.

Communication Process

See Chapter 3, "The Communication Process," Chapter 18, "Long, Formal Reports," and Chapter 19, "Proposals and Business Plans," for sample research papers and reports.

Performing Secondary Research Using the Communication Process

Writing a research report, as with other types of correspondence, is easiest when you follow a process. Prewrite (determine objectives and conduct research to gather your data), write a draft, and rewrite to ensure that you meet your goals successfully.

4. Prewrite using a variety of research techniques.

Prewriting Research Techniques

Use these techniques to help you prewrite for documents that require research.

Select a General Topic (or the Topic You Have Been Asked to Study)

Your topic may be a technical term, phrase, innovation, or dilemma. If you're in biomedical technology, for example, you might want to focus on biohazards in the workplace. If you work in electronics, select a topic related to nanotechnologies. If your field is computer science, you could focus on developments in online gaming software.

Spot-Check Sources of Information

Check a library or online sources to find material that relates to your subject. A quick review of your library's online periodical databases, such as Research Library from ProQuest or other equivalent sources, will help you locate periodical articles. Your library will have an electronic edition of the *Reader's Guide to Periodical Literature* and other similar specialized periodical indexes. A keyword search of Internet metasearch engines will give you an idea of how much information will be available through the Internet.

Establish a Focus

After you have chosen a topic for which you can find available source material, decide what you want to learn about your topic. A focus statement can guide you. In other words, if you are interested in biomedical technology, you might write a focus statement such as the following:

EXAMPLE ▶

> I want to research current problems with biohazards in the workplace, including crisis management, disposal of hazardous materials, and training for employees.

For electronics, you could write the following:

EXAMPLE ▶

> I want to research the uses, expenses, and maintenance of nanotechnology.

If you are in computer science, you might write the following:

EXAMPLE ▶

> I want to discover the uses, impact on employment, and expenses of software for online gaming.

With focus statements such as these, you can begin researching your topic, concentrating on articles pertinent to your topic.

Research Your Topic Using Traditional Sources

5. Research using traditional sources.

You may feel overwhelmed by the prospect of research. But there are many sources that, once you know how to use them, will make the act of research less overwhelming.

Books

All books owned by a library are listed in online catalogs. Books can be searched in online catalogs in a variety of ways: title, author, subject, keyword, or using some combination of these.

Periodicals

Use electronic or print periodical indexes to find articles on your topic. Online indexes can be searched in a variety of ways: title, author, subject, keyword, or using a combination of these.

Indexes to General Popular Periodicals.   Three leaders in providing periodical material in a general database are Gale/Cengage's Academic OneFile, Ebsco's Academic Search Premier, and ProQuest's Research Library. All are general resources in that they present material on a wide range of subjects or disciplines. However, each of the three presents this content from scholarly/professional sources, popular magazines, and from news sources. The content can be segmented to represent these categories of periodicals. There are electronic and print counterparts for some of these. The electronic indexes provide the full text of many of the articles.

◀ EXAMPLE

- *Reader's Guide to Periodical Literature*
- *Research Library from ProQuest*
- *Academic Search Premier*
- *Newsbank*

Indexes to Specialized, Scholarly, or Technical Periodicals.   Many libraries provide access to one or more specialized indexes covering the literature of a particular subject area. There are electronic and print counterparts for most of these. The following electronic indexes provide the full text of many of the articles.

- **Applied Science & Technology Index.** Covers engineering, aeronautics and space sciences, atmospheric sciences, chemistry, computer technology and applications, construction industry, energy resources and research, fire prevention, food and the food industry, geology, machinery, mathematics, metallurgy, mineralogy, oceanography, petroleum and gas, physics, plastics, the textile industry and fabrics, transportation, and other industrial and mechanical arts
- **Business Periodicals Index.** Covers major U.S. publications in marketing, banking and finance, personnel, communications, computer technology, and so on
- **ABI/Inform.** Covers business and management
- **General Science Index.** Covers the pure sciences, such as biology and chemistry
- **Social Sciences Index.** Covers psychology, sociology, political science, economics, and other social sciences topics
- **ERIC (Education Resources Information Center).** Provides bibliography and abstracts about educational research and resources. Available for free through the Internet

- **PubMed.** Covers medical journals and allied health publications
- **PsycINFO.** Covers psychology and behavioral sciences
- **LexisNexis.** Includes the full text of newspaper articles, reports, transcripts, law journals, and legal reporters and other reference sources in addition to general periodical articles

> 6. Use the Internet to research source material.

Use the Internet to Research Sources

Millions of documents from countless sources are found on the Internet. You can find material on the Internet published by government agencies, organizations, schools, businesses, or individuals (see Table 16.1). The list of options grows daily. For example, nearly all newspapers and news organizations have online Web sites.

TABLE 16.1 Examples of Internet Research Sources

Search Engines, Directories, Metasearch Engines	Online References	Online Libraries	Online Newspapers	Online Magazines	Online Government Sites
Yahoo!	Webster's Dictionary	Library of Congress	New York Times	National Geographic	United Nations
Excite	Roget's Thesaurus	New York Public Library	CNN	Atlantic	The White House
Lycos	Britannica Online Encyclopedia	Cleveland Public Library	USA Today	The New Republic	The IRS
AltaVista	Encyclopedia Smithsonian	Gutenberg Project	Washington Post	U.S. News Online	U.S. Postal Service
MetaSearch	The Old Farmer's Almanac	Most city and university libraries	Kansas City Star	Time Magazine	USA.gov
MetaCrawler			Most city newspapers	Ebony Online	Most states' supreme courts, legislatures, executive offices, and local governments
Google				Slate	
Britannica.com					
Ask.com					
Dogpile					

FAQs: Ethical Considerations when Evaluating Online Sources

Q: Can I quote or paraphrase from any source? Aren't all sources on the Internet equally valid?

A: Not all information, even if it's published online, is valid. Before you quote or paraphrase information, ask yourself these questions:

- What are the author's credentials? Not everyone who publishes a blog entry, a Twitter comment, a Web site, or an editorial in an online newspaper necessarily is an expert in the field. Is the author a college professor, a noted scientist, a government official, or some unknown individual? If you are uncertain about the author's background, you can perform research about the author by going to a search engine, such as Google, to find out the author's level of education, affiliations, and other publications.
- Is the source of information reputable? Not all online publications are equal. Many blog entries, Twitter

comments, Facebook pages, YouTube videos, or Web sites do not go through editorial review. Just because someone publishes a comment on their blog doesn't mean it is a valid source of information. In contrast, online journals, magazines, and newspapers generally hire professional writers, conduct research, cite sources, and verify information before it is published.

- Is the information up to date? Some information on the Internet is dated. If content is old, it might no longer be valid. Check the dates of publication to determine whether the content is still relevant and valid.
- Is the information biased? Much of what a person writes has a bias. However, some publications are more biased than others. To determine whether your source of information is biased, check to see who might have funded the research, who publishes the journal or magazine, or what affiliations the author might have.

Online Newspapers and Magazines

Many periodicals provide a current look at articles but charge for past articles or have otherwise limited offerings available on their www sites. For example, the *Kansas City Star*'s stories generally reside on kansascity.com for 7 to 14 days from the date of posting or publication. After that time, they move to the library archives. There is a fee to read stories in the archives. Although many articles are available online, you will have to access other newspaper or magazine articles in hard-copy format.

Directories

Directories, such as Yahoo!, Librarians' Internet Index, Infomine, About.com, and Google Directory, let you search for information from a long list of predetermined categories, including the following:

Arts	Government	Politics and Law
Business	Health and Medicine	Recreation
Computers	Hobbies	Science
Education	Money and Investing	Society and Culture
Entertainment	News	Sports

To access any of these areas, click on the appropriate category and then drill down, clicking on each subcategory until you get to a useful site.

Search Engines

Search engines, such as Google, Ask.com, Yahoo! search, and Exalead, let you search millions of Web pages by keywords. Type a word, phrase, or name in the appropriate blank space and press the Enter key. The search engine will search through documents on the Internet for hits, documents that match your criteria.

Metasearch Engines

A metasearch engine, such as Ixquick and Clusty, lets you search for a keyword or phrase in a group of search engines at once, saving you the time of searching separately through each search engine. In addition to these sources, you can consult the following for help: U.S. government publications and databases and your reference librarian.

Take Notes by Quoting, Paraphrasing, and Summarizing

7. Take notes by quoting, paraphrasing, and summarizing.

Once you have researched and located a source (whether it is online or a hard-copy book, magazine, journal, or newspaper article), study the material. For a book, use the index and the table of contents to locate your topic. When you have found information about your topic, refer to the pages indicated and skim, reading selectively. For shorter documents, such as magazine or journal articles or online materials, you can read closely. Reading and rereading the source material is an essential step in understanding your researched information. After you have studied the document thoroughly, go through it page by page and find content to quote, paraphrase, and summarize. This information will help you develop your ideas.

Quoting

When you quote, type the words and punctuation exactly as they appear in the original source. You cannot haphazardly change a word, a punctuation mark, or the ideas conveyed. Whether you use a quote or a paraphrase, however, you must correctly cite the source (see Appendix B). Following are three simple rules for correct quoting:

- Quote the author word for word.

EXAMPLE ▶

> "Corporate giants such as Comcast, JetBlue Airways, Whole Foods Market and others are beefing up direct communications with customers through social-media tools such as Twitter, Facebook, and YouTube" (Swartz, "Businesses Get Cheap Help," B1).

- Delete unnecessary words by using ellipses—three spaced periods (. . .). Be sure not to change the author's content or correct grammatical structure if you delete text.

EXAMPLE ▶

> "Corporate giants . . . Comcast, JetBlue Airways, Whole Foods Market . . . are beefing up direct communications with customers through social-media tools such as Twitter, Facebook, and YouTube" (Swartz, "Businesses Get Cheap Help," B1).

- Use brackets to insert words for clarity.

EXAMPLE ▶

> "Comcast, JetBlue Airways, Whole Foods Market and others [members of the Fortune 500] are beefing up direct communications with customers through social-media tools such as Twitter, Facebook, and YouTube" (Swartz, "Businesses Get Cheap Help," B1).

Paraphrasing

When you paraphrase, restate the original information using your own word and sentence structure. Note that you do not need to quote proper nouns like people's names, states, organizations, and companies.

EXAMPLE ▶

> Many large corporations including Comcast, PepsiCo, JetBlue Airways, and Whole Foods Market use social media such as Twitter, Facebook, and YouTube to provide rapid customer service (Swartz, "Businesses Get Cheap Help," B1).

Summarize

When you summarize content from an article, book, chapter, blog, or Web site, do the following:

- Provide the works cited or references information (documentation) for the material that you're summarizing.
- Begin with an introduction clearly stating the author's primary focus.
- Explain the author's primary contentions and omit secondary side issues.
- Explain the author's contentions through pertinent facts and figures while avoiding lengthy technicalities.
- Organize your discussion section according to the author's method of organization.
- Conclude the summary by reiterating the author's primary contentions, reveal the author's value judgment, or state the author's recommendations for future action.

The following two examples illustrate an original source of information from an engineering company's Web site followed by a summary of the same information.

"In the very competitive arena of site design, GBA strives for innovative, cost-effective, and functional design solutions that begin with a thorough understanding of the client's needs and goals for the project."

"Our approach may include but is not limited to consideration of traffic impact and transportation access, environmental constraints and opportunities, parking, storm water runoff, and the cost of infrastructure improvements, depending on the needs of the project. We are also experienced in the design of sustainable site features including Best Management Practices (BMPs) for Low Impact Development (LID)."

"When it comes to development, it's all about location and site accommodations. We have the ability to follow the critical path approach to obtaining planning and environmental permits, which enables our clients to start construction earlier. GBA focuses on identifying development issues early and creating an action plan to resolve these issues, which are then coordinated and executed concurrently with development of a site design package ("Site Development." GBA Architects and Engineers. 2012. Web. 14 Feb. 2012)."

GBA Architects and Engineers focuses on their clients' goals when designing affordable, practical, and unique facilities. A site development that meets client needs and construction deadlines must factor in traffic, environmental laws, sustainability, cost considerations, and the acquisition of necessary permits. How does GBA achieve these goals? They create a plan of action that can identify problems and solve construction issues in a timely and professional manner ("Site Development." GBA Architects and Engineers. 2012. Web. 14 Feb. 2012).

Isolate the Main Point to Establish Focus

After you complete the analysis of the document, isolate the main points discussed in the books or periodicals. You will find that, of the major points in an article or book, sometimes only three or four ideas will be relevant for your topic. Choose the ones discussed repeatedly or those that most effectively develop the ideas you want to pursue.

Write a Statement of Purpose

Once you have chosen two to four main ideas from your research, write a purpose statement that expresses the direction of your research. For example, one student wrote the following statement of purpose after performing research on social media.

The purpose of this report is to reveal the future of social media in business including the application of YouTube instructional videos, tweets, and Facebook pages.

Create an Outline

After you have written a purpose statement, formulate an outline. An outline will help you organize your paragraphs and ensure that you stay on track as you develop your ideas through quotes and paraphrases. Figures 16.2 and 16.3 are examples of topic and sentence outlines.

Writing the Researched Document

You now are ready to write your research report.

8. Write the research document by organizing effectively.

FIGURE 16.2 Topic Outline

I. Sensors used to help robots move
  A. Light
  B. Sound
  C. Touch

II. Touch sensor technology (microswitches)
  A. For gripping
  B. For maintaining contact with the floor

III. Optical sensors (LED/phototransistors)—Like bowling alley foul-line sensors
  A. Less bulky/connected to computer interface
  B. Not just for gripper, but for locating objects by following this sequence:
    1. Scan gripper to locate object
    2. Move gripper arm left and right to center object
    3. Move gripper forward to grasp
    4. Close gripper
  C. Problem—What force to use for gripping?

IV. Force sensors
  A. Spring and microswitch
  B. Optical encoder discs—Microprocessors determine speed of discs to determine force necessary
  C. Integrated circuits with strain gauge and pressure-sensitive paint
  D. Pressure sensors built with conductive foam

V. Conclusion

The Writing Process

See Chapter 3, "The Communication Process," for information about prewriting, writing, and rewriting.

Organize Your Report Effectively

When you are ready to write, provide an introductory paragraph, discussion (body) paragraphs, a conclusion or recommendation, and your works cited or references page(s).

Introduction.   In one to three sentences or a short paragraph, tell your readers the purpose of your long report. This purpose statement informs your readers *why* you are writing or *what* you hope to achieve.

Discussion Section, Quoting, and Paraphrasing.   The number of discussion paragraphs will depend on the number of divisions and the amount of detail necessary to develop your ideas. Use quotes and paraphrases to develop your content. Students often ask how much of a research report should be *their* writing, as opposed to researched information. A general rule is to lead into and out of every quotation or paraphrase with your own writing. In other words,

- Make a statement (your sentence).
- Support this generalization with a quotation, paraphrase, or summary (referenced material from another source).
- Provide a follow-up explanation of the referenced material's significance (your sentence).
- Cite sources for all paraphrased and quoted content to avoid plagiarism.

Conclusion/Recommendations.   In a final paragraph, summarize your findings, draw a conclusion about the significance of these discoveries, and recommend future action.

FIGURE 16.3 Sentence Outline

I. Because robots must move, they need sensors. These sensors could include light sensors, sound sensors, and touch sensors.

II. Touch sensors can have the following technology:

    A. Microswitches can be used for gripping.

    B. Microswitches can also be used for maintaining contact with the floor. This would keep the robot from falling down stairs, for example.

III. Optical sensors might be better than microswitches.

    A. LED/phototransistors are less bulky than switches.

    B. When connected to a computer interface, optical sensors also can help a robot locate an object as well as grip it.

    C. Here is the sequence followed when using optical sensors:

        1. The robot's grippers scan the object to locate it.

        2. The robot moves its gripper arms left and right to center the object.

        3. The robot moves forward to grip the object.

        4. The robot closes the gripper.

    D. The only problem faced is what force should be used when gripping.

IV. Force sensors can solve this problem.

    A. A combination spring and microswitch can be used to determine the amount of force required.

    B. Optical encoder discs can be used also. A microprocessor determines the speed of the disc to determine the required force.

    C. Integrated circuits with strain gauges and pressure-sensitive paint can be used to determine force.

    D. Another pressure sensor can be built from conductive foam.

V. To conclude, all these methods of tactile sensing comprise a field of inquiry important to robotics.

FAQs: The Ethics of Correct Source Citations and Avoiding Plagiarism

Q: Why is correct citation of sources so important?

A: It's all about avoiding plagiarism. Readers need to know where you found your information and from which sources you are quoting or paraphrasing. Therefore, you must document this information. Correct documentation is essential for several reasons:

- You must direct your readers to the books, periodical articles, and online reference sources that you have used in your research report or presentation. If your audience wants to find these same sources, they depend only on your documentation. If your documentation is incorrect, the audience will be confused. You want your audience to be able to rely on the correctness and validity of your research.

- *Plagiarism* is the appropriation (theft) of some other person's words and ideas without giving proper credit. Communicators are often guilty of unintentional plagiarism.

This occurs when you incorrectly alter part of a quotation but still give credit to the writer. Even if you have cited your source, an altered quotation constitutes plagiarism.

- On the other hand, if you intentionally use another person's words and claim them as your own, omitting quotation marks and source citations, you have committed theft. This is dishonest and could raise questions about your credibility or the credibility of your research. Teachers, bosses, and colleagues will have little, if any, respect for a person who purposely takes another person's words, ideas, or visuals. It is essential, therefore, for you to cite your sources correctly.

- However, if you use boilerplate content (text that resides in your company's archive of research reports), you do not necessarily have to cite the source of the content.

Ethical professional communicators carefully cite the source of their material.

9. Document sources to avoid plagiarism.

Document Sources to Avoid Plagiarism

On a final page or in an appendix, provide an alphabetized list of your research sources. (We discuss this documentation later in Appendix B.) Your readers need to know where you found your information and from which sources you are quoting or paraphrasing. Therefore, you must document this information. Correct documentation is essential to avoid plagiarism.

Style Manuals

To document your research correctly, you must provide parenthetical source citations following the quote or paraphrase within the text. Then, at the end of your document, supply a Works Cited page if you are using the Modern Language Association (MLA) style manual. If you are using either the American Psychological Association (APA) or Council of Science Editors (CSE) style manuals, you will include a References page. See Appendix B for examples of how to use parenthetical source citations and how to document sources.

Alternative Style Manuals

Although APA, CSE, and MLA are popular style manuals, others are favored in certain disciplines. Refer to these if you are interested or required to do so.

- *U.S. Government Printing Office Style Manual,* 30th edition. Washington, DC: Government Printing Office, 2008.
- *The Chicago Manual of Style*, 16th edition. Chicago: University of Chicago Press, 2010.
- Turabian, Kate L. *A Manual for Writers of Term Papers, Theses, and Dissertations,* 7th edition. Chicago: University of Chicago Press, 2007.

TECHNOLOGY TIPS

Using Microsoft Word 2013 for Documentation

Microsoft Word 2013 provides students and business employees many new tools related to research and documentation. When you click on the **References** tab, you will find ways to create tables of contents, insert footnotes, insert citations, and create either bibliographies or works cited notations.

For example, to create parenthetical and bibliographic citations using APA, follow these steps:

1. Click on the **Insert Citation** down arrow and **Add New Source**.

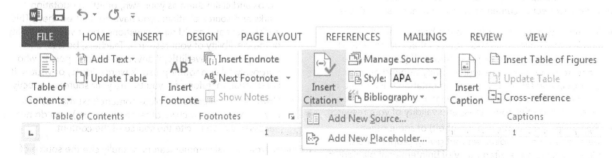

(Continued)

The following screen will pop up (the screen fields will be blank); we have added the appropriate information—author's name, title, publishing company, city, and date.

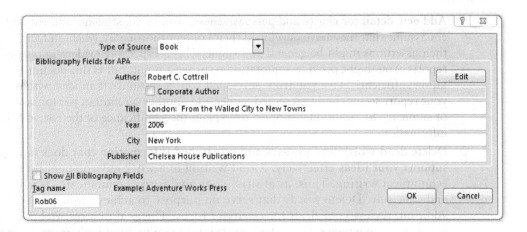

2. Click **OK** and the parenthetical source citation will be inserted.
3. To automatically insert the bibliographical information, click on the **Bibliography** down arrow and then on **Bibliography** to create your works cited information.

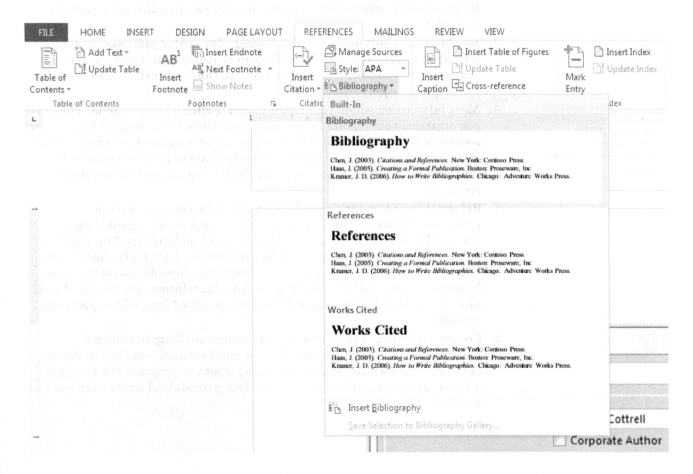

Rewriting Your Researched Document

As with all types of writing, drafting the text of your research report is only the second stage of the writing process. To ensure that your report is effective, revise your draft as follows:

1. **Add new detail for clarity and persuasiveness.** Too often, students and employees assume that they have developed their content thoroughly when, in fact, their assertions might be general and vague. This is especially evident in research reports. You might provide a quotation to prove a point, but is this documentation sufficient? Have you truly developed your assertions? If an idea within your report seems thinly presented, either add another quotation, paraphrase, or summary for additional support or explain the significance of the researched information.

2. **Delete dead words and phrases and researched information that does not support your ideas effectively.** Good writing in a work environment is economical writing. Thus, as always, your goal is to communicate clearly and concisely. Delete words that serve no purpose to achieve conciseness. In addition, review your draft for clarity of focus. The goal of a research report is not to use whatever researched information you've found wherever it seems valid. Instead, you want to use quotations, paraphrases, and summaries only when they help develop your statement of purpose. If your research does not support your thesis, it is counterproductive and should be eliminated. In the rewriting stage, delete any documented research that is tangential or irrelevant.

3. **Simplify your words for easy understanding.** The goal of professional writing is to communicate, not to confuse. Write to be understood. Don't say *grain-consuming animal units* if you mean *chickens*. Don't call a stock market crash a *fourth-quarter equity retreat*.

4. **Move information within your report to ensure effective organization.** How have you organized your report? Did you use a problem/solution format? Did you use comparison/contrast or cause/effect? Is your report organized as a chronological narrative or by importance? Whichever method you've used, you want to be consistent. To ensure consistency, rewrite by moving any information that is misplaced.

5. **Reformat your text for reader-friendly ease of access.** Look at any professional journal. You will notice that the writers have guided their readers through the text by using headings and subheadings. You will also notice that many journals use graphics (pie charts, bar graphs, line drawings, flowcharts, etc.) to clarify the writer's assertions. You should do the same. To help your readers follow your train of thought, reformat any blocks of wall-to-wall words. Add headings, subheadings, itemized lists, white space, and graphics.

6. **Correct any errors.** This represents your greatest challenge in writing a research report. You not only must be concerned with grammar and mechanics, as you are when writing an e-mail message, report, or proposal, but also with accurate quoting, paraphrasing, summarizing, parenthetical source citations, and works cited.

Organizational Techniques

See Chapter 3, "The Communication Process," for more information about organization.

RESEARCH CHECKLIST

_____ 1. Have you considered your purpose and audience in choosing a research topic?

_____ 2. Have you limited the focus of your research topic?

_____ 3. Have you used primary and secondary research sources?

_____ 4. Have you researched using secondary sources, such as books, periodicals, or the Internet or primary sources, such as interviews and questionnaires?

_____ 5. Do you have enough material to sufficiently develop your topic?

_____ 6. Have you used an outline to organize your content?

_____ 7. Have you used direct quotations, paraphrases, and summaries?

_____ 8. Have you quoted and paraphrased correctly to avoid plagiarism?

_____ 9. Have you documented your sources according to either the APA, CSE, or MLA formats to avoid plagiarism?

_____10. Have you used Microsoft Word to create tables of contents, insert footnotes, insert citations, and create either bibliographies or works cited?

The Writing Process at Work

Tom Waters, manager of information solutions at Science Technology Inc., spends approximately 20–30 percent of his work time on research. Research allows Tom to make sure that the software solutions he and his team design and develop are correct. Research helps Tom ensure that his products are on target to meet the client's needs, will be timely and cost efficient, and meet national, international, and industry standards. Tom also relies on research to make certain that he can achieve customer satisfaction and provide value to his organization.

Prewriting

To convey content to his audience, Tom prewrites for his project by considering the following:

- Goal—research wireless Internet options for sales representatives
- Audience—corporate employees and supervisors
- Channels—e-mail messages, face-to-face interviews, and a questionnaire
- Data—drawn from primary research

Figure 16.4 shows how Tom used an online survey to plan his project.

FIGURE 16.4 Online Survey Used for Prewriting the Research Project

SALES MANAGEMENT

1. What is your timeline for this project?

[ ]

2. Each wireless card will cost approximately $30 to $80 per month. Does that fit into your budget?

Yes	No
○	○

(Continued)

FIGURE 16.4 (Continued)

3. Under what circumstances will you expect the reps to use the wireless technology?

4. Will there be consequences for reps who misuse the technology?

Yes	No
○	○

5. If you answered Yes above, what will the consequences be and who will enforce them?

6. What will the process be for replacing lost or damaged wireless cards?

IS PROJECT MANAGER RESPONSIBLE FOR CELL PHONES?

7. Sales Management has asked us to provide wireless Internet connections for the sales force. What vendors will we have to involve to get coverage for everyone?

8. What are the plusses and minuses of each vendor's technology?

IS PROJECT MANAGER RESPONSIBLE FOR STANDARD COMPUTER CONFIGURATION?

9. Sales Management has asked us to provide wireless Internet connections for the sales force. What impact might we expect this technology to have on our standard configuration?

10. What questions and/or recommendations do you have for this project?

Writing

After analyzing the survey responses, Tom wrote a rough draft in a wiki for comments from his colleague. Figure 16.5 shows Tom's rough draft.

FIGURE 16.5 Wiki Draft with Review Comments from a Colleague

Comparing versions of <u>Wireless Connection Rough Draft</u>

Showing changes between <u>December 19, at 11:14:27 am</u> (~~crossed-out~~) and <u>December 19, at 11:23:27 am</u> (<u>underlined</u>)

Subject: Recommendations for Wireless Internet Cards

~~After~~<u>[Before you get to this point, add information about delivery dates and means of shipment. Also explain how vendors were chosen and why we will be using more than one vendor.]After</u> gathering input from all the end of the ~~week.~~

~~RECOMENDATIONS FOR~~<u>week.[You'll also need to clarify how they'll get their devices connected. They need specific sources for help.]</u>

<u>RECOMENDATIONS[You've spelled this word incorrectly. It might be better to write Guidelines for Use" plus avoid the all caps.]FOR</u> WIRELESS INTERNET ~~CARDS~~<u>CARDS[add bullets below to avoid this lengthy and wordy paragraph.]</u>

Provide wireless internet technology sales representatives and managers. ~~Next,~~

<u>Next,</u> because no one vendor availability by sales territory. ~~Use~~

<u>Use</u> wireless technology only when in their home offices. ~~We'll~~

<u>We'll</u> need to also bill amount considered typical usage. ~~Finally, bill sales~~

<u>Finally, bill</u>sales reps and managers for or damaged wireless cards.

<u>[The tone is a bit stiff. Let's add more positive wordsand try a less formal tone.]</u>

Your input is ~~appreciated.~~<u>appreciated.[Passive voice sounds stiff here, for example.]</u>

Rewriting

After receiving input from his colleague, Tom rewrote the draft. See Figure 16.6 for the completed e-mail message.

FIGURE 16.6 Rewritten E-mail to Sales Associates

Tom says, "It's so important to encourage all team members to participate in important decisions in the workplace. Since I became a manager 20 years ago, I have moved away from an authoritarian approach to decision making to an inclusive approach that allows input from even the newest members of the team. You can't have a great 'team' if anyone feels ignored. The communication process is the best approach to writing because it allows for input from team members. I never rush my communications. I want to give my team time to input comments and suggestions. Only then do I feel confident that we can meet the needs of our customers."

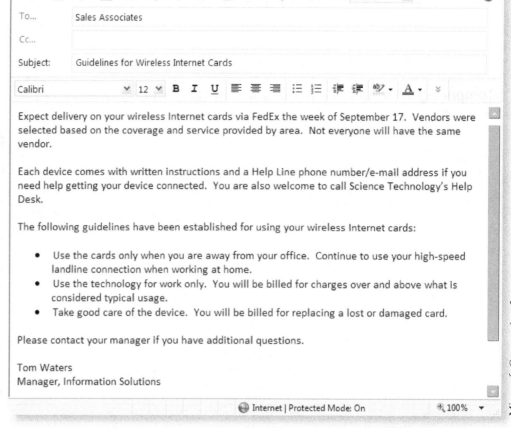

Send | To... Sales Associates | Cc... | Subject: Guidelines for Wireless Internet Cards

Calibri 12 B I U | Options... HTML

Expect delivery on your wireless Internet cards via FedEx the week of September 17. Vendors were selected based on the coverage and service provided by area. Not everyone will have the same vendor.

Each device comes with written instructions and a Help Line phone number/e-mail address if you need help getting your device connected. You are also welcome to call Science Technology's Help Desk.

The following guidelines have been established for using your wireless Internet cards:

- Use the cards only when you are away from your office. Continue to use your high-speed landline connection when working at home.
- Use the technology for work only. You will be billed for charges over and above what is considered typical usage.
- Take good care of the device. You will be billed for replacing a lost or damaged card.

Please contact your manager if you have additional questions.

Tom Waters
Manager, Information Solutions

Internet | Protected Mode: On   100%

Microsoft Corporation, Inc.

APPLY YOUR KNOWLEDGE

CASE STUDIES

1. The City of Oak Springs, Iowa, needs to improve a 10-mile stretch of Ironhorse Avenue. The winding, two-lane road was built in 1976. Since then, the city has grown with several businesses and new housing along this road, and the road no longer is sufficient for the increased traffic load. The road must be expanded to four lanes and straightened. To do so, new sewer lines, sidewalks, easements, esplanades, and lighting must be added. However, this construction will impact current homeowners (whose land, through eminent domain, can be expropriated without the owner's consent). In addition, part of the envisioned road construction will impact a wildlife refuge for waterfowl. After conducting primary and secondary research, the engineers can present a construction plan to Oak Springs' city management.

Assignment

a. Perform primary research. Create a questionnaire for interviewing city residents, as well as employees from the city's parks and recreation, police, fire, and transportation departments.

b. Conduct secondary research. Read statutes online for your city and state regarding eminent domain, real estate, environmental considerations, state wildlife refuges, zoning, planning, and land use. Summarize your findings for your professor either in an e-mail message or oral presentation.

2. You plan to create a new Minority and Women Owned business. Before you can incorporate your company, you must research the following:

- How you can become certified as an MWBD (Minority and Women Business Development).
- The standards you must uphold.
- The percentage of women and/or minorities that you must employ.
- The benefits of operating such a business.

Assignment

Research these topics either online for secondary research or interview an MWBD business owner for primary research. Then, write an e-mail message or memo to your professor about your findings.

ETHICAL CHALLENGE

Roberta Martin, an accountant, is compiling information for a quarterly report on a new client's profit and loss statement. Prior to using Roberta's accounting firm, the client had been with another accountant for many years. To facilitate Roberta's compilation of the report, the client gives Roberta access to prior records. When Roberta writes her report, she cuts and pastes many paragraphs from the original accountant's report. Roberta doesn't alter any words or cite that the content came from another writer.

Question

Is Roberta committing intellectual theft or plagiarism by using another accountant's content? Why or why not?

INDIVIDUAL AND TEAM PROJECTS

1. Summarize in one sentence any paragraph from this textbook. Provide a parenthetical source citation and documentation information (see Appendix B).

2. Practice direct quotations by quoting from a source and citing the source parenthetically.

3. Select a technical topic from your major field or your job and write a research report. You might want to consider an area of interest such as the greenhouse effect, crisis management, online gaming, or nanotechnologies.

PROBLEM-SOLVING THINK PIECES

1. Many communities have recycling projects that allow residents to recycle paper products, cans, and plastic. Not all businesses recycle, however. Research the benefits of recycling, determine how a business or businesses could implement a corporate recycling plan, and write a report recommending action based on your research.

2. In today's global economy, understanding and accommodating multiculturalism and cross-culturalism in business is important. Research the following:

   - The unique challenges that cultural diversity presents to businesses.
   - How companies have responded to these challenges.

   Write a report recommending why and how a business or businesses can help employees develop cultural awareness.

3. Many companies track the time their employees spend either surfing the Web or sending and receiving Twitter messages, posting blog entries, or viewing Facebook and YouTube. Research the following:

   - Software that companies use to track employee electronic communication usage
   - Corporate guidelines for employee use of company-owned electronic communication hardware and software
   - The legal and ethical ramifications of an employee's private use of corporate-owned e-mail and Internet access
   - The legal and ethical ramifications of an employer eavesdropping on an employee's Web usage

   Write a research report on your findings and provide a corporate guideline for both employee and employer electronic communication responsibilities.

4. Entrepreneurialism is one of the fastest-growing sectors in business. Many people are opening their own businesses. What does it take to open your own business? Before you can write an effective business plan and seek financing from a bank, you must research the project.

   Choose a new business venture, selling a product or service of your choice. What would it cost to open this business? What would be your best location, or should your business be online? What certifications or licensing is needed? How many personnel would you need? What equipment is necessary? Who would be your clientele?

   Based on research, write a proposal appropriate for presentation to a bank. In this proposal, present your business plan for a new entrepreneurial opportunity.

WEB WORKSHOP

1. USA.gov allows you to research a wide variety of topics, such as education and jobs, benefits and grants, consumer protection, environment and energy, science and technology, and public safety and health. This Web site also provides information on breaking news. Access USA.gov and research a topic relevant to your career goals. Write an e-mail message or report to your instructor summarizing your findings.

2. Go to an online news magazine, such as *Slate*, *Time*, or *U.S. News Online*. Type a topic of interest in the magazine's search engine. Research this topic and make a brief oral presentation to your class about the information you have gathered.

3. Using an online newspaper, such as *The New York Times* or *USA Today*, research business and technology news. Find the major news stories of the day which relate to your career path. Report your findings in an oral presentation to the class or in an e-mail message to your instructor.

CHAPTER SEVENTEEN

Short, Informal Reports

After completing this chapter, you will be able to

1. Learn what a report is
2. Understand the benefits of writing online reports
3. Distinguish among different types of reports
4. Follow the criteria for writing reports
5. Write an incident report
6. Write an investigative report
7. Write a trip report
8. Write a progress report
9. Write a lab or test report
10. Write a feasibility/recommendation report

In this scenario, a director of information technology must write a variety of reports, including trip reports, feasibility/recommendation reports, and progress reports.

Cindy Katz is director of information technology at **WIFINation**, an electronics communication company with branches in 35 cities and a home office in Austin, Texas.

Cindy has been traveling to Austin from her office in Miami, Florida, one week each month for a year. She is part of a team being trained on WIFINation's new corporate software. Her team consists of staff members from human resources, information technology, accounting, payroll, and administrative services. The software they are learning will be used to manage WIFINation's electronic communication systemwide. It will allow for

- Electronic payroll reports
- Employee benefits
- Paystubs
- Record of cumulative sick leave and vacation days available
- Corporate blogging

- WIFINation's intranet
- WIFINation's Web site
- WIFINation's e-mail system

In addition, with the help of the new software, WIFINation is creating an intracorporate Web site called "WIFILand." This social media site based on a Facebook model will provide WIFINation employees an online location to share stories, update their personal profiles, and inform their colleagues about community activities. The site will also allow for posting instructional videos and videos of corporate activities. WIFILand will be an online meet and greet site designed to build corporate rapport. Cindy's team is responsible for this new social media venture. In her travels to Austin, she is gathering information from her coworkers about what they want to build into this social media site.

Cindy must document her activities monthly and biannually. First, as the project team's recording secretary, Cindy records her travel expenses and the team's achievements, necessitating monthly trip reports submitted to her manager. As a member of her team's technology impact task force subcommittee, Cindy also has been asked to study technology options and emerging technology concerns. This means that she will write a feasibility/recommendation report following her study to justify the implementation of the new technology system the team decides on.

Finally, when her project is completed, Cindy will collaborate with her team members to write a progress report for WIFINation's board of directors. Though Cindy's area of expertise is computer information systems, her job requires much more than programming or overseeing the networking of her corporation's computer systems. Cindy's primary job has become communication with colleagues and administrators. Writing reports is a major component of this job requirement.

1. Learn what a report is.

Criteria for Reports

See Chapter 18, "Long, Formal Reports," for additional information about content in long, formal reports.

Report Definition

Reports come in different lengths and levels of formality, serve different and often overlapping purposes, and can be conveyed to an audience using different communication channels. You could write a short, informal progress report that conveys information about job-related projects. For a short report, an e-mail format might be sufficient, directed to either a high-tech or low-tech, internal audience. However, you might write a longer, more formal progress report, in a letter format for an external lay audience, which provides facts, analyzes these findings, and recommends follow-up action.

Your reports will satisfy one or all of the following needs:

- Supply a record of work accomplished
- Record and clarify complex information for future reference
- Present information to a large number of people with different skill levels
- Record problems encountered
- Document schedules, timetables, and milestones
- Recommend future action
- Document current status
- Record procedures

Long? Short? Formal? Informal? To clarify unique aspects of different kinds of reports, look at Table 17.1.

TABLE 17.1 Unique Aspects of Reports

Report Features	Distinctions	Definition of Unique Characteristics
Length and Scope	Short	A typical short report is limited to 1–5 pages. Short reports focus on topics with limited scope. This could include a limited timeframe covered in the report, limited financial impact, limited personnel, and limited impact on the company.
	Long	Long reports are more than 5 pages long. If a topic's scope is large, including a long timeframe, significant amounts of money, research, many employees, and a momentous impact on the company, a long report might be needed.
Formality (tone)	Informal	Most short reports are informal, routine messages written as letters, memos, or e-mail.
	Formal	Formal reports are usually long and contain standardized components, such as a title page, table of contents, list of illustrations, abstract, appendices, and works cited/references.
Audience	Internal (high tech or low tech)	Colleagues, supervisors, or subordinates within your company are an internal audience. Usually you would write an e-mail or memo report.
	External (multiple audience levels)	An external audience is composed of vendors, clients, customers, or companies with whom you are working. Usually you would write a report in letter format.
	Internal and external	If a report is being sent to both an internal and external audience, you would write either an e-mail, memo, or letter report.
Purpose	Informational	Informational reports focus on factual data. They are often limited in scope to findings: "Here's what happened."
	Analytical	Analytical reports provide information but analyze the causes behind. Then, these analytical reports draw conclusions, based on an interpretation of the data: "Here's what happened and why this occurred."
	Persuasive	Persuasive reports convey information and draw conclusions. Then, these reports use persuasion to justify recommended follow-up action: "Here's what happened, why this occurred, and what we should do next."
Communication Channels	E-mail	E-mail reports, written to internal and external audiences, are short and informal.
	Memo	Memo reports are written to internal audiences and are usually short and informal.
	Letter	Letter reports are sent to external audiences. These reports can be either long or short, formal or informal, depending on the topic, scope, purpose, and audience.
	Electronic (online)	Many reports can be accessed via a company's Web site. These reports can be downloaded and printed out and often are boilerplate—text that can be used repeatedly. Electronic reports also provide interactivity, allowing end users to fill out the report online and submit it to the intended audience.

Online Reports

With a new generation of technologically informed employees and citizens, many organizations have created opportunities for online and on-the-go report writing. Online report writing provides numerous benefits to both the organization and the individual: speed and convenience from a laptop or smartphone; financial savings (no postage or paper); decreased need for filing system storage space; immediate confirmation on a report's

> 2. Understand the benefits of writing online reports.

submittal; predetermined fields to organize a report; online help systems for instructions to help writers complete their reports.

Mobile apps have been created for a variety of reports, including

- Travel and expense reports allowing employees to write reports from the road, while waiting in lobbies, or while flying from various work-site destinations
- Sales reports to help sales personnel manage and update data, collaborate on accounts with team members, and document new sales
- Police reports for officers on the road or at sites where incidents have occurred

Following are examples of academic institutions and federal and local governments that offer online report writing:

- Social Security Payee Report
- City of Milwaukee Police Accident Report
- University of California Berkeley Travel Reimbursement Report
- U.S. Food and Drug Administration Medical Status Report
- National Fraud Information Center Online Incident Report Form

3. Distinguish among different types of reports.

Types of Reports

Many reports fall into the following categories:

- Incident reports
- Investigative reports
- Trip reports
- Progress or status reports
- Lab or test reports
- Feasibility/recommendation reports
- Research reports
- Proposals

Formatting Proposals

See Chapter 19, "Proposals and Business Plans," for additional information.

This chapter will focus on short, informal reports (incident reports, investigative reports, trip reports, progress reports, lab or test reports, and feasibility/recommendation reports).

4. Follow the criteria for writing reports.

Criteria for Writing Reports

Although there are many different types of reports and individual companies have unique demands and requirements, certain traits, including format, development, audience, and style, are basic to all report writing.

Organization

Every short report should contain five basic units: identification lines, headings and talking headings, introduction, discussion, and conclusion/recommendations.

Identification Lines.   Identify the date on which your report is written, the names of the people to whom the report is written, the names of the people from whom the report is sent, and the subject of the report. As discussed in Chapter 9, the subject line should contain a *topic* and a *focus*. In a short internal report, the identification lines will look like the following example.

Date: March 15, 2016
To: Rob Harken
From: Stacy Helgoe
Subject: Report on Usenet Conference

◀ EXAMPLE

Headings and Talking Headings. To improve page layout and make content accessible, use headings and talking headings. Headings—words or phrases such as "Introduction," "Discussion," "Conclusion," "Problems with Employees," or "Background Information"— highlight the content in a particular section of a document. Talking headings, in contrast, are more informative than headings. Talking headings, such as "Human Resources Committee Reviews 2016 Benefits Packages," informatively clarify the content that follows.

Headings

See Chapter 7, "Document Design," for information about creating a hierarchy of headings.

Introduction. The introduction supplies an overview of the report. It can include three or more optional subdivisions, such as the following:

- **Purpose.** Topic sentence(s) explaining why you are submitting the report (rationale, justification, objectives) and the subject matter of the report.
- **Personnel.** Names of others involved in the reporting activity.
- **Dates.** What period of time the report covers.

In this introductory section, use headings or talking headings to summarize the content. These can include headings for organization, such as "Overview" or "Purpose" or more informative talking headings, such as "HVAC Conference Dates Set for 2016."

◀ EXAMPLE

Introduction

Report Objectives: I attended the Southwest Regional Conference on Workplace Communication in Fort Worth, TX, to learn more about how our company can communicate effectively. This report addresses the workshops I attended, consultants I met with, and pricing for training seminars.

Conference Dates: August 5–8, 2016

Committee Members: Susan Lisk and Larry Rochelle

Some businesspeople omit the introductory comments in reports and begin with the discussion. They believe that introductions are unnecessary because the readers know why the reports are written and who is involved. These assumptions are false for several reasons. First, it is false to assume that readers will know why you're writing the report, when the activities occurred, and who was involved. Perhaps if you are writing only to your immediate supervisor, there's no reason for introductory overviews. Even in this situation, however, you might have an unanticipated reader for the following reasons:

- **Immediate supervisors change.** They are promoted, fired, retire, or go to work for another company.
- **Immediate supervisors aren't always available.** They're sick for the day, on vacation, or off-site for job-related travel.

Second, avoiding introductory overviews assumes that your readers will remember the report's subject matter. This is false because reports are written not only for the present, when the topic is current, but also for the future, when the topic is history. Reports go on file—and return at a later date. At that later date, the following may occur:

- You won't remember the particulars of the reported subject matter.

- Your colleagues, many of whom weren't present when the report was originally written, won't be familiar with the subject.
- You might have external, lay readers who need additional detail to understand the report.

An introduction, which seemingly states the obvious, satisfies multiple and future readers.

Discussion.   The discussion section of the report can summarize many topics, including your activities, the problems you encountered, costs of equipment, warranty information, and more. This is the largest section of the report requiring detailed development, illustrated in the different report types discussed throughout this chapter.

Conclusion/Recommendations.   The conclusion section of the report allows you to sum up, to relate what you have learned, or to state what decisions you have made regarding the activities reported. The recommendation section allows you to suggest future action, such as what the company should do next. Not all reports require recommendations.

EXAMPLE ▶

Headings, such as "Benefits of the Conference" and "Proposed Next Course of Action" provide focus.

The conclusion shows how the writer benefited.

The recommendation explains what the company should do next and why.

> **Conclusion/Recommendation**
>
> **Benefits of the Conference**
>
> The conference was beneficial. Not only did it teach me how to improve my technical communication but also it provided me contacts for technical communication training consultants.
>
> **Proposed Next Course of Action**
>
> To ensure that all employees benefit from the knowledge I acquired, I recommend hiring a consultant to provide technical communication training.

| Objectives |
See Chapter 4, "Objectives in Technical Communication," for additional information.

Development

Now that you know what subdivisions are traditional in short reports, the next questions might be, "What do I say in each section? How do I develop my ideas?"

First, answer the reporter's questions.

- **Who** did you meet or contact, who was your liaison, who was involved in the accident, who was on your technical team, and so on?
- **When** did the documented activities occur (dates of travel, milestones, incidents, etc.)?
- **Why** are you writing the report and why were you involved in the activity (rationale, justification, objectives)? Or, for a lab report, for example, why did the electrode, compound, equipment, or material act as it did?
- **Where** did the activity take place?
- **What** were the steps in the procedure, what conclusions have you reached, or what are your recommendations?

Second, when providing information, *quantify*. Do not be vague or imprecise. Specify to the best of your abilities with photographic detail.

The following justification is an example of vague, imprecise writing.

BEFORE

Installation of the machinery is needed to replace a piece of equipment deemed unsatisfactory by an Equipment Engineering review.

Which machine are we purchasing? Which piece of equipment will it replace? Why is the equipment unsatisfactory (too old, too expensive, too slow)? When does it need to be replaced? Where does it need to be installed? Why is the installation important? A department supervisor will not be happy with the preceding report. Instead, supervisors need information *quantified,* as follows:

The *exposure table* needs to be installed by *9/10/16* so that we can *manufacture printed wiring products with fine line paths and spacing (down to 0.0005 inch).* The table will replace the *outdated printer* in *Department 76.* Failure to install the table *will slow the production schedule by 27 percent.*

Note that the italicized words and phrases provide detail by quantifying.

Audience

Since reports can be sent both internally and externally, your audience can be high-tech, low-tech, lay, or include multiple readers. Before you write your report, determine who will read your text. This will help you decide if terminology needs to be defined and what tone you should use. In a memo report to an in-house audience, you might be writing simultaneously to your immediate supervisor (high-tech), to his or her boss (low-tech), to your colleagues (high-tech), and to a CEO (low-tech). In a letter report to an external audience, your readers could be high-tech, low-tech, or lay. To accommodate multiple audiences, use parenthetical definitions, such as cash in advance (CIA) or continuing property records (CPR).

In reports, audience determines tone. For example, you cannot write directive reports to supervisors mandating action on their part. It might seem obvious that you can write directives to subordinates or a lay audience, but you should not use a dictatorial tone. You will determine the tone of your report by deciding if you are writing vertically (up to management or down to subordinates), laterally (to coworkers), or to multiple readers.

Style

Style includes conciseness and highlighting techniques. Achieve conciseness by eliminating wordy phrases. Use *consider* rather than *take into consideration;* use *now* (or provide a date) rather than *at this present time.* Headings, subheadings, and graphics can be used to help communicate content. Note how the following text in the "Before" example fails to communicate effectively.

Johnson County is expected to add 157,605 persons to its 1990 population of 270,269 by the year 2016. That population jump would be accompanied by a near doubling of the 96,925 households the county had in 1990. The addition of 131,026 jobs also is forecast for Johnson County by 2016, more than doubling its employment opportunity.

The information is difficult to access readily. We are overloaded with too much data. Luckily, the report provided a table for easier access to the data. Through highlighting techniques (tables, white space, headings), the demographic forecast is made accessible at a glance.

Table 1 Johnson County Predicted Growth by 2016

	Population	Households	Employment
1990	270,269	96,925	127,836
2016	427,874	192,123	258,862
% change	+58.3%	+98.2%	+102%

5. Write an incident report.

Incident Reports

An *incident report* documents an unexpected problem that has occurred. This could be an automobile accident, equipment malfunction, fire, robbery, injury, or even problems with employee behavior.

Purpose and Examples

If a problem occurs within your work environment that requires analysis (fact-finding, review, study, etc.) and suggested solutions, you might be asked to prepare an incident report (also called a *trouble report* or *accident report*) as follows:

- **Biomedical technology.** A computed tomography (CT) scan in the radiology department is not functioning correctly. This has led to the department's inability to read X-rays. To avoid similar problems, you need to report this incident.
- **Hospitality management.** An oven in your restaurant caught fire. It not only injured one of your cooks but also damaged the oven, requiring that it be replaced with more fire-resistant equipment.
- **Retail.** One of your retail locations has experienced a burglary. The police have been contacted, but as site manager, you believe the problem could have been avoided with better in-store security. Your incident report will document the event and show how to avoid future problems.

Criteria

To write an incident report, include the following components:

1. Introduction
   Purpose. In this section, document when and where the incident occurred.
   Personnel. *Who* was involved, and *what* role do you play in the report? That could entail listing all of the people involved in the accident or event. These might be people injured, as well as police or medical personnel answering an emergency call.

   In addition, *why* are you involved in the activity? Are you a supervisor in charge of the department or employee? Are you a police officer or medical personnel writing the report? Are you an employee responsible for repairing the malfunctioning equipment?

2. Discussion (body, findings, agenda, work accomplished)
   Using subheadings or itemization, quantify what you saw (the problems motivating the report). Organize your content using problem/solution, chronology, cause/effect, or other modes of organization. Develop your content by including the following information:
   - Make or model of the equipment involved
   - Police departments or hospitals contacted
   - Names of witnesses
   - Witness testimonies (if applicable)
   - Extent of damage—financial and physical
   - Graphics (sketches, schematics, diagrams, layouts, etc.) depicting the incident visually
   - Follow-up action taken to solve the problem

3. Conclusion/recommendations
   Conclusion. Explain what caused the problem.
   Recommendations. Relate what could be done in the future to avoid a similar problem.

E-mail

See Chapter 9, "Routine Correspondence," to learn more about e-mail criteria.

Figure 17.1 presents an example of an incident report, written as an e-mail.

FIGURE 17.1   E-mail Incident Report

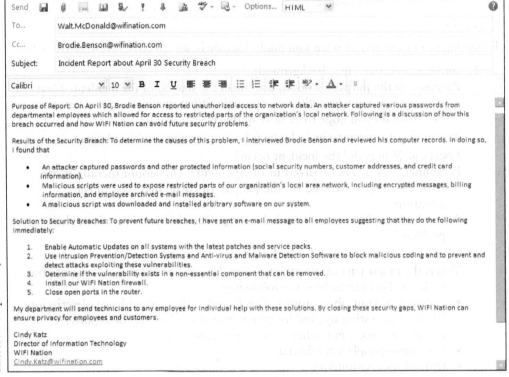

Microsoft Corporation, Inc.

Bullets are used to itemize causes of the problem. In contrast, numbers are used to sequence procedural steps.

The writer relies on a "problem/solution" analytical pattern of organization for this short, informal, internal e-mail report.

The writer includes a "signature file" that acts like an online business card, clearly identifying the writer and providing contact information.

Investigative Reports

6. Write an investigative report.

As the word *investigate* implies, an *investigative report* asks you to examine the causes behind an incident.

Purpose and Examples

This report does not just document the incident. It focuses more on why the event occurred. You might be asked to investigate causes leading up to a problem in the following instances:

- **Security.** You work in a bank's security department. You are responsible for investigating theft, burglary, fraud, vandalism, check forging, and other banking illegalities. One of your customers, a college student at the local university, reports losing her purse at a campus party. Within hours of the theft, checks bearing her name are showing up across the city. Your job now is to investigate the incident and report your findings.

- **Engineering.** A historic 100-year-old bridge crossing your city's river is buckling. The left lane is now two inches higher than the right lane, and expansion joints are separating beyond acceptable specifications. You must visit the bridge site, inspect the damage, and report on the causes for this construction flaw.

- **Computer technology.** You work in a college's technology department. The college requires that all student grades be kept on a newly installed campuswide database and then be submitted electronically when the semester ends. For some reason, faculty cannot access their students' records for grade inputting. You must investigate the causes behind this technology glitch and solve the problems—*now!* The semester grades are due within 24 hours.

Criteria

Following is an overview of what you might include in an effective investigative report.

1. Introduction (overview, background)

   Purpose. In the purpose section, document the date(s) of the incident. Then comment on your objectives or rationale. What incident are you reporting on and what do you hope to achieve in this investigation?

   You might also want to include the following optional subheadings:

   Location. Where did the incident occur?

   Personnel. Who was involved in the incident? This could include those with whom you worked on the project or those involved in the situation.

   Authorization. Who recommended or suggested that you investigate the problem?

2. Discussion (body, findings, agenda)

   This is the major part of the investigation. Using subheadings, document your findings. This can include the following:

   - A review of your observations. This includes physical evidence, descriptions, lab reports, testimony, and interview responses. Answer the reporter's questions: who, what, when, where, why, and how.
   - Contacts—people interviewed.
   - Difficulties encountered.
   - Techniques, equipment, or tools used in the course of the investigation.
   - Test procedures followed and organized chronologically.

3. Conclusion/recommendations

   Conclusion. What did you accomplish? What did you learn? What discoveries have you made regarding the causes behind the incident? Who or what is at fault?

   Recommendations. What do you suggest next? Should changes be made in personnel or in the approach to a particular situation? What training is required for use with the current technology, or should technology be changed? What is the preferred follow-up for the patient or client? How can the problem be fixed?

Figure 17.2 illustrates an investigative report written in letter format to an external audience. This example is written at a high-tech level because the city council needs to understand the scientific information about the incident.

In contrast to Figure 17.2, the high-tech investigative report, Figure 17.3 is written at a lay level for an external audience consisting of the elementary school administration, the teachers, and the parents of the school children affected by the incident in Frog Creek. The highly technical and scientific information is omitted, the material is condensed, and the writer includes sufficient details to alleviate the concerns of the audience. In addition, the writer emphasizes actions taken and assurances of how and why this incident will not recur.

FIGURE 17.2   Investigative Report in Letter Format (high-tech level)

Frog Creek Wastewater Treatment Plant

9276 Waveland Blvd.
Bowstring City, UT 86554

September 15, 2016

Bowstring City Council
Arrowhead School District 234
Bowstring City, UT 86721

Subject: Investigative Report on Frog Creek Wastewater Pollution

On September 7, 2016, teachers at Arrowhead Elementary School reported that over a five-day period (September 2–6), approximately 20 students complained of nausea, lightheadedness, and skin rashes. On the fifth day, the Arrowhead administration called 911 and the Arrowhead School District (ASD 234) in response to this incident.

Committee Findings

Impact on Schoolchildren: Arrowhead Elementary School administrators reported the following:

- Monday, September 2—two children reported experiencing nausea.
- Tuesday, September 3—two children reported experiencing nausea, and one child experienced lightheadedness.
- Wednesday, September 4—three children experienced skin rashes.
- Thursday, September 5—two children complained of nausea, one child was lightheaded, and two children showed evidence of skin rashes.
- Friday, September 6—two children reported nausea, three reported skin rashes, and two lightheadedness.

Report on Frog Creek Pollutants: Wastewater Engineers Sue Cottrell and Thomas Redburn took samples of Frog Creek on September 7–10 and found the following:

Typical Frog Creek Readings	Readings from September 7–10
Low alkalinity (generally <30 mg/l) (milligrams per liter)	High alkalinity readings: <45 mg/l
Low inorganic fertilizer nutrients (phosphorous and nitrogen): <20 mg/l	High phosphorous and nitrogen readings: <25 mg/l
Limited algae growth: 1 picometer	High algae readings: 4 picometers

Explanation of Findings: Despite normally low readings, in late summer, with heat and rain, these readings can escalate. Higher algae-related odors above the 3–6 picometer thresholds, along with increased alkalinity (<50 mg/l) can create health problems for youth, elderly, or anyone with respiratory illnesses.

Follow letter format when you write a short, informal report for an external audience.

The subject line provides a topic (Frog Creek Wastewater Pollution) and a focus (Investigative Report).

In the letter's introduction, reporter's questions clarify when and where the event occurred, who was involved, and why the report is being written.

The findings use chronological organization to document the incidents.

(Continued)

Include a new-page notation on every page after page one. In this notation, provide the writer's name, page number, and date. You can use a header to create this notation.

Mike Moore
Page 2
September 15, 2016

These studies showed that algae, alkalinity, and fertilizer nutrients were higher than usual. The above elevated readings were caused by three factors (heat, rain, and northeasterly winds).

Follow-up Studies: On September 11–14, our engineers rechecked Frog Creek, finding that the chemical levels had returned to a normal, acceptable range.

Conclusion about Incident

Frog Creek normally has acceptable levels of algae, alkalinity, and fertilizer nutrient levels. The heat and higher water levels temporarily led to elevated pollutant readings. These levels subsequently returned to normal. Wind directions during the school incidents also had an impact on the children's illnesses. On follow-up questionnaires, FCWTP employees found that the school children's ailments had subsided.

The findings not only investigate the causes of the incident but also document with specific details. To achieve a readable format, the text is made accessible through highlighting techniques—underlined and italicized subheadings, bulleted details, and a comparison/contrast table.

The Arrowhead Elementary School situation appears to have been an isolated incident due to atmospheric changes.

Recommendations for Future Course of Action

Though we constantly monitor Frog Creek for safety, FCWTP HAZMAT employees would be happy to work with parents and teachers to provide additional health information. In a one-hour workshop, presented during the school day or at a Parent-Teacher Organization meeting, FCWTP could offer the following information:

- Scientific data about stream and creek pollutants
- The effects of rain, wind, and heat on creek chemicals
- Useful preventive medical emergency techniques

Please let us know if you would like to benefit from this free-to-the-public workshop. We would be happy to schedule one at your convenience.

The letter's conclusion provides options for the readers and a positive tone appropriate for the intended audience.

Mike Moore

Mike Moore
Frog Creek Wastewater Treatment Plant Director of Public Relations

Enclosure

FIGURE 17.3 Investigative Report for Lay Audience in Letter Format

Frog Creek Wastewater Treatment Plant
"Safety is Our Number 1 Concern."
9276 Waveland Blvd.
Bowstring City, UT 86554

September 15, 2016

Attention: Parents and Arrowhead Elementary School Teachers and Administrators

Subject: Report on September 2–6 Frog Creek Incident

Last week, your children at Arrowhead Elementary School experienced nausea, lightheadedness, and skin rashes. This was due to an unusual environmental situation at Frog Creek, a rare case of airborne pollutants caused by high winds and rain.

How Did This Happen?

Algae is a good thing. The crayfish, snails, and minnows that your children love seeing in Frog Creek thrive on algae and lichen (the small, green plants that are the food base for most marine life). However, when temperatures rise above 84 degrees, algae can grow to an unhealthy level.

The same thing applies to the acid level in water. When acid is regulated by alkalinity, algae growth is controlled. When acid levels rise, however, algae can bloom or marine creatures can die. Both of these problems lead to unusual odors.

That's what happened last week. Our studies showed that algae and alkalinity were higher than usual. The causes were increased heat, rain, and wind.

What Can We Do to Help?

Could heat, rain, and wind lead to similar situations in the future? Yes. But . . . the incidents from last week were very rare. Please do not expect a repeat occurrence any time soon. Tell your children to enjoy Frog Creek for its beauty and natural resources.

We would be happy to meet with you and your children to explain the science of this environmental event. Plus, we'd like to provide techniques for managing simple ailments like nausea, skin rashes, and lightheadedness.

Because this letter report is sent to multiple audiences, no reader address is given. Instead, an attention line is used.

To communicate effectively with the public, the writer uses pronouns and a pleasant tone. Words like "unusual" and "rare" are used to lessen the parents' concerns.

For headings, the writer uses questions that an audience of concerned parents and teachers might have.

Because the audience is low-tech, the writer uses simple sentence structure ("Algae is a good thing"; "The same applies to acid levels in water.") and avoids complex discussion of science, focusing on temperature, rain levels, and wind direction.

(Continued)

The writer emphasizes the positive by using words such as "please," "enjoy Frog Creek for its beauty," and "We would be happy to meet with you."

The writer focuses on the audience's future concerns and suggests age-appropriate material to explain the incident to the elementary school children. This attempt at community outreach connects the wastewater treatment plant to concerned constituents.

page 2
September 15, 2016

In a one-hour workshop, presented during the school day or at a Parent-Teacher Organization meeting, Frog Creek Wastewater Treatment Plant (FCWTP) could offer the following information:

- Scientific data about stream and creek pollutants—with hands-on tutorials for your students.
- The effects of rain, wind, and heat on creek chemicals—complete with graphics and an age-appropriate PowerPoint presentation.
- Useful preventive medical emergency techniques—which every parent and child should know.

This information would explain real-world applications for science classes, as well as provide valuable health tips for parents and teachers. We have enclosed for your review information about our proposed training sessions.

Please let us know if you would like to benefit from this free-to-the-public workshop. We would be happy to schedule one at your convenience. FCWTP wants to assure you that your child's "safety is our number 1 concern."

Mike Moore

Mike Moore
Frog Creek Wastewater Treatment Plant Director of Public Relations

Enclosure

TECHNOLOGY TIPS

Creating Headers and Footers in Microsoft Word 2013

To create headers and footers (useful for new page notations in reports), follow these steps:

1. Click on the **Insert** tab on your toolbar. You will see the following ribbon.

(Continued)

2. Click on either **Header** or **Footer**. When you click on your choice of either **Header** or **Footer**, you will see a drop-down menu, such as the following.

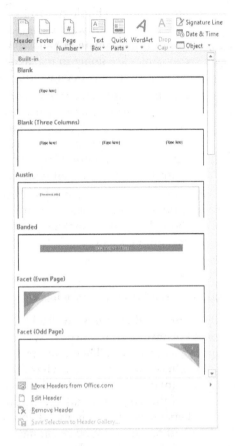

3. Choose the type of **Header** or **Footer** you want to use in your document and type your content (date, name, page number, etc.) as shown in the following example.

| Mary Subicas | Page 2 | April 21, 2015 |

Microsoft Corporation, Inc.

Trip Reports

A *trip report* allows you to report on job-related travel.

> 7. Write a trip report.

Purpose and Examples

When you leave your work site and travel for job-related purposes, your supervisors not only require that you document your travel expenses (food, mileage, hotel) and time while off-site, but also they want to be kept up to date on your work activities. For example, you might be engaged in work-related travel as follows:

- **Information technology.** You go to a conference to learn about the latest hardware and software technologies for the workplace. There you meet with vendors, participate in hands-on technology workshops, and learn what other companies are doing to manage their technology needs. When you return, you write a trip report documenting your activities.

- **Heating, ventilating, and air conditioning.** One of your clients is building an office site. Your company has been hired to install their heating, ventilating, and air conditioning (HVAC) system. You travel to your client's home office to meet with other contractors (engineering and architectural) so all team members can agree on construction plans. At the conclusion of your job-related travel, you will write a trip report about your meeting.

- **Biomedical equipment sales.** Four days a week, every week, you are on the road making sales calls. Each month, you must document your job-related travel to show that you are making your quota and to receive compensation for travel expenditures.

Criteria

Following is an overview of what you will include in an effective trip report.

1. Introduction (overview, background)
   Purpose. In the purpose section, document the date(s) and destination of your travel. Then comment on your objectives or rationale. What motivated the trip, what did you plan to achieve, and why were you involved in job-related travel? You might also want to include the following optional subheadings:
   Personnel. With whom did you travel?
   Authorization. Who recommended or suggested that you leave your work site for job-related travel?

2. Discussion (body, findings, agenda)
   Using subheadings, document your activities. This can include a review of your observations, contacts, seminars attended, or difficulties encountered.

3. Conclusion/recommendations
   Conclusion. What did you accomplish—what did you learn, whom did you meet, what sales did you make, what of benefit to yourself, colleagues, or your company occurred?
   Recommendations. What do you suggest next? Should the company continue on the present course (status quo) or should changes be made in personnel or in the approach to a particular situation? Would you suggest that other colleagues attend this conference in the future, or was the job-related travel not effective? In your opinion, what action should the company take?

Figure 17.4 is an example of an informal trip report written in memo format. The memo format providing identification lines (date, to, from, subject) is written to an internal audience and creates a hard-copy document.

FIGURE 17.4   Trip Report in Memo Format

Date: February 26, 2016
To: Debbie Rulo, Director of Education Technology
From: Oscar Holloway, Educational Technologist
Subject: Trip Report—21st Century Educational Technology Conference

Introduction

Purpose of the Meeting: On Tuesday, February 23, 2016, I attended the 21st Century Educational Technology Conference, held in Ruidoso, New Mexico. My goal was to find new ways to engage today's students and acquire hands-on instruction in courseware technologies, including the following:

- MOOCs (massive open online courses)
- Social media
- Mobile apps
- Game-based learning

Conference Participants: My coworkers Bill Cole and Gena Sebree also attended the conference.

Discussion

Presentations at the Conference:

Gena, Bill, and I attended the following sessions:

- *MOOCs*

  This two-hour workshop was presented by Dr. Peter Tsui, a noted instructional expert from Texas State University, San Marcos, Texas. During Dr. Tsui's presentation, we reviewed the concept behind massive open online courses, an online gathering of people who jointly exchange ideas. MOOCs are a unique and creative approach to education in two key ways:

  1. Classes depend on crowd-sourced interaction—peer reviews and group work
  2. Feedback is conveyed through online assessments and mLearning (mobile education)

  Though MOOCs allow for large enrollments and seek to engage students through innovation and customization, MOOCs have a downside. Usually, these course deliverables do not offer credit, nor do students need to enroll in the traditional way.

- *Social Media*

  This hour-long presentation was facilitated by Debbie Gorse, an employee of Xenadon E-Learning Inc. (Colorado Springs, Colorado).

(Continued)

The subject line presents the topic (technology conference) and the focus (trip report).

The introduction section answers the reporter's questions—who, what, when, where, and why.

Different heading levels and highlighting techniques are used to make the information more accessible and to help the readers navigate text.

FIGURE 17.4   Trip Report in Memo Format *(Continued)*

A new-page notation helps your readers avoid losing or misplacing pages.

Oscar Holloway
Page 2
February 26, 2016

She focused on the following benefits:

1. Wikis used for collaborative writing. She showed a variety of wiki sites and examples of classroom writing and rewriting.
2. Blogs used for journaling. Through WordPress, for example, students can create free blogs to share their thoughts regarding topics in various disciplines.
3. Google Sites used for creating Web-based e-folios. Students could use Web sites to showcase their achievements (certificates, writing samples, graphics, online resume, etc.).
4. Twitter used for secondary research. Twitter hashtags (#) allow students to follow newsworthy trends.

- *Mobile Apps*

  Dr. Randy Towner and Dr. Karen Pecis led this hour-long presentation. Both are professors at the University of Nevada, Las Vegas. Their workshop focused on ways in which mobile apps can enhance course development, instructional methodologies, student engagement, and real-world relevance. The professors provided workbooks and hands-on learning opportunities.

- *Game-based learning*

  Denise Pakula, Canyon E-Learning, Tempe, Arizona, spoke about her company's proprietary game-based learning software. The value of her products is that they enhance collaboration, interactivity, problem solving, situational analysis, creativity, critical thinking, and communication skills related to teamwork. Students using her company's software can develop video games, engage in virtual simulations, and see real-time results of actions taken.

Conclusion

Presentation Benefits:

Every presentation we attended was beneficial. However, the following information will clarify which workshop(s) would benefit our college the most:

The conclusion focuses on the primary findings to give the audience direction.

- Dr. Tsui's program was the most useful and informative. His interactive presentation skills were outstanding and included hands-on activities, small-group discussions, and individual instruction. He would charge only $90 per person (other people we researched charged at least $150 per person). Dr. Tsui's fees should fit our faculty training budget. However, the challenge is how our college can monetize this educational concept since MOOCs are generally non-credit.

- Social media obviously has value to education. However, we would not benefit from inviting Ms. Gorse to the college for additional guidance. Her training techniques are outdated. Videos and overhead projections will not create the interactivity our students require.

FIGURE 17.4   (Continued)

Oscar Holloway
Page 3
February 26, 2016

- The Towner/Pecis workshop was excellent. Mobile apps would be ideal for our needs. Mobile apps are very affordable. Plus, they allow our faculty and students to acquire information in the classroom and in dispersed locations through their smartphones. We will be able to customize instruction and provide up-to-date information about a wide range of topics. Furthermore, Drs. Towner and Pecis used informative workbooks and hands-on learning opportunities in their presentation.

- Ms. Pakula's presentation on game-based learning focused solely on her company's products. Perhaps if faculty members are interested in this topic, we could research freeware or ask our IT personnel to create game-based learning software, customized to our college's needs.

Recommendations

Gena, Bill, and I suggest that you invite Dr. Tsui to our site for further consultation. We also think you might want to contact Drs. Towner and Pecis for more information on their training.

Attachments: Travel receipts (mileage, food, and hotel)

The recommendation suggests the next course of action.

Progress Reports

8. Write a progress report.

A *progress report* lets you document the status of an activity, explaining what work has been accomplished and what work remains.

Purpose and Examples

Supervisors want to know what progress you are making on a project, whether you are on schedule, what difficulties you might have encountered, and what your plans are for the next reporting period. Because of this, your audience might ask you to write progress (or activity or status) reports—daily, weekly, monthly, quarterly, or annually.

- **Hospitality management.** The city's convention center is considering new catering options. Your job has been to compare and contrast catering companies to see which one or ones would best be suited for the convention center's needs. The deadline is arriving for a decision. What is the status? Whom have you considered, what are their prices and food choices, what additional services do they offer, and so forth? You need to submit a progress report so management can determine what the next steps should be.

- **Project management.** Your company is renovating its home office. Many changes have occurred. The supervisor wants to know when the renovations will be concluded. You need to write a progress report to quantify what has occurred, what work is remaining, and when work will be finished.

- **Automotive technology.** Your company recently suffered negative publicity due to product failures. As manufacturing supervisor, you have initiated new procedures for automotive manufacturing to improve your product quality. How are these procedural changes going? Your company CEO needs an update. To provide this information, you must write a progress report.

Criteria

Following is an overview of what you will include in an effective progress report.

1. Introduction (overview, background)
   Objectives. These can include the following:
   - Why are you working on this project (what's the rationale)?
   - What problems motivated the project?
   - What do you hope to achieve?
   - Who initiated the activity?

   Personnel. With whom are you working on this project (i.e., work team, liaison, contacts)?

   Previous activity. If this is the second, third, or fourth report in a series, remind your readers what work has already been accomplished. Bring them up to date with background data or a reference to previous reports.

2. Discussion (findings, body, agenda)
   Work accomplished. Using subheadings, itemize your work accomplished either through a chronological list or a discussion organized according to importance.

   Problems encountered. Inform your reader(s) of any difficulties encountered (late shipments, delays, poor weather, labor shortages) not only to justify possibly being behind schedule but also to show the readers where you'll need help to complete the project.

   Work remaining. Tell your reader what work you plan to accomplish next. List these activities, if possible, for easy access. A visual aid, such as a Gantt chart or a pie chart, fits well after these two sections. The chart will graphically depict both work accomplished and work remaining.

3. Conclusion/recommendations
   Conclusion. Sum up what you've achieved during this reporting period and provide your target completion date.

   Recommendations. If problems were presented in the discussion, you can recommend changes in scheduling, personnel, budget, or materials that will help you meet your deadlines.

   Figure 17.5 is an example of a progress report.

> **Visual Aids in Reports**
> See Chapter 8, "Visual Aids," for information about using graphics.

FIGURE 17.5 Progress Report in Memo Format

To: Buddy Ramos
From: Pat Smith
Date: April 2, 2016
Subject: First Quarterly Report—Project 80 Construction

Purpose of Report

In response to your December 20, 2015, request, following is our first quarterly report on Project 80 Construction (Downtown Airport). Department 93 is in the start-up phase of our company's 2016 build plans for the downtown airport and surrounding site enhancements. These construction plans include the following:

FIGURE 17.5   *(Continued)*

Pat Smith
Page 2
April 2, 2016

- *Airport construction*—terminals, runways, feeder roads, observation tower, parking lots, maintenance facilities.
- *Site enhancements*—northwest and southeast collecting ponds, landscaping, berms, and signage.

Work Accomplished

In this first quarter, we have completed the following:

- *Subcontractors*: Toby Summers (Project Management) and Karen Kuykendahl (Finance) worked with our primary subcontractors (ApexEngineering and Knoblauch and Sons Architects). Toby and Karen arranged site visitations and confirmed construction schedules. This work was completed January 12, 2016.
- *Permits*: Once site visitations were held and work schedules agreed upon, Toby and Wilkes Berry (Public Relations) acquired building permits from the city. They accomplished this task on January 20, 2016.
- *Core Samples*: Core sample screening has been completed by Department 86 with a pass/fail ratio of 76.4 percent pass to 23.6 percent fail. This meets our goal of 75 percent. Sample screening was completed January 30, 2016.
- *Shipments*: Timely concrete, asphalt, and steel beam shipments this quarter have provided us a 30-day lead on scheduled parts provisions. Materials arrived February 8, 2016.
- *EPA Approval*: Environmental Protection Agency (EPA) agents have approved our construction plans. We are within guidelines for emission controls, pollution, and habitat endangerment concerns. Sand cranes and pelicans nest near our building site. We have agreed to leave the north plat (40 acres) untouched as a wildlife sanctuary. This will cut into our parking plans. EPA approval occurred on February 15, 2016.

Problems Encountered

Core samples are acceptable throughout most of our construction site. Unfortunately, when Anderson Brothers began dredging, they hit rock, which had to be removed with explosives.

Since this northwest pond is near the sand crane and pelican nesting sites, EPA told us to wait until the birds were resettled. The extensive rock removal and wait for wildlife resettlement have slowed our progress. We are behind schedule on this phase. This schedule delay and increased rock removal will affect our budget.

Work Remaining

To complete our project, we need to accomplish the following:

- *Advertising*: Our advertising department is working on brochures, radio and television spots, and highway signs.

The introduction explains *why* the report has been written and *what* topic will be discussed.

The discussion provides quantified data and dates for clarity, such as "76.4 percent pass" and "January 30, 2016." The discussion also clarifies who worked on the project and lists other primary contacts.

A problems encountered section helps justify delays and explain why more time, personnel, or funding might be needed to complete a project.

Italicized sub-headings help readers access the information more readily. This is especially valuable when an audience needs to refer to documents at a later date.

(Continued)

FIGURE 17.5   Progress Report in Memo Format *(Continued)*

Pat Smith
Page 3
April 2, 2016

- *Signage*: Transportation is working with advertising on signage designs for the downtown airport's two entrances. These signs will juxtapose the city's symbol (a flying pelican) with an airplane taking off. The goal is to create a logo that simultaneously promotes the preservation of wildlife and suggests progress and community growth.
- *Landscaping*: Our architectural design team, led by Fredelle Schneider, is selecting and ordering plants, as well as directing a planting schedule. Anderson Brothers also is in charge of the berms and pond dredging. Fredelle will be our contact person for this project.
- *Construction*: The entire airport must be built. Thus, construction comprises the largest remaining task.

Project Completion/Recommendation

Though we have just begun this project, we have completed approximately 15 percent of the work. We anticipate a successful completion, especially since deliveries have been timely.

Only the delays at the northwest pond site present a problem. We are two weeks behind schedule and $3,575.00 over cost. With approximately ten additional personnel to speed the rock removal and with an additional $2,500, we can meet our target dates. Darlene Laughlin, our city council liaison, is the person to see about corporate investors, city funds, and big-ticket endowments. With your help and Darlene's cooperation, we should meet our schedules.

The Gantt chart in Figure 1 clarifies our status at this time.

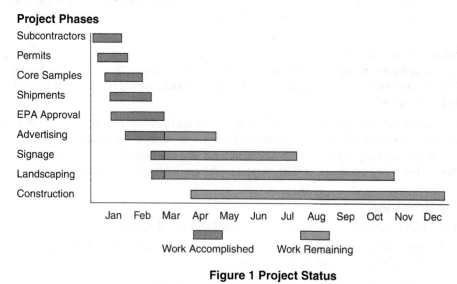

Figure 1 Project Status

The conclusion sums up the overall status of the project: "15 percent" complete.

The recommendation explains how the problems discussed in the problems encountered section can be solved and what is needed to complete the job: "additional personnel" and "increased funds." It also states who to contact for help, "Darlene Laughlin."

A graphic not only adds interest to the report but also helps quantify information and clarify details visually.

Lab Reports (also referred to as test reports)

9. Write a lab or test report.

A *lab or test report* lets you document the status of and findings from a laboratory experiment, procedure, or study.

Purpose and Examples

The knowledge acquired from a laboratory activity must be communicated to colleagues and supervisors so they can benefit from your discoveries. You write a lab report after you have performed the lab activity to share your findings.

- **Biomedical technology.** You have performed a pathology study on tissue, reviewed a radiological scan, or drawn blood. What have you found? To help nurses and doctors provide the best patient care, you must write a lab report documenting your findings.
- **Electronics.** Your company manufactures global positioning systems (GPSs) to correctly inform a user of an exact location. The GPS receiver must compare the time a signal is transmitted by a satellite with the time it is received. Your company's receptors are malfunctioning as are the units' electronic maps. Why? Your job is to study the electronic systems on randomly selected GPS units and write a lab report documenting your findings.
- **Information technology.** Customers are calling your company's 1–800 hotline almost daily, complaining about hard drive error readings. This is bad for business and profitability. To solve these hard drive malfunctions, you must study units to find the problem. Then, you will write a test report to document your discoveries.

You write a lab report after you've performed a laboratory test to share with your readers

- Why the test was performed
- How the test was performed
- What the test results were
- What follow-up action (if any) is required

Criteria

The following are components of a successful lab report.

1. Introduction (overview, background)
    Purpose. Why is this report being written? To answer this question, provide any or all of the following:
    - Rationale (What problem motivated this report?)
    - Objectives (What does this report hope to prove?)
    - Authorization (Under whose authority is this report being written?)
2. Discussion (body, methodology)
    How was the test performed? To answer this question, provide the following:
    - Apparatus (What equipment, approach, or theory have you used to perform the test?)
    - Procedure (What steps—chronologically organized—did you follow in performing the test?)
3. Conclusion/recommendations
    Conclusion. The conclusion of a lab report presents your findings. Now that you've performed the laboratory experiment, what have you learned or

discovered or uncovered? How do you interpret your findings? What are the implications?

Recommendations. What follow-up action (if any) should be taken?

You might want to use graphics to supplement your lab report, as shown in Figure 17.6.

FIGURE 17.6 Lab Report in Memo Format

Date: April 12, 2016
To: Dr. Lee Wang
From: Cassidy Poston
Subject: Lab Report on Anti-Cancer Characteristics of Anthocynanins in SW480 Cancer Cells

Introduction

Biosystems Inc. is testing new drugs in different applications to prevent colon cancer. This lab report will evaluate colon cancer cell growth rate and cell cycle apoptosis in response to varying concentrations of anthocynanin. The study was authorized by Dr. Amanda Wharton, laboratory supervisor at Biosystems, and conducted by me, Andrew Boston, and Cynthia Ruiz, laboratory technicians at Raston Pharmaceuticals.

Discussion

Apparatus—To conduct the lab work, we used the following chemicals and equipment:

Chemicals	Equipment
• Anthocynanin • Peonidin 3-glucose • Cyanidin 3-glucose • SW480 colon cancer cells • Fetal bovine serum • Penicillin	• Spectroscope • Petri dishes • 96-well microtiter platet

Procedure—To conduct the lab work, we followed this procedure:

1. *HPLC analysis*: Ms. Ruiz purified anthocynanin samples extracted from a sweet potato using HPLC (high-performance liquid chromatographic) analysis.
2. *Cells cultured*: SW480 colon cancer cells were cultured in a standard growth medium supplemented with 10 percent fetal bovine serum and 1 percent penicillin. The cells were then placed in 96-well microtiter plates for drug administration; approximately 2.0×10^4 cell density per well.
3. *Anthocynanin application*: The standard anthocynanins along with the sweet potato extracts were applied to the cell cultures in varying concentrations: 0.5, 1.0, 5.0 and 10.0 microgram/ml.
4. *Results*: Cancer cell growth inhibition was measured using spectroscopy, absorbance at 490nm with a 96-well plate reader. The analysis of anthocynanin standards were recorded graphically. We are awaiting results on the cells treated with sweet potato extract. (See graphs below.)

FIGURE 17.6 *(Continued)*

Cassidy Poston
Page 2
April 12, 2016

Colon Cancer Cell Growth vs. Anthocyanin Concentration

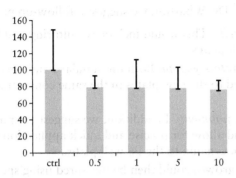

Standard: Peonidin 3-glucose

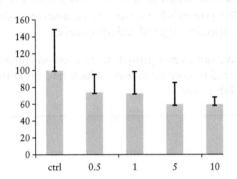

Standard: Cyanidin 3-glucose

Conclusion

The application of the anthocynanin standards gave undesirable results. The cell growth inhibition was similar for all concentrations of drug treatment; 0.5 to 10 microgram/ml. Our team of lab techs suspects that more concentrated applications would have a greater inhibitory effect on colon cancer cell growth or at least varying growth recordings for each concentration. We believe this discrepancy may be due to differences in cell densities per well. We did not invert our cell suspension often or follow a randomized application. This delayed analysis of the cell cultures treated with sweet potato extract.

(Continued)

FIGURE 17.6   Lab Report in Memo Format *(Continued)*

Cassidy Poston
Page 3
April 12, 2016

Recommendation

With the permission of Dr. Wharton, we suggest a follow-up procedure:

1.   *Reapplication of cells*: This would include re-culturing and inoculating SW480 cells on the 96 well plates.
2.   *Reapplication of anthocynanin*: Both the standard and sweet potato extracts should be reapplied to the new plates in the same concentrations discussed above.
3.   *Purchase of new equipment*: In addition, we suggest the purchase of a digital pipette. This would allow for precise and quick application of cell suspension, medium, and anthocynanin to the 96 well plates.
4.   *Analysis*: The cell growth could then be measured using spectrometry with the same conditions of prior trials.
5.   *Cytoflowmetry*: Once all values have been recorded for cell growth, the samples should be sent to Biosystems's laboratory to be assessed using cytoflowmetry. This will detect the specific stage of cell apoptosis.

Although our results have not been optimal, treated colon cancer cells did have inhibited growth compared to control values (which had no anthocynanin applied). This is an encouraging lab result.

Feasibility/Recommendation Reports

10. Write a feasibility/ recommendation report.

A *feasibility/recommendation report* accomplishes two goals. First, it studies the practicality of a proposed plan. Then, it recommends action.

Purpose and Examples

Occasionally, your company plans a project but is uncertain whether the project is feasible. Will the plan work, does the company have the correct technology, will the idea solve the problem, or is there enough money? One way a company determines the viability of a project is to perform a feasibility study to document the findings and then to recommend the next course of action.

- **Manufacturing.** Your company is considering the purchase of new equipment but is concerned that the machinery will be too expensive, the wrong size for your facilities, or incapable of performing the desired tasks. You need to research and analyze the options, determining which equipment best suits your company's needs. Then, you will recommend purchase.
- **Accounting.** Your company wants to expand and is considering new locations. The decision makers, however, are uncertain whether the market is right for expansion. Are interest rates good? Are local property taxes and sales taxes too high? Will the city provide tax rebate incentives for your

company's growth? You need to study the feasibility of expansion and report your recommendations.

- **Web design.** Your company wants to create a Web site to market your products and services globally. The CEO wants to be sure that online checkout is easy, that pricing is cost effective, that products are depicted in a visually appealing way, and that the site loads quickly. How will you make your Web site stand out from the competition? You must write a feasibility report to present the options as well as to offer your recommendations.

Criteria

One way a company determines the viability of a project is to perform a feasibility study and then write a feasibility report documenting the findings. The following are components of an effective feasibility report.

1. Introduction (overview, background)
   Objectives. Under this subheading, answer any of the following questions:
   - What is the purpose of this feasibility report? One of your responsibilities is to provide background data. To answer the question regarding the report's purpose, provide a clear and concise statement of intent.
   - What problems motivated this study? To clarify for your readers the purposes behind the study, explain what problems cause doubt about the feasibility of the project (i.e., is there a market, is there a piece of equipment available that would meet the company's needs, or is land available for expansion?). You can also explain what problems led to the proposed project (i.e., current equipment is too costly or time consuming, current facilities are too limited for expansion, current net income is limited by an insufficient market).
   - Who initiated the feasibility study? List the name(s) of the manager(s) or supervisor(s) who requested this report.

   Personnel. Document the names of your project team members, your liaison between your company and other companies involved, and your contacts at these other companies.

2. Discussion (body, findings)
   Under this subheading, provide accessible and objective documentation.
   Criteria. State the criteria upon which your recommendation will be based. Criteria are established so you have a logical foundation for comparison of personnel, products, vendors, costs, options, schedules, and so on.
   Analysis. In this section, compare your findings against the criteria. You might want to use a visual such as a table to organize the criteria and to provide easy access to information.

3. Conclusion/recommendations
   Conclusion. In this section, state the significance of your findings. Draw a conclusion from what you have found in your study. For example, state that "Tim is the best candidate for director of personnel" or "Site 3 is the superior choice for our new location."
   Recommendations. Once you have drawn your conclusions, the next step is to recommend a course of action. What do you suggest that your company do next? Which piece of equipment should be purchased, where should the company locate its expansion, or is there a sufficient market for the product?

Figure 17.7 presents an example of a feasibility/recommendation report.

FIGURE 17.7 Feasibility/Recommendation Report

From: Cindy Katz, Director of Information Technology
To: Shamir Rammalah, Accounts Payable
Date: August 13, 2016
Sbject: Feasibility Study for Technology Purchases

Purpose of the Report

The purpose of this report is to study which technology will best meet your communication needs and budget. After analyzing the feasibility of various technologies, we will recommend the most cost-effective technology options.

Vendor Contacts

Our vendor contacts for the laptops, printers, and software are as follows:

Electek	**Tech On the Go**	**Mobile Communications**
Steve Ross	Jay Rochlin	Karen Allen
stever1@electek.com	jrochlin@tog.com	karen.allen@mobcom.net

Criteria for Vendors

The following criteria were considered to determine which communication technology would best meet your department's needs:

- *Maintenance*—We need to purchase equipment and software complete with either a quarterly or biannual service agreement (at no extra charge).
- *Service Personnel*—The service technicians should be certified to repair and maintain whatever hardware we purchase. In addition, the vendors must also be able to train our personnel in hardware usage.
- *Warranties*—The warranties should be for at least one year with options for renewal.
- *Cost*—The total budgeted for your department is $15,000.

Vendor Evaluation

- **Electek**—Having been in business for ten years, this company is staffed by highly trained technicians and sales staff. All Electek employees are certified for software training. The company promises a biannual maintenance package and subcontracted personnel if employees cannot repair hardware problems. They offer manufacturers' guarantees with extended service warranties costing only $100 a year for up to five years. Electek offers 20 percent customer incentives for purchases of over $2,000.
- **Tech On the Go (TOG)**—This company has been in business for two years. TOG provides only subcontracted service technicians for hardware repair. TOG's employees are certified in software training. The owners do not offer extended warranty options beyond manufacturers' guarantees. No special customer pricing incentives are offered though TOG sells retail at a wholesale price.
- **Mobile Communications**—Having been in business for five years, Mobile has certified technicians and sales representatives. All repairs are provided in house. The company offers quarterly maintenance at a fee of $50 ($200 per year). Mobile offers a customer incentive of 10 percent discounts on purchases over $5000.

The purpose reminds the reader why this report has been written and what the report's goal is.

The vendor section provides contact information (names and e-mail addresses).

The criteria states the topics used to research the report and includes precise details. In this way, the audience can understand the rationale for later decisions.

The organizational mode comparison/contrast is used to analyze the strengths and weaknesses of each vendor.

The discussion provides specific details to prove the feasibility of the plan or project.

FIGURE 17.7 *(Continued)*

Cindy Katz
Page 2
August 13, 2016

The following table compares the three vendors we researched on a scale
of 1–3, 3 representing the highest score.

Graphics depict the findings
more clearly and more
concisely than a paragraph
of text.

Table 1: Criteria Comparison

Criteria	Electek	TOG	Mobile
Maintenance	3	2	3
Personnel	3	3	3
Warranties	3	2	2
Cost	3	2	2
Total	12	9	10

Recommended Action

Given the combination of cost, maintenance packages, warranties, and service
personnel, Electek is our best choice.

We suggest the following options for printers and laptops: purchasing five laptops
instead of ten; purchasing one additional printer instead of three; and/or sharing
printers with nearby departments.

The conclusion sums up
the findings, explaining the
feasibility of a course of
action—why a plan should
or should not be pursued.

The recommendation
explains what should
happen next and provides
the rationale for this
decision.

SHORT, INFORMAL REPORT CHECKLIST

_____ 1. Have you chosen the correct communication channel (e-mail, letter, or memo) for your short, informal report?

_____ 2. Does your subject line contain a topic and a focus?

_____ 3. Does the introduction explain the purpose of the report, document the personnel involved, or state when and where the activities occurred?

_____ 4. When you write the discussion section of the report, do you quantify what occurred by supplying accurate dates, times, calculations, and problems encountered?

_____ 5. Is the discussion accessible? To create reader-friendly ease of access, use highlighting techniques, such as headings, boldface, underlining, itemization, and graphics.

_____ 6. Have you selected an appropriate method of organization in your discussion, such as chronology, importance, comparison/contrast, or problem/solution?

_____ 7. Does your conclusion present a value judgment regarding the findings presented in the discussion?

_____ 8. In your recommendations, do you tell your reader what to do next or what you consider to be the appropriate course of action?

_____ 9. Have you effectively recognized your audience's level of understanding (high-tech, low-tech, lay, management, subordinate, colleague), multiple, internal, or external and written accordingly?

_____ 10. Is your report accurate? Correct grammar and calculations make a difference.

The Writing Process at Work

The best way to accomplish a writing task is by following the writing process. Linda Walkman, an accountant, used prewriting, writing, and rewriting to write a successful report.

Prewriting

Linda used a simple topic outline to gather data and determine her objectives (Figure 17.8).

FIGURE 17.8   Outline of Linda's Letter Report

I. Administrative changes in the company
  A. Third-party administrator hired to handle profit-sharing plan
  B. Change in third-party administrator of profit sharing plan
  C. Company administrative assistant did not notify owner of IRS notice
  D. Company administrative assistant fired
  E. New administrative assistant did not know rules

II. IRS sent second notice
  A. Company contacted former third-party administrator
  B. Original accounting firm had also undergone changes
  C. New accounting firm fired

III. New accounting firm hired
  A. Owner made changes to handling of correspondence and filing of all tax reports and returns

IV. Request for waiving of penalties due to
  A. Unfortunate failure of office assistant
  B. Third-party administrator was supposed to file an extension request
  C. Third-party administrator has been terminated

Writing

In Linda's problem-to-solution rough draft (Figure 17.9), she requested that the IRS waive a penalty assessed against her client. Then, she asked a colleague to provide feedback.

Rewriting

After reviewing the suggestions from her colleague, she sends the letter report to her company's production staff. Their job is to assemble, type, and save the report in the company's files. The production staff makes sure that the report meets the company's standards for font, margin, and design. See Figure 17.10 for Linda's letter report.

FIGURE 17.9   Linda's Rough Draft Letter Report with Colleague Comments

September 29, 2016

Internal Revenue Service
Ogden, UT 84201

RE: **PhotoFinish, Inc. Profit Sharing Plan**
         3900021
         Year ended: April 30, 2015

Dear Sirs:

PhotoFinish has received notices indicating that there is an outstanding penalty for late filing of the profit sharing plan return. They have asked for our assistance in resolving this issue.

PhotoFinish has been through several administrative changes over the past several years. Several years ago, a profit sharing plan was established by PhotoFinish. A third-party administrator was hired to handle the necessary calculations, recordkeeping, preparation of returns, and other administrative functions. This third-party administrator handled these functions for the plan for several years, without incident. PhotoFinish made a decision in the year 2014 to terminate this profit sharing plan and utilize another vehicle for employee savings. The plan was terminated as of December 31, 2015, and the third-party administrator handled the filing of the final returns, and the transfer of funds to another qualified account.

During this same time period, PhotoFinish underwent several personnel changes. The assistant handled the daily bookkeeping, correspondence, and phone calls. During 2015, the IRS issued a notice indicating that a late filing penalty had been assessed regarding the profit sharing plan for the April 30, 2014 year. The assistant received this correspondence, but did not call this to the owner's attention, or call on the third-party administrator for assistance. Due to job performance problems, this assistant was terminated, and another office worker was hired to take her place. The new assistant was unaware of the full extent of any late filings, or other open issues, and was only able to locate small portions of related correspondence regarding this late filing over a period of several months.

Subsequently, in 2016, another notice was issued by the IRS, indicating that the penalty was still outstanding, along with corresponding interest charges. The company contacted their former third-party administrator for assistance at that time. The accounting firm that had handled this filing had also undergone significant personnel changes. The staff who had serviced this account were no longer there. While the accounting office's records indicated that the Form 5500's had been filed timely each year, the actual mailing receipts and original extension applications could not be located. By this time, the accounting firm's involvement with administering this plan had been completed for slightly over four years, so it was difficult to provide any further investigation into the problem.

Sincerely,

Replace "Dear Sirs" with a more informative subject line.

Linda, let's break this text up with headings.

I notice some repetition such as "several" (shown in red). Also, phrases like "made a decision in the year 2012" can be revised for conciseness. Let's check for these style concerns throughout the letter.

Let's avoid excessively long paragraphs and make the information more accessible. How about trying bullets?

This would be a good place for a heading, something like "Corrective Action Taken."

Linda, my English teachers always told me to avoid using passive voice (like the sentence shown in red).

Some of your sentences are too long and difficult to understand.

FIGURE 17.10   Revised Letter Report

Anderson, Hubert, and Walkman, LLC
7701 College Blvd., Suite 150
Overland Park, KS 66210

September 29, 2016

Internal Revenue Service
Ogden, UT 84201

Subject: PhotoFinish, Inc. Profit Sharing Plan
         Case Number: 3900021
         Year ended: April 30, 2015

PhotoFinish received notices indicating that there is an outstanding penalty for late filing of the profit sharing plan return. They have asked for our assistance in resolving this issue.

History of PhotoFinish's Profit Sharing Plan

PhotoFinish has been through several administrative changes over the past five years. Ten years ago, a profit sharing plan was established by PhotoFinish. A third-party administrator was hired to handle the necessary calculations, recordkeeping, preparation of returns, and other administrative functions. PhotoFinish decided in December 31, 2014, to terminate this profit sharing plan and utilize another vehicle for employee savings. The third-party administrator handled the filing of the final returns, and the transfer of funds to another qualified account.

During this same time period, PhotoFinish underwent two personnel changes. The sequence leading to PhotoFinish's problems is as follows:

- The office administrative functions have been handled by the owner along with an assistant. The owner was primarily responsible for sales calls, purchasing, and marketing responsibilities. The assistant handled the daily bookkeeping, correspondence, and phone calls.
- During 2015, the IRS issued a notice indicating that a late filing penalty had been assessed regarding the profit sharing plan for the April 30, 2014, year. The assistant received this correspondence, but did not call this to the owner's attention, or call on the third-party administrator for assistance.

- Due to job performance problems, this assistant was terminated, and another office worker was hired to take her place. The new assistant, unaware of the full extent of any late filings or other open issues, was only able to locate small portions of related correspondence regarding this late filing over a period of several months.

FIGURE 17.10 *(Continued)*

Page 2   September 29, 2016

Subsequently, in 2016, another notice was issued by the IRS, indicating that the penalty was still outstanding, along with corresponding interest charges. The accounting firm that had handled this filing had also undergone significant personnel changes. The staff who had serviced this account was no longer there. While the accounting office's records indicated that the Form 5500's had been filed timely each year, the actual mailing receipts and original extension applications could not be located.

Corrective Action Taken by PhotoFinish

To provide better office procedures and office controls for PhotoFinish, the owner consulted a new accounting firm for assistance with their accounting review, tax return preparation, and other consulting work.

PhotoFinish's Request for Penalty Waiver

PhotoFinish asks you to waive the filing penalty for the following reasons:

1. They did not have a history of late filings for their profit sharing plan Form 5500.
2. They had been using the services of a reputable third-party administrator to handle the return filings, extension applications, and maintain the necessary documentation.
3. Due to the failure of the office assistant to take action on the notices, or alert anyone of the notices, PhotoFinish was unaware of this issue until five years had passed. As soon as they were aware of the outstanding penalty, they took appropriate action.
4. PhotoFinish believed that an extension was properly filed, due to the normal course of action taken by their former third-party administrator.
5. PhotoFinish has hired a new accounting firm to prevent future omissions in dealing with compliance filings and other paperwork.

PhotoFinish has made reasonable efforts to solve this filing issue, and reasonable cause should apply in this situation. We ask for abatement of the late filing penalty as well as any related interest charges. Thank you for helping us with this problem.

Sincerely,

Linda A. Walkman, C.P.A.

Linda says, "Following the communication process leads to my success as a written communicator. Prewriting, writing, and rewriting allow me to create well-written, carefully edited documents. I have a 95 percent 'win ratio' when communicating with the IRS. What a great testament to my success as a writer."

CASE STUDIES

1. You manage an engineering department at Acme Aerospace. Your current department supervisor is retiring. You must recommend the promotion of a new supervisor to the company's executive officer, Kelly Adams. You know that Acme seeks to promote individuals with the following traits:

   - Familiarity with modern management techniques and concerns, such as teamwork, global economics, crisis management, and the management of hazardous materials.
   - An ability to work well with colleagues (subordinates, lateral peers, and management).
   - Thorough knowledge of engineering.

   Using the information provided about the following candidates and the criteria for feasibility/recommendation reports discussed in this chapter, write a report recommending a new supervisor.

   a. **Pat Jefferson.** Pat has worked for Acme for 12 years. Pat has worked up to a position as a lead engineer after starting as an assembler, then working in test equipment, quality control, and environmental safety and health (ESH). As an engineer in ESH, Pat was primarily in charge of hazardous waste disposal. Pat has completed two years of college coursework and one class in management techniques. Pat is well liked by all colleagues and is considered to be a team player.

   b. **Kim Kennedy.** Kim has worked for ACME for two years. Kim was hired directly out of college after earning an MBA from the Mountaintop College School of Management. As such, Kim is familiar with modern management techniques. Kim's undergraduate degree was a BS in business with a minor in engineering. Currently, Kim works in the engineering department as a departmental liaison, communicating the engineering department's concerns to Acme's other departments. Kim has developed a reputation as an excellent coworker who is well liked by all levels of employees.

   c. **Chris Clinton.** Chris has a BS in engineering from Poloma College and an MBA from Weatherford University. Prior to working for Acme, Chris served on the IEEE (Institute of Electrical and Electronic Engineering) Commission for Management Innovation, specializing in global concerns and crisis management. Hired by Acme in 2000, Chris is now lead engineer in the engineering department. Chris has earned high scores on every yearly evaluation, especially regarding knowledge of engineering. Chris's only negative points on evaluations have resulted from difficulties with colleagues.

Assignment

Write your feasibility/recommendation report explaining whom you would hire.

2. Your company, Telecommunications R Us (TRU), has experienced a 45 percent increase in business, a 37 percent increase in warehoused stock, and a 23 percent increase in employees. You need more room. Your executive officer, Polina Gertsberg, has asked you to research existing options. To do so, you know you must consider these criteria:

   - **Ample space for further expansion.** Gertsberg says TRU could experience 150 percent growth. You need to consider room for parking, warehouse space, additional offices, and a cafeteria—approximately 20,000 square feet total.
   - **Cost.** Twenty million dollars should be the top figure, with a preferred payback of five years at 10 percent.

- **Location.** Most of your employees and customers live within 15 miles of your current location. This has worked well for deliveries and employee satisfaction. A new location within this 15-mile radius is preferred.
- **Aesthetics.** A beautiful site could improve employee morale and increase productivity.

After research, you've found three possible sites. Based on the following information and on the criteria for feasibility/recommendation reports discussed in this chapter, write your report recommending a new office site.

a. **Site 1 (11717 Grandview).** This four-story site, located 12 miles from your current site, offers three floors of finished space equaling 18,000 square feet. The fourth floor is an unfinished shell equaling an additional 3,000 square feet. As is, the building will sell for $19 million. If the current owner finishes the fourth floor, the addition would cost $4 million more. For the building as is, the owner asks for payment in five years at 12 percent interest. If the fourth floor is finished by the owner, payment is requested in seven years at 10 percent. The building has ample parking space but no cafeteria, although a building next door has available food services. Site 1 is nestled in a beautifully wooded area with hiking trails and picnic facilities.

b. **Site 2 (808 W. Blue Valley).** This one-story building offers 21,000 square feet that includes 100 existing offices, a warehouse capable of holding 80 storage bins that measure 20 feet tall × 60 yards long × 8 feet wide, and a full-service cafeteria. Because the complex is one story, it takes up 90 percent of the lot, leaving only 10 percent for parking. Additional parking is located across an eight-lane highway that can be crossed via a footbridge. The building, located 18 miles from your current site, has an asking price of $22 million at 3.75 percent interest for five years. Site 2 has a cornfield to its east, the highway to its west, a small lake to its north where flocks of geese nest, and a strip mall to its south.

c. **Site 3 (1202 Red Bridge Avenue).** This site is 27 miles from your current location. It has three stories offering 23,000 square feet, a large warehouse with four-bay loading dock, and a cafeteria with ample seating and vending machines for food and drink. Because this site is located near a heavily industrialized area, the asking price is $15 million at 3.5 percent interest for five years.

Assignment

Which site do you recommend? Write your feasibility/recommendation report stating your choice.

ETHICAL CHALLENGE

Jason White is the purchasing agent for a computer software company. All of the over 100 computer consultants who sell the company's software travel throughout the country to meet with clients. The consultants need smartphones to stay in touch with management, coworkers, and customers.

Jason knows that he can purchase these smartphones from many sources. Usually, Jason's company requires a minimum of three bids for major purchases. However, a good friend of his has promised to sell Jason the smartphones at "the lowest price possible." Jason wants to save time, save money, and do his friend a favor. To explain his decision, Jason will write a recommendation report to his boss favoring his friend.

Question

Is it ethical for Jason to go against company policy, even if it saves money and time? Why or why not?

INDIVIDUAL AND TEAM PROJECTS

1. Write a progress report. The subject of this report can involve a project or activity at work. If you haven't been involved in job-related projects, write about the progress you're making in this class or another course you're taking. Write about the progress you're making on a home improvement project (refinishing a basement, constructing a deck, painting and papering a room). Write about the progress you're making on a hobby (rebuilding an antique car, constructing a computer, or making model trains). Follow the criteria presented in this chapter regarding progress reports.

2. Write a lab report. The subject of this report can involve a test you're running at work or in one of your classes. Use the criteria regarding lab reports presented in this chapter to help you write the report.

3. Write a feasibility/recommendation report. You can draw your topic either from your work environment or home. For example, if you and your colleagues were considering the purchase of new equipment, the implementation of a new procedure, the expansion to a new location, or the marketing of a new product, you could study this idea and then write a report on your findings. If nothing at work lends itself to this topic, then consider plans at home. For example, are you and your family planning a vacation, the purchase of a new home or car, the renovation of your basement, or a new business venture? If so, study this situation. Research car and home options, study the market for a new business, and get bids for the renovation. Then write a feasibility/recommendation report to your family documenting your findings. Follow the criteria for feasibility/recommendation reports provided in this chapter to help you write the report.

4. Write an incident report. You can select a topic either from work or home. If you have encountered a problem at work, write an incident report documenting the problem and providing your solutions to the incident. If nothing has happened at work lending itself to this topic, then look at home. Has your car broken down, did the water heater break, or did you or any members of your family have an accident of any sort? Consider such possibilities, and then write an incident report documenting the incident. Follow the criteria for incident reports provided in this chapter.

PROBLEM-SOLVING THINK PIECES

1. Angel Guerrero, computer information technologist at HeartHome Insurance, has traveled from his home office to a branch location out of town. While on his job-related travel, he encountered a problem with his company's remanufactured laptop computer. He realized that the problem had been ongoing not only for this laptop but also for six other remanufactured laptops that the company had recently purchased.

   Angel thinks he knows why the laptops are malfunctioning and plans to research the issue. When he returns to his home office from his business trip, he needs to write a report. What type of report should he write? Explain your answer based on the information provided in this chapter.

2. Minh Tran is a special events planner in the marketing department at Thrill-a-Minute Entertainment Theme Park. Minh and her project team are in the middle of a long-term project. For the last eight months, they have been planning the grand opening of the theme park's newest sensation ride—The Horror—a wooden roller coaster that boasts a 10-gravity drop.

   During one of its weekly project meetings, the team has hit a roadblock. The rap group Bite R/B Bit originally slated to play at the midnight unveiling has cancelled at the last minute. The team needs to get a replacement band. One of Minh's teammates has researched the problem and presented six alternative bands (at varying prices and levels of talent) for consideration. Minh needs to write a report to her supervisor. What type of report should she write? Explain your decision based on the criteria for reports provided in this chapter.

3. Toby Hebert is human resource manager at Crab Bayou Industries (Crab Bayou, Louisiana), the world's largest wholesaler of frozen Cajun food. She and her staff have traveled to New Orleans to attend meetings held by five insurance companies that planned to explain and promote their employee benefits packages. During the trip, Toby's company van was sideswiped by an uninsured driver. When Toby returns to Crab Bayou, she has to write a report. What type of report should she write? Base your decision on the criteria for reports provided in this chapter.

4. Bill Baker, claims adjuster for CasualtyU Insurance Company (CUIC), traveled for a site visit. Six houses insured by his company were in a neighborhood struck by a hailstorm and 70 mile-per-hour winds. The houses suffered roof and siding damages. Trees were uprooted, with one falling on a neighbor's car. Though no injuries were sustained, over $50,000 worth of property damage occurred. When he returns to CUIC, what type of report should he submit to his boss? Defend your decision based on criteria for reports provided in this chapter.

WEB WORKSHOP

More and more, companies and organizations are putting report forms online. The reason for doing this is simple—ease of use. Go online and access online report forms. Use any Internet search engine, and type in "online _____ report form." (In the space provided, type in "trip," "progress," "incident," "investigative," or "feasibility/recommendation.") Once you find examples, evaluate how online reports differ from hard-copy reports. Share your findings with others in your class, either through oral presentations or written reports.

CHAPTER EIGHTEEN

Long, Formal Reports

After completing this chapter, you will be able to

1. Learn why to write a long, formal report

2. Distinguish among informative, analytical, and recommendation reports

3. Recognize the major components of long, formal reports

4. Write the front matter of a long, formal report, including a title page, cover letter, table of contents, list of illustrations, and abstract or executive summary

5. Write the long, formal report's text, including an introduction, discussion, and conclusion/recommendation

6. Write the long, formal report's back matter, including a glossary, works cited or reference page, and appendix

7. Use research in a long, formal report

In the following scenario, an information technology consulting company needs to write a long, formal report recommending a backup data storage facility for a client.

Alpha/Beta Consulting (A/B/C) is going to work with Nitrous Solutions to help them choose an off-site backup data storage facility that will be safe, secure, dependable, prompt, and able to grow as the company grows. Nitrous Solutions, a multi-discipline architectural design company, is a relatively new but successful business with no off-site backup data storage location.

Nitrous knows that off-site data backup is important for many reasons. First, after the 9/11 attacks, many companies realized that their data, if only housed on site, could be endangered. What companies need is a defense-in-depth strategy to ensure data security. Off-site data storage provides companies remote backup for worst-case scenarios, such as terrorism. With off-site data storage, even if a company's primary facility is damaged, data stored in another location is preserved.

In addition, weather problems, highlighted by the Hurricane Sandy catastrophe, proved to companies that multiple levels of data redundancy were needed to secure their files. High waters and rain damage caused many businesses in eastern states to lose valuable data. Off-site storage, in contrast, would provide a higher level of security.

Data loss isn't just due to terrorism or natural disasters. Nitrous also realizes that their files are endangered from simple causes like software and hardware failure. When disasters occur, a company can lose days, weeks, or months of past work. Recovering these files takes more time, negatively impacts the company's credibility with clients, and can lead to financial losses.

Nitrous needs to factor in one more consideration when choosing an off-site storage option. The company does not want to trust couriers to transport files from their home site to a separate location. Nitrous doesn't want an employee to put data tapes in a car and drive hundreds or thousands of miles. This kind of transport could lead to additional problems—time, car wrecks, theft, etc. Instead, Nitrous wants a secure means of electronic file transfer that will be quick, safe, and efficient.

Management has hired A/B/C to help them determine their data storage needs. A/B/C has met with management and key employees at Nitrous to study their client's unique needs. Now, A/B/C will write a long, formal report to inform, analyze, and recommend. The report will provide factual information about backup solutions for archiving architectural materials and critical communications for Nitrous.

The long, formal report will then analyze options, focusing on pros and cons of various solutions. In this analysis, A/B/C will draw conclusions about pricing, support, scalability, reliability, ease of use, and archiving choices.

Based on their findings and analysis of the options, A/B/C will draw a conclusion. Then, the consulting company will recommend the best options to help Nitrous fulfill their company's mission statement.

Why Write a Long, Formal Report?

In some instances, your subject matter might be so complex that a short report will not thoroughly cover the topic. For example, your company asks you to write a report about the possibility of an impending merger. This merger will require significant commitments regarding employees, schedules, equipment, training, facilities, and finances. Only a long report, complete with research, will convey your content sufficiently and successfully. Long reports are not simply extensions of short reports. Long, researched reports require time, resources of people and money, and have far-reaching effects.

The following are titles of long, formal, research reports:

- The Effect of Light Rail on Wyandotte County: An Economic Impact Study on Infrastructure
- The Increasing Importance of Mobile Communications: Security, Procurement, Deployment, and Support
- Managing Change in a Technologically Advancing Marketplace at EBA Corp.
- Information Stewardship at The Colony, Inc.: Legal Requirements for Protecting Information
- Remote Access Implementation and Management for New Hampshire State University

1. Learn why to write a long, formal report.

Short Reports

See Chapter 17, "Short, Informal Reports," for additional information about a variety of short reports.

Types of Long, Formal Reports: Informative, Analytical, and Recommendation

2. Distinguish among informative, analytical, and recommendation reports.

Reports can provide information, be analytical, and/or recommend a course of action. Occasionally, you might write a report that only informs. You might write a report that just analyzes a situation. You could write a report to recommend action persuasively. Usually, however, these three goals will overlap in your long, formal reports.

For example, let's say that your company is expanding globally and will need a wide area network (WAN) that spans a large geographical area. In a long, formal report to management, you might first write to inform the audience that this WAN will meet the strategic goals of storing and transmitting information to your global coworkers. The report also will analyze the ways in which this network will be safe, reliable, fast, and efficient. Finally, the report will recommend the best designs for providing secure communications to clients, partners, vendors, and coworkers. Ultimately, you must persuade the audience, through research, that your envisioned network's design, hardware specifications, software, and estimated budget will satisfy both internal and external needs. Such a report would have to be informative, analytical, and persuasive to convince the audience to act on the recommended suggestion.

Information

When you provide information to your audience, focus on the facts. These facts will help your readers better understand the situation, the context, or the status of the topic. For example, Nitrous Solutions asked A/B/C to report on backup data storage options for software archives (see the opening scenario on page 438). To do so, A/B/C needed to present factual information about the importance of backup storage. In a long, formal report, A/B/C informed its client about the following: the threats to data, the cost of lost data, how often data should be backed up, and what data should be archived. (See Figure 18.1.)

Analysis

When you analyze for your audience, you begin with factual information. However, you expand on this information by interpreting it and then drawing conclusions. Once A/B/C presented the informational findings about backup data storage facilities, they followed this information with a more in-depth analysis of backup options and drew a conclusion for their client (see Figure 18.2).

Recommendation

Visual Aids

See Chapter 8, "Visual Aids," for additional information about graphics.

After providing information and analysis, you can recommend action as a follow-up to your findings. The recommendation allows you to tell the audience why they should purchase a product, use a service, choose a vendor, select a software package, or follow a course of action. A/B/C presented findings and analyzed information for its client Nitrous Solutions. Based on the analysis, the consulting company then made a recommendation (see Figure 18.3).

Major Components of Long, Formal Reports

3. Recognize the major components of long, formal reports.

A long, formal report includes the following:

- **Front matter.** Title page, cover letter, table of contents, list of illustrations, and an abstract or executive summary
- **Text.** Introduction including purpose, issues, background, and problems; discussion; and conclusion/recommendation
- **Back matter.** Glossary, works cited or references page, and an optional appendix

Figure 18.4 shows the components of a long, formal report.

FIGURE 18.1   Information about Backup Data Storage

Description of Backup Data Storage

The following information provides facts about permanent storage to help Nitrous Solutions maintain a high degree of security in an off-site backup data storage location.

Facts

Due to events such as 9/11 and Hurricane Sandy, the importance of maintaining mission-critical electronic data and backups has become evident for companies that wish to avoid a catastrophic outcome. Moving data to a separate off-site location dilutes risk and protects against data loss when, for example, the company's IT infrastructure or its critical electronic information is damaged by the following:

Threats to Your Data	Dangers and Costs of Data Loss
➤ Fire	➤ Loss of mission-critical files
➤ Flood	➤ Corrupted database
➤ Hurricane or tornado	➤ Corrupted operating system
➤ Lightning strike	➤ Loss of all files on hard drive
➤ Earthquake	➤ Laptop damage, theft or loss
➤ Heat, sunlight	➤ Damaged computer hardware
➤ Humidity, moisture, and spilled liquids	➤ Backup media loss or theft
➤ Smoke, dust, and dirt	➤ Total loss of all data at your site
➤ Electrical surge or power failure	➤ Compromised data security
➤ Media failure	➤ Competitor access to your data
➤ Hard drive failure	➤ Lost business records

When and What Data Should be Backed Up?

Some basic questions that are commonly asked when determining when and what data should be backed up are as follows:

How often should desktop PCs or Macs be backed up?

- Data on desktop PCs and Macs should be backed up at a minimum of at least once a week. This will prevent anyone losing more than one week's worth of information.

Should everything on my PC or Mac be backed up once a week?

- This is not necessary. Only data needs to be backed up once a week. Software, operating systems, and even static data do not need to be backed up that often. Backing up everything on your PC or Mac adds time to the process and requires substantially more backup tapes, backup disks, or other types of media.

How long should I keep computer backups?

- Some companies keep original backups up to seven years and have a set schedule that includes five different increments of backups: daily, weekly, monthly, quarterly, and yearly.

This is a topic for discussion with your key management, legal aides, information systems technicians, records management, and key personnel. Balancing the needs of a company's computer system protection and good records management takes informed decisions on the part of all these departments.

Information is presented in a variety of ways. The consulting company details data in bulleted lists and explains facts and findings in paragraphs.

The writer uses a Q & A format to provide answers to typical customer questions.

FIGURE 18.2   Analysis of Backup Data Storage Options

Analysis of Backup Data Storage Options

Industry analysts claim that two out of five businesses that experience a disaster will go out of business within five years of the event due to information and service lost as a consequence of disasters.

This is an alarming statistic considering the high cost of not expecting the unexpected. In the event of a disaster, a corporation could lose hundreds, thousands, or even millions of dollars through lost productivity. At worst the corporation could go out of business without the possibility of a second chance.

The two main types of off-site data storage include the following:

Off-site Cloud Storage	Off-site Facility Storage
Off-site Internet cloud data storage has quickly become one of the fastest growing arenas in the records and information management industry. A business can simply select the files and the time that it wishes to have its files backed up with the off-site backup software. Its files are compressed and encrypted and sent to a backup server over an existing Internet connection. The business can upload or download your files as often as needed, with usually no additional charges.	Tapes, CDs, DVDs, or hard drives are sent to a predetermined off-site location for security and disaster recovery purposes. This allows a business to be safe in the knowledge that it will be able to recover quickly and seamlessly from any disaster. With simply a walk across the street or a drive downtown to the storage facility, the business can quickly begin to rebuild from the data it has stored.

Off-site Location for Data Storage Options

Off-site Cloud Storage	Off-site Facility Storage
Costs	
• Storage fees are calculated in gigabytes (ranging from .05 to .08 cents a month).	• $5–$650 per month depending on space and features
Reliability	
• Data are accessible when needed. • Save and archive as much as you need with no restrictions or extra fees.	• Insurance on data is automatic with contract. • Data are fully accessible during normal business hours, after hours by emergency only.

To clarify the distinction between storage facility options, the report uses a comparison/contrast format.

To help the audience understand the topic, the report analyzes unique features of storage facility options.

FIGURE 18.2 (Continued)

Security	
• More secure than leaving your data in your office building • Significant backup systems in place	• Vault and/or building monitored by 24-hour surveillance and alarm systems • Security patrolled

Scalability	
• Can store as little or as much data as needed, with no restrictions, guidelines, or rules	• Just like the Internet storage, as your company grows, so can data storage. This increases monthly fees.

Ease of Use	
• Easy to set up and manage • Step-by-step instructions available • Easy billing, access, deactivation, and authorization	• Easy to set up and manage • Step-by-step instructions available • Some companies offer free technical help.

Conclusion

Regardless of how much or how little a business uses a computer, your company will create important and unique data. This unique data can include financial and project budget records, digital images, client profiles, and marketing sales. This data is priceless and constantly at risk.

Data loss is likely to occur for many reasons:

- Hardware failure = 42%
- Human error = 35%
- Software corruption = 13%
- PC viruses = 7%
- Hardware destruction = 3%

> When analyzing information, A/B/C presents facts and figures, pros and cons, and specific details.

Front Matter

A long, formal report prefaces the report's text with the following components: title page, cover letter, table of contents, list of illustrations, and an abstract or executive summary.

> 4. Write the front matter of a long, formal report, including a title page, cover letter, table of contents, list of illustrations, and abstract or executive summary.

Title Page

The title page tells your reader the

- Title of the long report (thereby providing clarity of intent)
- Name of the company, writer, or writers submitting the long report
- Date on which the long report was completed

If the long report is being mailed outside your company to a client, you also might include on the title page the audience to whom the report is addressed. If the long report is being

FIGURE 18.3   Analysis and Recommendation

Off-site Storage Options Analysis and Recommendation

Analysis:

The following table analyzes the differences between two storage options.

Type	Off-site Cloud Storage	Off-site Facility Storage
Criteria	*Rating	*Rating
Cost	5	3
Performance	5	3
Reliability	5	4
Security	5	4
Scalability	5	5
Ease of use	5	4
Total	30	23

*Rating: The weighted analysis is based on a 1 to 5 scale: 1 is poor, 5 is excellent.

Recommendation:

Because Nitrous Solutions does not currently have a network infrastructure in place, the data storage decision should be considered a high priority since all new data that the company creates will be significant to the company's overall success. Alpha/Beta Consulting recommends that Nitrous Solutions use the off-site cloud data storage option due to the significant advantage it has over a professional storage facility.

Based on information and an analysis of the findings, provided in a weighted table, A/B/C recommends a backup solution for the client.

FIGURE 18.4   Components of a Long, Formal Report

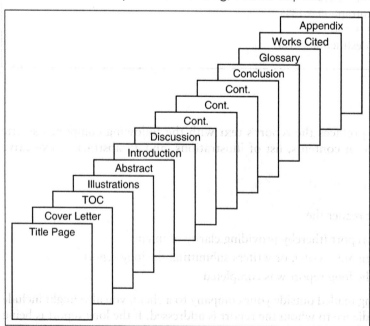

submitted within your company to peers, subordinates, supervisors, or owners, you might want to include a routing list of individuals who must sign off or approve the report.

Following are two sample title pages. Figure 18.5 is for a long report with routing information. Figure 18.6 is for a long report without routing information.

FIGURE 18.5   Title Page for Long Report (with routing information)

Report on Multicultural Workforce at StartCo Insurance

Prepared by: _____ Date: _____
            Pete Niosi
            Assistant Director, Human Resources

Reviewed by: _____ Date: _____
            Leah Workman
            Manager, Accounting

Recommended by: _____ Date: _____
            Greg Foss
            Department Supervisor, Customer Service

Recommended by: _____ Date: _____
            Shirley Chandley
            Director, Human Resources

Approved by: _____ Date: _____
            Ralph Houston
            Vice President

FIGURE 18.6   Title Page for a Long Report (without routing information)

Computer Networking Report:
The Need to Upgrade Intranet Capabilities   Title the report.

for
Acme Products, Inc.   Provide the reader's
2121 New Line Avenue   name(s), title, and company
San Antonio, TX 78666   name.

Submitted by
Thomas Brasher
CEO, Technology Upgrades   Give the writer's name(s),
3254 West King's Highway   title, and company.
San Antonio, TX 78221

August 13, 2016

Cover Letter

Cover Letters

See Chapter 9, "Routine Correspondence," for additional information about writing letters.

Your cover letter prefaces the long report and provides the reader an overview of what is to follow. It tells the reader

- Why you are writing
- What you are writing about (the subject of this long report)
- What exactly of importance is within the report
- What you plan to do next as a follow-up
- When the action should occur
- Why that date is important

Table of Contents

Long reports are read by many different readers, each of whom will have a special area of interest. For example, the managers who read your reports will be interested in cost concerns, timeframes, and personnel requirements. Technicians, in contrast, will be interested in technical descriptions and instructions. Not every reader will read every section of your long report.

Your responsibility is to help these different readers find the sections of the report that interest them. One way to accomplish this is through a table of contents. The table of contents should be a complete and accurate listing of the main and minor topics covered in the report. In other words, you don't want just a sketchy outline of major headings. This could lead to page gaps; your readers would be unable to find key ideas of interest. In the "before" table of contents, the discussion section contains approximately 16 pages of data. What is covered in those 16 pages? Is anything of value discussed? We don't know.

BEFORE

This poorly written table of contents omits second-level headings and has too many page gaps.

Table of Contents

In contrast, an effective table of contents fleshes out this detail so your readers know exactly what is covered in each section. By providing a thorough table of contents, you will save your readers time and help them find the information they want and need. The "After" example is a successful table of contents.

Table of Contents

Second-level and third-level headings help the audience access all parts of the report in any order.

This sample table of contents follows the IEEE (Institute of Electrical and Electronics Engineers) numbering style (1.0, 1.1, 1.2, 2.0, etc.) to number headings. See the sample proposal in Chapter 19 for an alternative approach to headings on the table of contents without numbers.

Creating a Hierarchy of Headings Using Microsoft Word 2013

When writing your long, formal reports, include headings and subheadings. To help your readers navigate the text, create a clear hierarchy of headings (also called *cascading headings*) that distinguishes among first-level headings, second-level headings, third-level headings, and fourth-level headings.

Microsoft Word 2013 will help you create this hierarchy of headings. You can apply a style to your entire document in Word's "formatting text by using styles" tool. For example, let's say that you want your headings and subheadings to look as follows:

Hierarchy of Headings	Description
HEADING ONE	First-level headings: Arial, all caps, boldface, 16 pt. font.
Heading two	Second-level headings: Times New Roman, italics, boldface, 14 pt. font.
Heading three	Third-level headings: Times New Roman, boldface, 12 pt. font.
Heading four	Fourth-level headings: Times New Roman, italics, 12 pt. font.

To create a hierarchy of headings, click on the Home tab and choose the option that works best for you.

Microsoft Corporation, Inc.

List of Illustrations

If your long report contains several tables or figures, provide a list of illustrations. This list can be included after your table of contents, if there is room on the page, or on a separate page. As with the table of contents, your list of illustrations must be clear and informative. Don't waste your time and your reader's time by providing a poor list of illustrations like in the "Before" example. Instead, provide a list as shown in the "After" example.

BEFORE	AFTER

Abstract

As mentioned earlier, a number of different readers will be interested in your long report. Because your readers are busy with many different concerns and might have little technical knowledge, they need your help in two ways: They need information quickly, and they need it presented in low-tech terminology. You can achieve both these objectives through an abstract or executive summary. The abstract is a brief overview of the proposal or long report's key points geared toward a low-tech reader. To accomplish the required brevity, limit your abstract to approximately one to two pages.

Each long report you write will focus on unique ideas. Therefore, the content of your abstracts will differ. Nonetheless, abstracts might focus on the following: the *problems* necessitating your report, your suggested *solutions*, and the *benefits* derived when your suggestions are implemented. For example, let's say you are asked to write a formal report suggesting a course of action (limiting excessive personnel, increasing your company's workforce, improving your corporation's physical facilities, etc). First, your abstract should specify the problem requiring your planned action. Next, mention the action you are planning to implement. This leads to a brief overview of how your plan would solve the problem, thus benefiting your company.

Another approach to an abstract for a long report is to summarize the content being analyzed in the report, setting the stage for the major issues to be discussed. In this abstract, highlight all the main points in the long report, omitting any supporting facts and documentation (to be developed in the discussion section of the report). You not only want to be brief, focusing on the most important issues, but also you should avoid high-tech terminology and concepts. The purpose of the abstract is to provide your readers with an easy-to-understand summary of the entire report's focus. Your readers want the bottom line, and they want it quickly. Therefore, either avoid all high-tech terminology completely, define your terms parenthetically, or refer readers to a glossary.

Executive Summary

An executive summary, similar to an abstract but generally longer, is found at the beginning of either a formal report or a proposal and summarizes the major topics covered in the document. By reading the executive summary, your audience (the decision makers) gets an overview of the much longer report or proposal. To write an effective and concise executive summary, use only the most important details and supporting statistics or information. Omit details or technical content which your reader either does not need to know or would unnecessarily confuse the reader. Details are used in the discussion section of the report or the proposal. To limit the length of your executive summary to one or two pages, choose the most important elements, and omit those that are secondary. An effective summary can include the following:

- Purpose and scope of the report or proposal, stating the problem or need and your ability to improve the situation
- Research or methods used to develop your content
- Conclusions about your analyses of the topic
- Your qualifications showing that you can resolve the issue
- A project management plan and timetable
- The total project budget
- Recommendations based on your findings

Text

The text of a long, formal report includes an introduction, discussion, and conclusion/recommendation.

> 5. Write the long, formal report's text, including an introduction, discussion, and conclusion/recommendation.

Introduction

Your introduction should include the following: purpose, background, and problem.

Purpose. In one to three sentences or a short paragraph, tell your readers the purpose of your long report. This purpose statement informs your readers *why* you are writing or *what* you hope to achieve. This statement repeats your abstract to a certain extent. However, it's not redundant; it's a reiteration. Although numerous people read your report, not all of them read each line or section of it. They skip and skim.

The purpose statement, in addition to the abstract, is another way to ensure that your readers understand your intent. It either reminds them of what they have just read in the abstract or informs them for the first time if they skipped over the abstract. Your purpose statement is synonymous with a paragraph's topic sentence, an essay's thesis, the first sentence in a letter, or the introductory paragraph in a shorter report.

Background and Problem. Whereas the purpose statement should be limited to one to three sentences or a short paragraph for clarity and conciseness, your discussion of the problem must be much more detailed. For example, if you are writing a report about the need for a new facility, explain that your company's current work and storage space are too limited. Your company has a problem that must be solved.

Your introduction's focus on the problem and background, which could average one to two pages, is important for two reasons. First, it highlights the importance of your report and establishes a context for the reader. The introduction emphasizes for your readers the report's priority. In this section, you persuade your readers that a problem truly exists and needs immediate attention.

Second, by clearly stating the problem and background information, you also reveal your knowledge of the situation. This section reveals your expertise. Thus, after reading this section of the introduction, your audience should recognize the importance of the issue and trust you to solve it or understand the complexity of the topic being discussed.

Discussion

The discussion section of your long report constitutes its body. In this section, you develop the detailed content of the long report. As such, the discussion section represents the major portion of the long report, perhaps 85 percent of the text.

What will you focus on in this section? Because every report will differ, we can't tell you exactly what to include. However, your discussion can contain any or all of the items shown in Table 18.1.

TABLE 18.1 Discussion Items

- Analyses
  - Existing situation
  - Solutions
  - Benefits
- Product specifications of mechanisms, facilities, or products
- Comparison/contrast of options
- Assessment of needs
- Features of the systems or products
- Process analyses of the systems or products
- Instructions for completing tasks
- Optional approaches or methodologies for solving the problems
- Managerial chains of command (organizational charts)
- Biographical sketches of personnel
- Corporate and employee credentials
  - Years in business
  - Satisfied clients
  - Certifications
  - Previous accomplishments
- Schedules
  - Implementation schedules
  - Reporting intervals
  - Maintenance schedules
  - Delivery schedules
  - Completion dates
  - Payment schedules
  - Projected milestones (forecasts)
- Cost analyses
- Profit and loss potential
- Documentation and researched material
- Survey results
- Lab report results
- Warranties
- Maintenance agreements
- Online help
- Training options
- Impact on the organization (time, personnel, finances, customers)

You will have to decide which of these sections will be geared toward high-tech readers, low-tech readers, or a lay audience. Once this decision is made, you will write accordingly, defining terms as needed. In addition to audience recognition, enhance your discussion with figures and tables for clarity, conciseness, and cosmetic appeal.

Conclusion/Recommendation

Sum up your long report in a page or so. The conclusion can restate the problem, the important implications of your analysis, your solutions, and the benefits to be derived. In doing so, remember to quantify. Be specific—state percentages and amounts.

Your recommendation will suggest the next course of action. Specify when this action will or should occur and why that date is important. The conclusion/recommendation section can be made accessible through highlighting techniques, including headings, subheadings, underlining, boldface, itemization, and white space.

Back Matter

The long, formal report's back matter can include a glossary, works cited or references page, and an appendix.

6. Write the long, formal report's back matter, including a glossary, works cited or reference page, and appendix

Glossary

Because you will have numerous readers with multiple levels of expertise, you must be concerned about your use of high-tech language (abbreviations, acronyms, and terms). Although some of your readers will understand your terminology, others won't. However, if you define your terms each time you use them, two problems will occur: You might insult high-tech readers, or you will delay your audience as they read your text. To avoid these pitfalls, use a glossary. A glossary is an alphabetized list of high-tech terminology placed after your conclusion/recommendation.

A glossary is invaluable. Readers who are unfamiliar with your terminology can turn to the glossary and read your definitions. Those readers who understand your word usage can continue to read without stopping for unneeded information.

Recognizing Audience

See Chapter 5, "Audience Recognition," for additional information.

Ethical Considerations when Documenting Sources in a Long, Formal Report

If you use research to write your long report, include a works cited or references page. This page(s) documents the sources (books, periodicals, interviews, computer software, Internet sites, etc.) you have researched and from which you have quoted or paraphrased. Correct documentation and source citations are essential in your long report to enhance your credibility and demonstrate your ethical behavior. Remember that boilerplate content and templates, already created material in your company's document library, do not necessarily require documentation. However, if you research the material on the Internet or from any other published source, you must document this material or be guilty of unethical behavior.

Ethics

See Chapter 6, "Ethical Considerations," Chapter 16, "Research," and Appendix B, "Parenthetical Source Citations and Documentation," for additional information.

Appendix

A final, optional component is an appendix. Appendices allow you to include any additional information (survey results, tables, figures, previous report findings, relevant letters or memos, etc.) that you have not built into your long report's main text. The contents of your appendix should not be of primary importance. Valuable data (proof, substantiation, or information that clarifies a point) should appear in the text where it is easily accessible.

Using Research in Long, Formal Reports

You can use researched material, quotes, and paraphrases to support and develop content in your long, formal report. Technical communicators often ask how much of a long, formal report should be *their* writing, as opposed to researched information. A general rule is to lead into and out of every quotation or paraphrase with your own writing. In other words,

- Make a statement (your sentence).
- Support this generalization with a quotation, paraphrase, or graphic (referenced material from another source).
- Provide a follow-up explanation of the referenced material's significance (your sentences).

Research Includes Secondary and Primary Sources

To conduct *secondary research*, you rely on printed and published information taken from sources including books, periodicals, newspapers, encyclopedias, reports, proposals, or other business documents. You might also rely on information taken from a Web site or a blog. All of this secondary research requires parenthetical source citations. In some instances, however, you will use already existing (boilerplate) material from your company's document files. This in-house material does not necessarily require documentation.

Primary research (field research) is research performed or generated by you. You do not rely on books or periodicals. Instead you create original research by preparing a survey or a questionnaire targeting a group of respondents, networking to discover information from other individuals, visiting job sites, or performing lab experiments. With primary research, you will be generating the information based on data or information from a variety of sources that might include observations, tests of equipment, interviews, networking, surveys, and questionnaires.

Preparing Questionnaires

See Chapter 16, "Research," for information about questionnaires.

LONG, FORMAL REPORT CHECKLIST

_____ 1. Have you included the major components for your long, formal report (front matter, text, and back matter)?

_____ 2. In your long, formal report, have you written to inform, analyze, and/or recommend?

_____ 3. Have you used research effectively, incorporating quotes, paraphrases, and graphics successfully?

_____ 4. Have you used primary and/or secondary research?

_____ 5. Did you correctly give credit for the source of your research?

_____ 6. Have you correctly used a hierarchy of headings to help your audience navigate the text?

_____ 7. Did you use tables and/or figures to develop and enhance content?

_____ 8. Have you met your audience's need for definitions of acronyms, abbreviations, and high-tech terms by providing a glossary?

_____ 9. Have you achieved clarity and conciseness?

_____ 10. Is your text grammatically correct?

The Writing Process at Work

To communicate effectively, Jon Russell, Assistant City Manager/City Clerk of Round Rock, Alabama, says that his city's engineers, police officers, fire chiefs, financial advisors, and parks and recreation employees often write formal reports to city council members and the city's mayor. These long reports help the city council members "make informed decisions" that impact the city's residents and business owners. To write his reports, Jon uses prewriting, writing, and rewriting.

Prewriting

To get started writing his report, Jon uses brainstorming to list the project requirements (Figure 18.7).

Save and Close   Mark Complete   📎   Repeat...   Forward   ✕   ▥ ▾

Subject: Prioritized List of Needs

Start date: None   ▾   Status: Not started   ▾

Due date: None   ▾   Priority: Normal   ▾

Date completed: None   ▾   % complete: 0   ▾

☑ Reminder: Mon 3/01/2016   ▾   8:00 AM   ▾   ☐ Private

Owner: Jon Russell

Fewer details ☆

Total work: Hours   ▾   Actual work: Hours   ▾

Mileage: Billing:

Companies:

Prioritized List of Needs

1. End-user ease of use, complete with extensive online help for inexperienced end users
2. Ability to accomplish specific finance tasks
    a.   automated interest allocations
    b.   pooled cash
    c.   seamless purchasing card interaction
    d.   superior budget module
    e.   project accounting
3. Lotus Notes compatibility
4. Seamless interaction with Excel and Word

Writing

Figure 18.8. shows the rough draft of Jon's executive summary.

FIGURE 18.8   Rough Draft of Executive Summary with Suggestions for Revision

Re: Procurement of new financial software

The financial software (ACS) that the Finance Department is currently utilizing is 16 years old. The current software is not Windows based and is extremely cumbersome to accomplish the simplest tasks. Many staff hours are wasted retyping information into Excel spreadsheets for reporting purposes; dual and even triple data entry is common due to the shortcomings of the current software. Additionally, the current software currently resides on an AS/400. Many technology experts believe the AS400 is becoming obsolete. In their management letter for year ended December 31, 2014, the City's auditors, Cottrell and Kreisler, LLC made several references to the inadequacy of the current software. Attached is an excerpt from the management letter referencing these weaknesses. They site multiple internal control issues, inability to track vital information, and redundant data entry. During the 2015 budget discussions, a Decision Package was approved for $90,000 for hardware or software that might be required for a new financial software package. It was determined at that time to wait to make the decision on how to proceed.

"Extremely" is misspelled.

Avoid using the word "current" repetitively.

Jon, add headings to break the text into more manageable units.

(Continued)

Check the data. Some of the figures and dates do not look accurate according to my research.

Requests for proposals for new financial software were sent to seven software vendors. The following table illustrates the five responses received on March 14, 2015.

New Land	Utica, NY	$54,295
MEGS	St. Louis, MO	$59,538
MoonRay	Jupiter Beach, FL	$129,884
ACS	Springfield, IA	$23,800
Roedell	Taos, NM	$19,318

The paragraphs are just too long. No one can wade through so many sentences without any breaks.

Avoid lengthy sentences. Check for grammar and spelling errors throughout the text. Explain why the cost analysis goes $70,000 over budget.

After a review of the responses, three vendors were selected to provide on-site demonstrations. Staff from finance and from other departments spent many hours comparing these vendor's products. The highest priorities were identified as end-user ease of use, ability to accomplish specific finance tasks, and Lotus Notes compatibility. MEGS software met all of the department's needs including the following: automated interest allocation, pooled cash, seamless purchasing card interaction, superior budget module, project accounting, etc. Unfortunately, MEGS is $70,000 over our allotted budget of $90,000. MEGS has offered to divide the payment over two years so the remaining $70,000 would be included in the General Fund in the 2016 budget. With the implementation of the software, the Finance Department will have some cost reductions related to the constant need for temporary help to accomplish current requirement for redundant data entry.

Rewriting

Jon revised and rewrote the draft by considering his colleague's revision suggestions. See Figure 18.9 for Jon's rewritten executive summary.

Long, Formal Report Sample

See Figure 18.10 for a sample long, formal report based on research.

FIGURE 18.9 Rewritten Executive Summary for City Manager and City Council Members

Executive Summary

Background of Finance Department's Software Requirements

The Finance Department's current ACS software is 16 years old. Because the software is outdated, it has the following problems:

- The software is not Windows based.
- The software is not time efficient. Many staff hours are wasted retyping information into Excel spreadsheets for reporting purposes. Dual and even triple data entry is common due to the shortcomings of the current software.
- The ACS financial software resides on an AS/400 hardware system. Many technology experts believe the AS400 is becoming obsolete.

FIGURE 18.9 *(Continued)*

- In their management letter for year ended December 31, 2014, the City's auditors, Cottrell and Kreisler, LLC, made several references to the inadequacy of the current software, including the following: multiple internal control issues, inability to track vital information, and redundant data entry.

During the 2015 budget discussions, a Decision Package was approved for $90,000 for hardware or software that might be required for a new financial software package.

Comparison of Vendor Proposals

Requests for proposals for new financial software were sent to seven software vendors. Table 1 illustrates the five responses received on March 14, 2015.

Table 1 Vendor Comparison		
Vendors	Locations	Proposed Fees
New Land	Utica, NY	$254,295
MEGS	St. Louis, MO	$159,538
MoonRay	Jupiter Beach, FL	$129,884
ACS	Springfield, IA	$123,800
Roedell	Taos, NM	$109,318

Jon says, "Working with a colleague on my rough draft helped me visualize how the reader would respond. The addition of bullets, headings, shorter paragraphs, and a table will enhance the summary's readability."

FIGURE 18.10 Long, Formal Report Recommending a Linux Operating System

Linux Desktop in Enterprise Settings at Design International, Inc.

Submitted by
John Staples, Manager of Information Technology
March 2, 2016

(Continued)

Date: March 2, 2016
To: Tiffany Steward, CEO
From: John Staples, Manager Information Technology
Subject: Recommendation for Migration to a Linux Operating System

Ms. Steward, thank you for allowing me to report on a possible solution to Design International's operating system challenges. This report is in response to a survey my team conducted and the Information Technology Department's focus on improving internal and external communication.

Currently, we use Microsoft Windows as our computing system. However, Windows has a few inherent challenges. These include limited security, limited customization, and increased cost. Based on substantial research, my team and I would like to present an option—Linux. This report will provide you the following information:

- A definition of Linux   page 1
- A comparison/contrast of Linux vs. Windows   page 2
- An explanation of how Linux works   page 2
- The benefits of migrating to Linux   page 8

Information Technology believes migrating to a Linux Operating System can benefit Design International. I appreciate your thoughts on this report and look forward to answering any questions you might have. If you could assess this report by March 20, we would be able to initiate a Linux migration before the next quarter begins.

FIGURE 18.10 *(Continued)*

Table of Contents

List of Illustrations

Summarizing the report in the Abstract allows the reader to get a "snapshot" of the content. Since most people think that Microsoft Windows is the only option for desktop computing, this report discusses an alternative—Linux.

The IT manager lets the reader see that a recommendation will be made based on the facts in the report.

Highlighting problems in the introduction shows the importance of your topic and the need to consider other options.

Embedding secondary research helps to develop the analysis and gives authority to your argument.

Abstract

The Microsoft Windows desktop has established a dominant presence in the enterprise desktop marketplace. Currently, the Windows desktop has nearly total control over the enterprise desktop market primarily due to

- an aggressive marketing strategy
- its intuitive ease of use
- its apparent ease of installation and administration
- readily available support

Windows, however, is not the only viable desktop solution in the enterprise marketplace. Because of potential risks and benefits associated with any technology product, as IT manager I must be alert for alternatives that are secure, reliable, and cost effective.

This report provides technical information regarding the issue of migration to a Linux enterprise desktop solution, analyzes the pros and cons of migrating to a Linux solution, and recommends appropriate action for our company, Design International, Inc.

Introduction

Purpose

We need to consider changing from a Windows-based desktop to a Linux-based desktop for Design International.

Challenges Created by Relying Only on Windows

Though Microsoft Windows is the standard in the marketplace, it presents the following challenges:

- Security—In Windows, the operating system and key applications are combined, which can lead to security issues for e-mail clients. In addition, users of Windows are dependent upon the vendor for recognition and correction of security issues, as well as the implementation of security patches.
- Customization—Prepackaged IT systems, such as Windows, disallow Design International to customize our computerization.
- Cost—Windows requires that Design International pay costly software licensure fees. This cost is compounded by the fact that Windows is platform-dependent, limiting server selection.

Discussion

What the Linux Desktop Environment Is

The current industry standard for the computing desktop is the Microsoft operating system. This system, like all other Microsoft products, is distributed to consumers, including enterprise consumers, in machine code only. The source code is kept as a tightly guarded secret. The Microsoft operating system comes to the consumer as an end-to-end solution that includes operating system, desktop, common applications, and utilities all bundled into one package.

In contrast, Linux is not necessarily distributed as machine code, nor is it necessarily an end-to-end solution. "Strictly speaking, Linux refers to the kernel maintained by Linus Torvalds and distributed under the same name through the main repository and various mirror sites. This code base includes only the kernel and no utilities whatsoever. The kernel provides the core system facilities" (Yaghmour).

1

FIGURE 18.10   *(Continued)*

The Linux kernel is open source. As such, a custom operating system may be built to suit a particular need based on the kernel. However, most distributions of Linux include common user applications and are marketed by one of several vendors in the open source community such as RedHat, SuSe, Mandrake, or Caldera. The products currently available come with differing levels of utilities, installation tools, support, and documentation. Linux is available either by way of machine code or in its native source code format. In this manner, and by virtue of the general public license, the user can modify the operating system, so long as the source code remains open.

Linux desktop environments including KDE and GNOME are utilities that are available for the user to interface with the other utilities that are driven by the Linux kernel. These tools also provide a graphic user interface (GUI) for the user to interact with the Linux operating system. Because a number of GUI desktop solutions are available in the marketplace, the user is free to choose from the ones that come with their Linux distribution or to search from the other available alternatives.

Linux vs. Windows Design Issues

The Microsoft Windows operating system follows the monolithic design. Most of the features of the system, including the GUI by way of the Internet Explorer (IE), are built into one single unit, the kernel (Petreley 20).

While the Linux kernel is also characterized in the literature as being monolithic in relation to traditional Unix microkernel design, fewer of the utilities and drivers are part of the kernel. What further differentiates the design of the Windows operating system and the design of Linux is that Linux is fully modular (Petreley 21). Microsoft developed the Internet Explorer browser as an integrated part of the operating system, ostensibly to provide for more convenience for the user. However, because of the interdependencies between IE and the other parts of the operating system, the entire system is exposed to any flaws that exist in the interdependent system (Petreley 23).

The Linux operating system does not for the most part allow graphics drivers to run inside the kernel. "A bug in a graphics driver may cause the graphical desktop to fail, but not cause the entire system to fail. If this happens, one simply restarts the graphical desktop" (Petreley 25).

Another architectural issue has been that Windows is heavily reliant on the remote procedural call (RPC), a situation where one computer initiates a call on another computer, over a network. While this is sometimes not possible to avoid, it is a procedure that should be kept to a bare minimum. This is the case with Linux (Petreley 27).

A final architectural issue that has rendered security issues with Windows is the high level of reliance on the GUI for performing administrative tasks. When the administrator is performing tasks as the administrative user, the GUI being used is essentially an instance of IE. Because this browser is built so closely to the operating system, the underlying kernel, and in the case of a server, the entire network, is exposed to security risks that are inherent with the browser (Petreley 22).

How Linux Works

Proprietary software primarily exists for the purpose of generating revenue for the owner of the copyright to the software. Its use is permitted in accordance with a licensing agreement. According to Kenston, software licenses serve three basic functions:

2

Comparing and contrasting facts about the options helps to explain the topic to the reader. By focusing on pros and cons, you analyze the topic, clarifying the issue and supporting the conclusions you will draw later.

(Continued)

1. They create profit for the developer.

2. They provide a means for controlling the distribution of the product.

3. They protect the underlying source code from modification. (6)

Raymond characterized software developed under the cathedral model as "carefully crafted by individual wizards or small bands of magi working in splendid isolation, with no beta released before its time" (29).

These concepts of business and development have two primary advantages:

1. They protect the vendor's profit motive.

2. They provide for centralized control of changes and development, along with a mechanism for funding and sustaining ongoing research and development.

Open source software exists for the benefits of the user community. It is distributed freely under the General Public License (GPL), which states that anyone may use or distribute the source code. The GPL says, "You may modify your copy or copies of the Program or any portion of it, thus forming a work based on the Program, and copy and distribute such modifications or work under the terms of Section 1 above" ("GNU").

In general, open source software products, including Linux, are developed by groups of users collaborating together to improve systems and applications. Raymond provides a description of the culture in which open source software is developed. Some of the cornerstones of this culture that were described by Raymond are as follows:

- Software development starts with some personal need of the developer.
- The developer remains in charge of the project, until it is passed along to some other competent developer.
- Reuse and improvement of code is preferable to coding from scratch.
- Treating users as co-developers helps to harness their creative energy.
- Releasing beta code often and listening to feedback leads to rapid debugging. (Raymond 27-28)

File Systems for Linux

The Second Extended File system (ext2fs) was developed specifically for Linux. According to Dalheimer et al., "The ext2fs is one of the most efficient and flexible file systems" (35). This file system was designed to provide an implementation of UNIX file commands and offers advanced file system features. This system is currently the most common file system found in Linux distributions. Some of the features that are supported by ext2fs are

- Standard UNIX file types, directories, device files, and links
- Large file systems, up to 32TB
- Long file names, usually 255 characters, but extendable to 1,012
- Root user reserved files
- Root user block sizes that can improve I/O performance
- File system state tracking (Dalheimer et al., 6)

Highlight facts and rely on research to emphasize important parts of the topic and strengthen the analysis.

3

FIGURE 18.10 *(Continued)*

In addition to ext2fs, Linux supports a wide range of non-Linux file systems including

- Minix-1 and XENIX
- Windows NTFS (Windows 2000 and NT)
- VFAT (Windows 95/98),
- FAT (MS-DOS)
- Macintosh
- HPFS [OS/2]
- Amiga
- ISO 9660 [CD-ROM]
- Novell
- SMB
- Apple McIntosh
- BFS [SCO Unix]
- UFS

(Dalheimer et al., 6)

Network Support for Linux

The Linux kernel provides for a full implementation of the TCP/IP networking protocols. According to Dalheimer et al., some of the features included with the kernel are

- Device drivers for a wide range of Ethernet cards;
- Parallel Line Internet Protocol (PLIP)
- IPv6 protocol suite
- DHCP
- Appletalk
- IRDA
- DECnet
- AX.25 for packet radio networks
- FTP
- Telnet
- NNTP
- Simple Mail Transfer Protocol (SMTP)

(Dalheimer et al., 42)

How Linux Addresses the Enterprise Desktop Need at Design International
The User and the Enterprise Desktop

Two examples of desktops are emerging as the front runners for enterprise use: KDE and GNOME. Both provide a colorful GUI environment that is highly customizable, even to the extent that it can be made to look a lot like the familiar Windows environment. Support for other features including drop and drag, multiple languages, and session management is now available.

4

Because long report topics are complex, help your reader navigate the text easily. To create a readable style in your long report, use short sentences, short paragraphs, bulleted lists, and headings and subheadings.

(Continued)

Enterprise Desktop Management

While the Linux desktop has evolved in the past few years, until recently, there was no true enterprise desktop management solution for the Linux environment. While installing a Linux desktop in a network environment has been easy, it is much more difficult to manage an entire network of desktops. Even the most basic desktop management issues such as common configuration management, software installation, software updates, printer configurations, network file configuration, and authentication management become a resource drain without an enterprise management solution.

What Are Design International's Options for Accessing Linux?

We can load Linux systems on our computers at Design International in the following ways (Freedman 69):

- **Live CD**: The most risk-free approach to accessing Linux is achieved by inserting CDs in all our computers, rebooting, and circumventing Windows once the system restarts. Computers started this way will function as Linux machines. This solution, however, is a quick fix and not advisable for long-term functionality.

- **Parallel installation**: We can download and install Linux software and run it parallel to Windows. This way, our end users can choose between the two operating systems at start-up. This solution does not allow users to access both systems synchronously.

- **Windows replacement**: If we choose to use Linux exclusively, we must remove all Windows files and install Linux. Then, if we would want access to Windows, we would need to perform a complete Windows reinstall.

Use of a Linux Environment at Design International

I conducted an online survey of our 3,000 nationwide employees, asking for their input on issues such as security, printer usage, budget, installation of hardware, networking, archiving of files, and more. They ranked their responses on the following four-point scale:

___ 4 = absolutely critical
___ 3 = important
___ 2 = somewhat important
___ 1 = not important

I received a 75 percent response rate, equaling 2,250 responses to the survey (see Appendix). The results of this survey are shown in Figure 1.

5

FIGURE 18.10 *(Continued)*

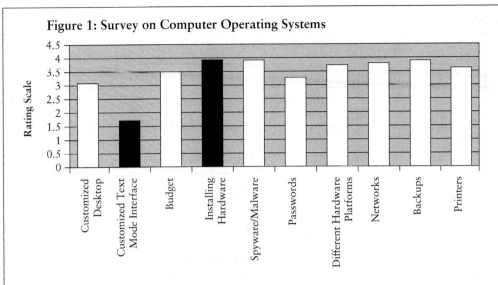

Figure 1: Survey on Computer Operating Systems

Conducting primary research through a survey is an excellent way to add content that is customized for your particular audience and topic.

Explanation of Survey Results

Based on the ten questions asked, the results strongly suggest that our employees would profit from a migration to a Linux environment. Table 1 compares and contrasts key facts regarding Windows and Linux, as represented in the survey:

A key component of analysis is interpretation. Informing allows you to convey data. In analysis, you explain what this data means to provide the audience a better understanding of the consequences of your findings.

Table 1 Comparison/Contrast of Key Facts Regarding Windows vs. Linux	
Windows	**Linux**
Graphical User Interface	
Windows offers some customization of the GUI.	Linux allows advanced users to customize GUIs to their liking.
Text Mode Interface	
Each version of Windows has a single command interpreter.	Linux supports multiple command interpreters that can present a highly customized user interface employing the user interface and input/output facilities.
Cost	
Windows is expensive. Microsoft allows a single copy of Windows to be used on only one computer.	Linux is very cheap and/or free. Once you have purchased Linux, you can run it on any number of computers for no additional charge.

Table 1 uses comparison/contrast to analyze Windows vs. Linux. This analysis provides information and then draws conclusions.

6

(Continued)

Table 1 Continued	
Windows	**Linux**
Getting the Operating System	
Windows comes with an operating system.	PC vendors sell Linux-based machines only as servers, not to consumers, so you must buy a new computer with the operating system pre-installed or as a consumer, you must install the Linux system yourself.
Viruses and Spyware	
Spyware is the worst problem affecting Windows-based computers.	One of the major reasons for using Linux is that viruses and spyware are far less troublesome.
Users and Passwords	
Windows XP allows the omission of the password. Windows software is not designed to be used by a restricted user. Therefore, viruses and malware can access Windows machines easily.	Linux supports root and restricted users, which limits viruses and malware.
Hardware the OS Runs On	
Windows does not run on different hardware platforms.	Linux runs on many different hardware platforms, even very old personal computers such as 486-based machines.
Clustering, Multiple Users, and Networking	
Windows is designed to be used by one person at a time.	Linux has been used to make enormous clusters of computers. Linux is a multi-user system, designed to handle multiple concurrent users.
User Data	
Windows allows programs to store user information (files and settings) anywhere. This makes it hard to back up user data files and settings and to switch to a new computer.	Linux stores all user data in the home directory making it much easier to migrate from an old computer to a new one.
Printer Drivers	
Windows supports many printers.	Linux does not support as many printers as Windows.

7

FIGURE 18.10 *(Continued)*

Migration Planning to the Linux Environment

Migration planning is critical to deploying the Linux desktop. This is the process of identifying how the organization will move from its current desktop environment to the Linux environment. The literature advocates a divide and conquer approach. According to Almond et al., the migration plan will include the assessment of user patterns, establishing functional continuity, dealing with user issues, and retraining considerations.

Table 2 Client Segmentation				
Fixed Function	**Technical Workstation**	**Transactional Workstation**	**Basic Office**	**Advanced Office**
Limited use of business apps	Limited use of business apps	Applications drive business process	Applications drive business process	Applications drive business process
Limited office productivity	Simple office productivity	Simple office productivity	Simple office productivity	Advanced office productivity
No e-mail	Simple e-mail	Simple e-mail	Simple e-mail	Advanced e-mail
No instant messaging	No instant messaging	No instant messaging	No instant messaging	Instant messaging
Simple browser access	Simple browser access	Simple browser access	Simple browser access	Advanced browser applications
Filepoint systems management; network access; host emulation	Filepoint systems management; network access; host emulation	Filepoint systems management; network access; host emulation	Filepoint systems management; network access; host emulation	Filepoint systems management; network access; host emulation

(Almond et al., 28)

A major concern with altering desktop solutions is how such a change will affect the end user. This table highlights key points of the analysis and clarifies pros and cons.

Solutions to Migration Issues

According to Almond et al., six potential technological solutions for solving migration issues are as follows:

- Bridging applications or applications that work in both the old and new environment.
- Similar applications or applications that work in the new environment that have the look and feel of the application used in the old environment.
- Server-based applications or applications that run solely on the server.
- Independent Software Vendor applications or custom software purchased for the new environment.

8

(Continued)

- Web-based functional equivalents or applications that run in the browser, regardless of the operating system.

- Custom porting of custom applications or modifying the code of custom applications so that they operate in the new environment as they did in the old. (33-36)

An example of establishing a migration solution by way of bridging applications is shown in Table 3:

Table 3 Bridging Applications in a Migrations Solution	
Application in Windows	**Linux/Open Source Equivalent**
Internet Explorer	Mozilla.org Firefox
Outlook	Mozilla.org Thunderbird
MS Office Word	OpenOffice.org Writer
MS Office Excel	OpenOffice.org Spreadsheet
MS Office Power Point	OpenOffice.org Impress
PaintShop Pro	Gimp
Messenger Client	Gaim

The table helps your reader quickly see the comparison/contrast between moving from a Windows-based application to a Linux-based environment.

(Almond et al., 31)

Managing User Satisfaction in Migration Planning

Perhaps the most critical issue in migration planning will be managing user satisfaction. Some strategies suggested by Almond et al. that can be used to help move users forward during the migration process are as follows:

- **Communicate.** Make sure that users understand what changes are upcoming, what the schedule is, and how the changes will have a positive impact on the organization.

- **Use bridging applications.** Some Linux-based software such as OpenOffice runs well on Microsoft systems. Where possible, give users the opportunity to gradually become acclimated to the new environment prior to the migration.

- **Provide a similar look and feel.** The GNOME and KDE desktops can be customized to provide the user with a familiar Windows lookalike work environment.

- **Use similar actions.** Linux customization features allow desktop actions (for example, mouse double clicks) to be changed so that they can more closely resemble the familiar Windows actions.

- **Move file systems in a contiguous manner.** A common issue with users is the new Linux file structure. When moving file systems, the user's new file structure can be customized to closely resemble the familiar Windows file structure.

- **Provide users with the opportunity to try it before the migration.** Many Linux distributions are available in the form of a bootable CD. Users can test the operating system on their own computers without actually installing it by merely booting from the CD (Almond et al., 34-36).

To ensure enthusiastic acceptance of migration to Linux, draw the reader's attention to the benefits.

9

FIGURE 18.10   *(Continued)*

Benefits of Using Linux at Design International

According to Dalheimer et al., some of the key attributes that make Linux a viable operating system for Design International include the following:

- **Linux is free.** Organizations using the Linux desktop can eliminate the need for paying software licensure fees. In fact, an independent study by Enterprise Management Associates concluded that "Linux may, in many cases, be substantially less expensive to own than Windows" (Galli 16).

- **Linux is reliable.** Linux systems rarely crash.

- **Linux is gaining popularity.** Linux is supported by user groups by way of the Internet through the open source community. In addition, commercial software vendors have seen the value of making software that is compatible with Linux.

- **Linux is powerful.** The Linux operating system is efficient and makes good use of hardware.

- **Linux is compatible.** Linux can exist on the same computer as Windows and even access Windows files that are in a different partition on the same disk as the Linux installation.

- **Linux is supported.** Support and documentation for Linux users are widely available online.

- **Linux provides for efficiency through customization.** The user can make the Linux system into exactly what the organization needs, thus making the most efficient use of equipment (Dalheimer et al., 11-19).

Focusing on the key benefits to using Linux helps your readers understand the implications of your topic and substantiates the recommendation.

Conclusion

The Microsoft Windows desktop has established a dominant presence in the enterprise desktop marketplace. Nonetheless, as the IT manager at Design International, I must be alert for opportunities to improve service delivery that is safe, secure, and cost effective. Among the options that are currently available in the marketplace is the range of Linux-based desktop solutions.

Recommendation

We should consider migrating to a Linux-based desktop system at Design International. A further study should be conducted to consider the economic impact of migration on our budget, resources, and personnel.

10

(Continued)

Glossary

Term, Acronym, or Abbreviation	Definition
Beta Testing	The process used to identify the correctness, completeness, security, and quality of computer software.
Browser	A user interface that allows navigation. For example, a Web browser provides access to the World Wide Web; a Wiki browser lets you read articles from Wikipedia; a file browser manages files and related objects; a code browser allows you to access the source code of a computer program.
Drivers	A driver is an electronic component used to control another electronic component. The term refers to a specialized chip that controls high-power transistors.
GNOME	GNOME is an international project to create an easy-to-use computing system built from free software.
GPL	The General Public License allows anyone to use or distribute source code.
GUI	Graphical User Interface. GUI, pronounced like "gooey," uses graphical images to represent information or actions available to the user.
I/O	Input/Output. Input allows a computer system user to manipulate the system; Output allows the system to produce what the user has asked for.
KDE	K Desktop Environment is a free desktop environment that uses GUIs.
Kernel	The kernel is the central component of most operating systems (OS). It manages the system's resources and allows for communication between hardware and software.
Open Source Software	Open source software is computer software that permits users to change and to redistribute the software in the revised form.
OS	A computer's operating system manages the computer's hardware and software.
RPC	Remote Procedure Call—a situation where one computer initiates a call on another computer or network.
Source Code	A program's source code is its programming language.
Utilities	Utility software helps manage the computer's hardware and operating system.

Use a glossary to define high-tech terms, abbreviations, and acronyms. By using a glossary, you do not have to burden the report's text with definitions or your readers slow down. Readers unfamiliar with terms can look for definitions in the glossary, while readers familiar with the terms can continue to read without interference.

11

FIGURE 18.10 *(Continued)*

Works Cited

Almond, Chris, et al. "Linux Client Migration Cookbook." *IBM Redbooks*, 20 Apr. 2005, www.ibmredbooks/almond.

Dalheimer, Matthias, et al. *Running Linux*. O'Reilly Safari Books, 2015.

Freedman, David H. "Tech geeks have long praised open-source software. Now's the time to see what the fuss is about." *Inc.*, vol. 28, no. 6, June 2016, pp. 69-71.

Galli, Peter. "OSDL Answers Microsoft Claims; Study: Linux is cost-effective alternative to Windows." *eWeek*, vol. 23, no. 7, 13 Feb. 2016, p. 16.

"GNU General Public License Version 2." *Free Software Foundation*, 10 Mar. 2015, www.freesoft/GNU.

Kenston, Geoff. "Software Licensing Issues." *Faulkner Information Services 2014*, 14 Feb. 2015, www.faulknerinfo/Kenston.

Petreley, Nicholas. "Security Report: Windows vs. Linux." *The Register*, October, 2014, pp. 20-28.

Raymond, Eric. *The Cathedral and the Bazaar*. O'Reilly Safari Books, 2014.

"Work Groups." *Free Standards Group*, 20 Mar. 2015.

Yaghmour, Karim. *Building Embedded Linux Systems*. O'Reilly Safari Books, 2015.

12

APPLY YOUR KNOWLEDGE

CASE STUDY

Alpha/Beta Consulting (A/B/C) helps clients assess their computing needs, provides training, installs required peripherals to expand computer capabilities, and builds e-commerce applications, such as Web sites and corporate blogs. Their client base includes international companies and academic institutions. A/B/C works extensively with FIRE industries (finance, insurance, and real estate).

A new client, Home and Hearth Security Insurance Company, has asked A/B/C to build its e-commerce opportunities. Since the Web site was created, Home and Hearth has added new services, including insurance coverage for electronic commerce, employment-related practices liability, financial institutions, management protection, and medical professional liability. In addition, Home and Hearth has expanded to the Far East (Japan, Taiwan, and Singapore), the Mid East (Dubai, Jordan, and Israel), and South Africa. Home and Hearth's new services and new service locations are not evident on the current Web site.

An additional challenge for Home and Hearth is client contact. Currently, Home and Hearth depends on a hardcopy newsletter. However, corporate communications at Home and Hearth wants A/B/C to build a corporate blog to connect with its client base.

Assignments

1. Write the abstract that will preface A/B/C's long, formal report to Home and Hearth Security Insurance Company.
2. Write the introduction that will preface A/B/C's long, formal report to Home and Hearth Security Insurance Company.
3. Create a questionnaire asking Home and Hearth employees what they would like to see on the company's Web site and corporate blog.

ETHICAL CHALLENGE

Jordan Benjamin is a marketing manager for an online distributor of healthcare-related equipment and supplies. As manager of 300 employees worldwide, she has to write a report documenting employee use of company-owned PCs. This report will be delivered to both management and company stockholders at the annual meeting. Jordan decides that the only way to obtain information about PC use is by e-mailing a questionnaire to her employees. Some of the questions she asks in the questionnaire are as follows:

1. Do you use your PC for personal activities?
2. Do you maintain a personal blog?
3. Do you send and receive personal e-mails over your PC?
4. How much time each day do you spend on your PC for personal use?
5. Do you order from any online stores using your company-owned PC?

When Jordan writes her report, she bases her analyses on the 77 percent response rate she received from employees. She also used many of their verbatim comments in her written report.

Question

Is it ethical for Jordan to require employees to respond to personal questions in a questionnaire she disseminated through the company? By e-mailing the questionnaire to the employees, does Jordan suggest that the company "expects" employees to respond? Is this ethical? Is it ethical for Jordan to use comments solicited from the employees on the questionnaire? Why or why not?

INDIVIDUAL AND TEAM PROJECTS

1. Read a long, formal report. It might be one your instructor provides you or one you obtain from a business or online (see the links provided in this chapter's Web Workshop). If the report contains an abstract or executive summary, is it successful, based on the criteria provided in this chapter? Explain your decision. If the abstract or executive summary could be improved upon, revise it to make it more successful. If the report does not have an abstract or executive summary, write one.
2. Read the following abstract from A/B/C, written for their client Nitrous Solutions. It needs revision. How would you improve its layout and content?

EXAMPLE ▶

> Nitrous Solutions, a multi-discipline architectural design firm, is a start-up business with plans to begin operation in October 2016. The company currently has no information technology network for internal and external communication. Alpha/Beta Consulting will recommend a server to meet Nitrous Solutions' unique needs, a backup solution for archiving architectural materials and critical communications, software for digital asset management, and a proposed network design. After implementing these suggestions, Nitrous Solutions will have an IT network that will be safe, secure, dependable, and prompt. The system that Alpha/Beta Consulting is recommending also can be restructured with simplicity as the company's needs grow. By installing the IT network recommended in this report, Nitrous Solutions will be able to fulfill their company's mission statement.

3. Alpha/Beta Consulting, as agreed upon in a meeting with its client Nitrous Solutions, determined that its research for a contracted long, formal recommendation report would take no less than 350 hours and no more than 400 hours. As detailed in the following time sheet, A/B/C's four employees assigned to the project worked 376 hours. Using the information provided in the time sheet, create a line graph that tracks the employees' hours and a pie chart showing the percentage of time each employee worked.

TABLE 1   Time Sheet

Staff	Hours Worked Each Week														Total Hours for Project
	wk 1	wk 2	wk 3	wk 4	wk 5	wk 6	wk 7	wk 8	wk 9	wk 10	wk 11	wk 12	wk 13	wk 14	
Kenyon Patel	6.0	10.5	7.0	7.5	7.0	8.5	6.0	11.0	8.0	8.0	7.0	6.5	12.0	8.0	**113.00**
Randy Butler	6.5	5.5	5.5	3.0	5.5	8.0	4.0	5.5	6.0	8.0	7.0	9.5	15.0	8.0	**97.00**
Maria Villatega	0.0	0.0	3.5	4.0	4.5	5.0	3.0	1.5	10.0	17.0	8.0	9.0	8.0	4.0	**77.50**
James Soto	4.5	3.5	4.0	5.5	5.0	5.0	3.0	2.0	14.0	12.0	8.5	10.5	7.5	3.5	**88.50**
Total	**17.00**	**19.50**	**20.00**	**20.00**	**22.00**	**26.50**	**16.00**	**20.00**	**38.00**	**45.00**	**30.50**	**35.50**	**42.50**	**23.50**	**376.00**

At the conclusion of their long, formal report, the employees at A/B/C wrote a recommendation. Their recommendation, though excellent in terms of its content, needed to be reformatted for easier access and better emphasis of key points. Read the employees' recommendation and improve its layout.

◄ EXAMPLE

Recommendations to Meet Nitrous Solutions' Network Needs
To ensure a secure and dependable network that can be restructured to meet Nitrous Solutions' developing needs, Alpha/Beta Consulting recommends the following:

You can improve your computing system's "Application Layer." The applications software programs that Alpha/Beta Consulting recommends consist of the Adobe Creative Suite 6 for imaging, editing, illustrations, file sharing, Web designs, and digital processing; Microsoft Office, for spreadsheets, word processing documents, project timelines and maintenance; and Internet communication utilities, such as Web browsers and FTP clients, to guarantee rapid deployment of company communications and contract sales. All of these applications will be run on both the current Mac and Windows operating systems.

You also need to add a "Data Layer" to your new computing system. The data transferred on the new network will consist of large graphic files and fonts being accessed from a centralized server, postscript printing, voice, and messaging. TCP/IP will be the main communication protocol, but Appletalk may also be used.

To meet Nitrous Solutions' network infrastructure requirements, Alpha/Beta Consulting recommends Category 6 (Cat 6) cables connected to lGbps switches. These constitute your "Network Layer." The logical star topology network will enable a higher degree of performance for all users associated with the production of architectural designs, meet all organizational duties and communications requirements, and provide a centralized server as the main point of access for all the client machines

Finally, once installed, the physical network must have a "Technology Layer." This will enable the company to perform at a competitive level. This communication system will be a vital piece of the overall success of the company. The server will provide quick and reliable data storage, meet Nitrous Solutions' data backup needs, and allow for streamlined work flow.

PROBLEM-SOLVING THINK PIECES

Read the following text from a long, formal report written by A/B/C. Based on the explanation in this chapter, decide whether the text informs, analyzes, and/or recommends. Explain your decisions.

EXAMPLE ▶

Purpose Statement

This report will recommend the design parameters, hardware specifications, and estimated expenses needed to build and install external and internal computer network technology to meet Nitrous Solutions' communication needs.

This report's recommendation will focus on the following key areas:

- A solution for Nitrous Solutions' server needs
- Recommendations for a backup solution
- Possible software for digital asset management
- A proposed network design

Needs Analysis

1. *Company Background*

Nitrous Solutions, a multi-discipline architectural design firm located in Raleigh, North Carolina, is a start-up company, Nitrous Systems will occupy a 2,000-square-foot office space. They plan to begin operations in October 2014. The company will provide industrial, interior, landscape, and green architecture services for an international client base. Communication is a major component of their business plan. Thus, a reliable, efficient, creative communications/IT technology system is mandatory.

2. *Basic IT Requirements*

Currently, Nitrous Solutions has no network. The company needs an infrastructure plan that allows for the following:

- An IT platform that will be compatible with Mac and Windows.
- Eight Windows work stations and four Mac platforms.
- A network based on WiMAX impact.
- An internal and external network that is safe, secure, reliable, and fast.
- An internal and external network that can store, execute, and transmit architectural materials.
- A network that will allow for successful communications to clients, partners, and vendors.
- A data backup solution.
- Digital asset management software.

3. *Detailed IT Requirements*

In addition to the basic network requirements for Nitrous Solutions, the company also has asked for recommendations to meet the following micro network needs:

- Information about the most efficient server operating system for reliable communication in a cross-platform environment.
- Requirements for different file and storage server hardware and software.

- An analysis of which FTP server software would transfer secure data reliably to external clients.
- The need for scripts to scan recently modified files and back them up to a server.
- Reliable backup solutions and backup media types.
- A comparison of different software applications that would prevent data duplication in the permanent backup archives.
- A comparison of software that could catalog backup storage.
- A software solution allowing Microsoft Word documents to be converted and catalogued into one PDF file.

WEB WORKSHOP

You can find many examples of long, formal reports online. Compare the format of the online reports with the criteria for long, formal, researched reports discussed in this chapter. Explain where the online reports are similar to or different from this chapter's criteria. Do the online reports successfully communicate information and analyze issues and problems? Check out these sites, for example:

- Reporting Student Progress: Policy and Practice, British Columbia Ministry of Education
- The Use of Radio Frequency Identification by the Department of Homeland Security to Identify and Track Suspicious Individuals
- The United States Government Accountability Office's Report on "Women's Participation in the Sciences"
- The June 2004 Progress Report on the Federal Building and Fire Safety Investigation of the World Trade Center
- The World Health Organization's Report on HIV/AIDS

CHAPTER NINETEEN

Proposals and Business Plans

After completing this chapter, you will be able to

1. Learn why to write a proposal
2. Understand when to write internal proposals within your company
3. Understand when to write external proposals to sell products or services
4. Follow the criteria for writing proposals
5. Write the front matter of a proposal including the title page, cover letter, table of contents, list of illustrations, and abstract
6. Write the text of the proposal including an introduction, discussion section, and conclusion/recommendation
7. Use business plans to persuade people to address issues and provide resources

COMMUNICATION AT WORK

In this scenario, small business owner Elle Whittier writes proposals to market her company.

Elle Whittier is the owner of **ProCom, Inc.** "I'm also the receptionist, the accountant, the marketing department, and the janitor. I'm a one-person company, so I do it all. The title 'Owner' or 'President' sounds pretentious under the circumstances. Anyway, the most important thing I do is help clients accomplish the goals they have for their print and electronic materials. I'm not sure what the title is for that."

Whatever her title, Elle creates a variety of materials for her clients, including requirements documentation, user manuals, training materials, job aids, newsletter articles, marketing copy, proposals, presentations, and more. Her clients are just as varied as the materials. She has worked with accounting firms, architectural companies, retail establishments, and more. In every instance, she has created documents to ensure continued success of the businesses and often to make existing customer bases grow.

On any project, the client's goal is Elle's primary focus. Before getting started, she encourages clients to be very clear about who their audience is and what the objectives of the finished materials are. She relies on meetings with the clients during which she asks questions or has questionnaires filled out so that she can obtain sufficient information. Then she uses her training and experience to create written documents that address the audience and meet the goals.

During her workday, if Elle is not in her home office writing or doing research, she is on the phone or in a client's office attending project meetings or interviewing subject matter experts. She also spends a good deal of time reading existing documentation in order to fully understand the subject matter of a project. Although tackling new topics can be a challenge at times, "It's what I love best," she says. "I'm always learning something new."

According to Elle, no matter what the project is, clear two-way communication is critical at every step. "Obviously, clients hire me to share my expertise with them, but if I don't listen to my clients carefully in return, I run the risk of creating materials that fail to get the job done. In that case, nobody is happy."

Her company has a small budget for advertising, so Elle cannot rely only on satisfied customers and word of mouth for new projects. She writes proposals to generate new business. Elle responds to requests for proposals (RFPs) and researches expanding companies in her region to see where growth opportunities might exist. Then, based on her research, she writes unsolicited proposals explaining how her company could help with corporate communication challenges. Elle's proposals have assured her a steady stream of revenue and a very successful business.

Why Write a Proposal?

When you write a proposal, your goal is to sell an idea persuasively. Consider this scenario: Your company is growing rapidly. As business increases, several changes must occur to accommodate this growth. For example, the company needs a larger facility. This new building could be located in your city's vibrant new downtown expansion corridor, in a suburban setting, or entail the expansion of your current site. A new building or expansion should include amenities to improve recruitment of new employees, such as workout facilities, day care, and restaurant options. Finally, as part of new employee recruitment, the company must increase its diversity hiring practices.

> 1. Learn why to write a proposal.

Internal Proposals

How will you convey these ideas to upper level management? The topic is large and will require extensive financial obligations, time for planning, and a commitment to new staffing. A short, informal report will not suffice. In contrast, you will have to write a type of longer, formal report—an *internal proposal* for your company's management.

Additional examples of internal proposals include the following:

- Your company needs to improve its mobile communication abilities for employees who work at diverse locations. To accomplish this goal, you write an internal proposal requesting the purchase of WiFi-compatible laptops, smartphones, and Mobile WiFi (MiFi) routers to give smartphones mobile WiFi capabilities.
- Your company's insurance coverage is skyrocketing. As a member of the human resources staff, you have researched insurance carriers and now will propose insurance options or opt-out options to upper-level management.

> 2. Understand when to write internal proposals within your company.

- Your company is migrating to a new software platform. Employees will need training to use the software. You propose consulting companies which can offer the training, optional schedules, funding sources, and post-training certification.

3. Understand when to write external proposals to sell products or services.

External Proposals

Whereas *internal proposals* are written to management within your company, *external proposals* are written to sell a new service or product to an audience outside your company. Your biotechnology company, for example, has developed new software for running virtual cell cultures. The software simulates cell runs and displays synchronous strip charts for sterile monitoring. Data from the runs are graphed for comparison purposes. Not only will your company sell the software, but also the company provides consulting services to train clients in the software use. Your responsibility is to write an *external proposal* selling the benefits of this new corporate offering to a prospective client.

Requests for Proposals

Many external proposals are written in response to *requests for proposals* (RFPs). Often, companies, city councils, and state or federal agencies need to procure services from other corporations. A city, for example, might need extensive road repairs. A governmental agency needs Internet security systems for its offices. A hospital asks engineering companies to submit proposals about facility improvements. An insurance company needs to buy a fleet of cars for its adjusters. To receive bids and analyses of services, organizations will write an RFP, specifying the scope of its needs. Competing companies will respond to this RFP with an external proposal.

Primary and Secondary Research

See Chapter 16, "Research," for more information.

In each of these instances, you ask your readers to make significant commitments regarding employees, schedules, equipment, training, facilities, and finances. Only a proposal, complete with research, will convey your content sufficiently and successfully.

4. Follow the criteria for writing proposals.

Criteria for Proposals

To guide your readers through a proposal, provide the following:

- Front matter
  o Title page
  o Cover letter (or cover e-mail message for electronic submission of proposals)
  o Table of contents
  o List of illustrations
  o Abstract
- Text
  o Introduction
  o Discussion (the body of the proposal)
  o Conclusion/Recommendation
- Back matter
  o Glossary
  o Works cited (or references) page
  o Appendix

For information about title pages, cover letters, table of contents, list of illustrations, discussion section, glossaries, works cited, and appendices, see Chapter 18. Each of these components, typical of long formal reports, is thoroughly covered in that chapter. Following is information specifically related to your proposal's abstract, introduction, and conclusion/recommendation.

Front Matter

The front matter of a proposal includes its title page, cover letter, table of contents, list of illustrations, and abstract.

5. Write the front matter of a proposal including the title page, cover letter, table of contents, list of illustrations, and abstract.

Abstract

Your audience for the proposal will be diverse. Accountants might read your information about costs and pricing, technicians might read your technical descriptions and process analyses, human resources personnel might read your employee biographies, and shipping/delivery might read your text devoted to deadlines. One group of readers will be management—supervisors, managers, and highly placed executives. How do these readers' needs differ from others? Because these readers are busy with management concerns and might have little technical knowledge, they need your help in two ways: They need information quickly, and they need it presented in low-tech terminology. You can achieve both these objectives through an abstract or executive summary.

Executive Summaries

See Chapter 18, "Long, Formal Reports," for information about executive summaries.

The abstract, limited to approximately three to ten sentences, presents the *problems* leading to your proposal, the suggested *solutions*, and the *benefits* your audience will derive. The following is an example of a brief, low-tech abstract for an internal proposal.

◀ EXAMPLE

An effective abstract highlights the problem, possible solutions, and benefits in the proposal.

> **Abstract**
>
> Due to deregulation and the recent economic recession, we must reduce our workforce by 12 percent.
>
> Our plan for doing so involves
>
> - Freezing new hires
> - Promoting early retirement
> - Reassigning second-shift supervisors to our Desoto plant
> - Temporarily laying off third-shift line technicians
>
> Achieving the above will allow us to maintain production during the current economic difficulties.

Text

The text of a proposal includes an introduction, discussion section, and conclusion/recommendation.

6. Write the text of the proposal including an introduction, discussion section, and conclusion/recommendation.

Introduction

Your introduction should include two primary sections: (1) purpose and (2) problem.

Purpose.   In one to three sentences, tell your readers the purpose of your proposal. Your purpose statement clarifies the proposal's context. The following is an effective purpose statement.

◀ EXAMPLE

> **Purpose Statement**
>
> The purpose of this report is to propose the immediate installation of the 102473 Numerical Control Optical Scanner. This installation will ensure continued quality checks and allow us to meet agency specifications.

Problem (needs analysis). To clarify for the audience why this proposal is important, explain the problems leading to your suggestions. For example, computer viruses are attacking your company's workstations. This is leading to a decline in productivity, compromised security, and corrupted documents. Your proposal highlights these problems to explain why new computer security measures are needed. One way to help your readers understand the problem is through the use of highlighting techniques, especially headings and subheadings. See Figure 19.1 for a sample introduction.

FIGURE 19.1 Introduction with Purpose Statement and Needs Analysis

---

Introduction

Purpose Statement

This is a proposal for a storm sewer survey for Yakima, Washington.

Needs Analysis

Increased Flooding

Residential basement flooding in Yakima has been increasing. Fourteen basements were reported flooded in 2015, whereas 83 residents reported flooded basements in 2016.

Property Damage

Basement flooding in Yakima results in thousands of dollars in property damage to washers, dryers, freezers, furniture, and furnaces. Major appliances cannot be repaired after water damage. Flooding also can result in expensive foundation repairs.

Indirect Costs

Flooding in Yakima is receiving increased publicity. Flood areas, including Yakima, have been identified in newspapers and on local newscasts. Until flooding problems have been corrected, potential residents and businesses may be reluctant to locate in Yakima.

Special-Interest Groups

Citizens over 55 years old represent 40 percent of the Yakima population. In city council meetings, senior citizens with limited incomes expressed their distress over property damage. Residents are unable to obtain federal flood insurance and must bear the financial burden of replacing flood-damaged personal and real property. Senior citizens (and other Yakima residents) look to city officials to resolve this financial dilemma.

Provide specific details to explain the problem. Doing so shows that you understand the reader's needs and highlights the proposal's importance.

Discussion

When writing the text for your proposal, sell your ideas persuasively, develop your ideas thoroughly through research, observe ethical technical communication standards, organize your content so the audience can follow your thoughts easily, and use graphics.

Communicating Persuasively. A successful proposal will make your audience act. Writing persuasively is especially important in an *unsolicited* proposal since your audience has not asked for your report. A *solicited* proposal, perhaps in response to an RFP, is written

Research Techniques and Persuasion

See Chapter 11, "Communicating to Persuade," and Chapter 16, "Research," for additional information.

to meet an audience's specific request. Your audience wants you to help them meet a need or solve a problem. In contrast, when you write an unsolicited proposal, your audience has not asked for your assistance. Therefore, in this type of proposal, you must convincingly persuade the audience that a need exists and that your proposed recommendations will benefit the readers.

To write persuasively, accomplish the following:

- Arouse audience involvement—focus on your audience's needs that generated this proposal.
- Refute opposing points of view in the body of your proposal.
- Give proof to develop your content through research and proper documentation.
- Urge action—motivate your audience to act upon your proposal by either buying the product or service or adopting your suggestions or solutions.

Researching Content for Proposals.　As in any long formal report, consider developing your content through research. This can include primary and secondary sources such as the following:

- Interviewing customers, clients, vendors, and staff members
- Creating a survey and distributing it electronically or as hard-copy text
- Visiting job sites to determine your audience's needs
- Using the Internet to locate sources of documentation such as articles
- Reading journals, books, newspapers, and other hard-copy text
- Using social media (Facebook, Twitter, LinkedIn, and more) for trends, customer commentary, and up-to-date information

Communicating Ethically in Proposals.　When you write a proposal, your audience will make decisions based on your content. They will decide what amounts of money to budget, how to allocate time, what personnel will be needed to complete a task, and if additional equipment or facilities will be required. Therefore, your proposal must be accurate and honest. You cannot provide information in the proposal that dishonestly affects your decision makers. To write an ethical proposal, provide accurate information about credentials, pricing, competitors, needs assessment, and sources of information and research. When using research, for example, cite sources accurately to avoid plagiarism as discussed in Chapter 16.

Ethics

See Chapter 6, "Ethical Considerations," for additional information about ethical use of research.

Organizing Your Content.　Your proposal will be long and complex. To help your audience understand the content, use modes of organization. These can include the following:

- **Comparison/contrast.** Rely on this mode when offering options for vendors, software, equipment, facilities, and more.
- **Cause/effect.** Use this method to show what created a problem or caused the need for your proposed solution.
- **Chronology.** Show the timeline for implementation of your proposal, reporting deadlines to meet, steps to follow, and payment schedules.
- **Analysis.** Subdivide the topic into smaller parts to aid understanding.

See Table 19.1 for organization and key components of a proposal's discussion section.

TABLE 19.1 Key Components of the Proposal's Discussion Section

Analyses of the existing situation, your suggested solutions, and the benefits your audience will derive	Spatial descriptions of mechanisms, tools, facilities, or products	Process analyses explaining how the product or service works	Chronological instructions explaining how to complete a task
Comparative approaches to solving a problem	Comparing and contrasting purchase options	Managerial chains of command	Chronological schedules for implementation, reporting, maintenance, delivery, payment, or completion
Corporate and employee credentials	Years in business	Testimonials from satisfied clients	Certifications
Analyses of previous accomplishments	Biographical sketches of personnel	Chronological listing of projected milestones (forecasts)	Comparative cost charts

Visual Aids

See Chapter 8, "Visual Aids," for information about tables and figures.

Using Graphics. Graphics, including tables and figures, can help you emphasize and clarify key points. For example, note how the following graphics can be used in your proposal's discussion section:

- **Tables.** Your analysis of costs lends itself to tables.
- **Figures.** The proposal's main text sections could profit from the following figures:
  - **Line charts.** Excellent for showing upward and downward movement over a period of time. A line chart could be used to show how a company's profits have decreased, for example.
  - **Bar charts.** Effective for comparisons. Through a bar or grouped bar chart, you could reveal visually how one product, service, or approach is superior to another.
  - **Pie charts.** Excellent for showing percentages. A pie chart could help you show either the amount of time spent or amount of money allocated for an activity.
  - **Line drawings.** Effective for technical descriptions and process analyses.
  - **Photographs.** Effective for technical descriptions and process analyses.
  - **Flowcharts.** A successful way to help readers understand procedures.
  - **Organizational charts.** Excellent for giving an overview of managerial chains of command.

Conclusion/Recommendations

Sum up your proposal, providing your readers closure. The conclusion can restate the problem, your solutions, and the benefits to be derived. Your recommendation will suggest the next course of action. Specify when this action will or should occur and why that date is important. The following example is a conclusion/recommendation from an internal proposal.

Solutions for Problem

Our line capability between San Marcos and LaGrange is insufficient. Presently, we are 23 percent under our desired goal. Using the vacated fiber cables will not solve this problem because the current configuration does not meet our standards. Upgrading the current configuration will improve our capacity by only 9 percent and still present us the risk of service outages.

Recommended Actions

We suggest laying new fiber cables for the following reasons. They will

- Provide 63 percent more capacity than the current system
- Reduce the risk of service outages
- Allow for forecasted demands when current capacity is exceeded
- Meet standard configurations

If these new cables are laid by September 1, 2016, we will predate state tariff plans to be implemented by the new fiscal year.

◀ **EXAMPLE**

Summarize the key elements of the proposal.

Recommend follow-up action and show the benefits derived.

PROPOSAL CHECKLIST

Have you included the following in your proposal?

_____ **1.** Title page (listing title, audience, author or authors, and date)

_____ **2.** Cover letter or e-mail cover message (stating why you're writing and what you're writing about; what exactly you're providing the readers; what's next—follow-up action)

_____ **3.** Table of contents (listing all major headings, subheadings, and page numbers)

_____ **4.** List of illustrations (listing all figures and tables, including their numbers, titles, and page numbers)

_____ **5.** Abstract (stating in low-tech terms the problem, solution, and benefits)

_____ **6.** Introduction (providing a statement of purpose and a lengthy analysis of the problem)

_____ **7.** Discussion (solving the readers' problem by discussing topics such as procedures, specifications, timetables, materials/equipment, personnel, credentials, facilities, options, and costs)

_____ **8.** Conclusion (restating the benefits and recommendation for action)

_____ **9.** Glossary (defining terminology)

_____**10.** Appendix (optional additional information)

Business Plan

A unique type of proposal is the business plan. These long reports serve as guides to help prospective business creators turn ideas into reality. Like proposals, a business plan is used to persuade decision makers to address issues and provide resources. Business plans include many of the same components as proposals, such as title pages, cover letters, and tables of content. However, as Elisa Waldman, consultant for the Kansas Small Business Development Center, says, "The differences between business plans and proposals include audience and very precise discussion elements, mandated by banks before they'll provide loans."

7. Use business plans to persuade people to address issues and provide resources.

Audience for a Business Plan

Internal proposals might be read by corporate managers; external proposals might be read by clients. Business plans, in contrast, are generally read by bank executives. A prospective business creator writes his or her business plan to gain financial support for an idea. Thus, the primary audience will be a bank's commercial loan officer. If this person believes the business plan has merit, the loan officer will convey the plan to the bank's loan review committee for approval. Next, this committee will send the business plan to the bank's underwriters, whose job is to crunch numbers and determine if the plan is financially feasible. If the business plan makes sense monetarily, the commercial loan officer will secure the funds for a loan. If the review committee and underwriter deny the plan, the loan officer might ask the writer of the business plan for more details. Then, the process could begin again until the loan is secured. Other audiences could include the Small Business Association (SBA) or individual investors.

Discussion Components

The audience, whether banks, the SBA, or investors, will judge the business plan's merits on standard discussion requirements, including the following.

Executive Summary. Like a long report's abstract, the executive summary provides a brief overview of the business plan's objectives. Most executive summaries include

- The purpose of the plan—an operating guide and/or a request for financing
- The proposed business structure—sole proprietorship, partnership, limited liability, and so on
- Names of business operators and their percentage of ownership
- Reasons why this proposed product or service will succeed
- Amount of funds needed
- How the money will be used
- How and when the money will be repaid
- Collateral offered

Description of the Business. Once the writer has provided an overview of the proposed project, it's time to clarify. Business plans include discussion of what could be called the *5Ps—Product/Service, Place, Personnel, Price, and Promotion.*

Product/Service. To convince a bank to loan money, the writer must explain in detail what he or she is selling. What are the features of your product or service? Are you marketing a new restaurant, home healthcare service, mobile app, green construction business, boutique wine and beer, landscaping service, line of salsas and dips, remodeling company, and so on? What makes your company unique? Are you beginning a new business or expanding an old one? What's your competitive edge?

Place. Where will your business be located? Banks will want to know if your company will be online or at a physical site. If the business plans to be located on property, the bank will need to know where this location is; why you've chosen this site; what the area looks like in terms of neighborhood and surrounding businesses; and whether you'll buy or lease. Part of this discussion will focus on zoning laws and leasing permits. If the business will be online, what hardware and software are needed to run the company's Web site ancillaries (security, online payments, ordering, shipping, and so on)?

Personnel. The business plan must focus on management and operations. The bank will need to know the business owner's or manager's credentials, including education, certifications, and past experience. In addition, you'll need to describe your preferred employees' credentials.

Price. Every business costs money and earns money. Banks need to know what your start-up costs are, including management salaries, employee salaries and benefits, facilities purchases, leasing arrangements, insurance, and licensing fees, for example. In addition, since the bank wants you to succeed so you can pay back loans, what are your projected earnings? This is called your company's *projected financials*. Before approving your plan, the bank's underwriter must see

- A capital equipment list
- Break-even analysis
- Weekly, monthly, quarterly, and yearly projected expense expectations
- Weekly, monthly, quarterly, and yearly projected income expectations
- Cash flow plans

These will include graphics, such as the following figure and table:

◀ EXAMPLE

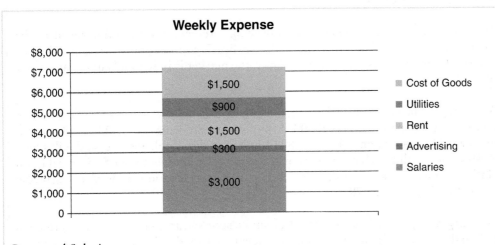

Personnel Salaries

The company will have a staff of six:

Personnel Salaries			
Personnel	Year 1	Year 2	Year 3
Operations manager	$37,000	$38,500	$40,000
Buyer/marketing	$28,000	$29,500	$32,000
Customer service	$27,500	$28,200	$29,400
Web site developer/manager	$32,000	$33,500	$35,000
Online order staff (2)	$12.00 an hour = $960/wk. $11,520/yr.	$11,520	$11,520
Total payroll	$136,020	$141,220	$147,920

Promotion. The business plan will also focus on how and to whom you plan to market your product or service. This includes the following considerations:

- Who will buy your goods? You can't say, "Everyone will want my product or service." The bank will not accept this generalization. They want to know exactly who your envisioned customer is. That includes age, gender, geography, income, and why the client needs your goods.

- How will you research this audience? The Small Business Development Center's SBDCNet.org Web site can help. This site will actually provide you a report of customer profiles, demographics, annual consumer expenditures for a geographic area, industry trend overview, market statistics, and psychographic study (for lifestyle and consumer interests). The findings are drawn from the Internal Revenue's tax returns.

- How do you plan to reach the consumer? You should discuss marketing through the use of fliers, brochures, social media, radio or television, neighborhood canvassing, and the Internet, for example. Following is a sample market analysis.

4.0 Market Analysis

This company projects a market of college-age students for an online social media site to sell products related to an individual college's goods (game tickets, used books, college sweatshirts, test banks, etc.). Thus, the market is individuals in the age range of 18–23.

According to the U.S. Census Bureau, this population in 2016 was 44.5 million, which represents 13.48 percent of the total U.S. population. Students age 18–23 spend approximately $315 billion annually, have $410 billion in disposable income, and will spend $110 billion online. This population segment spent an average of $157 weekly on clothing, entertainment, food, and technology. This generation is the fastest-growing demographic under age 70.

Specialty items, like college sweatshirts, t-shirts, sportswear, and sports gear, represent a billion dollar niche. In addition, college students buy textbooks annually, totaling over $5 billion. The market for used books is enormous and growing by 25 percent annually, according to e-marketing.com.

The Internet is an ever-accessible tool for this generation's shopping, by desktop, laptop, tablet, and smartphone.

The Writing Process at Work

To write an effective proposal, Elle Whittier, owner of ProCom Inc., followed the writing process.

Prewriting

Elle had to write a proposal to show how she was going to solve a potential client's problem. For her prewriting, she created a questionnaire. In the questionnaire, she interviewed employees about challenges they faced when using a manual to complete a job-related task. With this research, she was able to gather information from the client. See Figure 19.2 for Elle's questionnaire.

FIGURE 19.2 Online Questionnaire for Planning a Proposal

Proposal Planning

1. Are instructions for completing tasks set off clearly from the other text and written in a numbered, step-by-step format?

Never	Sometimes	Always
○	○	○

2. Are graphics and screen shots used when appropriate to enhance the instructions?

Never	Sometimes	Always
○	○	○

3. Does the material have a table of contents and/or index to help readers find the information they need?

Never	Sometimes	Always
○	○	○

4. Are terms and acronyms clearly defined?

Never	Sometimes	Always
○	○	○

5. What tasks do you perform without making mistakes?

```

```

6. How often do you perform those tasks?

Monthly	Weekly	Daily
○	○	○

7. What tasks do you struggle with?

```

```

(Continued)

FIGURE 19.2   Online Questionnaire for Planning a Proposal *(Continued)*

8. How often do you perform those tasks?

Monthly	Weekly	Daily
○	○	○

9. Do you refer to the training manual when you are having trouble?

Never	Sometimes	Always
○	○	○

10. If you do, can you find the information you need easily?

Never	Sometimes	Always
○	○	○

Done

Writing

Elle's draft of part of the discussion section of her proposal is in Figure 19.3.

FIGURE 19.3   Draft of "Findings" Section with Colleague Suggestions in Track Changes

Add a Heading, such as "Findings"

Employees perform 80% of the tasks in Application X with 100% accuracy. Each of these tasks is performed at least once a week if not daily. Employees also consistently identified five tasks that they have trouble completing. Three of these tasks are complicated, multi-step tasks with costly consequences when errors are made. Two of the tasks are performed only once a quarter.

> **Comment [JS1]:** Hala, you need to include more details about goals and recommendations. That way we can clarify what we hope to achieve in the proposal.

Next, the instructions in the existing training manual are well-written. Graphics and screen shots are used often to enhance the clarity of the instructions. Because of this, employees found the manual to be very thorough and helpful in the classroom training sessions.

> **Comment [JS2]:** Try using bullets and subheadings to break the information into smaller chunks for easy access.

However, the manual does not have an index, and the table of contents is skeletal. Employees do not use the manual as a reference when they are having trouble with a task because the manual is cumbersome to handle and instructions are hard to find.

> **Comment [JS3]:** The content is good, but we need to make this more persuasive. Let's highlight the errors that employees were making. Doing so will remind the readers exactly what prompted this proposal and how important our solutions are.

Rewriting

After drafting the document, Elle revised the proposal excerpt based on input from her colleague. Figure 19.4 is Elle's revised proposal excerpt.

FIGURE 19.4 Proposal Excerpt—Four Sections of the Total Proposal

Client Request

TechnoLand (client) has asked ProCom, Inc. to submit a proposal for the redesign of an existing training manual. The manual is used to train new finance department employees on Application X, the company's expense reporting application.

Employees are currently making costly mistakes, and the client believes that the manual is failing to communicate what employees need to know or do. The client's request for proposal (RFP) does not indicate what information the client relied on when deciding to have the manual revised.

Project Goal

The goal of the project is to eliminate the mistakes employees make when using Application X.

Findings

In response to TechnoLand, ProCom asked for and received permission to conduct a front-end analysis to confirm the cause of the mistakes being made in the application.

The findings include the following:

- Employees perform 80 percent of the tasks in Application X with 100 percent accuracy. Each of these tasks is performed at least once a week if not daily.
- Employees consistently identified five tasks that they have trouble completing. Three of these tasks are complicated, multi-step tasks with costly consequences when errors are made. Two of the tasks are performed only once a quarter.
- The instructions in the existing training manual are well-written. Graphics and screen shots are used often to enhance the clarity of the instructions.
- Employees found the manual to be very thorough and helpful in the classroom training sessions.
- The manual does not have an index, and the table of contents is skeletal.
- Employees do not use the manual as a reference when they are having trouble with a task because the manual is cumbersome to handle and instructions are hard to find.

Recommendation

Although the manual would benefit from the addition of an index and a more complete table of contents, a complete revision of the manual is unnecessary. Instead, ProCom recommends the creation of job aids for those tasks that are complicated, multi-step processes or for those tasks that are performed infrequently.

Elle says, "I always try to break text into small, readable amounts of content so that the lengthy proposal is easy to read. I know how busy my clients are, and I want to make the documents I submit to them pleasant to read—the less time they have to spend reading the proposal, the more likely it is that we'll be doing business together. The writing process helps me write effective proposals. Our business has increased over 40 percent since last year, so I know our writing is successful."

Sample Internal Proposal

Figure 19.5 shows a sample internal proposal.

FIGURE 19.5   Sample Internal Proposal for a New Mobile App

Include a proposal title, audience, author, and date of submission on your title page.

YourU
A Proposed Mobile Application Suite for College Students

For
Anderson Stolper, CEO
NanoTech Software Development, Inc.

By
Adriana Niemhoff, Manager
NanoTech Software Development Department

Date
March 22, 2016

FIGURE 19.5 (Continued)

Date: March 22, 2016
To: Anderson Stolper
From: Adriana Niemhoff
Subject: Proposal for New Mobile Application Suite

According to Kristen Purcell of Pew Internet Research, mobile technology has led to the development of an "apps culture." People with handheld devices are downloading billions of apps daily, and over half of the U.S. population owns mobile devices with app downloadable capabilities.

This represents huge potential and a growing market for our company. The software development department is proposing a new application suite, suitable for the largest audience of application users—college students. Our proposed app suite, entitled YourU, consists of the following *customizable* features:

- ConnectU. An app that provides students an all-in-one site for their friends and families' contact information (e-mail address, phone numbers, Facebook sites, Twitter links, and more)

- MoneyU. An app that helps students manage their finances

- FunU. An app that lets students buy concert and athletic event tickets, organize parties, make restaurant reservations, download music, and more

- PlanU. An app that allows students to organize their lives for test dates, work scheduling, and other calendar events

This proposal will provide you research about the suite's potential impact on the market, product information about each application within the suite, software development costs, return on investment, and our team's credentials. We believe that this application suite will add value to our company's product line. Once you have reviewed the proposal, please contact me by e-mail or phone so that our team can answer your questions. Thank you for your consideration.

The introductory paragraph explains the importance of the topic.

Itemizing the customizable features emphasizes the benefits and usefulness of this new app.

To persuade the reader of the proposal's value, the writer uses numerous pronouns and positive phrases, such as "huge potential," "growing market," and "add value."

(Continued)

FIGURE 19.5   Sample Internal Proposal for a New Mobile App *(Continued)*

Table of Contents

Headings, subheadings, and page numbers help the audience find information and navigate the text.

List of Illustrations

Figure and table numbers plus titles allow the readers to find the visuals quickly.

iii

FIGURE 19.5   *(Continued)*

Abstract

Problem

Our company has been searching for a new product idea with the potential for sales growth in a targeted audience. Currently, our mobile application products have been generic with broad but unfocused appeal. In addition, our mobile applications have been limited to static, off-the-shelf products with limited growth potential. Therefore, sales have been steady, but return on investments has diminished.

Solution

A key to any product's success is targeting an audience and meeting their needs. Mobile apps represent a growing market. The largest client base for mobile apps is young adults ages 18–30, many of whom are college students. Therefore, the YourU app suite, geared toward college student needs, is an ideal product for our company to develop.

Benefits

Our proposed YourU app suite addresses those problems as follows:

- Meets the needs of a niche market versus a generic market
- Provides end users creative options to customize their purchase
- Organizes many capabilities into one app

Stating the problem, solution, and benefits gives focus to the proposal. The abstract emphasizes the problems that generated the proposal. Next, the abstract shows how the proposal can solve the problems and highlights the benefits derived by implementing the proposal's suggestions.

iv

(Continued)

FIGURE 19.5   Sample Internal Proposal for a New Mobile App (*Continued*)

Introduction

Purpose

This report proposes the development of a new mobile application suite geared toward a niche market of college students.

Needs Assessment

Marketing Generic Products Does Not Meet the Needs of a Targeted Audience:
Our current product line is broad but without focused appeal. Currently, we sell mobile apps that offer our end users games, weathercasts, global positioning system (GPS) mapping capabilities, sports news updates, online dictionaries, and financial updates (among other choices). These apps are successful options for a mass audience. However, our company's research and development department (R&D) has found that marketing such diverse product lines is challenging.

For example, the demographic market for games (boys and men ages 14–24) is not the same as the demographic for financial updates (men and women ages 45–65). The demographic market for sports news (men ages 20–30) is not the same as the demographic for weathercasts (men and women ages 35–55).

With such diverse demographic ranges, we cannot effectively target our advertising. For men and women ages 45–65, we have found that our preferred marketing options are radio, television, and newspapers. However, online venues such as Facebook, Twitter, YouTube, and blogs appeal to youth, men, and women ages 14–35. See Figure 1 for a breakdown of advertising channels we currently use for various markets.

Figure 1 Percent of Market Based on Advertising Channel

42

58

■ TV, Radio, Newspaper   ■ Facebook, Twitter, YouTube, and Blog

To more successfully target our advertising, we need a product that appeals to a niche audience so we can saturate a precise consumer base. The YourU app suite accomplishes this goal by allowing us to market primarily in social media.

1

To highlight the importance of this proposal, the writer assesses how the app will meet marketing needs. These include targeted audiences, frequency of app usage, and profitability.

The figure adds visual appeal and makes content readily accessible to the readers.

FIGURE 19.5   *(Continued)*

Off-the-Shelf Products Create Problems with Revisitation and Profit Loss:
Currently, our mobile app software consists of commercially available off-the-shelf (COTS) products. These have been cost effective to produce, but they present our company with two challenges:

- Static product abilities without customization lead to user disinterest. Clients always look for new experiences online, new reasons to return to an app. If their apps don't meet their needs for new content, they stop using the product. Figure 2 presents the results of an R&D survey regarding software revisitation patterns.

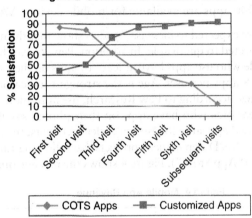

Figure 2 Software Revisitation Patterns

- Steady COTS app sales have not equaled profit. Our app products have produced steady sales. However, our return on investment (ROI) actually has diminished when we consider increases in business expenses (marketing, taxes, salaries, etc.). As seen in Figure 3, R&D compares sales with profit decreases due to business expenditures.

Figure 3 COTS Sales versus Decreased Profits

Figure 3 shows that though our sales have increased each quarter, our profits are steadily decreasing. We need a product line that provides us an opportunity for increased profits.

2

(Continued)

FIGURE 19.5   Sample Internal Proposal for a New Mobile App *(Continued)*

Discussion

Research Proving Financial Value of Mobile Apps

Why should our company pursue a new line of mobile apps? R&D has found the following regarding the growing audience of app users:

- Half of all Americans own a smartphone that can support software applications. Even more potential clients own PDAs, cell phones, and handheld computers with app capabilities.
- More than 10 billion apps are downloaded each month.
- Approximately 500,000 apps are available for mobile devices (Wexford).

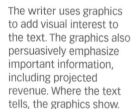

The writer adds depth to the content and persuades the audience through primary and secondary research.

These impressive numbers are expected to increase. Studies show that the number of users accessing mobile apps is expected to quadruple. In addition, Pew Research shows that on average, 93 percent of people who use apps have at least 10 apps on their mobile devices, and users regularly access 18 apps per user. Age is the strongest indicator of app usage. Young adults with cell phones, according to Pew Research, use apps more than other age cohorts. "79% of 18–29 year-olds who have apps on their phones say they use them" (Purcell). Because our proposed app suite focuses on college students, the age range highlighted in the Pew study, our R&D department projects the following future revenue in contrast to current app sales ("App Store"). Figure 4 shows the revenue our app can earn.

The writer uses graphics to add visual interest to the text. The graphics also persuasively emphasize important information, including projected revenue. Where the text tells, the graphics show.

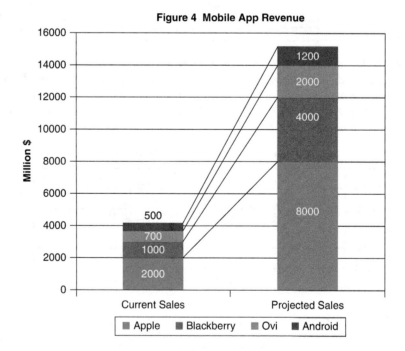

Figure 4 Mobile App Revenue

Finally, though many apps can be downloaded for free, 90 percent of app users are willing to pay for additional apps (Purcell). Therefore, based on current sales, projected app usage, and the college-aged clientele we are targeting, our company can profit from the expanding market of mobile apps.

3

FIGURE 19.5 *(Continued)*

Product Description

Features:

Our proposed YourU app suite will initially consist of the following four bundled apps.

- ConnectU—This app will provide students an all-in-one site for their friends', families', and professors' contact information. This will include, but not be limited to, e-mail addresses, phone numbers, street addresses, city/state/zip information, Facebook sites, Twitter links, and blog links. Having all contact information in one app helps students quickly access information without the fear of losing these URLs, sites, addresses, and contact numbers.
- MoneyU—This app helps students manage their finances. MoneyU will contain CalcU with the following features:
  - A calculator for basic math
  - A tip calculator
  - An interest calculator for purchases
  - A way to calculate GPAs

  MoneyU will also help students balance their checkbooks, keep track of their charges, and remind them when rent is due. We have installed a "money bing" tune that they can select from a list of sound options. This will warn them two days in advance of a rent due date, for example.
- FunU—This app lets students buy concert and athletic event tickets, organize parties, make restaurant reservations, and download music. Students also will be able to receive updates from local stores near their colleges when sales start and end.
- PlanU—This app allows students to organize their lives for test dates, work scheduling, and other calendar events. PlanU will include a list of important dates, such as parents' anniversaries, friends' birthdays, college game days and times, rock concert events, and more. Students will be able to select from our list, add new items, and then input pertinent date data. This app even includes "wake up bings," tunes they select for alarm clock functions.

These apps represent our basic components of the YourU app suite. After this suite is created and depending upon our sales success, we hope to add new apps to the suite to generate additional income.

Customization:

What distinguishes the proposed YourU app suite from our other app products is customization. Our current COTS apps are static. They cannot be personalized for each customer. In contrast, our proposed YourU app suite is programmed to allow end users a new level of personal ownership. The YourU app suite lets clients

- Add photos
- Add their own data for phone numbers, important dates, and contacts
- Customize their app by choosing tunes for their "wake up bings" and "money bings"
- Choose background images

4

Highlighting the app's features persuades the reader of the product's uniqueness and benefit to the consumer.

(Continued)

FIGURE 19.5   Sample Internal Proposal for a New Mobile App *(Continued)*

- Customize the app with their college colors. See Figure 5 for examples of this customization
- Access their apps and input customized data in their language of choice, including English, French, Italian, German, Spanish, Arabic, and Chinese

Figure 5 Color Customization by Client's College Color

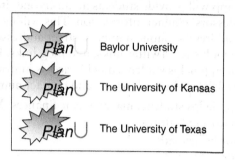

Customization capabilities will provide students a sense of ownership for and commitment to the app. This, we believe, will lead to higher usage rates, higher visitation numbers, increased customer satisfaction, and increased revenue.

Specifications:
Table 1 provides an overview of YourU app suite specifications.

Table 1 YourU App Suite Specifications	
Features	**Specifications**
Supporting mobile devices	iPhones, iPad, IPod, Android, Nokia, Samsung, Sony Ericsson, Palm, Blackberry, and Windows smartphone
Approximate file size per app	286 K
Graphics format	.jpg approximately 2,000 bytes 70 × 90 pixels per graphic

Software Development Costs

Based on prior app development benchmarks, we anticipate that each member of our software development team will spend a month on this project. Our team consists of five staff members. With each member devoting an eight-hour day to the project, five days a week, times four weeks, this will equal 800 work hours.

Each of our software development team members earns approximately $25 an hour. Therefore, at $25 × 800 hours, the upfront development costs will equal $20,000.

Return on Investment

Currently, our product line of apps ranges in price from as low as free downloads to as high as $39.99. After discussion with our accounting and marketing departments,

5

FIGURE 19.5 (Continued)

we have decided that a cost-effective price for the YourU app suite is $9.99. Figure 6 shows you the projected return on investment (ROI) based on this pricing structure.

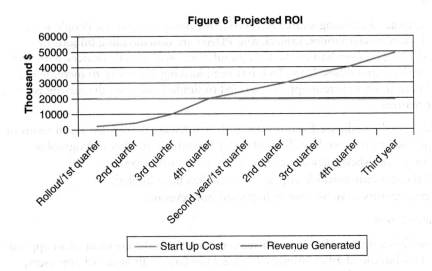

Figure 6 Projected ROI

An analytical breakdown of the ROI shows the reader not only that the writer of the proposal has thoroughly researched the topic but also how soon the company will turn a profit.

Considering national sales figures, we project a break-even ROI at the end of the first year. We project that our investment will more than double by the beginning of the third year.

Credentials

Our software development team consists of specialists ready to answer your questions. The following individuals have been assigned to the YourU app project:

- *Randy Draper, Team Lead*
  Randy (BS, Information Technology, South Central Texas University, 2009) has worked for NanoTech for five years. He was promoted to software development team lead after two years as an IT specialist. Randy has worked on over 60 software development projects, including ten app creations.

- *Ruth Bressette*
  Ruth (BS, Computer Information Systems, Northwest New Mexico State University, 2011) has worked for NanoTech for three years. Having received an additional CIS Certification for Excellence in App Development, Ruth has proven skills in the creation of apps.

- *Doug Loeb*
  Doug (double degree: BS, Computer Information Systems; BS, Accounting, Idaho Tech University, 2010) has worked for NanoTech for two years. Doug's two degrees from Idaho Tech make him especially valuable in the YourU app suite project.

- *Dana Anders*
  Dana (AS, Information Technology, Landview Community College, 2013) has worked for NanoTech for a year. Dana will be responsible for most of the coding.

Though Randy will be in charge of this project, as manager of the software development department, I will oversee all stages of app development.

Qualifications and experience help persuade the reader of staff members' expertise and ability to do the job.

6

(Continued)

FIGURE 19.5   Sample Internal Proposal for a New Mobile App *(Continued)*

Conclusion/Recommendation

Conclusion

Mobile apps are a growing source of revenue for many companies. People with mobile devices (smartphones, tablets, and PDAs) are downloading billions of apps daily. In fact, over half of the U.S. population owns mobile devices with app downloadable capabilities. NanoTech has been looking for a way to increase our revenue stream. Our current app line has led to limited sales and decreasing revenues for four reasons:

- Because the audience for our off-the-shelf software is very diverse in terms of age and interest, we have had difficulty targeting a market demographic.
- Our off-the-shelf software does not allow for customization.
- Without a customizable app, revisitation has been limited.
- These factors have negatively impacted our revenue.

Recommendation

The NanoTech software development team recommends the creation of an app suite, YourU. This bundle of applications will target the largest audience of app users, college students. In addition, we propose a customizable suite that will empower users, personalize their apps, and lead to increased visitation. Our proposed suite will benefit NanoTech as follows:

A Targeted Audience + Customization + Revisitation = **Revenue Growth**

Glossary

The alphabetized glossary defining jargon and technical terms helps both a low-tech and a lay audience understand the text.

Term	Definition
Application suite	Multiple applications bundled together with related functions and features that can interact with other apps within the suite.
Commercially available off-the-shelf (COTS) software	A Federal Acquisition Regulation (FAR) term to define software that is commercially available to both the public and government.
Mobile apps	Application software developed for small handheld devices such as PDAs, tablets, and smartphones.
Niche market	A subset of a larger market. In a niche market, product features strive to satisfy a specific market's needs, price range, and demographics.
Revisitation	The number of times a client visits a software application.
ROI	Return on investment (a calculation that compares initial cost versus revenue generated)

7

FIGURE 19.5   *(Continued)*

Works Cited

"App Store." *Wikipedia*, 27 Mar. 2016, en.wikipedia.org/wiki/App_Store_(iOS).
Purcell, Kristen, et al. "The Rise of Apps Culture." *Pew Internet,* 14 Sep. 2015,
    www.pewinternet.org/2010/09/14/the-rise-of-apps-culture/.
Wexford, Charles. "Mobile Apps in the Workplace." *Time,* 12 Aug. 2016,
    www.time.com/wexford/mobile.

8

APPLY YOUR KNOWLEDGE

CASE STUDY

You own **Buzz Electronics Co.**, 4256 Blue Mountain Blvd., Raleigh, North
Carolina 65221. Mr. and Mrs. Allan Thibodeux, 3876 Spanish Moss Drive,
Bayside, North Carolina 65223, have asked you to bid on electrical work for
a new family room.

The room, 18' (east to west) by 15' (north to south), should have
four 110 V outlets for three lamps, an iPod and MP3 player, and a high-
definition, flat-screen, 46" television. The family wants the four 110 V
outlets to be placed equidistant throughout the room.

The client wants two 220 V outlets. One 220 V outlet will go by
the southwest window on the west wall where the family plans to put a
window air conditioning unit. The other 220 V outlet must be placed on
the east wall, where the family plans to put home office equipment (computer, printer, scanner,
and fax machine).

The family wants electrical wiring for a fan with a light. In addition to this light, the family
also wants a light mounted on the east wall above the desk area, so wiring is needed there,
approximately 5–6' up from the floor.

The family wants two light switches in the room: one by the door and one on the north wall,
approximately 6' in from where the north and west walls meet. They want both to be double
switches, one to control the fan and ceiling lights; the other to control additional floor and ceiling
lights in the room. All light switches need rheostats for dimming. Finally, the Thibodeuxs plan
to have a whole-house vacuum system installed in the walls, and they have asked you if you can
provide this service.

Buzz Electronics has been in business since 1995. Buzz has long-standing contracts for service
with Acme Construction, J&L Builders, Food-to-Go Groceries, the City of Piedmont, North
Carolina, and Ross and Reed Auto Showroom.

As owner of Buzz Electronics, you have an Associate's degree in electronics from Sandy
Shoal Community College, Sandy Shoal, North Carolina. You are ETA-I (Electronics Technicians
Association International) Certified; NASTec (National Appliance Service Technician) Certified;
and a Certified Industrial Journeyman. You have eight employees, all of whom also are Certified
Industrial Journeymen.

Assignment

Write an external proposal—bid for contract. To do so, study the Thibodeux's electrical needs, list the parts you will need to complete the job, estimate the time for your labor—including setup, work performed, and cleanup. Then, provide a price quote. Follow the guidelines provided in this chapter.

ETHICAL CHALLENGE

Candice Miller, manager of Human Resources, is preparing an internal proposal to promote new training classes for her company's employees. To discover where employees can improve their job skills, Candice conducts primary research. She interviews staff and tells them how their comments will be used. After completing her research but before she writes the proposal, Candice gets the employees' approval and has them sign a release statement: "I give the company permission to use my statements in this proposal."

Months later, Candice is invited to attend the national convention of human resource managers and is asked to make a PowerPoint presentation about her company's new training options. In this presentation, she uses the interviews from her earlier proposal. She did not get permission from her fellow employees since they had already signed release forms.

Question

Is it ethical for Candice to use her colleagues' interview comments in her speech at the convention? Why or why not?

INDIVIDUAL AND TEAM PROJECTS

1. Write an external proposal. To do so, create a product or a service and sell it through a proposal. Your product can be an improved radon detection unit, a new MP3 player, safety glasses for construction work, bar codes for pricing or inventory control, a piece of biotechnology equipment for monitoring blood work, computer graphics for an advertising agency, and so on. Your service may involve dog grooming, automobile servicing, computer maintenance, home construction (refinishing basements, building decks, room additions, and so on), freelance professional or business writing, at-home occupational therapy, or home theater installation and maintenance. The topic is your choice. Draw from your job experience, college coursework, or hobbies. To write this proposal, follow the process provided in this chapter.

2. Write an internal proposal. You can select a topic from either work or school. For example, your company or department is considering a new venture. Research the project by reading relevant information. Interview involved participants or survey a large group of people. Once you have gathered your data, document your findings and propose to management the next course of action. If you choose a topic from school, you could propose a day care center, on-campus bus service, improved computer facilities, tutoring services, coed dormitories, pass/fail options, and so on. Write an internal proposal to improve your company's Web site, expand or improve the security of your company's parking lot, create or improve your company's physical site security in light of post 9/11 issues, improve policies for overtime work, improve policies for hiring diversity, or improve your company's policies for promotion.

    Research your topic by reading relevant information or by interviewing or surveying students, faculty, staff, and administration. Once you have gathered your data, document your findings and recommend a course of action.

PROBLEM-SOLVING THINK PIECES

1. Stinson, Heinlein, and Brown Accounting, LLC, employs over 2,000 workers, including accountants, computer information specialists, a legal staff, paralegals, and office managers. The company requires a great deal of written and oral communication with customers, vendors, governmental agencies, and coworkers. For example, a sample of their technical communication includes the following:

   - Written reports to judges and lawyers
   - Letters and reports to customers
   - E-mail and memos to coworkers
   - Oral communication in face-to-face meetings, videoconferences, and sales presentations
   - Maintaining the company's social media sites (Facebook, Twitter, LinkedIn, and blog)

   Unfortunately, not all employees communicate effectively. The writing companywide is uneven. Discrepancies in style, grammar, content, and format hurt the company's professionalism. The same problems occur with oral communication.

   George Hunt, a mid-level manager, plans to write an internal, unsolicited proposal to the company's principal owners, highlighting the problems and suggesting solutions. What must Mr. Hunt include in his proposal—beyond the obvious proposal components (a title page, table of contents, abstract, introduction, and so forth)—to persuade the owners to accept his suggestions? Suggest ways in which the problem can be solved.

2. Roger Hebert is sales manager at Crab Bayou Industries (Crab Bayou, Louisiana). In his position, Toby manages a sales staff of 12 employees who travel throughout Louisiana, Texas, Arkansas, and Mississippi. Currently, the sales staff members use their own cars to make sales calls, and CBI pays them 31 cents per mile for travel expenses. Each staff member currently travels approximately 2,000 miles a month, with cars getting 20 miles per gallon.

   Gasoline prices, at the moment, are over $3.00 a gallon. With gasoline and car maintenance costs higher than ever, the current rate of 31 cents per mile means that CBI's sales employees are losing money. Something must be done to solve this problem. Roger has met with his staff, and they have decided to write a proposal to Andre Boussaint, CBI's CEO.

   What must Roger and his staff include in the proposal—beyond the obvious proposal components (a title page, table of contents, abstract, introduction, and so forth)—to persuade the CEO to accept the suggestions? Suggest ways in which the problem can be solved.

WEB WORKSHOP

1. By typing "RFP," "proposal," "online proposal," or "online RFP" in an Internet search engine, you can find tips for writing proposals and requests for proposals, software products offered to automatically generate e-proposals and winning RFPs, articles on how to write proposals, samples of RFPs and proposals, and online RFP and proposal forms.

   To perform a more limited search, type in phrases like "automotive service RFP," "computer maintenance RFP," "desktop publishing RFP," "Web design RFP," and many more topics. You will find examples of both proposals and RFPs from businesses, school systems, city governments, and various industries. To enhance your understanding of business and industry's focus on proposal writing, search the Web for information on RFPs and proposals. Using the criteria in this chapter and your knowledge of effective technical communication techniques, analyze your findings. What are some of the industries that are requesting proposals, and what types of products or services are they interested in?

   a. Report your findings, either in an oral presentation or in an e-mail message.
   b. Respond to an online RFP by writing a proposal. To complete this assignment, go online to research any information you need for your content.

2. Research "business plans" on the Internet. Access companies you are interested in to see if they post any business plans. Also access the Small Business Administration Web site for information about writing successful business plans.

CHAPTER TWENTY

Oral Presentations

After completing this chapter, you will be able to

1. Distinguish between verbal and nonverbal communication

2. Develop effective listening skills

3. Understand the importance of oral communication in the workplace

4. Learn everyday oral communication skills

5. Communicate effectively in informal oral presentations

6. Make formal oral presentations successfully

7. Know the parts of a formal oral presentation

8. Use the communication process to make an effective oral presentation

9. Use visual aids to enhance the oral presentation

10. Understand the benefits of PowerPoint slides in an oral presentation

COMMUNICATION AT WORK

In this scenario, a recruitment specialist talks about the importance of oral communication in her new job.

Estella Alvarez has just been hired as recruitment specialist in the **human resources department for Monroe, Louisiana.** She will be responsible for coordinating the recruitment and hiring of city employees. She also will administer the city's retirement system, unemployment compensation, family medical leave programs, and the city's drug and alcohol abuse policy. She will have to keep up to date on new federal, state, and local regulations related to hiring employees. In order to do so, she will have to take some online courses and attend seminars presented by other human resource specialists in the area.

As a recruitment specialist, Estella will be involved in staffing merit system positions. This will include ensuring that employee recruiting, advertising, testing, certification, and applicant record-keeping comply with legal, professional, and merit system standards. Maintaining high ethical standards is an important function of her position in human resources and recruiting.

She will meet these responsibilities as follows:

- Advertising and recruiting
- Reviewing approximately 120,000 applications a year to determine whether or not the applicants meet position requirements
- Verifying experience and education by contacting former employers, checking social media sites, and performing credit checks
- Establishing, maintaining, and certifying eligibility lists for vacant city positions
- Being informed about legal issues related to hiring

The largest part of her job will be making presentations to city employees and conducting new-employee orientations each month. In this capacity, she will prepare PowerPoint slides and speak in front of small and large groups, in both formal and informal presentations. She will also meet individually with employees to work with them regarding their benefits, on-the-job certification needs, and education requirements record keeping. For example, when employees complete college classes or attend off-site training, Estella reminds them to update their records since additional education enhances their opportunities for raises and promotions.

Estella is well educated with a degree in communications from the University of Texas at Austin where she gave many classroom presentations and participated in debate. As a member of the debate team, she traveled nationwide to participate in debate tournaments and helped her team win numerous awards in competition.

Although she has confidence about her abilities to meet her job requirements, like most employees, she knows that much of her job success will be based on how she presents important information to audiences. She says, "I have to learn how to determine the makeup of my audience. Who are the participants? How much do they know about my subject matter? What's the most effective way to hold their interest during an hour-long presentation? What would be the benefit of using a highly visual approach to content rather than simply talking about the topic? Should I rely on handouts? These are some of the things I have to consider as I perfect my communication skills to be effective in my new job."

The Role of Verbal and Nonverbal Communication

> 1. Distinguish between verbal and nonverbal communication.

Here is a fact: Excellent communication does not just depend on the words you say and write. Communication depends greatly on the tone of your voice, your body language, and other verbal and nonverbal cues you give your audience.

How Important Is Verbal and Nonverbal Communication?

Some researchers suggest that up to 90 percent of communication is body language (Warfield 1). Albert Mehrabian, a University of California at Los Angeles professor and researcher, states that three elements combine to express messages. He calls it the *3 Vs*—verbal, vocal, and visual. His studies determined that words account for 7 percent of communication, tone of voice for 38 percent, and body language equals 55 percent of the message (Mehrabian).

What Is Verbal Communication?

What you say is either positively or negatively impacted by *how* you say it. In fact, "after your physical appearance, your voice is the first thing people notice about you" (Clarke). Verbal communication includes your pace (rate of speech) and modulation.

Pace.   A key concern is pace, the speed with which you speak. Generally, we speak about 150–200 words a minute. Speeding up this rate can have several meanings. You could imply that you are happy, fearful, angry, or surprised. In addition, if you talk too fast, not only will your coworkers have trouble following your train of thought but also they might think you're nervous or impatient. If you are in a meeting, for instance, and you speak too rapidly, your voice might be telling your coworkers it's time to move on to the next topic. You are saying, "I don't have time for you" or "I want to be doing something else (anything else)."

Talking too slowly causes problems also. Slowing down your pace could convey sadness, boredom, or lack of interest. People also might feel you are talking down to them. A very slow delivery style could imply condescension—"I assume that you can't keep up with my thoughts, so I'll speak *Reallllly Slowllllly*."

Modulation.   Modulation is the loudness, tone, and pitch of your voice as you speak. A very loud voice will be perceived as dominating, while a very quiet voice potentially will make you sound meek. Mumbling can destroy your credibility. A monotone voice, one with no change in pitch, will put your listener to sleep. Modulation, varying your pace and pitch, adds interest to your comments.

What Is Nonverbal Communication?

Nonverbal communication is another phrase for body language. Basically, nonverbals entail both conscious and unconscious body movements that convey an emotional message. Nonverbal cues can include any of the following.

Eye Contact.   Good eye contact suggests openness, confidence, and interest in your audience. However, too much of anything is potentially bad. Do not overdo eye contact; staring and glaring will make any audience uncomfortable. Many people believe that eye contact should last no longer than three to five seconds. Any more than that might be unsettling.

Facial Expressions.   Smiling, frowning, yawning, wide-eyed bewilderment, open-mouthed shock, lip-curling angry glares—each of your facial expressions conveys a message. For example, if you are listening to a colleague speak and you are simultaneously wrinkling your nose, arching your eyebrows, scrunching your face, or furrowing your forehead, you are conveying a very negative message. You are showing your coworker your uncertainty or disbelief. In contrast, a pleasant smile shows your involvement and support.

Posture.   Imagine you are in a meeting. A coworker is speaking, but you are fidgeting, slouching, placing your hands behind your head, or leaning to the side. You are reflecting boredom and disinterest. On the other hand, if you lean forward, you show you are interested and actively participating in a team meeting. Nodding your head up and down shows your affirmation.

Proximity.   What does proximity to your audience show? If you stand very close to a spouse or child, you show your love. However, if you stand very close to a coworker, you will encroach upon his or her personal space. This suggests a threatening or domineering behavior. In *The Silent Language*, Edward T. Hall discusses four main proximity zones of personal space (see Table 20.1).

TABLE 20.1 Proximity Zones of Personal Space

Zone	Distance	Who Gets Access	What Occurs Within the Zone
Intimate	6–18 inches (15–45 centimeters)	Spouse, children, and significant others	Intimate relationships, such as touching and hugging
Personal	1½–4 feet (45–125 centimeters)	Family members and close personal friends	Within an arm's length, less intimate but still very close encounters
Social	4–12 feet (1.25–3.60 meters)	Friends and colleagues	Social and business relationships
Public	Over 12 feet (+3.6 meters)	Strangers/general public	Little interaction

Gestures. We speak a lot with our hands. A thumbs-up gesture in the United States signifies that everything is A-OK. A thumbs-down gesture denotes a negative opinion. These two gestures are commonly understood in Western culture. Other gestures, however, are subtler. Making a fist shows anger, while rubbing the back of your neck or your forehead shows stress. If you place your hands in front of your mouth, you might be showing secretiveness and deception. Finger pointing might help you make a point, but it is also very assertive, even threatening. We all know how annoying a person's rhythmic drumming can be. This suggests that the individual is impatient and bored.

Listening Skills

2. Develop effective listening skills.

In business, you must also learn to listen effectively. If you are not listening while others are speaking, no communication can occur. Collaboration demands that all voices be heard. Successful workplaces value everyone's input.

Barriers to Active Listening

What gets in the way of your active listening abilities? Following are barriers to active listening.

Multitasking. Envision this scenario: You take your seat at the meeting table. You are still on your smartphone, speaking to a customer. You are also checking your online calendar to see when you can meet with this client. Meanwhile, you are thinking about the deadline you have for submitting a proposal. You are wondering if your graphic artist has completed the sketch you need for the upcoming meeting. You are juggling many tasks at the same time. Now, the meeting you are currently in has begun, and your coworker is speaking. How can you also pay attention to this colleague, when you have so many other tasks to complete and so many other conflicting responsibilities?

Preconceived Notions. If you believe that you already know what a coworker will say in a meeting, then you will not hear his or her comments objectively. Assumptions like this are a barrier to successful teamwork. Preconceived notions will diminish the colleague's comments and hinder your contributions to the team effort.

Focusing on Your Response. While your colleague is speaking, you do not wait to hear all he or she says. You only focus on your follow-up response. You can barely wait to jump in with a clever retort, a witty comeback, an oppositional point of view, or a debunking doubt. This does your colleague and your team a disservice. By thinking only of what you might say in response, you risk not hearing all that is said in the meeting.

Interrupting the Speaker. Interrupting a speaker is rude. A person should have the right to his or her say. If you interject your comments abruptly, then your coworker cannot complete his or her thoughts. Waiting to respond is not only good manners; it is also good business.

External Distractions. Think of all the possibilities:

- The room is too hot or too cold.
- It is 11:30 A.M. and you are hungry.
- It is Friday and you have exciting plans for the weekend.
- A road crew is repairing the street outside your office complex, complete with the sounds of a jackhammer and the annoying beeping of a truck backing up.
- You get a tweet from an acquaintance about an upcoming activity.

These external distractions might disallow you from focusing on what your coworkers are saying.

Keys to Effective Listening

Effective listening is critical to a company's goal of continuous improvement. Employees can best represent a company by actively listening to vendors, customers, and coworkers. When you hear what others say, then you also can contribute to discussions more effectively and provide useful feedback. Following are some keys to effective listening.

Stop What You Are Doing; Concentrate on the Task at Hand. In a meeting, turn off your smartphone. Leave your unfinished work in your office. Focus on the team's project and on the speaker.

Do Not Talk. It is hard to speak and listen at the same time. Sure, you might have valuable things to add, but wait your turn. Let the speaker have his or her say. Once that person has spoken, you can.

Make Eye Contact. When someone in the group is speaking, look at this person. Do not look out the window, or at your watch, your shoes, or your fingernails. Making eye contact with the speaker will help you focus your attention.

Take Notes. Another way to stay focused is by taking notes. Write down key ideas for future reference. You also can jot down any thoughts you might have for follow-up discussion. This way, you will not feel compelled to interrupt the speaker. The notes will help you remember what you had wanted to say, once the speaker has completed his or her thoughts.

Be Objective. Without a doubt, a colleague might say something with which you disagree. That's OK. We all have different points of view. However, remember that these differences are exactly what make cross-functional teams successful. Be open to alternatives; be willing to accept differences in opinion. Remember to judge the content of the comment instead of the speaker. Often, you will be asked to work with someone you might not like. Maybe you do not like his or her politics, work ethic, or personality. Still, that individual might make valuable comments.

Ask Questions. Once someone has spoken, then review your notes and ask questions. Be sure your questions are sincere and not oppositional. The goal of a question should be clarity. Then, after the speaker answers your question, repeat what he or she has said to confirm your understanding.

Control Your Reactions. If you hear something you do not agree with, avoid just shouting out your opposition. Your goal is to *respond*, not *react*. When you *react*, you tend to do so defensively. Immediate reactions are based on impatience, maybe even aggression. A more effective approach to successful listening is to calmly *respond*, after a brief delay—five to ten seconds. This delay acts as a filter or a buffer. Even a short delay will allow you to digest the issue, consider the person's point of view, organize your thoughts, and then respond more professionally.

The Importance of Oral Communication

3. Understand the importance of oral communication in the workplace.

Many people, even the seemingly most confident, are afraid to speak in front of others. You may have to communicate orally with your peers, your subordinates, your supervisors, and the public. Oral communication is an important component of your business success because you will be required to speak formally and informally on an everyday basis.

Everyday Oral Communication

4. Learn everyday oral communication skills.

"Hi. My name is Bill. How may I help you?" Think about how often you have spoken to someone today or this week at your job. You constantly speak to customers, vendors, and coworkers face-to-face, on the telephone, in meetings, or by leaving messages on voice mail.

- If you work a helpline, your primary job responsibility is oral communication.
- When you return the dozens of calls you receive or leave voicemail messages, each instance reveals your communication abilities.
- As an employee, you must achieve rapport with your coworkers. Much of your communication to them will be verbal and nonverbal. What you say and your body language impact your working relationships.

Every time you communicate orally, you reflect something about yourself and your company. The goal of effective oral communication is to ensure that your verbal and non-verbal skills make a good impression and communicate your messages effectively.

Telephone and Voicemail

You speak on the telephone dozens of times each week. When speaking on a telephone, make sure that you do not waste either your time or your listener's time.

Ten Tips for Telephone and Voice Mail Etiquette

1. Know what you are going to say before you call.
2. Speak clearly and enunciate each syllable.
3. Avoid rambling conversationally.
4. Avoid lengthy pauses.
5. Leave brief messages.
6. Avoid communicating bad news.
7. Repeat your phone number twice including the area code.
8. State your e-mail address as an option.
9. Sound pleasant, friendly, and polite.
10. If a return call is unnecessary, say so.

Informal Oral Presentations

As a team member, manager, supervisor, employee, or job applicant, you often will speak and meet with a coworker, a group of colleagues, or a committee. You will need to communicate orally in an informal setting for several reasons:

- Your boss needs your help preparing a presentation. You conduct research, interview appropriate sources, and prepare reports. When you have concluded your research, you might be asked to share your findings with your boss in a brief, informal oral presentation.
- Your company is planning corporate changes (staff layoffs, mergers, relocations, or increases in personnel). As a supervisor, you want to provide your input in an oral briefing to a corporate decision maker.
- At a departmental meeting, you are asked to report orally on the work you and subordinates have completed and to explain future activities.
- Your company is involved in a project with coworkers, contractors, and customers from distant sites. To communicate with these individuals, you participate in a teleconference, videoconference, or webinar, orally sharing your ideas.

Video and Teleconferences

If you are communicating with a dispersed group located in different cities or countries and you want to hear what everyone else is saying simultaneously, video or teleconferences are an answer. Consider using a video or teleconference when three or more people at separate locations need to talk.

In a video or teleconference, you want all participants to feel as if they are in the same room facing each other. With expensive technology in place, such as cameras, audio components, coder/decoders, display monitors, and user interfaces, you want to avoid wasting time and money with poor communication.

Ten Tips for Video and Teleconferences

1. Inform participants of conference date, time, time zone, and expected duration.
2. Make sure participants have printed materials before the teleconference.
3. Ensure that equipment has good audio quality.
4. Choose your room location carefully for quiet and privacy.
5. Consider arrangements for hearing impaired participants. You might need a text telephone (TTY) system or simultaneous transcription in a chat room.
6. Introduce all participants.
7. Direct questions and comments to specific individuals.
8. Do not talk too loudly, too softly, or too rapidly.
9. Turn off smartphones.
10. Limit side conversations.

Video Chat Platform

Video chat platforms, like Skype, Facetime, Google Hangouts, Google Voice, or Tango, might provide your company a more cost-effective means of oral communication. You and your audience need a computer, webcam, microphone or headset, and speakers. You can even use video chat platforms from your smartphone, allowing for mobile, anywhere, anytime oral communication. Most of these video chat platforms are free and allow for easy collaboration through group chats and conference calls.

Skype

See Chapter 2, "Digital Communication," for more information about Skype.

Webconferencing

Due to economic hardships, rising costs of airline tickets, time-consuming travel, and the need to complete projects or communicate information quickly, many companies and organizations are using webconferencing to communicate with their personnel, customers, prospective employees, and vendors. Webconferencing, sometimes called *webinars* or *virtual meetings,* include web-based seminars, lectures, presentations, or workshops transmitted over the Internet. A unique feature of a live webconference or webinar is interactivity. In a webconference, the presenter and audience can send, receive, and discuss information. In contrast to webconferences, a podcast is a digital sound or video file that can be downloaded.

Following are instances where a webconference would be appropriate:

- **Train employees.** Companies are using webconferences as a training tool not only to save travel expenses but also because of the interactivity a webconference allows.
- **Make sales presentations.** With webconferences, a sales team can make numerous sales calls without leaving the office. Through webconferencing, sales presentations can be enhanced with product demonstrations, questions and answer sessions, text messaging, and interactive polling to gather client and customer information.
- **Conduct quarterly or annual meetings.** An online conference room allows corporations to make reports to dispersed employees and investors.
- **Hold press conferences.** Share corporate information with news agencies at diverse locations.
- **Enhance online collaboration with colleagues at diverse locations.** Along with e-mail, teleconferencing, and videoconferencing, webconferencing provides companies another option when face-to-face meetings are difficult due to cost and distance.

Ten Tips for Webconferencing

1. Limit webconferences to 60 to 90 minutes.
2. Limit a webconference's focus to three or four important ideas.
3. Start fast by limiting introductory comments.
4. Keep it simple. Instead of using too many Internet tools, limit yourself to simple and important features like polling and messaging.
5. Plan ahead. Make sure that all webconference participants have the correct Internet hardware and software requirements; know the correct date, time, agenda, and Web log-in information for your webconference; and have the correct Web URL or password.
6. Before the webconference, test your equipment, hypertext links, and PowerPoint slide controls.
7. Use both presenter and participant views. One way to ensure that all links and slides work is by setting up two computer stations. Have a computer set on the presenter's view and another computer logged in as a guest. This will allow you to view accurately what the audience sees and how long displays take to load.
8. Involve the audience interactively through questions and/or text messaging.
9. Personalize the presentation. Introduce yourself, other individuals involved in the presentation, and audience members.
10. Archive the presentation.

Formal Presentations

You might need to make a formal presentation for the following reasons:

- Attend a civic club meeting and provide an oral presentation to maintain good corporate or community relations.
- Represent your company at a city council meeting. You will give an oral presentation explaining your company's desired course of action or justifying activities already performed.
- Represent your company at a local, regional, national, or international conference by giving a speech.
- Make an oral presentation promoting your service or product to a potential customer.

Types of Formal Oral Presentations

Three types of formal oral presentations include the following:

The Memorized Speech. The least effective type of oral presentation is the memorized speech. This is a well-prepared speech which has been committed to memory. Although such preparation might make you feel less anxious, too often these speeches sound mechanical and impersonal. They are often stiff and formal, and allow no speaker-audience interaction.

The Manuscript Speech. In a manuscript speech, you read from a carefully prepared manuscript. The entire speech is written on paper. This may lessen speaker anxiety and help you to present information accurately, but such a speech can seem monotonous, wooden, and boring to the listeners.

The Extemporaneous Speech. Extemporaneous speeches are probably the best and most widely used method of oral communication. You carefully prepare your oral presentation by conducting necessary research, and then you create a detailed outline. However, you avoid writing out the complete presentation. When you make your presentation, you rely on notes or PowerPoint slides with the major and minor headings for reference. This type of presentation helps you avoid seeming dull and mechanical, allows you to interact with the audience, and still ensures that you correctly present complex information. (See Figure 20.1 for the oral presentation process.)

Parts of a Formal Oral Presentation

A formal oral presentation consists of an introduction, a discussion (or body for development), and a conclusion.

Introduction

The introduction should welcome your audience, clarify your intent by providing a verbal road map, and capture your audience's attention and interest. This is the point in the presentation where you are drawing in your listener, hoping to generate enthusiasm and a positive impression.

To create a positive impression, set the table. Address your audience politely by saying "Good morning" or "Good afternoon." Tell the audience your name and the names of others who might be speaking. Welcome the audience and thank them for inviting you to speak.

FIGURE 20.1 Oral Presentation Process

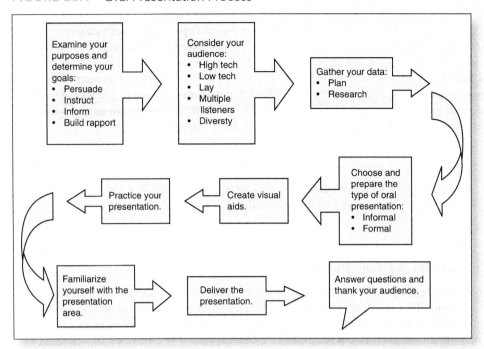

- **Opening comments for goodwill.** Begin with a pleasant, welcoming statement.

"Good evening. Thank you for allowing us to speak to you tonight. I'm (name and title). You'll also hear from (name and title)."

Next, provide a road map, clarifying what points you'll discuss and laying out the order of the topics.

- **Road map (thesis statement).** Briefly summarize key points in your presentation.

"What we are going to talk about, and in this order, are the following key points:

- The issues that led us to consider road improvements
- Challenges to this construction task
- Optional approaches
- Costs
- A timeframe for your consideration."

You can use a variety of openings to capture your audience's attention, such as the following:

- **Word pictures.** Anecdotes (specific in *time, place, person, and action*), quotes, and data (facts and figures) can be used to interest the listener.

Word Pictures to Arouse Your Audience's Interest

"From November 2015 to February 2016, our police department received over 75 reports of problems regarding Elm Street. These ranged from injuries related to hill jumping, cars sliding into the street's ditches, increased rush hour traffic, and tight turn lanes. One stretch of the road, from grid 39 to grid 47, is too narrow for snow removal crews to clean effectively. And these issues promise to get worse with residential growth anticipated to increase by 39 percent. As Sgt. Smith of the police department stated, 'Elm Street is a disaster waiting to happen.'"

- **A question or a series of questions.** Asking questions involves the audience immediately. A training facilitator could begin a workshop as follows:

Questions to Involve Your Audience

"How many reports do you write each week or month? How often do you receive and send e-mail messages to customers and colleagues? How much time do you spend on the telephone? Face it; technical communication is a larger part of your engineering job than you ever imagined."

These three questions are both personalized and pertinent. Through the use of the pronoun *you*, the facilitator speaks directly to each individual in the audience. By focusing on the listeners' job-related activities, the questions directly lead into the topic of conversation—technical communication.

- **A quotation from a famous person.** The training facilitator in the previous example could have begun his speech with a quote from Warren Buffett, a famous businessperson.

A Quote to Arouse Audience Interest

"How important is effective technical communication? Just listen to what Warren Buffett has to say on the topic:

For more than forty years, I've studied the documents that public companies file. Too often, I've been unable to decipher just what is being said or, worse yet, had to conclude that nothing was being said. . . . Perhaps the most common problem . . . is that a well-intentioned and informed writer simply fails to get the message across to an intelligent, interested reader. In that case, stilted jargon and complex constructions are usually the villain. . . . When writing Berkshire Hathaway annual reports, I pretend that I'm talking to my sisters. I have no trouble picturing them: Though highly intelligent, they are not experts in accounting or finance. They will understand plain English, but jargon may puzzle them. My goal is simply to give them the information I would wish them to supply me if our positions were reversed. To succeed,

I don't need to be Shakespeare; I must, though, have a sincere desire to inform.

That's what I want to impart to you today: Good writing is communication that
is easy to understand. If simple language is good enough for Mr. Buffett, then that
should be your goal."

(Buffett 1-2)

Discussion (or Body)

After you have aroused your listeners' attention and clarified your goals, you have to
prove your assertions. In the *discussion* section of your formal oral presentation, provide
details to support your thesis statement. You can develop your content in a variety of ways,
including the following:

- **Quotes, testimony, anecdotes.** You can find this type of information through
  primary and secondary research. For example, primary research, such as a
  survey or questionnaire, will help you substantiate content. By using people's
  direct comments, you help your audience to relate to the content. You also can
  find quotes, testimony, or anecdotes in secondary research, such as periodicals,
  newspapers, books, and online. This type of information validates your
  comments.
- **Data.** Facts and figures, again found through research or interviews, develop and
  support your content. Stating exactly how often your company's computer system
  has been attacked by viruses will support the need for improved firewalls. A
  statistical analysis of the increase in insurance premiums will clarify the need for
  a new insurance carrier. Showing the exponential increase in the cost of gasoline
  over the last 10 years will show why your company should consider purchasing
  hybrids.

To help your audience follow your oral communication, present these details using any
of the following modes of organization.

Comparison/Contrast. In your presentation, you could compare different makes of
office equipment, employees you are considering for promotion, different locations for a
new office site, vendors to supply and maintain your computers, different employee benefit
providers, and so forth. Comparison/contrast is a great way to make value judgments and
provide your audience options.

Problem/Solution. You might develop your formal oral presentation by using a problem-
to-solution analysis. For example, you might need to explain to your audience why your
division needs to downsize. Your division has faced problems with unhappy customers,
increased insurance premiums, decreased revenues, and several early retirements of top
producers. In your speech, you can then suggest ways to solve these problems ("We need to
downsize to lower outgo and ultimately increase morale"; "Let's create a Twitter account
to answer customer concerns"; "We should compare and contrast new employee benefits
packages to find creative ways to lower our insurance costs").

Argument/Persuasion. Almost every oral presentation has an element of argument/
persuasion to it—as does all good written communication. You will usually be persuading your audience to do something based on the information you share with them in the
presentation.

> **Persuasive Communication**
>
> See Chapter 11,
> "Communicating to
> Persuade," for additional
> information.

Importance. Prioritizing the information you present from least to most important (or most important to least) will help your listeners follow your reasoning more easily. To ensure the audience understands that you are prioritizing, provide verbal cues. These include simple words like *first, next, more important*, and *most important*. Do not assume that these cues are remedial or obvious. Remember, sometimes it is hard to follow a speaker's train of thought. Good speakers realize this and give the audience verbal signposts, reminding the listeners exactly where they are in the oral presentation and where the speaker is leading them.

Chronology. A chronological oral presentation can outline for your audience the order of the actions they need to follow. For example, you might need to prepare a yearly evaluation of all sales activities. Provide your audience with target deadlines and with the specific steps they must follow in their reports each quarter.

Coherence. To maintain coherence, guide your audience through your speech as follows:

- **Use clear topic sentences.** Let your listeners know when you are beginning a new, key point: "Next, let's talk about the importance of conciseness in your workplace writing."
- **Restate your topic often.** Constant restating of the topic is required because listeners have difficulty retaining spoken ideas. A reader can refer to a previously discussed point by turning back a page or two. Listeners do not have this option.

Furthermore, a listener is easily distracted from a speech by noises, room temperature, uncomfortable chairs, or movement inside and outside the meeting site. Restating your topic helps your reader maintain focus.

- **Use transitional words and phrases.** This helps your listeners follow your speech. Transitional words and phrases, like those shown in Table 20.2, aid reader comprehension, emphasize key points, and highlight your speech's organization.

TABLE 20.2   Purpose of Transitional Words

Purpose	Examples
Cause and effect	because, since, thus, therefore, for this reason, due to this, as a result of, consequently, in order to
Example	for instance, for example, another
Interpreting jargon	that is, in other words, more commonly called
Sequencing ideas	first, second, next, last, following, finally, above, below
Adding a point	furthermore, next, in addition, besides, not only . . . but also, similarly, likewise
Restating	in other words, that is, again, to clarify
Contrasting	but, instead, yet, however, on the other hand, nevertheless, in contrast, on the contrary, whereas, still
Emphasis	in fact, more importantly, clearly
Summarizing	to summarize, therefore, in summary, to sum up, consequently, therefore

Conclusion

Conclude your speech by restating the main points, by recommending a future course of action or by asking for questions or comments. A polite speaker leaves time for a few follow-up questions from the audience. Gauge your time well, however. You do not want to bore people with a lengthy discussion after a lengthy speech. You also do not want to cut short an important question-and-answer session. If you have given a controversial speech that you know will trouble some members of the audience, you owe them a chance to express their concerns.

The Communication Process

Use the communication process to make an effective oral presentation.

8. Use the communication process to make an effective oral presentation.

Prewrite by Considering the Purpose

Determine why you are making an oral presentation. Ask yourself questions similar to the following:

- Are you selling a product or service to a client?
- Do you want to inform your audience of the features in your newly created software?
- Are you persuading your audience to increase corporate spending to enhance a benefits package?
- Are you giving a speech for one of your college classes?
- Has your boss asked for your help in preparing a presentation? After you research the content, will you have to present your information orally?
- Are you a supervisor justifying workforce cuts to your division?
- Did a customer request information on solutions to a problem?
- Are you representing your company at a conference by giving a speech?
- Are you running for an office on campus and giving a speech about your candidacy?
- At the division meeting, are you reporting orally on the work you and your team have completed and the future activities you plan for the project?

Determining the purpose of a presentation will ensure that you choose the appropriate content.

Inform. For example, when your speech is to *inform*, you want to update your listeners. Such a speech could be about new tax laws affecting listeners' pay, new management hirings or promotions, or budget constraints or cutbacks. Speeches that inform do not necessarily require any action on the part of your audience. Your listeners cannot alter tax laws, change hiring or promotion practices, or prevent cutbacks. The informative speech keeps your audience up to date.

Persuade. On the other hand, some speeches *persuade*. You will speak to motivate listeners. For instance, you might give an oral presentation about the need to hold more regular and constructive meetings. You might tell your audience that teamwork will enhance productivity. Maybe you are giving an oral presentation about the value of quality controls to enhance product development. In each instance, you want your audience to leave the speech ready and inspired to act on your suggestions.

Instruct. You might speak to *instruct*. In an instruction, you will teach an audience how to follow procedures. For instance, you could speak about new sales techniques, ways to handle customer complaints, implementation of software, manufacturing procedures, or how to prepare for on-campus interviews. When you instruct, your goal is both to inform and persuade. You will inform your audience how to follow steps in the procedure. In addition, you will motivate them, explaining why the procedure is important.

Build Trust.   Finally, you might give an oral presentation to *build trust*. Let's say you are speaking at an annual meeting. Your goal not only might be to inform the audience of your company's status but also to instill the audience with a sense of confidence about the company's practices. You could explain that the company is acting with the audience's best interests in mind. Similarly, in a departmental meeting, you might speak to build rapport. As a supervisor, you will want all employees to feel empowered and valued. Speaking to build trust will accomplish this goal.

Consider Your Audience.   When you plan your oral presentation, consider your audience. Ask yourself questions such as the following:

Audience Recognition

See Chapter 5, "Audience Recognition," for more information.

- Is your audience high-tech, low-tech, or lay?
- Are you speaking up to supervisors?
- Are you speaking down to subordinates?
- Are you speaking laterally to peers?
- Are you speaking to the public?
- Are you addressing multiple audience types (supervisors, subordinates, and peers)?
- Is your audience friendly and receptive or hostile?
- Are you speaking to a captive audience (one required to attend your presentation)?
- Is your audience diverse in terms of culture, gender, or age?
- Will you need translators for those with hearing impairments?

Considering your audience's level of knowledge and interest will help you prepare your presentation. You should consider whether or not your audience needs terms defined and what tone you should take. You cannot communicate effectively if your audience fails to understand you or if your tone offends or patronizes them. Plan how you will design your content and style to communicate most effectively with your audience.

Presentation Plan.   A presentation plan, like the example shown in Figure 20.2, can help you accomplish your goals for the oral presentation.

Gather Information.   The best delivery by the most polished professional speaker will lack credibility if the speaker has little of value to communicate. You must study and research your topic thoroughly before you write about it.

You can rely on numerous sources when you research a topic for an oral presentation. For example, you could use any of the following sources:

Additional Information

See Chapter 16, "Research," for information on surveys and questionnaires. See Appendix B, "Parenthetical Source Citations and Documentation."

- Interviews
- Questionnaires/surveys
- Visits to job sites
- Participant observation
- Conversations in meetings or on the telephone
- Company reports
- Internet research
- Library research including periodicals and books
- Market research
- Social media search for trends

FIGURE 20.2   Presentation Plan

Topic: _____

Objectives:

- What do you want your audience to believe or do as a result of your presentation?

- Are you trying to persuade, instruct, inform, build trust, or a combination of these?

Development:
What main points are you going to develop in your presentation?

1.

2.

3.

4.

5.

Organization:
Will you organize your presentation using analysis, comparison/contrast, chronology, importance, or problem/solution?

Visuals:
Which visual aids will you use?

Technology:
Will you use PowerPoint, Prezi, Google Slides, or any other technology to make your presentation?

Using information from company reports or other sources such as the Internet, books, or periodicals requires that you read and document your research. Gathering information through interviews, questionnaires, or conversations, on the other hand, requires help from other people.

Writing the Presentation

After you obtain your information, your next step is to write a draft and consider visual aids for the presentation. The writing step in the communication process lets you use the research you gathered in the prewriting stage. When you organize your information, you will determine whether or not additional material is needed or if you can delete some of the material you have gathered.

Avoid writing the complete text of the presentation. Often when people have the complete text in front of them, they rely too heavily on the written words. They end up reading most of the paper to the audience rather than speaking more conversationally. Instead of writing out a complete copy of the presentation, use an outline or note cards to present your speech.

Outline.   You may want to write a detailed outline focusing on your speech's major units of discussion and supporting information. A skeleton speech outline (Figure 20.3) provides you a template for your presentation.

FIGURE 20.3 Skeleton Speech Outline

Skeleton Outline
Title:
Purpose:

I. Introduction
   A. Attention getter
   B. Focus statement

II. Body
   A. First main point
      1. Documentation/Subpoint
        a. Documentation/Subpoint
        b. Documentation/Subpoint
      2. Documentation/Subpoint
      3. Documentation/Subpoint
   B. Second main point
      1. Documentation/Subpoint
      2. Documentation/Subpoint
        a. Documentation/Subpoint
        b. Documentation/Subpoint
      3. Documentation/Subpoint
   C. Third main point
      1. Documentation/Subpoint
      2. Documentation/Subpoint
      3. Documentation/Subpoint
        a. Documentation/Subpoint
        b. Documentation/Subpoint
        c. Documentation/Subpoint

III. Conclusion
   A. Summary of main points
   B. Recommended future course of action

Note Cards. If you decide that presenting your speech from an outline will not work for you, consider writing highlights on 3" × 5" note cards. Avoid writing complete sentences or filling in the cards from side to side. If you write complete sentences, you will be tempted to read the notes rather than speak to the audience. If you fill in the cards from side to side, you will have trouble finding key ideas. Write short notes (keywords or short phrases) that will aid your memory when you make your presentation.

Practice. Practice your speech including manipulation of your visual aids so you use them at appropriate times and places during the presentation. As you practice, you will grow more comfortable and less dependent on your note cards or outline.

You will find that the more you practice, the more comfortable you feel. Practicing will help you achieve the following:

Decrease your fear	Pronounce troublesome words	Rearrange your content
Process your thoughts	Decide what to emphasize and how to emphasize it	Add further details
Become more comfortable with the topic	Enhance verbal and nonverbal cues	Know when and how to use your visual aids

The Presentation

In the rewriting or revision step of the communication process, consider all aspects of style, delivery, appearance, body language, and gestures. Then, most importantly, practice. Even if you have excellent visual aids and well-organized content, if you fail to deliver effectively, your audience could miss your intended message.

Style. As with good writing, effective oral communication demands clarity and conciseness. To achieve clarity, stick to the point. Your audience does not want to hear about your personal life or other irrelevant bits of information. You need to maintain focus on the topic. Concise oral presentations depend on the same skills evident in concise writing—word and sentence length. Trim your sentences of excess words (12–15 words per sentence is still the preferred length).

Remember to speak so that your audience can understand you and your level of vocabulary. You should speak to communicate rather than to impress your listeners.

Delivery. Effective oral communicators interact with and establish a dynamic relationship with their audiences. The most thorough research will be wasted if you are unable to create rapport and sustain your audience's interest. Although smaller audiences are usually easier to connect with, you can also establish a connection with much larger audiences through a variety of delivery techniques.

Eye Contact. Avoid keeping your eyes glued to your notes. You will find it easy to speak to one individual because you will naturally look him or her in the eye. The person will respond by looking back at you.

With a larger audience, whether the audience has 20 or 200 plus people, keeping eye contact is more difficult. Try looking into different people's eyes as you move through your presentation (or look slightly above their heads if that makes you more comfortable). Most of the audience has been in your position before and can sympathize.

Rate. Because your audience wants to listen and learn, you need to speak at a rate slow enough to achieve those two goals. Determine your normal rate of speech, and cut it in half. *Slow* is the best rate to follow in any oral presentation. You could speed up your delivery when you reach a section of less interesting facts. Slow down for the most important and most interesting parts. Match your rate of speaking to the content of your speech, just as actors vary their speech rate to reflect emotion and changes in content.

Enunciation. Speak each syllable of every word clearly and distinctly. Rarely will an audience ask you to repeat something even if they could not understand you the first time. It is up to you to avoid mumbling. Remember to speak more clearly than you might in a conversational setting. Slowing down your delivery will help you enunciate clearly.

Pitch.   When you speak, your voice creates high and low sounds. That's *pitch*. In your presentation, capitalize on this fact. Vary your pitch by using even more high and low sounds than you do in your normal, day-to-day conversations. Modulate to stress certain keywords or major points in your oral presentations.

Pauses.   One way to achieve a successful pace is to pause within the oral presentation. Pause to ask for and to answer questions, to allow ideas to sink in, and to use visual aids or give the audience handouts. These pauses will not lengthen your speech; they will only improve it.

A well-prepared presenter will allow for pauses and will have budgeted time effectively. Know in advance if your speech is to be five minutes, ten minutes, or an hour long. Then plan your speech according to time constraints, building pauses into your presentation. Practice the speech beforehand so you can determine when to pause and how often.

Emphasis.   You will not be able to underline or boldface comments you make in oral presentations. However, just as in written communication, you will want to emphasize key ideas. Your body language, pitch, gestures, and enunciation will enable you to highlight words, phrases, or even entire sentences.

Interaction with Listeners.   You might need your audience to be active participants at some point in your oral presentation, so you will want to encourage this response. Your attitude and the tone of your delivery are key elements contributing to an encouraging atmosphere.

Conflict Resolution.   You might be confronted with a hostile listener who either disagrees with you or does not want to be in attendance. You need to be prepared to deal with such a person. If someone disagrees with you or takes issue with a comment you make, try these responses:

- "That is an interesting perspective."
- "Thanks for your input."
- "Let me think about that some more and get back to you."
- "I have got several more ideas to share. We could talk about that point later, during a break."

The important point to remember is to not allow a challenging person to take charge of your presentation. Be pleasant but firm and maintain control of the situation. You will be unable to please all of your listeners all of the time. However, you should not let one unhappy listener destroy the effect of your presentation for the rest of the audience.

Appearance.   When you speak to an audience, they see you as well as hear you. Therefore, avoid physical distractions. For example, avoid wearing clothes or jewelry that might distract the audience. You might be representing your company or trying to make a good impression for yourself when you speak, so dress appropriately.

Body Language and Gestures.   During an oral presentation, nonverbal communication can be as important as verbal communication. Your appearance and attitude are important. In addition, the way you present your speech through your movements and tone of voice will affect your listeners. If you are enthusiastic about your topic, your listeners will respond enthusiastically. If you are bored or ambivalent, your tone and mannerisms will reflect your attitude. If you are negative, your tone will communicate negativity to your audience.

To communicate effectively, be aware of your body language and your gestures:

- Avoid standing woodenly. Move around somewhat, scanning the room with your eyes, stopping occasionally to look at one person. Remember to look at all parts of the room as you make the oral presentation.
- If you are nervous and your hands shake, try holding onto a chair back, lectern, the top of the table, or a pen.
- Use hand motions to emphasize ideas and provide transitions. For instance, you could put one finger up for a first point, two fingers up for a second point, and so forth.
- Avoid folding your arms stiffly across your chest. This projects a negative, defensive attitude.

Postspeech Question and Answers.   After your presentation, be prepared for a question-and-answer session. Politely invite your audience to participate by saying, "If you have any questions, I am happy to answer them."

When an audience member asks a question, make sure everyone hears it. If not, you can repeat the question. If you fail to understand the question, ask the audience member to repeat it and clarify it.

If you have no answer, tell the audience. You could say that you will research the matter and get back with them. Faking an answer will only harm your credibility and detract from the overall effect of your presentation. Another valid option is to ask the audience what they think regarding the question or if they have any possible solutions or answers to the question. This is not only a good way to answer the question but also to encourage audience interaction.

Visual Aids

Most speakers find that visual aids enhance their oral communication. Pie charts, bar charts, line graphs, photographs, tables, and schematics show what is difficult to express. With a variety of visual aids, you can enhance your communication and create interest in the audience to break up the monotony of words.

Although PowerPoint slide shows, tables, and figures are powerful means of communication, you must be the judge of whether visual aids will enhance your presentation. Avoid using them if you think they will distract from your presentation or if you lack confidence in your ability to create them and integrate them effectively. However, with practice, you probably will find that visuals add immeasurably to the success of most presentations.

Table 20.3 lists the advantages, disadvantages, and helpful hints for using visual aids. For all types of visual aids, practice using them before you actually make your presentation. When you practice your speech, incorporate the visual aids you plan to share with the audience.

> 9. Use visual aids to enhance the oral presentation.

Powerpoint Presentations

One of the most powerful oral communication tools is visual—Microsoft PowerPoint. Whether you are giving an informal or formal oral presentation, your communication will be enhanced by PowerPoint slides.

> 10. Understand the benefits of PowerPoint slides in an oral presentation.

TABLE 20.3   Visual Aids—Advantages and Disadvantages

Type	Advantages	Disadvantages	Helpful Hints
Chalkboards	Are inexpensive. Help audiences take notes. Allow audiences to focus on a statement. Help you be spontaneous.	Make you turn your back to the audience. Can be hard to see from a distance.	Have extra chalk. Stand to the side as you write. Print in large letters. Avoid talking with your back to the audience.
Chalkless whiteboards	Same as above.	Require unique, erasable pens. Some pens can be hard to erase if left on the board too long. Pens that run low on ink create light, unreadable impressions.	Use blue, black, or red ink. Cap pens to avoid drying out. Use pens made especially for these boards. Erase soon after use.
Flip charts	Can be prepared in advance. Can be reused. Are inexpensive. Are portable. Allow for spontaneity. Help audiences take notes.	Are limited by small size. Won't work well with large groups. Can have markers that run out of ink.	Have two pads. Have numerous markers. Use different colors for effect. Print in large letters. Turn pages when through with an idea so audience will not be distracted. Don't write on the back of pages where print bleeds through.
CDs	Are easy to use. Are entertaining. Can choose from many. Can be used for large groups.	Require equipment and outlets. Deny speaker–audience interaction. Can become dated.	Use up-to-date information. Avoid long presentations. Provide discussion time. Use to supplement the speech, not replace it.
Prezi presentations	Are entertaining—Prezi is animation intensive with swirling and zooming capabilities. Can be customized and updated. Can be used for large groups. Can be prepared in advance.	Does not lend itself to handouts. Requires Internet connectivity during the presentation if you want to use Prezi's ability to support YouTube. Has preformatted templates which limit your ability to change color, font, or style.	Practice with the equipment. Be prepared with a backup plan if the system crashes. Have the correct computer equipment (cables, monitors, screens, etc.). Practice your presentation.
PowerPoint presentations	Can be customized and updated. Can be used for large groups. Can be supplemented with handouts easily generated by PowerPoint. Can be prepared in advance. Can be reused. Can include animation and hyperlinks.	Require computers, screens, and outlets. Work better with dark rooms. Computers can malfunction. Can be too small for viewing. Can distance the speaker from the audience.	Practice with the equipment. Be prepared with a backup plan if the system crashes. Have the correct computer equipment (cables, monitors, screens, etc.). Practice your presentation.

Today, you will attend very few meetings where the speakers do not use PowerPoint slides. PowerPoint slides are used frequently because they are simple to use, economical, and transportable. Even if you have never created a slide show before, you can use the templates in the software and the autolayouts to develop your own slide show easily. An added benefit of PowerPoint slides is that you can print them and create handouts for audience members.

Benefits of PowerPoint

When you become familiar with Microsoft PowerPoint, you will be able to achieve the following benefits:

- Choose from many different presentation layouts and designs.

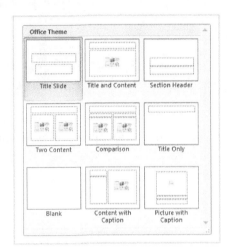

◀ EXAMPLE

- Create your own designs and layouts, changing colors and color schemes from preselected designs.

◀ EXAMPLE

- Add, delete, or rearrange slides as needed. By left-clicking on any slide, you can copy, paste, or delete it. By left-clicking between any of the slides, you can add a new blank slide for additional information.

EXAMPLE ▶

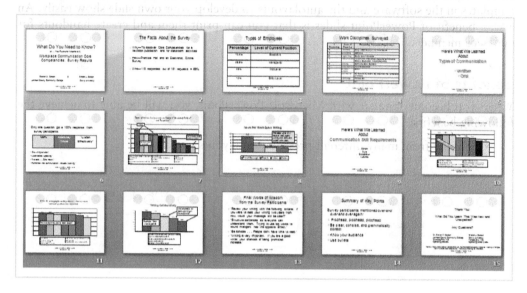

- Insert tables, charts, SmartArt, pictures from files, clip art, or media clips.

EXAMPLE ▶

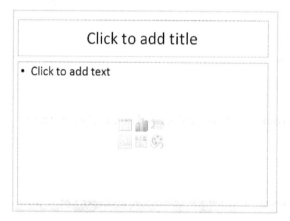

- Add hyperlinks either to slides within your PowerPoint presentation or to external Web links.

Tips for Using PowerPoint

To make it easy for you to add PowerPoint slides to your presentations, consider the following hints.

1. **Create optimal contrast.** Use dark backgrounds for light text or light backgrounds for dark text. Avoid using red or green text (individuals who are color blind cannot see these colors). You should use color for emphasis only.

2. **Choose an easy-to-read font size and style.** Use common fonts, such as Times New Roman, Courier, or Arial. Arial is considered to be the best to use because sans serif fonts (those without feet) show up best in PowerPoint. Use no more than three font sizes per slide. Use at least a 24-point font size for text and 36-point font for headings.

3. **Limit the text per slide.** Too many words per slide are not effective in an oral presentation. Use short sentences or phrases to ensure readability. You can accomplish this by creating a screen for each major point discussed in your oral presentation.

4. **Use headings for readability.** To create a hierarchy of headings, use larger fonts for a first-level heading and smaller fonts for second-level headings. Each slide should have at least one heading to help the audience follow your thoughts.

5. **Use emphasis techniques.** To call attention to a word, phrase, or idea, use color (sparingly), boldface, all caps, or arrows. Use a layout that includes white space. Include figures, graphs, pictures from the Web, or other line drawings.

6. **End with an obvious concluding screen.** Often, if speakers do not have a final screen that *obviously* ends the presentation, the speakers will click to a blank screen and say, "Oh, I didn't realize I was through," or "Oh, I guess that's it." In contrast, an obvious ending screen will let you as the speaker end graciously—and without surprise.

7. **Prepare handouts.** Give every audience member a handout, and leave room on the handouts for note taking.

8. **Avoid reading your screens to your audience.** Remember they can read and will become quickly bored if you read slides to them. Speakers lose their dynamism when they resort to reading slides rather than speaking to the audience.

9. **Elaborate on each screen.** PowerPoint should not replace you as the speaker. In contrast, PowerPoint should add visual appeal, while you elaborate on the details. Give examples to explain fully the points in your presentation.

10. **Leave enough time for questions and comments.** Instead of rushing through each slide, leave sufficient time for the audience to consider what they have seen and heard. Both during and after the PowerPoint presentation, give the audience an opportunity for input.

See Figures 20.4 through 20.9 for a sample effective PowerPoint slide presentation of a report.

FIGURE 20.4   Report Overview

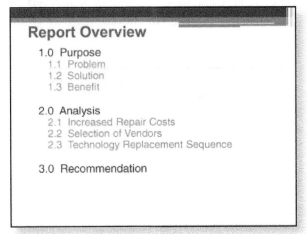

Sans serif font (such as Arial, Verdana, or Calibri) achieves a professional and readable look. Numbered headings and subheadings help readers navigate the report. White space breaks up the text.

FIGURE 20.5   Purpose

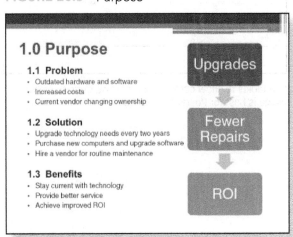

The SmartArt feature in PowerPoint allows the writer to emphasize a flow of ideas and add visual appeal. Bullets help to enhance the text.

FIGURE 20.6 Increased Repair Costs

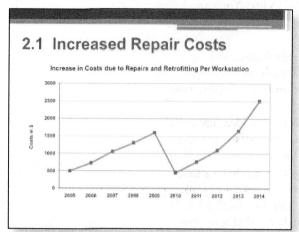

A line graph highlights increases and decreases in expenditures over a period of time. This figure will help persuade an audience to accept recommendations for change.

FIGURE 20.7 Selection of Vendors

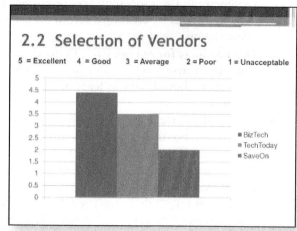

The bar chart, using a graded scale, compares options and clarifies which vendor should be selected.

FIGURE 20.8 Technology Replacement Sequence

2.3 Technology Replacement Sequence

Grey Team (even numbered years)	GreenTeam (odd numbered years)
Accounting	Manufacturing
Administrative Services	Personnel
Corporate Communication	Sales
Information Technology	Shipping and Receiving

Use tables to show comparison/contrast. Concise word usage is best for PowerPoint slides.

FIGURE 20.9 Recommendation

White space between text aids readability. The photograph adds visual appeal and personalizes the content.

Adding Hypertext Links to a PowerPoint Presentation

Microsoft Corporation, Inc.

To enhance your oral presentation, you can add hypertext links to PPT. Doing so will allow you to accomplish two goals: moving seamlessly between slides within the PPT presentation and linking to external sites.

To create hypertext links, follow these steps:

1. Highlight the word or words that you want to make a hypertext link.
2. Right click on that word or those words. A box will pop up, as shown in the following figure.

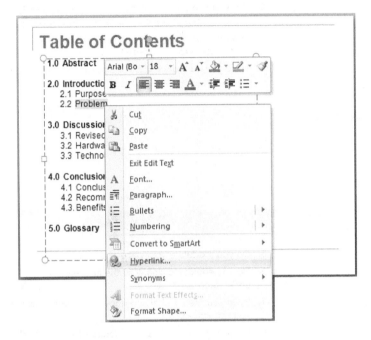

3. Scroll to **Hyperlink** and right click. You will see the following pop up.

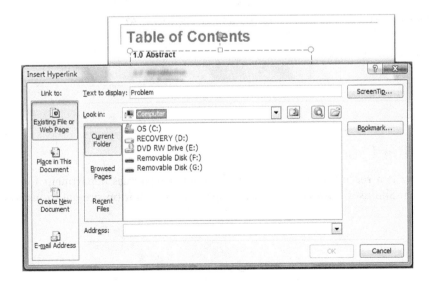

4. To create a hypertext link to screens within your PPT presentation, left click on **Place in This Document**.
5. To create a hypertext link to external Web sites, left click on **Existing File** or **Web Page**. Then, type the URL to the existing site in the Address line.

POWERPOINT SLIDES CHECKLIST

_____ 1. Does the presentation include headings for each slide?

_____ 2. Have you used an appropriate font size for readability?

_____ 3. Did you choose an appropriate font type for readability?

_____ 4. Did you limit yourself to no more than three different font sizes per screen?

_____ 5. Has color been used effectively for readability and emphasis, including font color and slide background?

_____ 6. Did you use special effects effectively versus overusing them?

_____ 7. Have you limited text on each screen?

_____ 8. Did you size your graphics correctly for readability, avoiding ones that are too small or too complex?

_____ 9. Have you used highlighting techniques (arrows, color, white space) to emphasize key points?

_____10. Have you edited for spelling and grammatical errors?

EFFECTIVE ORAL PRESENTATION CHECKLIST

_____ 1. Does your speech have an introduction,
- Arousing the audience's attention?
- Clearly stating the topic of the presentation?

_____ 2. Does your speech have a body,
- Explaining what exactly you want to say?
- Developing your points thoroughly?

_____ 3. Does your speech have a conclusion, suggesting
- What is next?
- When (due date) a follow-up should occur?
- Why that date is important?

_____ 4. Does your presentation provide visual aids to help you make and explain your points?

_____ 5. Do you modulate your pace and pitch?

_____ 6. Do you enunciate clearly so the audience will understand you?

_____ 7. Have you used verbal and nonverbal communication skills effectively?

_____ 8. Do you speak slowly and remember to pause so the audience can think?

_____ 9. Have you practiced with any equipment you might use?

_____10. Have you left time for Q/A?

The Writing Process at Work

Estella Alvarez is a recruitment specialist in the human resources department for Monroe, Louisiana. She uses PowerPoint to make new-hire orientation presentations to city employees.

Prewriting

To plan a presentation, Estella considered the following:

- Goal—to communicate human resources information and regulations
- Audience—new city employees
- Channels—oral presentation, PowerPoint slides, and handouts
- Data—details about governmental regulations illustrated by real-life experiences of other employees

Figure 20.10 shows how Estella used brainstorming to plan her communication.

FIGURE 20.10 Estella's Brainstorming List

Before the interview: Be prepared, find the best place and time for the interview, make up the interview questions, invite the panel of interviewers, get water and glasses, paper and pens.

During the interview: Create a comfortable environment, ask questions correctly (check with our HR legal staff to find out what we can and can't ask), use correct body language when speaking to the applicant, and be sure to manage time (don't go too long, so we have enough time for the next interview).

At the end of the interview: Thank the applicant, walk her or him out of the building (or to meet management), meet with the panel to discuss the interview, rate the applicant, and prepare documentation for records.

Writing

Estella drafted PowerPoint slides, such as the one shown in Figure 20.11.

Estella says, "After my first unsuccessful presentation, I realized I was too slide dependent. My slides had too much content and were way too boring, and I read them to my audience! No wonder I scored poorly on the effectiveness of the presentation."

Create a Comfortable Environment

Not every applicant will be comfortable in an interview. It's your job to make them comfortable. Conduct the interview in a pleasant, informal, and conversational manner. Here's what to do:

Welcome applicants and make them feel comfortable by creating a climate that shows trust. Introduce any panel members and explain how the interview will be conducted. Tell the applicant how many questions will be asked, who will ask them, the sequence of the questions, and let them know about note-taking resources available during the interview, like pen, paper, or reference materials. Let applicants know that they will be allowed to ask questions at the end of the interview if there's time and encourage them to organize their thoughts before responding to questions.

Create a positive body language by making eye contact and facing the applicants so that you seem caring and interested. Avoid negative body language. Don't frown, cross your arms, look at your watch, take phone calls, or yawn during the interview.

Rewriting

Figure 20.12 shows Estella's revised PowerPoint slide.

FIGURE 20.12   Revised PowerPoint Slide

Large heading to help reader navigation

Subheadings using Arial font for readability

Graphics and white space

Four lines of text, bulleted, with few words per line

Create a Comfortable Environment

In the interview, be pleasant, informal, and conversational.

- Welcome applicants by using their name.
- Introduce the interviewers.
- Allow applicants to ask questions.
- Avoid negative body language.

CASE STUDY

A customer is treated poorly by a salesperson at TechStop. The vice president of customer service, Shuan Wang, has received reports of other problems with customer-sales staff interaction, such as employee rudeness, unfamiliarity with store policy and products, taking personal calls while interacting with customers, or visiting with other employees instead of focusing on customer needs. Shuan needs to make a formal oral presentation to all sales personnel to improve customer service.

Assignment

Based on this scenario, outline the content for Shuan's oral presentation as follows:

- Determine what kind of introduction should be used to arouse the audience's interest.
- Provide a thesis statement.
- Develop information to teach employees sales etiquette and customer interaction, store policy regarding customer satisfaction, the impact on poor customer relations, and the consequences of failing to handle customers correctly. Research these topics to find your content.
- Organize the speech's content with appropriate transitions to aid coherence.
- Determine which types of visual aids would work best for this presentation.

ETHICAL CHALLENGE

Hank Reynolds, sales manager at RolledOut Advertising, is giving an oral presentation to a new client. He's behind schedule, however, and needs to put the presentation together quickly. To meet his timeline, he takes a shortcut. He uses several photographs taken from the Internet rather than using his company's staff photographer or accessing his company's photo files. None of the photographs he downloads from the Internet come from his advertising competitors, so he feels comfortable using the graphics.

Question

When, why, and how do you provide source information for graphics to ensure ethical behavior? Is Hank's use of online photographs ethical? Why or why not?

INDIVIDUAL AND TEAM PROJECTS

Evaluating an Oral Presentation

Listen to a speech. You could find one on television; at a student union; at a church, synagogue, or mosque; at a civic event, city hall meeting, or community organization; at a company activity; or in your classroom. Answer the following questions:

1. What type of introduction did the speaker use to arouse listener interest (anecdote, question, quote, or facts)? Give examples to support your decision.
2. What visual aids were used in the presentation? Were these visual aids effective? Explain your answers.
3. How did the speaker develop the assertions? Did the speaker use analysis, comparison/contrast, argument/persuasion, problem/solution, or chronology? Give examples to prove your point.
4. Was the speaker's delivery effective? Use the Presentation Delivery Rating Sheet to assess the speaker's performance by placing check marks in the appropriate columns.

PRESENTATION DELIVERY RATING SHEET

Delivery Techniques	Good	Bad	Explanation
Eye contact			
Rate of speech			
Enunciation			
Pitch of voice			
Use of pauses			
Emphasis			
Interaction with listeners			
Conflict resolution			

Giving an Oral Presentation

Research a topic in your field of expertise or degree program. The topic could include a legal issue, a governmental regulation, a news item, an innovation in the field, or a published article in a professional journal or public magazine. Make an oral presentation about your findings.

Creating a PowerPoint Slide Presentation

For the assignment "Giving an Oral Presentation," create PowerPoint slides to enhance your oral communication. To do so, follow the guidelines provided in this chapter.

Assessing PowerPoint Slides

After reviewing the following Microsoft PowerPoint slides, determine which are successful, which are unsuccessful, and explain your answers based on the guidelines provided in this chapter.

A Library Perspective on Distance Learning

1. Library online research sites are rapidly evolving.
2. Instructors engaged in online course development should visit with library staff for updates on resources.
3. Many traditional library services are available to serve distance learning students, including course reserves, interlibrary loan, reference, reciprocal borrowing policies, document delivery, access to online databases and collections.
4. Many new services, electronic books and journals, are available from libraries.
5. The Internet is not the online equivalent of an academic library.

A Technology Perspective on Online Education

Students—24-hour Call Centers answer student hardware/software needs

Faculty—The Tech Center helps faculty with course creations and tech resources

Comparison of Enrollment: Online vs. TV Classes

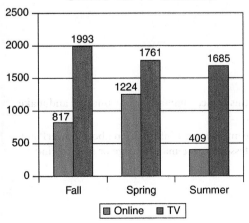

	Online	TV
Fall	817	1993
Spring	1224	1761
Summer	409	1685

PROBLEM-SOLVING THINK PIECES

After reading the following scenarios, answer these questions:

- Would your oral presentation be everyday, informal, or formal?
- Would you use a videoconference, Skype, teleconference, or face-to-face meeting?
- Which visual aids would work best for your presentation?
- Is your oral presentation goal to persuade, instruct, inform, or build trust?

In addition, complete any or all of these assignments:

- Write an outline for the presentation.
- Write a brief, introductory lead-in to arouse your listeners' interest.
- Create a questionnaire.

1. You work at FlashCom Electric. Your company has created a new interface for modems. Your boss has asked you to make a presentation to sales representatives from 20 potential vendors in the city. The oral presentation will explain to the vendors the benefits of your product, the sales breaks you will offer, and how the vendors can increase their sales.

2. After working for Friendly's, a major discount computer hardware and software retailer, for over a year, you have created a new organizational plan for the company's vast inventory. In an oral presentation, you plan to show the CEO and board of directors why your plan is cost effective and efficient.

3. You are the manager of an automotive parts supply company, Plugs, Lugs, & More. Your staff of 10 in-store employees lacks knowledge of the store's new merchandise, has not been meeting sales goals, and does not always treat customers with respect and care. It's time to address these concerns.

4. As CEO of your engineering/architectural firm (Levin, Lisk, and Lamb), you must downsize. Business is decreasing and costs are rising. To ensure third-quarter profitability, 15 percent of the staff must be laid off. That will amount to a layoff of over 500 employees. You now must make an oral presentation to your stockholders and the entire workforce (located in three states) at your company's annual meeting to report the situation.

WEB WORKSHOP

1. PowerPoint has proven to be a useful tool in business presentations. However, PowerPoint must be used correctly, as discussed in the chapter, to be effective. Access an online search engine and type in phrases such as follows (include any additional ones you create):

   - "using powerpoint in business"
   - "effective powerpoint use"
   - "tips for using powerpoint"
   - "powerpoint + business presentations"
   - "powerpoint + pros and cons"

   Once you have found articles discussing this topic, summarize your findings and make an oral presentation (using PowerPoint slides).

2. Research the following articles. Then, compare and contrast information about PowerPoint versus Prezi. Report your findings to your professor in an e-mail message or memo. Find additional articles that make this comparison.

   - Bird, Phil. "Prezi vs Powerpoint."
   - Wicks, David. "Prezi vs. PowerPoint."

APPENDIX A

GRAMMAR, PUNCTUATION, MECHANICS, AND SPELLING

Correct organization and development of your memos, letters, or reports is important for the success of your technical communication. However, no one will be impressed with the quality of your work, or with you, if your writing is riddled with errors in sentence construction or punctuation. Your written correspondence is often your first contact with business associates. Many people mistakenly believe that only English teachers notice grammatical errors and wield red pens, but businesspeople as well take note of such errors and may see the writer as less competent.

We were working recently with a young executive who is employed by a branch of the federal government. This executive told us that whenever his supervisor found a spelling error in a subordinate's report, this report was paraded around the office. Everyone was shown the mistake and had a good laugh over it, and the report was then returned to the writer for correction. Our acquaintance assured us that all of this was in good-natured fun. However, he also said that employees quickly learned to edit and proofread their written communication to avoid such public displays of their errors. He went on to say that his dictionary was well thumbed and always on his desk.

Your writing at work may not be exposed to such scrutiny by coworkers. Instead, your writing may go directly to another firm, and those readers will see your mistakes. To avoid this problem, you must evaluate your writing for grammar, punctuation, and spelling errors. If you don't, your customers, bosses, and colleagues will.

Grammar Rules

To understand the fundamentals of grammar, you must first understand the basic components of a sentence.

A correctly constructed sentence consists of a subject and a predicate (some sentences also include a phrase or phrases).

The meeting	began	at 4:00 A.M.
subject	predicate	phrase

◀ EXAMPLE

Subject: The *doer* of the action; the subject usually precedes the predicate.
Predicate: The *action* in the sentence

He	ran	to the office to avoid being late.
doer	action	phrases

◀ EXAMPLE

If the subject and the predicate (a) express a complete thought and (b) can stand alone, you have an *independent clause.*

The meeting began	at 4:00 P.M.
independent clause	phrase

◀ EXAMPLE

A *phrase* is a group of related words that does not contain a subject and a predicate and cannot stand alone or be punctuated as a sentence. The following are examples of phrases:

at the house
in the box
on the job
during the interview

If a clause is dependent, it cannot stand alone.

EXAMPLE ▶

> Although he tried to hurry, he was late for the meeting.
> dependent clause   independent clause

EXAMPLE ▶

> He was late for the meeting although he tried to hurry.
> independent clause   dependent clause

NOTE: When a dependent clause begins a sentence, use a comma before the independent clause. However, when an independent clause begins a sentence, do not place a comma before the dependent clause.

Agreement between Pronoun and Antecedent (Referent)

A pronoun has to agree in gender and number with its antecedent.

> *Susan* went on *her* vacation yesterday.
> The *people* who quit said that *they* deserved raises.

Problems often arise when a singular indefinite pronoun is the antecedent. The following antecedents require singular pronouns: *anybody, each, everybody, everyone, somebody,* and *someone.*

incorrect

Anyone can pick up *their* applications at the job placement center.

correct

Anyone can pick up *his* or *her* application at the job placement center.

Problems also arise when the antecedent is separated from the pronoun by numerous words.

incorrect

Even when the best *employee* is considered for a raise, *they* often do not receive it.

correct

Even when the best *employee* is considered for a raise, *he or she* often does not receive it.

Agreement between Subject and Verb

Writers sometimes create disagreement between subjects and verbs, especially if other words separate the subject from the verb. To ensure agreement, ignore the words that come between the subject and verb.

incorrect

Her *boss* undoubtedly *think* that all the employees want promotions.

correct

Her *boss* undoubtedly *thinks* that all the employees want promotions.

incorrect

The *employees* who sell the most equipment *is* going to Hawaii for a week.

correct

The *employees* who sell the most equipment *are* going to Hawaii for a week.

If a sentence contains two subjects (a compound subject) connected by *and,* use a plural verb.

incorrect

Joe and Tiffany *was* both selected employee of the year.

correct

Joe and Tiffany *were* both selected employee of the year.

incorrect

The bench workers and their supervisor *is* going to work closely to complete this project.

correct

The bench workers and their supervisor *are* going to work closely to complete this project.

Add a final *s* or *es* to create most plural subjects or singular verbs, as follows:

PLURAL SUBJECTS/VERBS	SINGULAR SUBJECTS/VERBS
bosses hire	a boss hires
employees demand	an employee demands
experiments work	an experiment works
attitudes change	the attitude changes

If a sentence has two subjects connected by *either . . . or, neither . . . nor,* or *not only . . . but also,* the verb should agree with the closest subject. This also makes the sentence less awkward.

Either the salespeople or the warehouse worker deserves raises.
Not only the warehouse worker but also the salespeople deserve raises.
Neither the salespeople nor the warehouse worker deserves raises.

◀ EXAMPLE

Singular verbs are used after most indefinite pronouns such as the following:

another	each	everything	nothing
anybody	either	neither	somebody
anyone	everybody	nobody	someone
anything	everyone	no one	something

Anyone who works here *is* guaranteed maternity leave.
Everybody wants the company to declare a profit this quarter.

Singular verbs often follow collective nouns such as the following:

class	department	organization	staff
corporation	group	platoon	team

The *staff is* sending the boss a bouquet of roses.

Comma Splice

A *comma splice* occurs when two independent clauses are joined by a comma rather than separated by a period or semicolon.

incorrect

Sue was an excellent employee, she got a promotion.

Several remedies will correct this error.

1. Separate the two independent clauses with a semicolon.

correct

Sue was an excellent employee; she got a promotion.

2. Separate the two independent clauses with a period.

correct

Sue was an excellent employee. She got a promotion.

3. Separate the two independent clauses with a comma and a *coordinating conjunction (and, but, or, for, so, yet).*

correct

Sue was an excellent employee, *so* she got a promotion.

4. Separate the two independent clauses with a semicolon (or a period), a conjunctive adverb, and a comma. *Conjunctive adverbs* include *also, additionally, consequently, furthermore, however, instead, moreover, nevertheless, therefore,* and *thus.*

correct

Sue was an excellent employee; therefore, she got a promotion.

or

Sue was an excellent employee. *Therefore*, she got a promotion.

5. Use a *subordinating conjunction* to make one of the independent clauses into a dependent clause. Subordinating conjunctions include *after, although, as, because, before, even though, if, once, since, so that, though, unless, until, when, where,* and *whether*.

correct

Because Sue was an excellent employee, she got a promotion.

Faulty or Vague Pronoun Reference

A pronoun must refer to a specific noun (its antecedent). Problems arise when (a) there is an excessive number of pronouns (causing vague pronoun reference) and (b) there is no specific noun as an antecedent. Notice that there seems to be an excessive number of pronouns in the following passage, and the antecedents are unclear.

Although Bob had been hired over two years ago, *he* found that *his* boss did not approve *his* raise. In fact, *he* was also passed over for *his* promotion. The boss appears to have concluded that *he* had not exhibited zeal in *his* endeavors for their business. Instead of being a highly valued employee, *he* was not viewed with pleasure by those in authority. Perhaps it would be best if *he* considered *his* options and moved to some other company where *he* might be considered in a new light.

The excessive and vague use of *he* and *his* causes problems for readers. Do these words refer to Bob or to his boss? You are never completely sure. To avoid this problem, limit pronoun usage, as in the following revision.

◀ EXAMPLE

Although Bob had been hired over two years ago, he found that his boss, Joe, did not approve his raise. In fact, Bob was also passed over for promotion. Joe appears to have concluded that Bob had not exhibited zeal in his endeavors for their business. Instead of being a highly valued employee, Bob was not viewed with pleasure by those in authority. Perhaps it would be best if Bob considered his options and moved to some other company where he might be considered in a new light.

To make the preceding paragraph more precise, we have replaced vague pronouns (*he* and *his*) with exact names (*Bob* and *Joe*).

Fragments

A *fragment* occurs when a group of words is incorrectly used as an independent clause. Often the group of words begins with a capital letter and has end punctuation but is missing either a subject or a predicate.

incorrect

Working with computers.
(lacks a predicate and does not express a complete thought)

The group of words may have a subject and a predicate but be a dependent clause.

incorrect

Although he enjoyed working with computers.

(has a subject, *he*, and a predicate, *enjoyed*, but is a dependent clause because it is introduced by the subordinate conjunction *although*)

It is easy to remedy a fragment by doing one of the following:

- Add a subject.
- Add a predicate.
- Add both a subject and a predicate.
- Add an independent clause to a dependent clause.

correct

Joe found that working with computers used his training.
(subject, *Joe,* and predicate, *found,* have been added)

correct

Although he enjoyed working with computers, he could not find a job in a computer-related field.

(independent clause, *he could not find a job,* added to the dependent clause, *Although he enjoyed working with computers*)

Fused Sentence

A *fused sentence* occurs when two independent clauses are connected with no punctuation.

incorrect

The company performed well last quarter its stock rose several points.

There are several ways to correct this error.

1. Write two sentences separated by a period.

correct

The company performed well last quarter. Its stock rose several points.

2. Use a comma and a coordinating conjunction to separate the two independent clauses.

correct

The company performed well last quarter, *so* its stock rose several points.

3. Use a subordinating conjunction to create a dependent clause

correct

Because the company performed well last quarter, its stock rose several points.

4. Use a semicolon to separate the two independent clauses.

correct

The company performed well last quarter; its stock rose several points.

5. Separate the two independent clauses with a semicolon, a conjunctive adverb or a transitional word or phrase, and a comma.

correct

The company performed well last quarter; *therefore*, its stock rose several points.

correct

The company performed well last quarter; *for example*, its stock rose several points.

The following are transitional words and phrases listed according to their use.

TO ADD

again	first	moreover	second
also	furthermore	next	still
besides	in addition		

TO COMPARE/CONTRAST

also	nevertheless
but	on the contrary
conversely	still
in contrast	

TO PROVIDE EXAMPLES

for example	of course
for instance	put another way
in fact	to illustrate

TO SHOW PLACE

above	here
adjacent to	nearby
below	on the other side
elsewhere	there
further on	

TO REVEAL TIME

afterward	second
first	shortly
meanwhile	subsequently
presently	thereafter

TO SUMMARIZE

all in all	last
finally	on the whole
in conclusion	therefore
in summary	thus

Modification

A *modifier* is a word, phrase, or clause that explains or adds details about other words, phrases, or clauses.

Misplaced Modifiers. A *misplaced modifier* is one that is not placed next to the word it modifies.

incorrect

He had a heart attack *almost* every time he was reviewed by his supervisor.

correct

He *almost* had a heart attack every time he was reviewed by his supervisor.

incorrect

The worker had to *frequently* miss work.

correct

The worker *frequently* had to miss work.

Dangling Modifiers. A *dangling modifier* is a modifier that is not placed next to the word or phrase it modifies. To avoid confusing your readers, place modifiers next to the word(s) they refer to. Don't expect your readers to guess at your meaning.

incorrect

While working, tiredness overcame them.
(Who was working? Who was overcome by tiredness?)

correct

While working, the staff became tired.

incorrect

After soldering for two hours, the equipment was ready for shipping. (Who had been soldering for two hours? Not the equipment!)

correct

After soldering for two hours, the technicians prepared the equipment for shipping.

Parallelism

All items in a list should be parallel in grammatical form. Avoid mixing phrases and sentences (independent clauses).

incorrect

We will discuss the following at the department meeting:

1. Entering mileage in logs (phrase)
2. All employees have to enroll in a training seminar. (sentence)
3. Purpose of quarterly reviews (phrase)
4. Some data processors will travel to job sites. (sentence)

We will discuss the following at the department meeting:

1. Entering mileage in logs
2. Enrolling in training seminars
3. Reviewing employee performance quarterly
4. Traveling to job sites

} phrases

At the department meeting, you will learn how to do the following:

1. All employees will enter mileage in logs.
2. Participants will enroll in training seminars.
3. New managers will review employee performances.
4. Sales personnel will travel to job sites.

} sentences

Punctuation

Apostrophe (')

Place an *apostrophe* before the final *s* in a singular word to indicate possession.

Jim's tool chest is next to the furnace.

◄ EXAMPLE

Place the apostrophe after the final *s* if the word is plural.

The employees' reception will be held next week.

◄ EXAMPLE

Don't use an apostrophe to make singular abbreviations plural.

The EXT's will be shipped today.

The EXTs will be shipped today.

Colon (:)

Use a *colon* after a salutation.

Dear Mr. Harken:

◄ EXAMPLE

In addition, use a colon after an emphatic or cautionary word if explanations follow.

Note: Hand-tighten the nuts.
Caution: Wash thoroughly if any mixture touches your skin.

◄ EXAMPLE

Use a colon after an independent clause to precede a quotation, list, or example.

EXAMPLE ▶

> She said the following: "No comment."
>
> These supplies for the experiment are on order: a plastic hose, two batteries, and several chemicals.
>
> The problem has two possible solutions: Hire four more workers, or give everyone a raise.

NOTE: In the preceding examples, the colon follows an independent clause.

A common mistake is to place a colon after an incomplete sentence. Except for salutations and cautionary notes, whatever precedes a colon *must* be an independent clause.

incorrect

The two keys to success are: earning money and spending wisely.

correct

The two keys to success are earning money and spending wisely.

or

The two keys to success are as follows: earning money and spending wisely.

or

The two keys to success are as follows:

1. Earning money
2. Spending wisely

Comma (,)

Writers often get in trouble with *commas* when they employ one of two common "words of wisdom."

- When in doubt, leave it out.
- Use a comma when there is a pause.

Both rules are inexact. Writers use the first rule to justify the complete avoidance of commas; they use the second rule to sprinkle commas randomly throughout their writing. On the contrary, commas have several specific conventions that determine usage.

1. Place a comma before a coordinating conjunction *(and, but, or, for, so, yet)* linking two independent clauses.

EXAMPLE ▶

> You are the best person for the job, *so* I will hire you.
>
> We spent several hours discussing solutions to the problem, *but* we failed to decide on a course of action.

2. Use commas to set off introductory comments.

EXAMPLE ▶

> First, she soldered the components.
>
> In business, people often have to work long hours.
>
> To work well, you need to get along with your coworkers.
>
> If you want to test equipment, do so by 5:25 P.M.

3. Use commas to set off sentence interrupters.

> The company, started by my father, did not survive the last recession.
> Mrs. Patel, the proprietor of the store, purchased a wide array of merchandise.

◀ EXAMPLE

4. Set off parenthetical expressions with commas.

> A worker, it seems, should be willing to try new techniques.
> The highway, by the way, needs repairs.

◀ EXAMPLE

5. Use commas after each item in a series of three or more.

> Prakash, Mirren, and Justin were chosen as employees of the year.
> We found the following problems: corrosion, excessive machinery breakdowns, and power failures.

◀ EXAMPLE

6. Use commas to set off long numbers.

> She earns $100,000 before taxes.

◀ EXAMPLE

NOTE: Very large numbers are often written as words.

> Our business netted over $2 million in 2017.

◀ EXAMPLE

7. Use commas to set off the day and year when they are part of a sentence.

> The company hired her on September 7, 2017, to be its bookkeeper.

◀ EXAMPLE

NOTE: If the year is used as an adjective, do not follow it with a comma.

> The 2017 corporate report came out today.

◀ EXAMPLE

8. Use commas to set off the city from the state and the state from the rest of the sentence.

> The new warehouse in Austin, Texas, will promote increased revenues.

◀ EXAMPLE

NOTE: If you omit either the city or the state, you do not need commas.

> The new warehouse in Austin will promote increased revenues.

◀ EXAMPLE

Dash (—)

A *dash,* typed as two consecutive hyphens with no spaces before or after, is a versatile punctuation mark used in the following ways.

1. After a heading and before an explanation

EXAMPLE ▶

Forecasting—Joe and Joan will be in charge of researching fourth-quarter production quotas.

2. To indicate an emphatic pause

EXAMPLE ▶

You will be fired—unless you obey company rules.

3. To highlight a new idea

EXAMPLE ▶

Here's what we can do to improve production quality—provide on-the-job training, salary incentives, and quality controls.

4. Before and after an explanatory or appositive series

EXAMPLE ▶

Three people—Sue, Luci, and Tom—are essential to the smooth functioning of our office.

Ellipses (...)

Ellipses (three spaced periods) indicate omission of words within quoted materials.

EXAMPLE ▶

"Six years ago, prior to incorporating, the company had to pay extremely high federal taxes."
"Six years ago, . . . the company had to pay extremely high federal taxes."

Exclamation Point (!)

Use an *exclamation point* after strong statements, commands, or interjections.

EXAMPLE ▶

You must work harder!
Do not use the machine!
Danger!

Hyphen (-)

A *hyphen* is used in the following ways.

1. To indicate the division of a word at the end of a typed line. Remember, this division must occur between syllables.
2. To create a compound adjective.

EXAMPLE ▶

He is a well-known engineer.
Until her death in 2017, she was a world-renowned chemist.
Tom is a 24-hour-a-day student.

3. To join the numerator and denominator of fractions.

EXAMPLE ▶

Four-fifths of the company want to initiate profit sharing.

4. To write out two-word numbers.

Twenty-six people attended the conference. ◀ EXAMPLE

Parentheses ( )

Parentheses enclose abbreviations, numbers, words, or sentences for the following reasons.

  1. To define a term or provide an abbreviation for later use

We belong to the Society for Technical Communication (STC). ◀ EXAMPLE

  2. To clarify preceding information in a sentence

The people in attendance (all regional sales managers) were proud of their accomplishments. ◀ EXAMPLE

  3. To number items in a series

The company should initiate (1) new personnel practices, (2) a probationary review board, and (3) biannual raises. ◀ EXAMPLE

Period (.)

A *period* must end a declarative sentence (independent clause).

I found the business trip rewarding. ◀ EXAMPLE

Periods are often used with abbreviations.

D.C.	Mr.	Ms.	P.M. or p.m.
e.g.	Mrs.	A.M. or a.m.	

It is incorrect to use periods with abbreviations for organizations and associations.

incorrect

Society for Technical Communication (S.T.C.)

correct

Society for Technical Communication (STC)

State abbreviations do not require periods if you use two capital letters.

incorrect

KS. (Kansas)	MO. (Missouri)	TX. (Texas)

correct

KS	MO	TX

Question Mark (?)

Use a *question mark* after direct questions.

EXAMPLE ▶

> Do the lab results support your theory?
> Will you work at the main office or at the branch?

Quotation Marks (" ")

Quotation marks are used in the following ways.

1. When citing direct quotations

EXAMPLE ▶

> He said, "Your division sold the most compressors last year."

NOTE: When you are citing a quotation within a quotation, use double quotation marks (" ") and single quotation marks (' ').

EXAMPLE ▶

> Kim's supervisor, quoting the CEO, said the following to explain the new policy regarding raises: "'Only employees who deserve them will receive merit raises.'"

2. To note the title of an article or a subdivision of a report.

EXAMPLE ▶

> The article "Robotics in Industry Today" was an excellent choice as the basis of your speech.
> Section III, "Waste Water in District 9," is pertinent to our discussion.

When using quotation marks, abide by the following punctuation conventions:

- Commas and periods always go *inside* quotation marks.

EXAMPLE ▶

> She said, "Our percentages are fixed."

- Colons and semicolons always go *outside* quotation marks.

EXAMPLE ▶

> He said, "The supervisor hasn't decided yet"; however, he added that the decision would be made soon.

- Exclamation points and question marks go inside the quotation marks if the quoted material is either exclamatory or a question. However, if the quoted material is not exclamatory or a question, then these punctuation marks go outside the quotation marks.

EXAMPLE ▶

> John said, "Don't touch that liquid. It's boiling!"

(Although the sentence isn't exclamatory, the quotation is. Thus, the exclamation point goes inside the quotation marks.)

EXAMPLE ▶

> How could she say, "We haven't purchased the equipment yet"?

(Although the quotation isn't a question, the sentence is. Thus, the question mark goes outside the quotation marks.)

Semicolon (;)

Semicolons are used in the following instances.

1. Between two independent clauses *not* joined by a coordinating conjunction.

The light source was unusual; it emanated from a crack in the plastic surrounding the cathode.

◄ EXAMPLE

2. To separate items in a series containing internal commas.

When the meeting was called to order, all members were present, including Susan Johnson, the president; Jewel Smith, the vice president; Harold Holbert, the treasurer; and Linda Hamilton, the secretary.

◄ EXAMPLE

Mechanics

Abbreviations

Never use an abbreviation that your reader will not understand. A key to clear technical writing is to write on a level appropriate to your reader. You may use the following familiar abbreviations without explanation: *Mrs., Dr., Mr., Ms.,* and *Jr.*

A common mistake is to abbreviate inappropriately. For example, some writers abbreviate *and* as follows:

I quit my job & planned to retire young.

◄ EXAMPLE

This is too colloquial for professional technical writing. Spell out *and* when you write.

Capital Letters

Capitalize the following:

1. Proper nouns

people	cities	countries	companies	schools	buildings
Susan	Houston	Italy	Bendix	Harvard	Oak Park Mall

◄ EXAMPLE

2. People's titles (only when they precede the name)

Governor Sally Renfro
or
Sally Renfro, governor
Technical Supervisor Wes Schneider
or
Wes Schneider, the technical supervisor

◄ EXAMPLE

3. Titles of books, magazines, plays, movies, television programs, and CDs (excluding the prepositions and all articles after the first article in the title).

EXAMPLE ▶

Mad Men
The Hobbit: The Battle of Five Armies
The Catcher in the Rye
The Hunger Games: Mockingjay, Part 1
The Walking Dead

4. Names of organizations.

EXAMPLE ▶

Girl Scouts
Phoenix, AZ, Regional Home Care Association
Kansas City Regional Council for Higher Education
Programs for Technical and Scientific Communication
American Civil Liberties Union
Students for a Democratic Society

5. Days of the week, months, and holidays.

EXAMPLE ▶

Monday   Thanksgiving
December

6. Races, religions, and nationalities.

EXAMPLE ▶

American Indian   Polish
Jewish

7. Events or eras in history.

EXAMPLE ▶

the Gulf War   World War II
the Vietnam War

8. North, South, East, and West (when used to indicate geographic locations).

EXAMPLE ▶

They moved from the North.
People are moving to the Southwest.

NOTE: Don't capitalize these words when giving directions.

EXAMPLE ▶

We were told to drive south three blocks and then to turn west.

9. The first word of a sentence.
10. Don't capitalize any of the following:

EXAMPLE ▶

Seasons—spring, fall, summer, winter
Names of classes—sophomore, senior
General groups—middle management, infielders, surgeons

Numbers

Write out numbers one through nine. Use numerals for numbers 10 and above.

10	536	2,093
104	12	5,550,286

Although the preceding rules cover most situations, there are exceptions.

1. Use numerals for all percentages.

2 percent   18 percent   25 percent

2. Use numerals for addresses.

12 Elm   935 W. Harding

3. Use numerals for miles per hour.

5 mph   225 mph

4. Use numerals for time.

3:15 A.M.

5. Use numerals for dates.

May 31, 2017

6. Use numerals for monetary values.

$45 $.95 $2 million

7. Use numerals for units of measurement.

14 feet   6 ¾ inches   16 mm   10 V

8. Do not use numerals to begin sentences.

incorrect

568 people were fired last August.

correct

Five hundred sixty-eight people were fired last August.

9. Do not mix numerals and words when writing numbers. When two or more numbers appear in a sentence and one of them is 10 or more, figures are used.

EXAMPLE ▶

> We attended 4 meetings over a 16-day period.

10. Use numerals and words in a compound number adjective to avoid confusion.

EXAMPLE ▶

> The worker needed six 2-inch nails.

Spelling

The following is a list of commonly misspelled or misused words. You can avoid many common spelling errors if you familiarize yourself with these words. Remember to run spell checks; also remember that spell check will not understand context. The incorrect word contextually could be spelled correctly.

accept, except	council, counsel	personal, personnel
addition, edition	desert, dessert	principal, principle
access, excess	disburse, disperse	quiet, quite
advise, advice	fiscal, physical	rite, right, write
affect, effect	forth, fourth	stationery, stationary
all ready, already	incite, insight	their, there, they're
assistants, assistance	its, it's	to, too, two
bare, bear	loose, lose	whose, who's
brake, break	miner, minor	your, you're
coarse, course	passed, past	
cite, site, sight	patients, patience	

APPENDIX B
PARENTHETICAL SOURCE CITATIONS AND DOCUMENTATION

To document research correctly, you must provide parenthetical source citations following the quote or paraphrase within the text. At the end of your document, supply a references page (American Psychological Association and Council of Science Editors) or a works cited page (Modern Language Association).

Parenthetical Source Citations

The American Psychological Association (APA), the Council of Science Editors (CSE), and the Modern Language Association (MLA) use a parenthetical form of source citations. If your boss or instructor requests footnotes or endnotes, you should still use these forms. However, the most modern approach to source citations according to MLA and APA requires only that you cite the source of your information parenthetically after the quotation or paraphrase. The CSE style manual uses a number after the quote or paraphrase to reference the source to the references page at the end of the document. CSE also uses an author-year sequence similar to APA.

APA Format

One Author.   If you do not state the author's name or the year of the publication in the lead-in to the quotation, include the author's name, year of publication, and page number in parenthesis, after the quotation.

> "Social media has helped companies quickly answer customer complaints" (Cottrell, 2011, p. 118).

(Page numbers are included for quoted material. The writer determines whether page numbers are included for source citations of summaries and paraphrases.)

Two Authors.   When you cite a source with two authors, always use both last names with an ampersand (&).

> "Line charts reveal relationships between sets of figures" (Gerson & Gerson, 2011, p. 158).

Three or more Authors.   When your citation has more than two authors but fewer than six, use all the last names in the first parenthetical source citation. For subsequent citations, list the first author's last name followed by *et al.* (Latin for "and others"), the year of publication, and for a quotation, the page number.

> "Employees require instantaneous access to crisis communication in the workplace" (Conners et al., 2011, p. 2).

Anonymous Works.   When no author's name is listed, include in the source citation the title or part of a long title and the year. Book titles are underlined or italicized, and periodical titles are placed in quotation marks.

> Flash drives have revolutionized data storage (*Electronic Databases,* 2011).

> Effective e-mail messages can be organized in three paragraphs ("Using Templates," 2011).

CSE Format

In-text citations for quoted or paraphrased material are in the form of superscript numbers. Sometimes, editors prefer that numbers are placed in parentheses or in brackets. The number refers to the numbered source citation on the references list at the end of the document. Many editors and publishers believe that a numbered form of citation is less intrusive to the reader than the method used by APA or MLA.

"Social media has helped companies quickly answer customer complaints."[1]

"Line charts reveal relationships between sets of figures" (2).

"Employees require instantaneous access to crisis communication in the workplace" [3].

MLA Format

One Author. After the quotation or paraphrase, parenthetically cite the author's last name and the page number of the information.

"Viewing the molecular activity required state-of-the-art electron microscopes" (Heinlein 193).

Note that the period follows the parenthesis, not the quotation. Also note that no comma separates the name from the page number and that no lowercase *p* precedes the number.

Two Authors. After the quotation or paraphrase, parenthetically cite the authors' last names and the page number of the information.

"Twitter has dramatically changed the way we write on the job" (Crider and Berry 292).

Three or More Authors. Writing a series of names can be cumbersome. To avoid this, if you have a source of information written by three or more authors, parenthetically cite one author's name, followed by *et al.* and the page number.

"The development of gaming software is a growing industry" (Norwood et al. 93).

Anonymous Works. If your source has no author, parenthetically cite the shortened title and page number.

"Robots are more accurate and less prone to errors caused by long hours of operation than humans" ("Useful Robots" 81).

Documentation of Sources

Parenthetical source citations are an abbreviated form of documentation. In parentheses, you tell your readers only the names of your authors and the page numbers on which the information can be found, or you provide a number that parallels the numbered source at the end of the document. Such documentation alone would be insufficient. Your readers would not know the names of the authors (in CSE numerical-sequence format), the titles of the books, the names of the periodicals, or the dates, volumes, or publishing companies. This more thorough information is found on the references page (APA) or works cited page (MLA), a listing of research sources alphabetized either by author's name or title (if anonymous). On the references page (CSE), you organize the citations numerically by the order in which the quote or paraphrase appeared in the text. This is the last page[s] of your research report.

Your entries should follow APA, CSE, and MLA standards. (Additional style manuals are available for many professions.) In contrast to previous editions, *The MLA Handbook* 8th ed. (2016) requires that each source citation include a URL (uniform resource locators) or doi (digital object identifier) to allow readers to more readily find the original source.

APA References

The APA style is commonly used in both engineering and scientific fields. The following are sample entries for the reference page which is placed at the end of the document. Include on the reference page only sources from which you cited in the document. For a comprehensive illustration of reference page entries, use the *Publication Manual of the American Psychological Association* (2009) and the *APA Style Guide to Electronic References* (2007).

A book with one author

Cottrell, R. C. (2006). *Smoke jumpers of the civilian public service in World War II*. London: McFarland and Co., Inc.

A book with two authors

Heath, C., & Heath, D. (2007). *Made to stick: Why some ideas survive and others die*. New York: Random House.

A book with three or more authors

Nadell, J., McNeniman, L., & Langan, J. (1997). *The Macmillan writer*. Boston: Allyn & Bacon.

A book with a corporate authorship

Corporate Credit Union Network. (2007). *A review of the credit union financial system: History, structure, and status and financial trends*. Kansas City, MO: U.S. Central.

A translated book

Phelps, R. (Ed.). (1983). *The collected stories of Colette* (M. Ward, Trans.). New York: Farrar, Straus & Giroux.

An entry in a collection or anthology

Hamilton, K. (2005). What's in a name? In R. Atwan (Ed.), *America now: Short readings from recent periodicals* (pp. 12–20). New York: Bedford/St. Martin's.

A signed article in a journal

Davis, R. (2007, April). Getting—and keeping good clients. *Intercom*, 8–12.

A signed article in a magazine

Rawe, J. (2007, May 28). A question of honor. *Time*, 59–60.

A signed article in a newspaper

Gertzen, J. (2007, March 29). University to go wireless. *The Kansas City Star*, p. C3.

An unsigned article

Effective communication with clients. (2009, September 23). *Technical Communication*, 22.

Encyclopedias and almanacs

Internet. (2000). *The world book encyclopedia*. Chicago: World Book.

Computer software

Drivers and Utilities [Computer software]. (2002–2004). Dell, Inc.

An article from an online database (or other electronic subscription service)

Pascal, J. (2005). Top ten qualities/skills employers want. *Job Outlook 2006 Student Version*. National Association of Colleges and Employers, 5 (12–16). Retrieved from ProQuest database.

E-mail

According to APA, do not include e-mail messages in the list of references. You should cite the message parenthetically in your text. (R. Schneider, personal communication, April 2, 2011).

Blog

McWard, J. (2011, May 31). Graphics on-line. Message posted to http://www.jmcward.net.

Personal Web Site

Mohr, E. (2011, Dec. 29). Home page of Ellen Mohr's Web Site. Retrieved from http://emohr.edu.

Professional Web Site

Johnson County Community College Writing Center. (2011, Jan. 5). Johnson County Community College. Retrieved from http://jccc.edu.

Posting to a discussion listserv

Tsui, P. (2011, Sep. 15). Questionnaire [Msg.16]. Message posted to http://groups.stc.com./ html.

APA References Page

Place the references page at the end of the document or in an appendix. The entries on the reference page are alphabetized by author's last name or title.

References

Corporate Credit Union Network. (2007). *A review of the credit union financial system: History, structure, and status and financial trends.* Kansas City, MO: U.S. Central.

Effective communication with clients. (2009, September 23). *Technical Communication*, 22.

Gertzen, J. (2007, March 29). University to go wireless. *The Kansas City Star*, p. C3.

Pascal, J. (2005). Top ten qualities/skills employers want. *Job Outlook 2006 Student Version*. National Association of Colleges and Employers, 5 (12–16). Retrieved from ProQuest database.

Tsui, P. (2011, Sep. 15). Questionnaire [Msg.16]. Message posted to http:// groups.stc.com./html.

CSE References

The CSE style guide shows two systems for organizing references at the end of the document. First, you can use the citation-sequence system that lists the numbered references in the order cited within the text. Second, you can follow the name-year system that lists references in alphabetical order by author's last name.

Personal e-mail messages, blog entries, personal web sites, and entries to listservs should not automatically be included on a references list. According to the CSE style manual, the decision to include such references is left to publishers and editors.

The CSE style of documentation is used in the fields of biology and medicine. Following are sample entries using the numerical system for the references list. For a comprehensive illustration of entries on the references list, use the *Scientific Style and Format: The CSE Manual for Authors, Editors, and Publishers* (2006).

A book with one author

Cottrell R. Smoke jumpers of the civilian public service in World War II. London: McFarland and Co., Inc.; 2006. p 27–28.

A book with two authors

Heath C., Heath D. Made to stick: Why some ideas survive and others die. New York: Random House; 2007. p 217–24.

A book with three or more authors

Nadell J., McNeniman L., Langan J. The Macmillan writer. Boston: Allyn and Bacon; 1997. p 224.

A book with a corporate authorship

Corporate Credit Union Network. Review of the credit union financial system: History, structure, and status and financial trends. Kansas City, MO: U.S. Central; 2007.

A translated book

Patel J. Technical communication and globalization. McWard J, translator. New York: Bedford/St. Martin's; 2006. p 15.

An entry in a collection or anthology

Hamilton K. What's in a name? R. Atwan, editor. America now: Short readings from recent periodicals. New York: Bedford/St. Martin's; 2005. p 12–20.

A signed article in a journal

Davis R. Getting—and keeping good clients. Intercom 2007 April: 8–12.

A signed article in a magazine

Rawe J. A question of honor. Time 2007 May 28: 59–60.

A signed article in a newspaper

Gertzen J. University to go wireless. The Kansas City Star 2007 Mar 16; Sect C: 3.

An unsigned article

Effective communication with clients. Technical Communication 2009 Sep 23; 22.

An article from an online database (or other electronic subscription service)

Pascal J. Re: top ten qualities/skills employers want. In: Job outlook 2006 student version. 2005. National Association of Colleges and Employers; 2006 May 5 [cited 2006 Apr 14]. Available from: ProQuest database.

Professional web site

Johnson County Community College Writing Center. Johnson County Community College. [cited 2011 Jan 5]. Available from: http://jccc.edu.

CSE References Page

The references page in CSE style is placed either at the end of the document or in an appendix. List the sources in the numerical order in which they appeared in the document.

References

1. Davis R. Getting—and keeping good clients. Intercom 2007; April: 8–12.
2. Heath C., Heath D. Made to stick: why some ideas survive and others die. New York: Random House; 2007. p 217–24.
3. Patel J. Technical communication and globalization. McWard J, translator. New York: Bedford/St. Martin's; 2006. p 15.
4. Pascal J. Re: top ten qualities/skills employers want. In: Job outlook 2006 student version. 2005. National Association of Colleges and Employers; 2006 May 5 [cited 2006 Apr 14]. Available from: ProQuest database.
5. Gertzen J. University to go wireless. The Kansas City Star 2007 Mar 16; Sect C: 3.

MLA Works Cited

MLA documentation format is used in the arts and humanities fields. Following are examples of entries on the works cited page. For a comprehensive illustration of MLA format for the works cited page, refer to the *The MLA Handbook*, 8th ed. (2016).

A book with one author

Cottrell, Robert C. *Smoke Jumpers of the Civilian Public Service in World War II*. McFarland and Co., Inc., 2006.

A book with two or three authors

Heath, Chip, and Dan Heath. *Made to Stick: Why Some Ideas Survive and Others Die*. Random House, 2007.

A book with four or more authors

Nadell, Judith, et al. *The Macmillan Writer*. Allyn and Bacon, 1997.

A book with a corporate authorship

Corporate Credit Union Network. *A Review of the Credit Union Financial System: History, Structure, and Status and Financial Trends*. U.S. Central, 2007.

A translated book

Phelps, Robert, editor. *The Collected Stories of Colette*. Translated by Matthew Ward, Farrar, Straus Giroux, 1983.

An entry in a collection or anthology

Hamilton, Kendra. "What's in a Name?" *America Now: Short Readings from Recent Periodicals*. Edited by Robert Atwan, Bedford/St. Martin's, 2005, pp. 12–20.

A signed article in a journal

Davis, Rachel. "Getting—and Keeping Good Clients." *Intercom,* April 2007, pp. 8–12.

A signed article in a magazine

Rawe, Julie. "A Question of Honor." *Time,* 28 May 2007, pp. 59–60.

A signed article in a newspaper

Gertzen, Jason. "University to go wireless." *The Kansas City Star,* 29 Mar. 2007, p. C3.

An unsigned article

"Effective Communication with Clients." *Technical Communication*, 23 Sep. 2009, p. 22.

Encyclopedias and almanacs

"Internet." *The World Book Encyclopedia.* 2000 ed., World Book.

Computer software

Drivers and Utilities. Computer software. Dell, Inc., 2002–2004.

An article from an online database (or other electronic subscription service)

Pascal, Janet. "Top Ten Qualities/Skills Employers Want." *Job Outlook 2006 Student Version,* National Association of Colleges and Employers, vol. 5, no. 1, 2015, pp. 12–16. *ProQuest,* www.joboutlook/studentversion/skills.

E-mail

Schneider, Max. "Re: Teaching Technical Communication." Message to Sharon Gerson, 2 Apr. 2016, E-mail.

Blog

McWardj. "Graphics On-line." *Myinfosite,* 31 May 2016, www.mcwardj/myinfosite/graphics.

Personal web site

Mohr, Ellen. "Home page." *JCCConline,* 29 Dec. 2015, www.mohr/home/jccconline.edu.

Professional web site

"Writing Center." *Johnson County Community College Writing Center,* 5 Jan. 2016, www.jccc.edu/student-resources/tutors-accessibility/writing-center/.

Posting to a discussion listserv

Ptsui [Peter Tsui]. "Questionnaire." *Organization for Technical Writing Listserv,* 15 Sep. 2015, www.OTWL/questionnaire/ptsui.

MLA Works Cited Page

At the end of the document, include a works cited page or place this page in an appendix.

Works Cited

Corporate Credit Union Network. *A Review of the Credit Union Financial System: History, Structure, and Status and Financial Trends.* U.S. Central, 2007.

"Effective Communication with Clients." *Technical Communication*, 23 Sep. 2009, p. 22.

McWardj. "Graphics On-line." *Myinfosite,* 31 May 2016, www.mcwardj/ myinfosite/graphics.

Pascal, Janet. "Top Ten Qualities/Skills Employers Want." *Job Outlook 2006 Student Version,* National Association of Colleges and Employers, vol. 5, no. 1, 2015, pp. 12–16. *ProQuest.* www.joboutlook/studentversion/skills.

Rawe, Julie. "A Question of Honor." *Time,* 28 May 2007, pp. 59–60.

CREDITS

TEXT CREDITS

CHAPTER 5

P. 110, Figure 5.8—Skype: "Use Skype in your business" website screenshot placement. Courtesy of Skype. The Skype name, associated trademarks, logos, the "S" logo and other marks, are trademarks of the Microsoft group of companies.

CHAPTER 6

P. 122, Figure 6.1—The National Society of Professional Engineers' Ethics Web Page. Reprinted by permission of the National Society of Professional Engineers (NSPE). www.nspe.org.

P. 124, Figure 6.2—Courtesy of STC Single Sourcing SIG.

CHAPTER 7

P. 149, Figure 7.13—GBA Architects/Engineers Web Page. Courtesy of George Butler Associates, Inc.

P. 150, Figure 7.14—GBA Architects/Engineers screenshot of Web Page: Market Sectors/Life Sciences, Courtesy of George Butler Associates, Inc.

CHAPTER 8

P. 174, Figure 8.15—Virtual Reality Drawing, Courtesy of Johnson County Community College.

P. 175, Figure 8.17—Courtesy of Andrew P. Buchwitz, PLA.

CHAPTER 10

P. 229, Figure 10.1—Screenshot of the Jobs by CareerBuilder's app. Copyright 2015 CareerBuilder, LLC—Reprinted with permission.

CHAPTER 13

P. 317, Figure 13.3—Web Site Home Page with Lead-in Information. Used by permission of Treanor Architects.

P. 319, Figure 13.4—GBA Architects/Engineers screenshot of Web Page: Market Sectors/Life Sciences, Courtesy of George Butler Associates, Inc.

CHAPTER 14

P. 337–338, Figure 14.1—Courtesy of Andrew P. Buchwitz, PLA.

REFERENCES

CHAPTER 1 AN INTRODUCTION TO TECHNICAL COMMUNICATION

"Achieving the Promise of the Mobile Enterprise." *Motorola: Position Paper*, Aug. 2005, www.motorolasolutions.com/.

Fisher, Lori, and Lindsay Bennion. "Organizational Implications of the Future Development of Technical Communication: Fostering Communities of Practice in the Workplace." *Technical Communication*, vol. 52, Aug. 2005, pp. 277–288.

Google Docs. 2012, www.google.com/docs/about/.

Hughes, Michael A. "Managers: Move from Silos to Channels." *Intercom*, Mar. 2003, pp. 9–11.

"The Importance of E-mail Continuity." An Osterman Research White Paper. Osterman Research, Inc., Dec. 2010, www.ostermanresearch.com/downloads.htm.

"Individual's and Teams' Roles and Responsibilities." *GOAL/QPC*, 24 Feb. 2003, www.goalqpc.com/index.htm.

Johnson, L. et. al. "Key Trends." 2011 *Horizon Report*, 16 Feb. 2012, http://eric.ed.gov/?id=ED515956.

Karch, Marziah. Personal Interview, 20 June 2011.

Mader, Stewart. "Using Wiki in Education." *The Science of Spectroscopy*, 16 Nov. 2008, www.scienceofspectroscopy.info/edit/index.php?title=Using_wiki_in_education.

Mader, Stewart. "Your Wiki Isn't Wikipedia: How to Use It for Technical Communication." *Intercom*, Jan. 2009, pp. 14–15.

Messina, Chris. "Hashtag." 2015, en.wikipedia.org/wiki/Hashtag.

Nesbitt, Pamela, and Elizabeth Bagley-Woodward. "Practical Tips for Working with Global Teams." *Intercom*, June 2006, pp. 25–30.

Robidoux, Charlotte, and Beth Hewett. "Is There a Write Way to Collaborate?" *Intercom*, Feb. 2009, pp. 4–9.

"Top Ten Skills for Job Candidates." National Association of Colleges and Employers, 3 Apr. 2013, www.naceweb.org/s10022013/job-outlook-skills-quality.aspx.

"TWiki—Enterprise Wiki & Collaboration Platform." *TWiki*, 4 Mar. 2007, www.twiki.org.

"Wiki." *Wikipedia*, 9 Jan. 2007, www.wikipedia.org.

"Writing: A Ticket to Work . . . Or a Ticket Out, a Survey of Business Leaders." *National Commission on Writing*, Sept. 2004, www.writingcommission.org.

"Writing Skills Necessary for Employment, Says Big Businesses." *National Commission on Writing*, 2007, www.writingcommission.org/pr/writing_for_employ.html.

CHAPTER 2 DIGITAL COMMUNICATION

Barnes, Nora, and Ava M. Lescault. "The 2014 Fortune 500 and Social Media: LinkedIn. Dominates as Use of Newer Tools Explodes." *Center for Marketing Research, UMass Dartmouth*, 2014, www.umassd.edu/cmr/socialmediaresearch/2014fortune500/.

"Business." *Skype*, 2012. 7 Mar. 2012, www.skype.com/en/meetings/.

Crowther, Don. "QR Codes—21 Ways to Use Them to Build Your Business." *Don Crowther.com*, 2014, doncrowther.com/marketing/qrcodes.

Google Docs. 2014, www.google.com/docs/about/.

Google Sites. 2014, www.google.com/sites/help/intl/en_GB/overview.html.

Karch, Marziah. Personal Interview, 20 June 2011.

Korhan, Jeff. "How QR Codes Can Grow Your Business." *SocialMediaExaminer*, 7 Feb. 2011, www.socialmediaexaminer.com/how-qr-codes-can-grow-your-business/.

Lenhart, Amanda. "Teens, Smartphones & Texting." *Pew Research Center*, 19 Mar. 2012, www.pewinternet.org/2012/03/19/teens-smartphones-texting/.

Pinkham, Ryan. "11 Reasons You SHOULD Be Using QR Codes at Your Place of Business." *SOL*, 28 Aug. 2012, solwebsolutions.com/qrcodes.

"QR Codes for Your Business." *Forbes*, 19 Aug. 2013, www.forbes.com/sites/thesba/2013/08/19/qr-codes-for-your-business/#4b0571333848.

"QR Code Generator." *GoQR*, 2014, goqr.me/.

Sutter, Brian. "5 Ways to Use QR Codes." *Wasp Buzz*, 2014, www.waspbarcode.com/buzz/5-ways-qr-codes/.

"Twiki Success Stories." *TWiki*, 2010, twiki.org/cgi-bin/view/Main/TWikiSuccessStories.

Welinske, Joe. "Developing User Assistance for Mobile Applications." *Intercom*, vol. 58, no. 9, 30 Nov. 2011, pp. 6–10.

CHAPTER 3   THE COMMUNICATION PROCESS

Albers, Michael J. "Single Sourcing and the Technical Communication Career Path." *Technical Communication,* vol. 50, Aug. 2003, pp. 336–343.

Carter, Locke. "The Implications of Single Sourcing for Writers and Writing." *Technical Communication,* vol. 50, Aug. 2003, pp. 317–320.

"Writing: A Powerful Message from State Government." *Report of the National Commission on Writing.* College Board, 2005.

CHAPTER 4   OBJECTIVES IN TECHNICAL COMMUNICATION

"GNOME Documentation Style Guide." *Gnome Dev Center,* 2005–2011, developer.gnome.org/gdp-style-guide/.

Labbe, J.R. "A Post-Literate World Will Leave Much to Be Desired." *The Kansas City Star,* Aug. 29 2007, p. B9.

Moore, Linda E. "Serving the Electronic Reader." *Intercom,* Apr. 2003, p. 17.

Online-Utility.org. 2012. www.online-utility.org/.

Perlin, Neil. "Technical Communication: The Next Wave." *Intercom,* Jan. 2001, pp. 4–8.

Readability Formulas. 2016, www.readabilityformulas.com/free-readability-formula-tests.php.

Rizzo, Tony. "Dropped Decimal Point Becomes a Costly Point of Contention." *The Kansas City Star,* 19 Feb. 2005, pp. A1, A6.

Self, Tony. "What If Readers Can't Read?" *Intercom,* Feb. 2009, pp. 10–14.

"Social Networking Websites Review." *Top Ten Reviews,* 20 Jan. 2011, www.toptenreviews.com/services/internet/best-social-networking-websites/.

U.S. Census Bureau. 2010. www.census.gov/.

CHAPTER 5   AUDIENCE RECOGNITION

Bannister, Linda. "Global Leaders May Benefit from Global Growth." *Edward Jones,* 2010, www.edwardjones.com.

Cardarella, Toni. "Business." *The Kansas City Star,* Sept. 2003, www.kansascity.com/mld/kansascity/business/6877765.htm?lc.

Cerner. 2012, www.cerner.com.

Courtis, John K., and Salleh Hassan. "Reading Ease of Bilingual Annual Reports." *Journal of Business Communication,* vol. 39, Oct. 2002, pp. 394–413.

"Demographics of the United States." *Wikipedia,* 2011, www.wikipedia.com.

"Diversity and inclusion." General Mills, 2014, www.generalmills.com.

Flint, Patricia, et al. "Helping Technical Communicators Help Translators." *Technical Communication,* vol. 46, May 1999, pp. 238–248.

Grimes, Diane Susan, and Orlando C. Richard. "Could Communication Form Impact Organizations' Experience with Diversity?" *Journal of Business Communication,* vol. 40, Jan. 2003, pp. 7–27.

Horton, William. "The Almost Universal Language: Graphics for International Documents." *Technical Communication,* vol. 40, Nov. 1993, pp. 682–693.

Jordan, Katrina. "Diversity Training: What Works, What Doesn't, and Why?" *Civil Right Journal,* Fall 1999, p. 1.

McInnes, R. "Workforce Diversity: Changing the Way You Do Business." *Diversity World,* 12 Feb. 2003, www.diversityworld.com/workforce_diversity.htm.

Melgoza, Cesar M. "Geoscape Findings Confirm Hispanic Market Growth." *Hispanic MPR.com,* 2007, www.hispanicmpr.com/2006/02/14/.2006.

Nethery, Kent. "Let's Talk Business." 2003, www.cuspomona.edu/-cljones/powerpoints/chap02/sld001.htm.

"New Census Bureau Report Analyzes Nation's Linguistic Diversity." *U.S. Census,* 27 Apr. 2010, http://www.census.gov/newsroom/press-releases/2015/cb15-tps16.html.

Rains, Nancy E. "Prepare Your Documents for Better Translation." *Intercom,* vol. 41, no. 5, Dec. 1994, p. 12.

Sanchez, Mary. "KC hospitals seek to overcome language barriers." *The Kansas City Star,* 7 Jan. 2003, pp. A1, A4.

Scott, Julie S. "When English Isn't English." *Intercom,* May 2000, pp. 20–21.

"Set Up Windows XP for Multiple Languages." *Microsoft,* 7 Sept. 2006, www.microsoft.com/en-us/.

Skype. 2012, www.skype.com.

St. Amant, Kirk R. "Communication in International Virtual Offices." *Intercom,* Apr. 2003, pp. 27–28.

SurveyMonkey. 2014, www.surveymonkey.com.

Swenson, Lynne V. "How to Make (American) English Documents Easy to Translate." *Proceedings: 34th International Technical Communication Conference,* 10 May 1987, pp. WE-193–195.

"Swiss Fight Encroachment of English." *The Kansas City Star,* 7 Dec. 2002, pp. A16.

Walmer, Daphne. "One Company's Efforts to Improve Translation and Localization." *Technical Communication,* vol. 46, May 1999, pp. 230–237.

Weiss, Edmund H. "Twenty-five Tactics to 'Internationalize' Your English." *Intercom*, May 1998, pp. 11–15.

"What Is the 'Business Case' For Diversity?" *Society for Human Resource Management*, 2003, www.shrm.org/diversity/businesscase.asp.

CHAPTER 6 ETHICAL CONSIDERATIONS

Adams, Rae, et al. "Ethics and the Internet." *Proceedings: 42nd Annual Technical Communication Conference*, 23–26 Apr. 1995, p. 328.

Barker, Thomas, et al. "Coming into the Workplace: What Every Technical Communicator Should Know—Besides Writing." *Proceedings: 42nd Annual Technical Communication Conference*, 23–26 Apr. 1995, pp. 38–39.

Bowman, George, and Arthur E. Walzer. "Ethics and Technical Communication." *Proceedings: 34th Annual Technical Communication Conference*, 10–13 May 1987, MPD–93.

Bremer, Otto A., et al. "Ethics and Values in Management Thought." *Business Environment and Business Ethics: The Social, Moral, and Political Dimensions of Management*, edited by Karen Paul, Ballinger, 1987, pp. 61–86.

"Circular 66, Copyright Registration for Online Works." *U.S. Copyright Office*, 13 Sept. 2012, www.copyright.gov/circs/circ66.html.

"Code for Communicators and Ethical Principles." Rocky Mountain Chapter, *Society for Technical Communication*, 10 June 2014, www.stermc.org/resources/resource_code.htm.

Gerson, Steven M., and Sharon J. Gerson. "A Survey of Technical Writing Practitioners and Professors: Are We on the Same Page." *Proceedings: 42nd Annual Technical Communication Conference*, 23–26 Apr. 1995, pp. 44–47.

Girill, T. R. "Technical Communication and Ethics." *Technical Communication*, vol. 34, Aug. 1987, pp. 178–179.

Guy, Mary E. *Ethical Decision Making in Everyday Work Situations*. Quorum Books, 1990.

Hartman, Diane B., and Karen S. Nantz. "Send the Right Messages About E-Mail." *Training & Development*, May 1995, pp. 60–65.

Johnson, Dana R. "Copyright Issues on the Internet." *Intercom*, June 1999, p. 17.

LeVie, Donald S. "Internet Technology and Intellectual Property." *Intercom*, Jan. 2000, pp. 20–23.

Li, Charlene. "Blogging: Bubble or Big Deal?" *Forrester*, 5 Nov. 2004, www.forrester.com/Research/Print/Document/0,7211,35000,00.html.

National Society of Professional Engineers, 2012, www.nspe.org.

Reyman, Jessica. "Rethinking Plagiarism for Technical Communication." *Technical Communication*, Feb. 2008, pp. 61–67.

Turner, John R. "Ethics Online: Looking Toward the Future." *Proceedings: 42nd Annual Technical Communication Conference*, 23 Apr. 1995, pp. 59–62.

U.S. Copyright Office. 2012, www.copyright.gov.

Wilson, Catherine Mason. "Product Liability and User Manuals." *Proceedings: 34th International Technical Communication Conference*, 10 May 1987, pp. WE-68–71.

CHAPTER 7 DOCUMENT DESIGN

Schriver, Karen A. "Quality in Document Design." *Technical Communication*, Second Quarter 1993, pp. 250–251.

CHAPTER 8 VISUAL AIDS

Horton, William. "The Almost Universal Language: Graphics for International Documents." *Technical Communication*, Nov. 1993, pp. 682–693.

Reynolds, Michael, and Liz Marchetta. "Color for Technical Documents." *Intercom*, Apr. 1998, pp. 5–7.

CHAPTER 9 ROUTINE CORRESPONDENCE

Bradley, Tony. "Policing Instant Messaging: Create a Policy to Govern IM Applications." *Processor*, 5 Aug. 2005.

Clark, Robert. Interview. June 2010.

Gerson, Steven M., et. al. "Core Competencies." Survey, Prentice Hall, 2004.

Hoffman, Jeff. "Instant Messaging in the Workplace." *Intercom*, Feb. 2004, pp.16–17.

"How to: Instant Messaging Security." *PC World*, 14 July 2006, www.pcworld.com.

Ollman, Gunter. "Instant Message Security." *Technical Info*, Mar. 2004.

Sieberg, Daniel. "How apps are changing our phones—and us." *CBS News*, 13 Mar. 2011, www.cbsnews.com.

Shinder, Deb. "Instant Messaging: Does it Have a Place in Business Networks?" *WindowSecurity.com*, 2005, www.windowsecurity.com/articles-tutorials/misc_network_security/Instant-Messaging-Business-Networks.html.

Toor, Amar. "200,000 Text Messages Are Sent Every Second, U.N. Agency Says." *Switched,* 20 Oct. 2010, www.switched.com.au/.

"Writing: A Ticket to Work . . . Or a Ticket Out, a Survey of Business Leaders." *National Commission on Writing,* Sept. 2004, www.writingcommission.org.

CHAPTER 10 THE JOB SEARCH

Bloch, Janel M. "Online Job Searching: Clicking Your Way to Employment." *Intercom,* Sept./Oct. 2003, pp. 11–14.

Brogan, Chris. "Write Your LinkedIn Profile for Your Future." *Chris Brogan,* 9 Aug. 2008, www.chrisbrogan.com.

CareerBuilder. 2012, www.careerbuilder.com.

Dixson, Kirsten. "Crafting an E-mail Resume." *BusinessWeek Online,* 13 Nov. 2001, www.businessweek.com.

Doyle, Alison. "LinkedIn Profile: Use Your LinkedIn Profile as a Resume." *About.Com,* 2011, www.about.com.

Hartman, Peter J. "You Got the Interview. Now Get the Job!" *Intercom,* Sept./Oct. 2003, pp. 23–25.

Isaacs, Kim. "Resume Critique Checklist." *MonsterTRAK Career Advice Archives,* 2012, www.monster.com.

Kallick, Rob. "Research Pays Off During Interview." *The Kansas City Star,* Mar. 23, 2003, p. D1.

Kendall, Pat. "Electronic Resumes." 9 Dec. 2003, www.reslady.com.

Khare, Shweta. "Adding QR Code to your Resume." *CareerBright,* 2014, careerbright.com/career-self-help/adding-qr-code-to-your-resume.

LinkedIn, 2014, www.linkedin.com.

Mantell, Ruth. "Make Your Résumé Stand Out From the Crowd." *MarketWatch,* 10 July 2012, www.marketwatch.com/story/.

Nale, Mike. "10 Things That Make Up a Good Video Resume." *eremedia,* 8 Feb. 2008, www.eremedia.com/ere/10-things-that-make-up-a-good-video-resume/.

Ralston, Steven M., et al. "Helping Interviewees Tell Their Stories." *Business Communication Quarterly,* vol. 66, Sept. 2003, pp. 8–22.

"Resume Guidelines." *Career Services Center,* Johnson County Community College, KS., 2003.

Schawbel, Dan. "7 Secrets to Getting Your Next Job Using Social Media." *Mashable,* 5 Jan. 2009, www.mashable.com/2009/01/05/job-search-secrets/#90sl9mBe2Eq9.

Schawbel, Dan. "10 iPhone Apps to Manage Your Job Search." *Mashable,* 26 Mar. 2009, www.mashable.com/2009/03/26/iphone-job-search-apps/#jNEhjgUoMuqD.

"SelectMinds Survey: 72% of Companies Use Social Media to Recruit Candidates." *SelectMinds.* 2012, www.oracle.com.

"72% of Employers Use Social Networking Sites to Research Job Candidates." *CareerBuilder,* 2010, www.careerbuilder.com.

Silverman, Rachel Emma. "No More Résumés, Say Some Firms." *The Wall Street Journal,* 25 Jan. 2012, www.wsj.com/articles.

Stafford, Diane. "Show Up Armed with Answers." *The Kansas City Star,* 3 Aug. 2003, p. L1.

Stafford, Diane. "LinkedIn: a must-have tool for job hunters." *The Kansas City Star,* Mar. 13, 2010, p. L1.

Vogt, Peter. "Your Resume's Look Is as Important as Its Content." *MonsterTRAK Career Advice Archives,* 19 Mar. 2009, www.monster.com.

"Writing: A Powerful Message from State Government." *Report of the National Commission on Writing,* College Board, 2005.

CHAPTER 11 COMMUNICATING TO PERSUADE

"CDC and CPSC Warn of Winter Home Heating Hazards." *Centers for Disease Control and Prevention,* 7 Jan. 2009, www.cdc.gov/nceh/pressroom/2006/COwarning.htm.

CHAPTER 12 SOCIAL MEDIA

Agarwal, Anil. "Google Plus Vs Facebook: Who Will Win This Social Media War?" *BloggersPassion,* 24 Apr. 2014, www.bloggerspassion.com/google-plus-vs-facebook.

"Bosses 'should embrace Facebook.'" *BBC News,* 29 Oct. 2008, www.news.bbc.co.uk/2/hi/business/7695716.stm.

Brewer, Melissa. "YouTube Marketing Basics—Know and Follow the Rules." *Ezine Articles,* 2011, www.ezinearticles.com/?YouTube-Marketing-Basics—Know-and-Follow-the-Rules&id=885792.

Brito, Michael. Presentation. Social Networking Conference. 25 June 2009.

Brogan, Chriss. "50 Ideas on Using Twitter for Business." *Chris Brogan,* 20 Apr. 2008, www.chrisbrogan.com/50-ideas-on-using-twitter-for-business/.

Brown, Andrew. "Should Facebook be worried about Google Plus?" *SocialFresh,* 10 Aug. 2012, www.socialfresh.com/facebook-google-plus/.

Ciortea, Marius. Presentation. Social Networking Conference. 25 June 2009.

"Facebook Newsroom." *Facebook,* 2014, www.facebook.com.

"Facing Up to Facebook." *British Trade Union,* Aug. 2007, www.scribd.com/document/52253906/When-Pervasive-Becomes-Invasive-Issues-Surrounding-e-

Monitoring-and-Surveillance-Technology-in-the-Modern-Organisation.

Foremski, Tom. "IBM is preparing to launch a massive corporate wide blogging initiative." *Silicon Valley Watcher,* 13 May 2005, www.siliconvalleywatcher.com/mt/archives/2005/05/can_blogging_bo.php.

"Gates backs blogs for businesses." *BBC News,* 21 May 2004, www.news.bbc.co.uk/2/hi/technology/3734981.stm.

Goforth, Alan. "Firms Find Friend in Facebook." *The Kansas City Star,* 19 May 2009, pp. D:15, D:17.

Jackson, Renee M. "Social Media Permeate the Employment Life Cycle." *The National Law Journal,* 18 Mar. 2011, www.nationallawjournal.com/id=1202437746082/Social-media-permeate-the-employment-life-cycle?slreturn=20160714152858.

Kazi, Sumaya. Presentation. Social Networking Conference. 25 June 2009.

Kelly, Heather. "Vine, Instagram and the rise of bite-sized video." *CNN,* 21 June 2013, www.cnn.com/2013/06/19/tech/social-media/social-video-trend/index.html.

Kharif, Olga. "Blogging for Business." *BusinessWeek Online,* 9 Aug. 2004, www.businessweek.com.

Kirkpatrick, Marshall. "Study: Fastest Growing US Companies Rapidly Adopting Social Media." *Read Write Web,* 15 Aug. 2008, www.readwrite.com/2008/08/15/study_fast_growing_us_companie/.

Li, Charlene. "Blogging: Bubble or Big Deal?" *Forrester,* 5 Nov. 2004, www.forrester.com/Research/Print/Document/0,7211,35000,00.html.

Mathews, Whitney. Interview. 4 May 2009.

Miller, Michael. "Using YouTube to Promote Your Business." *InformIT,* 28 Sept. 2007, www.informit.com.

Ostrander, Susan. "Google Plus vs. Facebook: Who's Winning?" *The Huffington Post,* 20 Nov. 2011, www.huffingtonpost.com.

Payne, Jarrod. "Micro-video: how brands should tackle 2014's hottest medium." *MarketingMagazine.* 22 Nov. 2014, www.marketingmagazine.com.

Pogue, David. "Twittering Tips for Beginners." *The New York Times,* 15 Jan. 2009, www.nytimes.com.

Pring, Cara. "100 More Social Media Statistics for 2012." *The Social Skinny,* 13 Feb. 2012, www.thesocialskinny.com/100-more-social-media-statistics-for-2012/.

Ranii, David. "More companies use social media for marketing." *The News & Observer,* 12 Dec. 2008, www.newsobserver.com/.

Ray, Ramon. "Blogging for Business." *Inc.com,* Sept. 2004, www.inc.com/partners/sbc/articles/20040929-blogging.html.

Reddick, Paul. Interview. 15 Oct. 2009.

"Statistics." *Facebook,* 2011, www.facebook.com.

Vascellaro, Jessica E. "Why Email No Longer Rules." *The Wall Street Journal,* 12 Oct. 2009, pp. R1, R3.

Wohlsen, Marcus. "Mining our Messages." *The Kansas City Star,* 13 Feb. 2012, p. A2.

Wuorio, Jeff. "Blogging for business: 7 tips for getting started." *Microsoft,* 9 Sept. 2007, www.microsoft.com/business/issues/marketing/online_marketing/5_ways_blog.htm.

Zahorsky, Darrel. "YouTube Your Business with Online Video." *About,* 20 Apr. 2009, www.about.com.

CHAPTER 13   WEB SITES AND ONLINE HELP

Arno, Christian. "Why Your Business Must Embrace the Foreign Language Internet." *Mashable,* 15 Nov. 2010, www.mashable.com/2010/11/15/business-foreign-language-web/#Njx.NfraMsqf.

Badminton, Nikolas. "Great Web Design for Mobile Devices." *The HuffingtonPost,* 10 May 2013, www.huffingtonpost.com.

Conner, Shannon. Interview. 11 July 2011.

"E-Commerce, Trust, and SSL." *Symantec,* 2011, www.symantec.com.

Eddings, Earl, Kim Buckley, Sharon Coleman Bock, and Nathaniel Williams. Software Documentation Specialists at PDA. Interview. 12 Jan. 1998.

Fogg, B. J. et al. "How Do People Evaluate a Web Site's Credibility?" *ACM Digital Library*, 2003, www.dl.acm.org/citation.

"508 Law." *Section 508.* 15 Aug. 2002, www.section508.gov/section508-laws.

Goldman, David. "HTML5: The future of the Web is finally here." *Money Magazine,* 17 Dec. 2012, www.money.cnn.com/2012/12/17/technology/html5/.

Hemmi, Jane A. "Differentiating Online Help from Printed Documentation." *Intercom,* July/Aug. 2002, pp. 10–12.

"Internet Accessibility." *STC AccessAbility SIG,* 9 June 2004, www.stc-access.org/category/focus/internet_access/.

"Internet Growth: Today's Road to Business and Trade." *Internet World Stats,* 2004, www.internetworldstats.com.

"Internet World Stats: Usage and Population Statistics." *Internet World Stats,* 2014, www.internetworldstats.com.

Karch, Marziah. Interview. 15 July 2011.

McGowan, Kevin S. "The Leap Online." *Intercom,* Sept./Oct. 2000, pp. 22–25.

Mielach, David. "Four Common Website Design Mistakes to Avoid." *BusinessNewsDaily,* 27 Aug. 2012, www.businessnewsdaily.com/3056-website-design-mistakes.html.

Olive, Eric G. "Usability: Making the Web Work." *Intercom*, Nov. 2002, pp. 8–10.

"OnGuard Online," *Federal Trade Commission,* 1 Aug. 2007, www.consumer.ftc.gov/features/feature-0038-onguardonline.

"Privacy Initiatives." *Federal Trade Commission,* 1 Aug. 2007, www.ftc.gov/tips-advice/business-center/privacy-and-security.

"Protecting Personal Information." *Federal Trade Commission,* 1 Aug. 2007, www.ftc.gov/tips-advice/business-center/guidance/protecting-personal-information-guide-business.

Smith, Aaron. "Mobile Access 2010." *Pew Internet.* 7 July 2010, www.pewinternet.org/2010/07/07/mobile-access-2010/.

CHAPTER 15 INSTRUCTIONS, USER MANUALS, AND STANDARD OPERATING PROCEDURES

Sharp, Roger A. "Incorporating Animation into Help Files." *Intercom,* Nov. 2007, pp. 8–12.

Tietjen, Phil. "Adventures in Screencasting." *Intercom,* July/Aug. 2008, pp. 8–10.

CHAPTER 16 RESEARCH

The American Psychological Association, 2012, www.apa.org.

Fall, Jason, and Erik Deckers. "How to Use Social Media for Research and Development." *Entrepreneur,* 7 Dec. 2011, www.entrepreneur.com/article/220812.

Gardner, Oli. "10 Social Media Research Strategies to Inject Your Next Blog Post With 'Roids.'" *Unbounce,* 20 July 2011, www.unbounce.com/social-media/10-social-media-research-strategies-to-enhance-your-next-blog-post/

Gray, Catherine. "Social Media: A Guide for Researchers." *Research Information Network,* 7 Feb. 2011, www.healthconnect-intl.org/FIRB_apr11.html.

Krum, Randy. *Cool Infographics,* 2007, www.coolinfographics.com/.

Messina, Chris. "Hashtag." 2015, en.wikipedia.org/wiki/Hashtag.

The Modern Language Association. MLA, 14 Aug. 2016, www.mla.org.

CHAPTER 19 PROPOSALS AND BUSINESS PLANS

App Store." *Wikipedia,* 27 Mar. 2016, en.wikipedia.org/wiki/App_Store_(iOS).

Purcell, Kristen, et al. "The Rise of Apps Culture." *Pew Internet,* 14 Sept. 2015, www.pewinternet.org/2010/09/14/the-rise-of-apps-culture/.

Wexford, Charles. "Mobile Apps in the Workplace." *Time,* 12 Aug. 2016, www.time.com/wexford/mobile.

CHAPTER 20 ORAL PRESENTATIONS

Buffett, Warren. "Preface." In *A Plain English Handbook: How to Create Clear SEC Disclosure Documents.* Office of Investor Education and Assistance, U.S. Securities and Exchange Commission, Aug. 1998, pp. 1–2.

Clarke, Robyn D. "Put your best voice forward—speech recommendations." *The CBS Interactive Business Network,* 2000 June, www.cbsinteractive.com/.

Hall, Edward T. *The Silent Language,* Random House, Inc. 1981.

Mahin, Linda. "PowerPoint Pedagogy." *Business Communication Quarterly,* vol. 67, June 2004, pp. 219–222.

Mehrabian, Albert. *Silent Messages: Implicit Communication of Emotions and Attitudes,* Wadsworth, 1981.

Murray, Krysta. "Web Conferencing Tips from a Pro." *EServer TC Library,* 1 July 2004, www.tc.eserver.org/31474.html.

Tilton, James E. "Adventures in public speaking: A guide for the beginning instructor or public speaker." *NCJRS,* Feb. 2002, www.ncjrs.gov/App/publications/abstract.aspx?ID=193657.

Warfield, Anne. "Do You Speak Body Language?" *The CBS Interactive Business Network,* Apr. 2001, www.cbsineractive.com/.

INDEX